TRANSFORMING BIODIVERSITY GOVERNANCE

Over fifty years of global conservation has failed to bend the curve of biodiversity loss, so we need to transform the ways we govern biodiversity. The UN Convention on Biological Diversity aims to develop and implement a transformative framework for the coming decades. However, the question of what transformative biodiversity governance entails and how it can be implemented is complex. This book argues that transformative biodiversity governance means prioritizing ecocentric, compassionate and just sustainable development. This involves implementing five governance approaches – integrative, inclusive, adaptive, transdisciplinary and anticipatory governance – in conjunction and focused on the underlying causes of biodiversity loss and unsustainability. *Transforming Biodiversity Governance* is an invaluable source for academics, policy makers and practitioners working in biodiversity and sustainability governance. This is one of a series of publications associated with the Earth System Governance Project. For more publications, see www.cambridge.org/earth-system-governance. This title is also available as Open Access on Cambridge Core.

INGRID J. VISSEREN-HAMAKERS serves as Professor and Chair of the Environmental Governance and Politics (EGP) group at Radboud University, Netherlands, and specializes in transformative global environmental governance. She aims to contribute to both academic and societal debates on how societies and economies can become sustainable. Her research focuses on governing the relationships between animal interests, biodiversity and food, among others.

MARCEL T. J. KOK is Programme Leader of the International Biodiversity Policy group at PBL Netherlands Environmental Assessment Agency. His research concentrates on global environmental governance and scenario analysis of global environmental problems, with a focus on biodiversity. He specializes in bottom-up governance approaches.

The **Earth System Governance Project** was established in 2009 as a core project of the International Human Dimensions Programme on Global Environmental Change. Since then, the Project has evolved into the largest social science research network in the area of sustainability and governance. The Earth System Governance Project explores political solutions and novel, more effective governance mechanisms to cope with the current transitions in the socioecological systems of our planet. The normative context of this research is sustainable development; earth system governance is not only a question of institutional effectiveness, but also of political legitimacy and social justice.

The **Earth System Governance series** with Cambridge University Press publishes the main research findings and synthesis volumes from the Project's first ten years of operation.

Series Editor

Frank Biermann, Utrecht University, the Netherlands

Titles in print in this series

Biermann and Lövbrand (eds.), *Anthropocene Encounters: New Directions in Green Political Thinking*

van der Heijden, Bulkeley and Certomà (eds.), *Urban Climate Politics: Agency and Empowerment*

Linnér and Wibeck, *Sustainability Transformations: Agents and Drivers across Societies*

Betsill, Benney and Gerlak (eds.), *Agency in Earth System Governance*

Biermann and Kim (eds.), *Architectures of Earth System Governance: Institutional Complexity and Structural Transformation*

Baber and Bartlett (eds.), *Democratic Norms of Earth System Governance*

Djalante, Siebenhüner (eds.), *Adaptiveness: Changing Earth System Governance*

Behrman and Kent (eds.), *Climate Refugees*

Lamalle andStoett (eds.), *Representations and Rights of the Environment*

TRANSFORMING BIODIVERSITY GOVERNANCE

Edited by
INGRID J. VISSEREN-HAMAKERS
Radboud University, Netherlands

MARCEL T. J. KOK
PBL Netherlands Environmental Assessment Agency

CAMBRIDGE
UNIVERSITY PRESS

University Printing House, Cambridge CB2 8BS, United Kingdom

One Liberty Plaza, 20th Floor, New York, NY 10006, USA

477 Williamstown Road, Port Melbourne, VIC 3207, Australia

314–321, 3rd Floor, Plot 3, Splendor Forum, Jasola District Centre, New Delhi – 110025, India

103 Penang Road, #05–06/07, Visioncrest Commercial, Singapore 238467

Cambridge University Press is part of the University of Cambridge.

It furthers the University's mission by disseminating knowledge in the pursuit of education, learning, and research at the highest international levels of excellence.

www.cambridge.org
Information on this title: www.cambridge.org/9781108479745
DOI: 10.1017/9781108856348

© Cambridge University Press 2022

This work is in copyright. It is subject to statutory exceptions and to the provisions of relevant licensing agreements; with the exception of the Creative Commons version the link for which is provided below, no reproduction of any part of this work may take place without the written permission of Cambridge University Press.

An online version of this work is published at doi.org/10.1017/9781108856348 under a Creative Commons Open Access license CC-BY-NC-ND 4.0 which permits re-use, distribution and reproduction in any medium for non-commercial purposes providing appropriate credit to the original work is given. You may not distribute derivative works without permission. To view a copy of this license, visit https://creativecommons.org/licenses/by-nc-nd/4.0

All versions of this work may contain content reproduced under license from third parties.

Permission to reproduce this third-party content must be obtained from these third-parties directly.

When citing this work, please include a reference to the DOI 10.1017/9781108856348

First published 2022

Printed in the United Kingdom by TJ Books Limited, Padstow Cornwall

The first print run of this book uses FSC-certified paper.

A catalogue record for this publication is available from the British Library.

Library of Congress Cataloging-in-Publication Data
Names: Visseren-Hamakers, Ingrid J., 1970– editor. | Kok, Marcel T. J., 1968– editor.
Title: Transforming biodiversity governance / edited by Ingrid J. Visseren-Hamakers, Radboud Universiteit Nijmegen; Marcel T.J. Kok, PBL Netherlands Environmental Assessment Agency.
Description: Cambridge, United Kingdom ; New York, NY : Cambridge University Press, 2022. | "The book evolved through presentations of draft chapters at ESG conferences, numerous discussions at workshops, CBD sessions, and meetings of the RBG network" – ECIP introduction. | Includes bibliographical references and index.
Identifiers: LCCN 2021056164 (print) | LCCN 2021056165 (ebook) | ISBN 9781108479745 (hardback) | ISBN 9781108790741 (paperback) | ISBN 9781108856348 (ebook)
Subjects: LCSH: Biodiversity conservation – Law and legislation – Congresses.
Classification: LCC K3488.A6 T73 2022 (print) | LCC K3488.A6 (ebook) | DDC 346.04/695–dc23/eng/20220331
LC record available at https://lccn.loc.gov/2021056164
LC ebook record available at https://lccn.loc.gov/2021056165

ISBN 978-1-108-47974-5 Hardback

Cambridge University Press has no responsibility for the persistence or accuracy of URLs for external or third-party internet websites referred to in this publication and does not guarantee that any content on such websites is, or will remain, accurate or appropriate.

Contents

List of Contributors	*page* viii
Preface	xv
Acknowledgments	xvii
List of Abbreviations	xviii

Part I Introduction — 1

1 The Urgency of Transforming Biodiversity Governance — 3
INGRID J. VISSEREN-HAMAKERS AND MARCEL T. J. KOK

Part II Unpacking Central Concepts — 23

2 Defining Nature — 25
HANS KEUNE, MARCO IMMOVILLI, ROGER KELLER, SIMONE MAYNARD, PAMELA MCELWEE, ZSOLT MOLNÁR, GUNILLA A. OLSSON, UNNIKRISHNAN PAYYAPPALLIMANA, ANIK SCHNEIDERS, MACHTELD SCHOOLENBERG, SUNEETHA M. SUBRAMANIAN AND WOUTER VAN REETH

3 Global Biodiversity Governance: What Needs to Be Transformed? — 43
JOANNA MILLER SMALLWOOD, AMANDINE ORSINI, MARCEL T. J. KOK, CHRISTIAN PRIP AND KATARZYNA NEGACZ

4 How to Save a Million Species? Transformative Governance through Prioritization — 67
INGRID J. VISSEREN-HAMAKERS, BENJAMIN CASHORE, DERK LOORBACH, MARCEL T. J. KOK, SUSAN DE KONING, PIETER VULLERS AND ANNE VAN VEEN

Part III Cross-Cutting Issues Central to Transformative Biodiversity Governance — 91

5 One Health and Biodiversity — 93
HANS KEUNE, UNNIKRISHNAN PAYYAPPALLIMANA, SERGE MORAND AND SIMON R. RÜEGG

6 Biodiversity Finance and Transformative Governance: The Limitations of Innovative Financial Instruments — 115
RICHARD VAN DER HOFF AND NOWELLA ANYANGO-VAN ZWIETEN

7 Emerging Technologies in Biodiversity Governance: Gaps and Opportunities for Transformative Governance — 137
FLORIAN RABITZ, JESSE L. REYNOLDS AND ELSA TSIOUMANI

8 Rethinking and Upholding Justice and Equity in Transformative Biodiversity Governance — 155
JONATHAN PICKERING, BRENDAN COOLSAET, NEIL DAWSON, KIMBERLY MARION SUISEEYA, CRISTINA Y. A. INOUE AND MICHELLE LIM

9 Mainstreaming the Animal in Biodiversity Governance: Broadening the Moral and Legal Community to Nonhumans — 179
ANDREA SCHAPPER, INGRID J. VISSEREN-HAMAKERS, DAVID HUMPHREYS AND CEBUAN BLISS

10 Industry Responses to Evolving Regulation of Marine Bioprospecting in Polar Regions — 200
KRISTIN ROSENDAL AND JON BIRGER SKJÆRSETH

Part IV Transforming Biodiversity Governance in Different Contexts — 219

11 Transformative Biodiversity Governance for Protected and Conserved Areas — 221
JANICE WEATHERLEY-SINGH, MADHU RAO, ELIZABETH MATTHEWS, LILIAN PAINTER, LOVY RASOLOFOMANANA, KYAW T. LATT, ME'IRA MIZRAHI AND JAMES E. M. WATSON

12	The Convivial Conservation Imperative: Exploring "Biodiversity Impact Chains" to Support Structural Transformation BRAM BÜSCHER, KATE MASSARELLA, ROBERT COATES, SIERRA DEUTSCH, WOLFRAM DRESSLER, ROBERT FLETCHER, MARCO IMMOVILLI AND STASJA KOOT	244
13	Transformative Biodiversity Governance in Agricultural Landscapes: Taking Stock of Biodiversity Policy Integration and Looking Forward YVES ZINNGREBE, FIONA KINNIBURGH, MARJANNEKE J. VIJGE, SABINA J. KHAN AND HENS RUNHAAR	264
14	Cities and the Transformation of Biodiversity Governance HARRIET BULKELEY, LINJUN XIE, JUDY BUSH, KATHARINA ROCHELL, JULIE GREENWALT, HENS RUNHAAR, ERNITA VAN WYK, CATHY OKE AND INGRID COETZEE	293
15	Transformative Governance for Ocean Biodiversity BOLANLE ERINOSHO, HASHALI HAMUKUAYA, CLAIRE LAJAUNIE, ALANA MALINDE S. N. LANCASTER, MITCHELL LENNAN, PIERRE MAZZEGA, ELISA MORGERA AND BERNADETTE SNOW	313

Part V Strategic Reflections — 339

16	Enabling Transformative Biodiversity Governance in the Post-2020 Era MARCEL T. J. KOK, ELSA TSIOUMANI, CEBUAN BLISS, MARCO IMMOVILLI, HANS KEUNE, ELISA MORGERA, SIMON R. RÜEGG, ANDREA SCHAPPER, MARJANNEKE J. VIJGE, YVES ZINNGREBE AND INGRID J. VISSEREN-HAMAKERS	341

Index — 361

Contributors

Nowella Anyango-van Zwieten
Wageningen University, Netherlands

Cebuan Bliss
Radboud University, Netherlands

Harriet Bulkeley
Durham University, UK and Utrecht University, Netherlands

Bram Büscher
Wageningen University, Netherlands

Judy Bush
University of Melbourne, Australia

Benjamin Cashore
National University of Singapore

Robert Coates
Wageningen University, Netherlands

Ingrid Coetzee
ICLEI Cities Biodiversity Centre, South Africa

Brendan Coolsaet
Lille Catholic University, France and University of East Anglia, UK

List of Contributors

Neil Dawson
Lille Catholic University, France and University of East Anglia, UK

Susan de Koning
Radboud University, Netherlands

Sierra Deutsch
University of Zurich, Switzerland

Wolfram Dressler
University of Melbourne, Australia

Bolanle Erinosho
University of Cape Coast, Ghana

Robert Fletcher
Wageningen University, Netherlands

Julie Greenwalt
Go Green for Climate, Netherlands

Hashali Hamukuaya
Benguela Current Convention, Namibia

David Humphreys
Open University, UK

Marco Immovilli
Wageningen University, Netherlands

Cristina Y. A. Inoue
University of Brasília, Brazil and Radboud University, Netherlands

Roger Keller
University of Zurich, Switzerland

Hans Keune
University of Antwerp and INBO, Belgium

Sabina J. Khan
Helmholtz Center for Environmental Research – UFZ, Germany

Fiona Kinniburgh
Technical University of Munich, Germany

Marcel T. J. Kok
PBL Netherlands Environmental Assessment Agency, Netherlands

Stasja Koot
Wageningen University, Netherlands

Claire Lajaunie
Inserm, France

Alana Malinde S. N. Lancaster
The University of the West Indies, Barbados

Kyaw T. Latt
Wildlife Conservation Society, Myanmar

Mitchell Lennan
University of Strathclyde, UK

Michelle Lim
Macquarie University, Australia

Derk Loorbach
DRIFT, Erasmus University Rotterdam, Netherlands

Kimberly Marion Suiseeya
Northwestern University, USA

Kate Massarella
Wageningen University, Netherlands

Elizabeth Matthews
Wildlife Conservation Society, USA

Simone Maynard
IUCN Commission on Ecosystem Management, Australia

Pierre Mazzega
CNRS, France and University of Strathclyde, UK

Pamela McElwee
Rutgers University, USA

Me`ira Mizrahi
Wildlife Conservation Society, Myanmar

Zsolt Molnár
Centre for Ecological Research, Hungary

Serge Morand
CNRS – CIRAD, France

Elisa Morgera
University of Strathclyde, UK

Katarzyna Negacz
Free University of Amsterdam, Netherlands

Cathy Oke
University of Melbourne, Australia

Gunilla A. Olsson
University of Gothenburg, Sweden

Amandine Orsini
Université Saint-Louis – Bruxelles, Belgium

Lilian Painter
Wildlife Conservation Society, Bolivia

Unnikrishnan Payyappallimana
Transdisciplinary University, India

Jonathan Pickering
University of Canberra, Australia

Christian Prip
The Fridtjof Nansen Institute, Norway

Florian Rabitz
Kaunas University of Technology, Lithuania

Madhu Rao
Wildlife Conservation Society, Singapore

Lovy Rasolofomanana
Wildlife Conservation Society, Madagascar

Jesse L. Reynolds
University of California, Los Angeles, USA

Katharina Rochell
Utrecht University, Netherlands

Kristin Rosendal
The Fridtjof Nansen Institute, Norway

Simon R. Rüegg
University of Zurich, Switzerland

Hens Runhaar
Utrecht University and Wageningen University, Netherlands

Andrea Schapper
University of Stirling, UK

Anik Schneiders
Research Institute for Nature and Forest (INBO), Belgium

Machteld Schoolenberg
PBL Netherlands Environmental Assessment Agency, Netherlands

List of Contributors

Jon Birger Skjærseth
The Fridtjof Nansen Institute, Norway

Joanna Miller Smallwood
University of Sussex, UK

Bernadette Snow
University of Strathclyde, UK

Suneetha M. Subramanian
United Nations University, Japan

Elsa Tsioumani
University of Trento, Italy

Richard van der Hoff
Federal University of Minas Gerais, Brazil

Wouter Van Reeth
Research Institute for Nature and Forest (INBO), Belgium

Anne van Veen
Radboud University, Netherlands

Ernita van Wyk
ICLEI Cities Biodiversity Centre, South Africa

Marjanneke J. Vijge
Utrecht University, Netherlands

Ingrid J. Visseren-Hamakers
Radboud University, Netherlands

Pieter Vullers
Erasmus University Rotterdam, Netherlands

James E. M. Watson
University of Queensland, Australia

Janice Weatherley-Singh
WCS EU, Belgium

Linjun Xie
University of Nottingham Ningbo China

Yves Zinngrebe
Helmholtz Center for Environmental Research – UFZ, Germany and University of Göttingen, Germany

Preface

The idea for this book was conceived in December 2016 during the 13th Conference of the Parties to the Convention on Biological Diversity (CBD COP13) in Cancun, Mexico. Several members of the Rethinking Biodiversity Governance (RBG) network, a network of social scientists and policy practitioners working on biodiversity governance, were chatting during a coffee break in the hallway in front of the meeting rooms.

Discussions on global biodiversity governance after the 2020 deadline for the Aichi targets were starting, and the development of the Global Assessment of the Intergovernmental Science-Policy Platform on Biodiversity and Ecosystem Services (IPBES) was underway – the time seemed right for a book on transforming biodiversity governance. It was clear from the start that we wanted the book to become part of the tradition of the Earth System Governance (ESG) book series, since many of us have been active members of the ESG community for years, and earlier volumes in the community's series have inspired and shaped our own work in countless ways.

The book evolved through presentations of draft chapters at ESG conferences, numerous discussions at workshops, CBD sessions and meetings of the RBG network. The book reflects the diversity of views, disciplines, philosophical perspectives, motivations and areas of expertise of the RBG network. It's only through this sense of community that we were able to together contribute to the discussion on an issue as complex as transforming biodiversity governance.

The book reflects and contributes to current thinking on sustainable development. It is increasingly recognized among policy practitioners and scholars that fundamental societal change is needed to achieve the sustainability goals established by the international community, including those on addressing biodiversity loss. A rich debate on such transformations, transformative change and transitions is ongoing. How and the extent to which these much-needed fundamental changes can be governed is an outstanding question. The book has set out to contribute to this question – hence its title *Transforming Biodiversity Governance*: Sustainability transformations also require transformations in and of governance.

On the cover you see an image of a beaver, a transformative animal in its own right through its role as ecosystem engineer in shaping its environment and thereby the landscape. Please bear with us as we take a Dutch perspective in explaining the cover image. Beavers became extinct in the Netherlands in 1826 due to hunting. They were reintroduced in the late twentieth century, and also found their way into the Netherlands themselves from Belgium and Germany. Once in the Netherlands, they benefited from rewilding efforts and the climate adaptation policy of creating "room for the river," and the number of beavers increased across the country. The population grew to the extent that now in some parts of the country beavers are starting to be seen as a problem. Special management plans are being put in place to resist the transformative powers of the beaver; hunting is unfortunately again taking place. This short, Dutch history of the beaver illustrates how we as humans have to rethink how we relate to nature and animals, and thereby also need to rethink biodiversity governance. Instead of trying to manage nature for humans, we can learn to live with, and as part of, nature.

The aim of this book was to inform the development and implementation of the CBD Post-2020 Global Biodiversity Framework (GBF). These negotiations, as well as the writing of this book, were severely challenged by the COVID-19 pandemic. The pandemic made us all more aware in so many ways of the inextricable link between nature, biodiversity conservation, increasing risks of pandemics and human wellbeing. At the time of writing, we expect the book to be published around the finalization of the negotiations of the CBD Post-2020 GBF. We hope that the analyses in this book may inform and inspire its further implementation, and support the efforts of actors around the world to enable the transformative change that is so urgently needed for the conservation, and sustainable and equitable use, of biodiversity.

Acknowledgments

This book truly represents a community effort. So many people contributed in so many different ways. We are grateful to all of you.

First of all, a huge thanks to all the authors of the book for the wonderful cooperation, sharing of ideas and excellent chapters you have written. Discussing these chapters at various moments in time has been one of the most rewarding aspects of our joint efforts. We would especially also like to note the role of the first authors of the chapters in extending the author teams to include diverse geographical and disciplinary perspectives.

Thanks also to Frank Biermann, editor of this series, who supported the idea for this book from the start.

We would like to genuinely thank the chapter reviewers for taking the time to go through the chapters critically. All chapters were peer reviewed by an author from a different chapter in the book and an anonymous reviewer. These reviews helped tremendously in improving the chapters and strengthening the line of argumentation of the book.

Sincere thanks also to Caro Dijkman of Caro Grafico, who designed the figures in Chapters 1 and 4.

We are deeply indebted to Cebuan Bliss. On top of her role as coauthor in several chapters, Cebby took on numerous coordinating roles in the development of the book, including copy-editing all draft chapter texts, organizing the peer review process, and overseeing administration and communication. This book would not have been published without you, Cebby!

We are grateful to the Nijmegen School of Management of Radboud University and the PBL Netherlands Environmental Assessment Agency for covering the open access fees for the book.

Last but not least we would like to thank our respective families for their patience during this book project, which ended up lasting for about five years.

Abbreviations

ABMTs	– area-based management tools
ABNJ	– areas beyond national jurisdiction
ABS	– access and benefit-sharing
ACP	– African, Caribbean and Pacific
AHTEG	– ad hoc technical expert group
ART	– architecture for REDD+ transactions
ATS	– Antarctic Treaty System
BBNJ	– Marine Biodiversity Beyond the Limits of National Jurisdiction
BCC	– Benguela Current Convention
BECCS	– bioenergy with carbon capture and sequestration
BIC	– biodiversity impact chain
BIOFIN	– Biodiversity Finance Initiative
BIOPAMA	– Biodiversity and Protected Areas Management Program
BPI	– biodiversity policy integration
C40	– C40 Climate Leadership Group
CAP	– Common Agricultural Policy
CBD	– Convention on Biological Diversity
CBFP	– Congo Basin Forest Partnership
CCA	– community conserved area
CDR	– carbon dioxide removal
CHM	– common heritage of mankind
CITES	– Convention on International Trade in Endangered Species of Wild Fauna and Flora
CO_2	– carbon dioxide
COBA	– community based group
COP	– Conference of the Parties

List of Abbreviations

CMS	– Convention on the Conservation of Migratory Species of Wild Animals
COST	– European Cooperation on Science and Technology
CRA	– environmental reserve quota (*Cota de Reserva Ambiental*)
CSCP	– Collaborating Centre on Sustainable Consumption and Production
DAC	– direct air capture
DSI	– digital sequence information
EC	– European Commission
EBSAs	– ecologically or biologically significant marine areas
EID	– emerging infectious disease
ES	– ecosystem services
ESG	– environmental, social and governance
EU	– European Union
FAO	– Food and Agriculture Organization
FAWC	– Farm Animal Welfare Council
FPIC	– free, prior and informed consent
GBF	– Global Biodiversity Framework
GCA	– Global Commission on Adaptation
GEF	– Global Environment Facility
GELOSE	– Gestion Locale Sécurisée (secured local managed forests law)
GM	– genetically modified
IAS	– invasive alien species
ICC	– International Chamber of Commerce
ICCAs	– Indigenous and community conserved areas
ICDP	– integrated conservation and development project
ICLEI	– International Council for Local Environmental Initiatives (now Local Governments for Sustainability)
IG	– integrative governance
IMET	– Integrated Management Effectiveness Tool
IPBES	– Intergovernmental Science-Policy Platform on Biodiversity and Ecosystem Services
IPCC	– Intergovernmental Panel on Climate Change
IPLC	– Indigenous peoples and local communities
IPRs	– intellectual property rights (including patents)
ITPGRFA	– International Treaty on Plant Genetic Resources for Food and Agriculture
IUCN	– International Union for the Conservation of Nature

IUU	– illegal, unreported and unregulated
KBA	– key biodiversity area
KIFCA	– Kyeintali Inshore Fisheries Co-Management Association
LBSAPs	– local biodiversity strategies and action plans
LED-R	– low emissions rural development
LME	– large marine ecosystem
LMMA	– locally managed marine areas
LMOs	– living modified organisms
MA	– Millennium Ecosystem Assessment
MARISMA	– Marine Spatial Management and Governance Project
MAT	– mutually agreed terms
METT	– Management Effectiveness Tracking Tool
MGR	– marine genetic resources
MPA	– marine protected area
MSP	– marine spatial planning
NAMA	– nationally appropriate mitigation action
NBSAPs	– national biodiversity strategies and action plans
NCP	– nature's contributions to people
NEOH	– Network for EcoHealth and One Health
NGFS	– Network for Greening the Financial System
NGO	– nongovernmental organization
ODA	– official development assistance
OECD	– Organisation for Economic Co-operation and Development
OECMs	– other effective area-based conservation measures
OIE	– [Office International des Epizooties] World Organisation for Animal Health
PA	– protected area
PAME	– Protected Area Management Effectiveness
PES	– payments for ecosystems services
PIC	– prior informed consent
PIP Framework	– Pandemic Influenza Preparedness Framework
PPP	– public–private partnerships
RAN	– Rainforest Action Network
REDD+	– Reducing Emissions from Deforestation and Forest Degradation
RFMOs	– regional fisheries management organizations
RoN	– Rights of Nature
RTQMM	– Rio Tinto QIT Madagascar Mineral
SBSTTA	– Subsidiary Body on Scientific, Technical and Technological Advice

List of Abbreviations

SDGs	– sustainable development goals
SOI	– Sustainable Ocean Initiative
SRM	– solar radiation modification
TBG	– transformative biodiversity governance
TEEB	– the economics of ecosystems and biodiversity
TRIPS	– trade-related aspects of intellectual property rights
UFF	– Unlocking Forest Finance
UN	– United Nations
UNCLOS	– United Nations Convention on the Law of the Sea
UNDP	– United Nations Development Programme
UNDRIP	– United Nations Declaration on the Rights of Indigenous Peoples
UNEP	– United Nations Environment Programme
UNESCO	– United Nations Educational, Scientific and Cultural Organization
UNFCCC	– United Nations Framework Convention on Climate Change
UNFSA	– United Nations Fish Stocks Agreement
UNGA	– United Nations General Assembly
WCPA	– World Commission on Protected Areas
WCS	– Wildlife Conservation Society
WEF	– World Economic Forum
WHO	– World Health Organization
WIPO	– World Intellectual Property Organization
WRI	– World Resources Institute
WTO	– World Trade Organization

Part I
Introduction

1

The Urgency of Transforming Biodiversity Governance

INGRID J. VISSEREN-HAMAKERS AND MARCEL T. J. KOK

1.1 Introduction: The Third Era in Global Biodiversity Governance

This book is written at a vital time for biodiversity around the world. Biodiversity is threatened more than ever before in human history, and nature and its vital contributions to people are deteriorating worldwide, as highlighted by various recent reports (CBD, 2020a; EEA, 2019; IPBES, 2019; WWF, 2020). This is not only a problem for these ecosystems and their inhabitants, but also for humans, since we depend on biodiversity for many vital processes such as food production and provision of natural resources. These risks of biodiversity loss are increasingly recognized among policymakers, academics and society at large (IPBES, 2019; WEF, 2021).

The worldwide deterioration of biodiversity is taking place despite over half a century of efforts to combat biodiversity loss by governments, civil society and, increasingly, business, at all levels of governance from the local to the global. Past and ongoing efforts are therefore not effectively supporting the conservation and sustainable and equitable use of biodiversity, and this worldwide failure to address biodiversity loss has created a growing consensus that fundamental, transformative changes are needed in order to reverse these trends, or "bend the curve of biodiversity loss" (IPBES, 2019; Mace et al., 2018).

This increasing attention for transformative change can be seen as the start of a new, third era in global biodiversity governance. During the first era, early nature conservation policies were developed in silos – the focus was on conserving biodiversity and developing and better managing protected areas. These older intergovernmental processes, such as the Convention on International Trade in Endangered Species of Wild Fauna and Flora (CITES), the Ramsar Convention on Wetlands and the Convention on the Conservation of Migratory Species of Wild Animals (CMS), date back to the 1970s.

The central intergovernmental biodiversity process, the Convention on Biological Diversity (CBD), was adopted in 1992 at the United Nations Conference on Environment and Development (UNCED), along with the United Nations Framework Convention on Climate Change (UNFCCC) and the UN Convention to Combat Desertification (UNCCD) (Le Prestre, 2002). The CBD has three main objectives, namely the conservation of biological diversity, the sustainable use of its components and the fair and equitable sharing of the benefits arising out of the utilization of genetic resources (CBD, 1992). In 2002, parties to the CBD agreed on targets to significantly reduce of the rate of biodiversity loss by 2010. After this target was not met, the CBD developed new targets for 2020, the Aichi

Table 1.1 *Overview of the Aichi Targets (CBD, 2010)*

Strategic goal	Target
A. Addressing the underlying causes of biodiversity loss	1. Raising awareness
	2. Integration of biodiversity values into national development policies
	3. Elimination of harmful incentives and development of positive incentives
	4. Sustainable production and consumption
B. Reducing the direct pressures on biodiversity	5. Loss of natural habitats
	6. Sustainable fish harvesting
	7. Sustainable agriculture, aquaculture and forestry
	8. Pollution
	9. IAS
	10. Coral reefs and other vulnerable ecosystems
C. Safeguarding ecosystems, species and genetic biodiversity	11. Protected areas
	12. Threatened species
	13. Genetic diversity of cultivated plants and farmed animals
D. Enhancing benefits	14. Ecosystem services
	15. Conservation and restoration of carbon stocks
	16. Nagoya Protocol
E. Enhancing implementation	17. NBSAPs
	18. Indigenous and local communities
	19. Knowledge, science base and technologies
	20. Financial resources

targets, as part of its Strategic Plan 2011–2020 (Table 1.1). With this strategic plan, a second era started as attention shifted toward mainstreaming biodiversity in the most relevant policy domains and sectors, such as forestry and fisheries. However, most of these targets, again, were not met (CBD, 2010; 2020b) (also see Chapter 3).

The adoption of the United Nations Sustainable Development Goals (SDGs) in 2015 can be seen as the start of the third biodiversity governance era. Biodiversity concerns are well integrated into the SDGs (See SDG 14 and 15 in Table 1.2), and are part of a broader transformative change agenda for sustainability and environmental justice. The focus of biodiversity policy has thus broadened over time, and the call for transformative change now recognizes the need for deepening such efforts. In this third era, all three strategies are recognized as vital: stepping up protection and restoration of nature, broadening biodiversity efforts across society and deepening effects to enable transformative change (as elaborated in Section 1.3 below). With the COVID-19 pandemic, discussions on the urgency of such transformative change and changing our relationship with nature have further intensified (see e.g. Platto et al., 2020 and Chapter 5).

Despite growing societal and academic interest in transformative change, it is far from clear how to enable, achieve or accelerate transformative change for biodiversity. This book aims to provide and further develop a governance perspective on achieving such

Table 1.2 *The United Nations Sustainable Development Goals (SDGs) (UN, 2015)*

SDG	Topic
1	No poverty
2	Zero hunger
3	Good health and wellbeing
4	Quality education
5	Gender equality
6	Clean water and sanitation
7	Affordable and clean energy
8	Decent work and economic growth
9	Industry, innovation and infrastructure
10	Reduced inequality
11	Sustainable cities and communities
12	Responsible consumption and production
13	Climate action
14	Life below water
15	Life on land
16	Peace and justice, strong institutions
17	Partnerships to achieve the goals

transformative change. The book captures the state-of-the-art knowledge on transformative biodiversity governance and further explores its practical implications in various contexts and issues relevant for the long-term biodiversity policy agenda.

The book is written against the backdrop of the development of the Post-2020 Global Biodiversity Framework (GBF), the new global framework following the CBD Strategic Framework 2011–2020 and its Aichi targets. At the time of writing, the GBF was expected to be adopted in 2022 at the 15th Conference of the Parties of the CBD (COP15) in Kunming, China. COP15 was originally due to be held in 2020 but was postponed because of the COVID-19 pandemic. The GBF represents the guiding policy framework for biodiversity action across societies and governments, and, in our view, should provide a global answer to shaping transformative change in the multilateral system, and through implementation at the national and subnational levels by state and nonstate actors. We hope that the book will contribute to transformative action for biodiversity in the implementation of the Post-2020 GBF around the world over the coming years.

This first chapter is organized as follows. We first set the stage by providing an overview of the current state of biodiversity, causes of biodiversity loss and its implications. We then introduce the concepts of transformative change and governance. The two final sections explain the book's logic and organization, and provide an overview of the book.

1.2 The Problem of Biodiversity Loss and the Potential for Transformative Change

According to the Global Assessment of the Intergovernmental Science-Policy Platform on Biodiversity and Ecosystem Services (IPBES GA),[1] "nature, and its vital contributions to people, which together embody biodiversity and ecosystem functions and services, are deteriorating worldwide" (Díaz et al., 2019: 10). Most indicators of the state of nature are declining, including the number and population size of wild species, the number of local varieties of domesticated species, the distinctness of ecological communities and the extent and integrity of many terrestrial and aquatic ecosystems. Around one million species are threatened with extinction. Biodiversity in areas owned, managed or used by Indigenous People and local communities (IPLC) is declining less rapidly than elsewhere (Díaz et al., 2019).

This biodiversity loss has accelerated over the past fifty years (the period analyzed by the IPBES GA), and is caused by the following *direct drivers*: land and sea use change, with agricultural expansion representing the most important form of land-use change; direct exploitation, and especially overexploitation, of animals, plants and other organisms, mainly through harvesting, logging, hunting and fishing; climate change, which is becoming an increasingly important driver; pollution and invasive alien species. Land-use change is the main direct driver in terrestrial areas, and direct exploitation is the most important one in marine systems. These trends in nature and its contributions to people are projected to worsen over the coming decades, unevenly in different regions. These direct drivers are influenced by *indirect drivers*, or underlying causes, which can be demographic (e.g. human population dynamics), sociocultural (e.g. consumption patterns), economic (e.g. production and trade), technological, or relating to institutions, governance, conflicts and epidemics. These indirect drivers are underpinned by societal values and behaviors (Díaz et al., 2019).

Biodiversity issues are an integral part of broader sustainable development debates, and are intertwined with many other sustainability issues, including climate change. Humans depend on nature and biodiversity for human health through the production of food, medicines and clean water, among others, and the provision of natural resources, such as timber. Nature also provides regulatory ecosystem services that are vital for humans, including regulating air quality and climate. Nature is thus essential for achieving the SDGs, and biodiversity loss and ecosystem degradation will undermine progress toward the vast majority of the SDG targets, as the capacity of nature to provide these services has declined significantly over the last decades.

In this context, it is important to address biodiversity loss coherently with climate change mitigation and adaptation, since there are both synergies and trade-offs among biodiversity and climate change efforts. Limiting climate change to well below 2 degrees Celsius is crucial to reducing the impacts on nature and ecosystem services, but some large-scale land-based climate change mitigation measures, such as large-scale afforestation and

---

[1] This section relies strongly on the IPBES GA because it represents the most recent and comprehensive global assessment of biodiversity-relevant knowledge. Both authors were involved in the GA.

reforestation or bioenergy crop development, will have negative impacts on biodiversity. Other efforts, such as ecosystem restoration or avoiding and reducing deforestation, can provide synergies between climate and biodiversity goals (Díaz et al., 2019; Pörtner et al., 2021).

As discussed above, biodiversity policy has so far not been able to deliver the intended results, and it is clear that conservation efforts need to be improved, broadened and deepened: "Goals for conserving and sustainably using nature and achieving sustainability cannot be met by current trajectories, and goals for 2030 and beyond may only be achieved through transformative changes across economic, social, political and technological factors" (Díaz et al., 2019: 14). Explorative scenario-projections, covering a wide range of plausible socioeconomic pathways and biodiversity policies, indicate that global biodiversity will continue to decline, even under optimistic socioeconomic pathways oriented toward sustainability. Only specific solution-oriented scenarios that step up ambition levels in conservation and restoration, address indirect drivers of biodiversity loss and capitalize on nature-based solutions, which use nature to address societal challenges, are able to bend the curve while also mitigating climate change (Kok et al., under review; Leclère et al., 2020). However, many of the social dimensions of such scenario analyses require further attention to evaluate the equity implications of these future pathways (Ellis and Mehrabi, 2019; Mehrabi et al., 2018; Otero et al., 2020; Schleicher et al., 2019). Transformative change is thus urgently needed.

1.3 Understanding, Shaping and Delivering Transformative Change and Governance

1.3.1 Transformative Change

As accurately noted by Otsuki (2015: 1): "Current debates on sustainable development are shifting their emphasis from the technocratic and regulatory fix of environmental problems to more fundamental and transformative changes in social-political processes and economic relations." However, discussions on societal transformations are of course not new (see for a detailed overview of the literature on sustainability transformations Linnér and Wibeck [2019]). The concept of social transformation generally "implies an underlying notion of the way society and culture change in response to such factors as economic growth, war, or political upheavals" (Castles, 2001: 15). Often-named examples include the "great transformation" (Polanyi, 1944) in Western societies brought about by industrialization and modernization, or more recent changes such as decolonization (Castles, 2001).

Scoones et al. (2020) distinguish structural, systemic and enabling approaches to conceptualizing transformations, with structural approaches focused more on societal change, systemic approaches on transitions in specific socioecological systems, and enabling approaches on developing capacities for change. Others differentiate between discussions on transformations and transitions (Grin et al., 2010), with the former focused on societal change and the latter on change in subsystems (e.g. the food, energy or mobility systems). These two approaches are also rooted in different literatures (Hölscher et al., 2018;

Loorbach et al., 2017). In our view, all these different approaches can be seen as complementary (see Chapter 4 for a more elaborate overview of the literatures on transformations and transitions, and their governance).

Transformative change can be differentiated from incremental or gradual change, which often occurs as a result of disturbances and is often aimed at resolving problems without changing existing systems or structures, although there are incremental changes that can contribute to transformations (Termeer et al., 2017). Transformative change incorporates both personal and social transformation (Chaffin et al., 2016; Otsuki, 2015), and includes shifts in values and beliefs, and patterns of social behavior (Chaffin et al., 2016).

Burch et al. (2019) highlight that transformations can be studied analytically, normatively and critically. Although debates among academics and policymakers on transformative change toward sustainability have often remained rather apolitical, a more critical perspective has emerged that incorporates politics, power and equity issues in the debates on transformation (see e.g. Chaffin et al., 2016; Lawhon and Murphy, 2012). Transformations include the making of "hard choices" by decision-makers (Meadowcroft, 2009: 326). Blythe et al. (2018) highlight the potential risks of apolitical approaches to transformative change, arguing that consideration of the politics of transformative change is necessary to address these risks, which include: shifting the burden of response onto vulnerable parties; the transformation discourse may be used to justify business-as-usual, pays insufficient attention to social differentiation and excludes the possibility of non-transformation or resistance; and insufficient treatment of power and politics can threaten the legitimacy of the discourse of transformation. In this book we recognize these risks and actually place them center stage by focusing on the governance of and for such transformations.

The IPBES GA defines transformative change as a fundamental, system-wide reorganization across technological, economic and social factors, including paradigms, goals and values (Díaz et al., 2019). Building on this definition, we here define *transformative change* as follows:

a fundamental, society-wide reorganization across technological, economic and social factors and structures, including paradigms, goals and values.

With this renewed definition, we emphasize changes in generic, societal structures. Such a society-wide transformation encompasses transitions in specific subsystems or sectors, and is necessary, since current societal structures inhibit sustainable development – they actually represent the underlying causes of biodiversity loss. Thereby, transformative change addresses both generic societal underlying causes and underlying causes in specific transitions (see Chapter 4 for an extensive discussion on the relationships between transformations, transitions, transformative change and transformative governance).

Transformative solutions are often synergistic: By focusing on the indirect drivers, they simultaneously address multiple sustainability issues, since the same indirect drivers simultaneously cause various problems. An example is the development of healthy and sustainable food systems, including through reducing production and consumption of animal products (especially in developed and newly industrialized countries), which can support progress on the majority of SDGs, and also addresses animal interests (Visseren-Hamakers, 2020). With this emphasis on the societal underlying causes of environmental problems, environmental policy becomes less "environmental" and increasingly integrated into mainstream policy and politics, becoming an integral part of discussions on the economy, innovation, development and societal values (also see Biermann, 2021).

While this book is focused on transforming biodiversity governance, we explicitly reflect on this issue as embedded in discussions on transformative change toward sustainability more broadly. We do so because biodiversity and other environmental and social justice issues are interwoven, and broader societal transformations are necessary to address all of these sustainable development issues.

1.3.2 Transformative Governance

While a burgeoning literature discusses transformative change, less research investigates how to govern such transformations (Chaffin et al., 2016; Patterson et al., 2017), and very few authors have specifically used the concept of transformative governance (Chaffin et al., 2016; Colloff et al., 2017; Visseren-Hamakers et al., 2021). Chaffin et al. (2016: 400) define transformative environmental governance as "an approach to environmental governance that has the capacity to respond to, manage, and trigger regime shifts in coupled socio-ecological systems at multiple scales." It thus has the capacity to shape nonlinear change. An important literature related to transformative governance is work on "transition management," defined as "the attempt to influence the societal system into a more sustainable direction, ultimately resolving the persistent problem(s) involved" (Grin et al., 2010: 108). The thinking on governing transformative change has thus so far focused on systemic – and not necessarily societal – change.

Hence, there is a difference between the concepts of transformative change and transformative governance, with change referring to the actual shift and governance to "steering" the shift, although some authors do not clearly differentiate between the two concepts (e.g. Chaffin et al., 2016). An important question is the extent to which the shift can actually be governed (Meadowcroft, 2009), with some authors noting that transformative sustainable development "is a contingent and creative process, which cannot be readily planned" (Otsuki, 2015: 4). Chaffin et al. (2016) list several constraints and opportunities for transformative governance, with constraints including: entrenched power relations, capitalism and dominant economic and political subsystems, and cognitive limits of humans; and opportunities including: law, formal institutions and governmental structure, previous success of adaptive governance, and human agency and imagination (Chaffin et al., 2016: 411). Interestingly, all of these opportunities and constraints are part of the underlying causes of biodiversity loss that need to be addressed through transformative change.

Transformative governance is deliberate (Chaffin et al., 2016), and inherently political (Blythe et al., 2018; Patterson et al., 2017), since the desired direction of the transformation is negotiated and contested, and power relations will change because of the transformation (Chaffin et al., 2016). Current vested interests (including in dominant technologies) are expected to inhibit, challenge, slow or downsize transformative change, among others, through "lock-ins" (see e.g. Blythe et al., 2018; Chaffin et al., 2016; Meadowcroft, 2009).

Transformative governance is about framing and agenda setting, and requires leadership, financial investment and capacity for learning. Also, the change needs to be increasingly institutionalized (Chaffin et al., 2016).

Literature on earth system governance has explored different ways of conceptualizing the governance of transformations. Burch et al. (2019) and Patterson et al. (2017) differentiate between the following conceptualizations of governing transformations:

- Governance for transformations (i.e. governance that creates the conditions for transformation to emerge from complex dynamics in socio-technical-ecological systems),
- Governance of transformations (i.e. governance to actively trigger and steer a transformation process),
- Transformations in governance (i.e. transformative change in governance regimes).

Based on these insights and earlier definitions on environmental governance (Biermann et al., 2010), we here define *transformative governance* as:

The formal and informal (public and private) rules, rule-making systems and actor-networks at all levels of human society (from local to global) that enable transformative change, in our case, towards biodiversity conservation and sustainable development more broadly

(Visseren-Hamakers et al., 2021: 21)

Since governing transformative change is inherently difficult because of its political character, transformative governance needs to take on board various lessons learned from the governance literature. We therefore propose that, based on Visseren-Hamakers et al. (2021), transformative governance includes five governance approaches, namely: integrative, inclusive, transdisciplinary, adaptive and anticipatory governance, which are based on various niches in the governance literature. These governance approaches have been studied separately in detail, and in the literature on sustainability transformations combinations of these approaches are often recognized as important (Linnér and Wibeck, 2019). We hypothesize that governance can only become transformative when the five governance approaches are (Visseren-Hamakers et al., 2021):

a) focused on addressing the underlying causes of unsustainability;
b) implemented in conjunction; and
c) operationalized in the following specific manners.

Thereby, in order to be transformative, governance needs to be:

1. *Integrative*, operationalized in ways that ensure solutions also have sustainable impacts at other scales and locations, on other issues and in other sectors (see e.g. Castán Broto et al., 2019; Chaffin et al., 2016; Visseren-Hamakers, 2015; 2018a; 2018b; Visseren-Hamakers et al., 2021; Wagner and Wilhelmer, 2017);
2. *Inclusive*, in order to empower and emancipate those whose interests are currently not being met and who represent values that constitute transformative change toward sustainability (see e.g. Biermann et al., 2010; Blythe et al., 2018; Chaffin et al., 2016; Li and Kampmann, 2017; Meadowcroft, 2009; Otsuki, 2015);

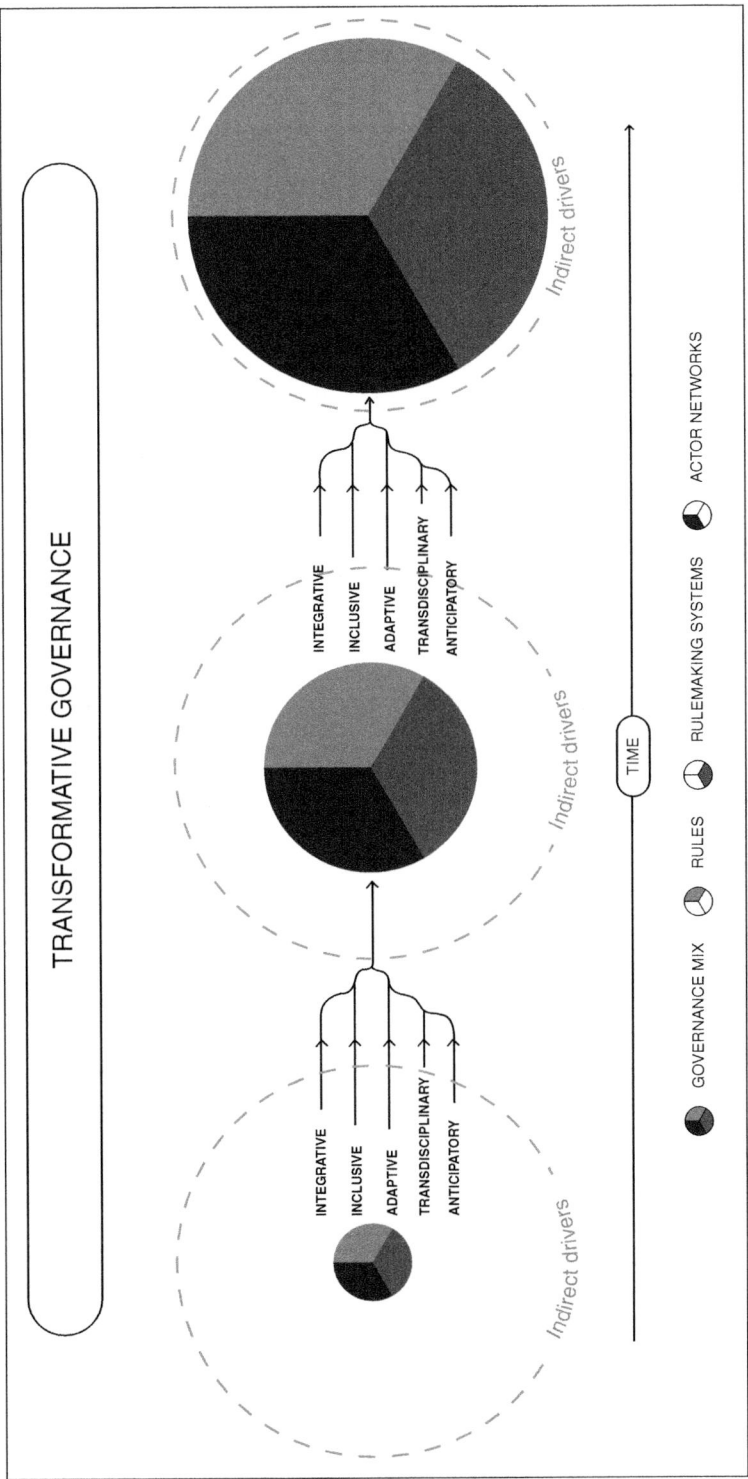

Figure 1.1 Transformative governance
Governance (including rules, rulemaking systems and actor networks) becomes transformative if integrative, inclusive, adaptive, transdisciplinary and anticipatory governance approaches are: 1) implemented in conjunction; 2) operationalized in a specific manner; and 3) focused on addressing the indirect drivers underlying sustainability issues. Over time, governance then becomes increasingly capable of addressing the indirect drivers (as indicated by the growth of the governance mix from left to right, i.e. over time). (adapted from Visseren-Hamakers et al., 2021).

3. *Adaptive*, since transformative change and governance, and our understanding of them, are moving targets, so governance needs to enable learning, experimentation, reflexivity, monitoring and feedback (see e.g. Blythe et al., 2018; Chaffin et al., 2016; Meadowcroft, 2009; Otsuki, 2015; Van den Bergh et al., 2011; Wagner and Wilhelmer, 2017; Wolfram, 2016);
4. *Transdisciplinary*,[2] in ways that recognize different knowledge systems, and support the inclusion of sustainable and equitable values by focusing on types of knowledge that are currently underrepresented (see e.g. Blythe et al., 2018; Colloff et al., 2017; Chaffin et al., 2016; Keitsch and Vermeulen, 2021; Moser, 2016; Scoones et al., 2020); and
5. *Anticipatory*, in ways that apply the precautionary principle when governing in the present for uncertain future developments, and especially the development or use of new technologies (see e.g. Burch et al., 2019; ESG, 2018; Guston, 2014).[3]

With this operationalization, transformative governance is focused on the underlying societal causes of unsustainability while being cognizant of relationships between issues, sectors, scales and places, aiming to emancipate those holding transformative sustainability values, governing through learning, incorporating different knowledge systems and taking a precautionary stance in situations of uncertainty. Any actor can contribute to transformative governance, and governance mixes can be polycentric in character, encompassing initiatives by actors operating in different places, sectors or at different levels of governance. All actors can regularly evaluate whether the governance mix includes the necessary governance instruments to address the indirect drivers underlying a specific sustainability issue, and governance mixes will need to evolve as sustainability transformations progress. Over time, governance will become increasingly transformative, and transformative governance will become easier, as societal structures increasingly become sustainable (see also Chapter 4).

As a whole, the book does not take a specific stance on the various academic and theoretical debates on transformative change and governance, but embraces the diversity of approaches. Although this first chapter highlights structural approaches to transformative change, given the definitions of transformative change and governance above, we see this structural change as embedding systemic and enabling approaches to transformations. The various chapters in the book can be positioned differently in the various approaches:

- Highlighting structural, systemic and/or enabling approaches to transformations;
- Studying transformations analytically, normatively and/or critically;
- Focusing on governance for transformations, governance of transformations and/or transformations in governance.

---

[2] We use the term transdisciplinary governance, instead of pluralist governance, as Visseren-Hamakers et al. (2021) do, in order to use a more generic term, instead of referring directly to the literature on pluralism.
[3] While Visseren-Hamakers et al. (2021) distinguish four governance approaches, based on chapter 6 of the IPBES GA (Razzaque et al., 2019), we have here added anticipatory governance as a fifth approach.

1.4 Characteristics, Aim and Research Questions of the Book

The *aim* of the book is to enhance our understanding of ways forward for transformative biodiversity governance. With this, the book aims to inform the development and implementation of transformative biodiversity policies and action.

The book addresses the following *research questions*:

What are lessons learned from existing attempts to:
a) Address the underlying causes of biodiversity loss?
b) Apply different approaches to, and instruments for, transformative governance (as operationalized in the above)?

The book is part of the Earth System Governance series at Cambridge University Press, which aims to draw lessons from the research of the global Earth System Governance Project, a global network of scholars in the social sciences and humanities working on governance and global environmental change. By drawing lessons from past, and explaining current, attempts for transformative biodiversity governance against the backdrop of the Post-2020 GBF, the book fits well into this series, especially since governance perspectives on biodiversity remain relatively underrepresented as compared to other sustainability issues such as climate change. One of the main added values of the book is its governance perspective on transformative change. As stated earlier, such a governance lens on transformative change is relatively new, and such insights from the perspective of the earth system governance community on transformative change for biodiversity conservation and sustainable development more broadly are urgently needed. Such a governance angle implies a multiactor perspective throughout the book, acknowledging and critically reflecting on the role of governmental, market and civil society actors in governing biodiversity. With this, the book builds on earlier contributions to the series, especially Linnér and Wibeck (2019), by further delving into the governance of transformative change.

The idea for a book on "transforming biodiversity governance" was born in discussions among members of the Rethinking Biodiversity Governance network, an informal network of academics and practitioners interested in biodiversity governance. Because our community includes both academics and policymakers, we have aimed to develop a book that is academic but policy-relevant.

1.5 Overview of the Book

The book is organized into five sections. Following this introductory part, Part II focuses on unpacking the central concepts of the book. Parts III and IV respectively focus on cross-cutting issues and key contexts that are vital to biodiversity conservation and its sustainable and equitable use. Part V strategically reflects on the insights developed throughout the book.

All chapters are built around broad, reflexive literature reviews. The chapters are focused on possible solutions, based on a critical reflection on past policies and practices. The book explicitly incorporates insights from different ontological, epistemological and theoretical perspectives to ensure coverage of various relevant literatures. The chapters include local, national, regional, global or multilevel lenses. All chapters have been peer reviewed by two reviewers.

In answering the research questions, each chapter focuses on one or multiple underlying causes of biodiversity loss, and/or one or more approaches to transformative governance. Each chapter includes an introduction of the issue (problem, main underlying causes), existing attempts to address the underlying causes of biodiversity loss, governance approach(es) and ways forward. In this way, the book provides rich insights into the diversity of current thinking on transforming biodiversity governance.

After this introductory chapter, Chapter 2 illustrates how nature has been defined in the context of shifts in biodiversity governance in recent decades, and how different stakeholders have engaged with these concepts. The chapter aims to show that nature is defined, and cannot be taken for granted as one objectifiable concept. The concepts of biodiversity, wilderness, intrinsic value and protected areas are introduced, and the concept of landscape is illustrated regarding ecosystem services and biocultural diversity. Furthermore, instrumental and relational values of nature are discussed. Conferring nature with legal rights (rights of nature) is introduced as a hybrid form of biodiversity governance merging Western and non-Western ontologies and definitions of nature. The chapter also discusses the importance of scenarios for nature in order to develop alternative pathways grounded on value pluralism. It concludes that defining nature is far from an objective and conflict-free exercise. Instead of reductionist approaches, the authors promote pluralistic approaches, highlight the importance of transparency and warn for the danger of treating concepts and approaches as truth-claims, making them less open to other perspectives.

Chapter 3 focuses on global biodiversity governance. The CBD is discussed as the main international treaty governing biodiversity. Its Post-2020 GBF aims to transform biodiversity governance to steer the necessary transformative change to halt biodiversity loss. For this undertaking, the CBD operates alongside multiple international conventions and international governmental and nongovernmental organizations at different scales that together form global biodiversity governance. The chapter presents what needs to be transformed within global biodiversity governance and discusses ways to achieve such transformation. It begins with a historical account of the evolution of global biodiversity governance. A "regime complex" lens is then used to show why biodiversity governance approaches have to intervene with sectors responsible for biodiversity loss such as agriculture, trade and development, and reflections are made on the implementation of global biodiversity law and policies. The conclusion considers how obstacles can be overcome to achieve true transformation.

Chapter 4 aims to understand why the current state of biodiversity is so fragile, despite over half a century of global conservation efforts, and develop insights for more effective ways forward. The chapter generates insights by integrating largely disconnected literatures that have sought to understand how to govern transformative change, transformations and transitions. It pays particular attention to the role of four distinct sustainability problem conceptions, namely commons, optimization, compromise and prioritization. Combining insights on transformations and transitions allows more focused attention to the generic societal underlying causes of sustainability issues and integrative governance of transitions. Through integrating problem type thinking, the chapter shows that treating biodiversity loss, and thereby ecocentric, compassionate and just sustainable development, as a priority is an essential part of transformative governance. Such prioritization radically changes governance: Governance mixes that combine instruments from all four problem conceptions will need to evolve over time for governance to become increasingly transformative.

The main aim of Chapter 5 is to discuss linkages between nature and generic health from a One Health as well as a transformative biodiversity governance perspective. The transformative governance ambitions of being integrative, inclusive, transdisciplinary, adaptive and anticipatory resonate quite well with One Health as an overarching concept for nature–health linkages. Especially during the COVID-19 pandemic, interest in One Health broadened. But what does, or can, it entail? What is the beauty of One Health in the eyes of different beholders? The chapter outlines different aspects and interpretations of One Health to illustrate both its potential and challenges. This includes integrative ambitions of including animal, human, plant and ecosystem health, as well as structural societal drivers of these "healths" and related complexity.

Chapter 6 critically discusses the role of innovative financial instruments in transformative biodiversity governance. These instruments are a subset of the broader spectrum of biodiversity finance instruments and directly mobilize financial resources for biodiversity conservation, compensate negative impacts of economic activity or manage risks of biodiversity loss. The chapter presents four general arguments: innovative financial instruments (1) conceptualize nature from an anthropocentric, mechanical and managerial perspective; (2) emphasize monetary values at the expense of others; (3) frame uncertainty as a manageable risk and (4) integrate different sectors, levels and stakeholders without challenging the foundations of existing systems and relations. These arguments underscore the limitations of innovative financial instruments in most dimensions of transformative governance (particularly inclusive and transdisciplinary governance), while offering some opportunities in others (i.e. integrative governance). The chapter's assessment of these instruments critically challenges their capacity for fostering transformative governance, although they may be useful as component of broader and more fundamental developments.

Chapter 7 discusses the relationships between biodiversity and emerging technologies. Emerging technologies have potentially far-reaching impacts on the conservation and sustainable and equitable use of biodiversity. Simultaneously, biodiversity increasingly serves as an input for novel technological applications. The chapter assesses the relationship between the CBD regime and the governance of three sets of emerging technologies: climate-related geoengineering (carbon dioxide removal and solar radiation modification), synthetic biology (including gene drives) as well as bioinformatics and digital sequence information. It presents an overview of relevant applications of these technologies, including potential positive and negative impacts on the CBD's objectives; explores the state of relevant deliberations under the CBD and other intergovernmental fora, including normative gaps and opportunities for action; and assesses the extent to which they could support transformative governance of technologies and biodiversity from the vantage points of adaptiveness, integration, anticipation, inclusion and transdisciplinarity.

Chapter 8 assesses how principles of justice and equity should be interpreted and upheld in efforts to pursue transformative biodiversity governance. Justice and equity are not only core social values but also key to addressing biodiversity decline. The chapter argues that the depth, scale and urgency of transformative change required demand heightened attention to both existing injustices and the advancement of multiple dimensions of justice, including procedural justice, recognition and distributive justice. It addresses questions of justice arising at three key stages of biodiversity governance: decision-making processes, resource mobilization and allocation, and implementation. Building on understandings of transformative governance as being both inclusive and integrative, the chapter highlights potential synergies and trade-offs between environmental sustainability and justice. The findings converge on the need for a "just transformation" of biodiversity governance.

Chapter 9 argues that transformative biodiversity governance requires mainstreaming the interests of the individual animal. Applying an integrative governance perspective, the chapter brings together debates from animal and biodiversity governance systems through a literature review and document analysis on animal rights and welfare, rights of nature (Earth jurisprudence), One Health and One Welfare, and compassionate conservation. It shows that, especially through rights-based approaches, moral and legal communities are expanding beyond humans to include nature and nonhuman animals. Since Earth jurisprudence does not explicitly recognize the interests of the individual animal, and the animal rights discourse does not include flora or natural objects, both approaches are necessary to complete the shift from dominant anthropocentric ontologies to a more holistic and ecocentric approach that includes recognition of individual animals. Such a shift is vital to enact the transformative change required for a biodiversity governance model in which justice between species is integral.

Chapter 10 focuses on bioprospecting. While it has potential to create high-value products in the pharmaceutical, cosmetics, food and other life science-based industries, bioprospecting the deep oceans beyond national jurisdictions is cost-intensive and receives significant state funding. Moreover, it is dominated by multinational companies from a few developed nations. This has spurred debate on whether some of the benefits derived from these genetic resources should be more equitably shared among the international community. Legal regulation of the use of genetic material from areas beyond national jurisdiction (ABNJ) is currently subject to negotiation in the UN Convention on the Law of the Sea (UNCLOS). Discussions are fueled by controversies over the principles of the freedom of the high seas versus principles stemming from the access and benefit-sharing regime that governs the use of genetic resources. This chapter examines variation in corporative response to the proposed regulations, thereby filling a gap, as commercial actors in bioprospecting are rarely studied academically.

Chapter 11 examines the need for transformative change in the governance of protected and conserved areas, with a focus on the Post-2020 GBF under the CBD. While progress has been made in designating sites under Aichi Target 11, this has not resulted in equitably and effectively managed or ecologically representative sites. Drawing from three case studies, the chapter proposes a new approach based on biodiversity and equity outcomes that incorporates integrative, inclusive and adaptive elements of transformative governance. Governance needs to go beyond including IPLC to focus on rights-based approaches and equity considerations. Adopting this type of approach at the global level will require a common understanding of biodiversity outcomes, redirecting of finance from high- to low-income countries, and complementary efforts by high-income countries to address the underlying causes of biodiversity loss by adopting sustainable trade and consumption patterns.

Increasingly heated debates concerning species extinction, climate change and global socioeconomic inequality reflect an urgent need to transform biodiversity governance. A central question in these debates is whether fundamental transformation can be achieved within mainstream institutional and societal structures. Chapter 12 argues that it cannot. Indeed, mainstream neoprotectionist and natural capital governance paradigms that do not sufficiently address structural issues, including an increase of authoritarian politics, might even set us back. The way out, the chapter contends, is to combine radical reformism with a vision for structural transformation that directly challenges neo-liberal political economy and its newfound turn to authoritarianism. Convivial conservation is a recent paradigm that promises just this. The chapter reviews convivial conservation as a vision, politics and set of governance mechanisms that move biodiversity governance beyond market mechanisms and protected areas. It further introduces the concept of "biodiversity impact chains" as one potential way to operationalize its transformative potential.

Current forms of agriculture are a major driver of biodiversity loss. Prevailing threats to biodiversity in agricultural landscapes are linked to management choices and habitat conversion. Biodiversity conservation in agricultural landscapes requires both setting aside valuable ecological areas (land-sparing) and radically changing agricultural practices (land-sharing). Chapter 13 employs the concept of biodiversity policy integration (BPI) to assess to what extent biodiversity is integrated into agricultural governance in developed and developing countries. The chapter finds that biodiversity policies are predominantly "add-on" and neither directly address biodiversity-threatening agricultural practices, nor specifically support more "nature-inclusive" agriculture. Thus, existing knowledge of biodiversity-sound agriculture is not reflected in dominant agricultural policies and practices. The chapter argues that political will can target the following leverage points to transform existing governance structures: (a) working toward a clear vision for sustainable agriculture; (b) building social capital; (c) integrating private sector initiatives; and (d) better integrating knowledge and learning in policy development and implementation.

Chapter 14 explores how the governance of urban nature is transforming in response to the increasing urgency of this agenda, and the extent to which it is in turn becoming transformative for the governance of biodiversity. The chapter finds that urban biodiversity governance is being transformed both in terms of its focus (moving from only a concern with reducing the threat of cities to biodiversity to also realizing their benefits) and in terms of the forms that governance is taking (through the growth of governance experimentation in cities and the growth in transnational governance networks). Nonetheless, there remain significant challenges to address in terms of how matters of biodiversity can become mainstream to urban development and how cities come to be positioned within biodiversity governance, which forms of urban nature come to count in the pursuit of urban sustainability and how issues of social inclusion and justice can be addressed.

Chapter 15 analyzes the major underlying causes of marine biodiversity loss and focuses specifically on the lessons learned for transformative ocean governance in the context of area-based management and spatial planning. It illustrates the broad recognition of the vital need for integrative, anticipatory, adaptive and inclusive governance of ocean biodiversity. Fundamentally, however, the chapter underscores the need for transdisciplinary governance in supporting integration, inclusion and learning in ocean affairs for transformative change. An alternative governance approach is proposed: Building on the interdependencies between human rights and marine biodiversity, a broader approach to fair and equitable benefit-sharing can support institutionalized shifts toward more transdisciplinary, integrative, inclusive and adaptive governance for the ocean at different scales.

Chapter 16 wraps up the edited volume. Based on the contributions of the different chapters, it takes a next step in operationalizing the key concepts of the book, namely transformative change, transformative governance, transformations and transitions. It then discusses opportunities and challenges for transformative biodiversity governance in the context of the Post-2020 GBF and its implementation. The GBF has the ambition to develop a transformative framework for the next stage in biodiversity governance. This requires prioritizing ecological, justice and equity concerns in addressing the underlying causes of biodiversity loss and developing governance arrangements to make this happen. We apply the book's transformative governance framework to further harness the transformative potential of a number of governance arrangements put forward for the GBF. We argue that in this manner, transformative biodiversity governance can contribute to ecocentric, compassionate and just sustainable development.

References

Biermann, F. (2021). The future of "environmental" policy in the Anthropocene: Time for a paradigm shift. *Environmental Politics* 30, 61–80. DOI: 10.1080/09644016.2020.1846958

Biermann, F., Betsill, M. M., Gupta, J., et al. (2010). Earth system governance: A research framework. *International Environmental Agreements: Politics, Law and Economics* 10, 277–298.

Blythe J., Silver J., Evans L., et al. (2018). The dark side of transformation: Latent risks in contemporary sustainability discourse. *Antipode* 50, 1206–1223.

Burch, S., Gupta, A., Inoue, C. Y., et al. (2019). New directions in earth system governance research. *Earth System Governance* 1, 100006.
Castán Broto, V., Trencher, G., Iwaszuk, E., and Westman L. (2019). Transformative capacity and local action for urban sustainability. *Ambio* 48, 449–462.
Castles, S. (2001). Studying social transformation. *International Political Science Review* 22, 13–32.
CBD. (1992). Convention on Biological Diversity. United Nations Treaty Series 1760, 79-307. Available from https://bit.ly/3s86Eam.
CBD. (2010). *The strategic plan for biodiversity 2011–2010 and the Aichi biodiversity targets*. UNEP/CBD/COP/DEC/X/2. Montreal: Convention on Biological Diversity.
CBD. (2020a). Global Biodiversity Outlook 5: Humanity at a crossroads. Available from www.cbd.int/gbo5.
CBD. (2020b). Aichi target pages. Available from www.cbd.int/aichi-targets/target/1.
Chaffin, B. C., Garmestani, A. S., Gunderson, L. H., et al. (2016). Transformative environmental governance. *Annual Review of Environment and Resources* 41, 399–423.
Colloff, M. J., Martín-López, B., Lavorel, S., et al. (2017). An integrative research framework for enabling transformative adaptation. *Environmental Science & Policy* 68, 87–96.
Díaz, S., Settele, J., Brondízio, E. S., et al. (2019). *Summary for policymakers of the global assessment report on biodiversity and ecosystem services of the Intergovernmental Science-Policy Platform on Biodiversity and Ecosystem Services*. Intergovernmental Science-Policy Platform on Biodiversity and Ecosystem Services. Bonn: IPBES secretariat.
EEA. (2019). The European environment – State and outlook 2020: Knowledge for transition to a sustainable Europe. European Environment Agency. Available from www.eea.europa.eu/soer/publications/soer-2020.
Ellis, E. C., and Mehrabi, Z. (2019). Half Earth: Promises, pitfalls, and prospects of dedicating half of Earth's land to conservation. *Current Opinion in Environmental Sustainability* 38, 22–30.
ESG. (2018). *Earth system governance: Science and implementation plan of the Earth System Governance Project*. Utrecht, the Netherlands.
Grin, J., Rotmans, J., Schot J., Geels, F., and Loorbach, D. (2010). *Transitions to sustainable development: New directions in the study of long term transformative change*. New York: Routledge.
Guston, D. H. (2014). Understanding "anticipatory governance." *Social Studies of Science* 44, 218–242.
Hölscher, K., Wittmayer, J. M., and Loorbach, D. (2018). Transition versus transformation: What's the difference? Environmental Innovation and Societal Transitions 27, 1–3.
Intergovernmental Science-Policy Platform on Biodiversity and Ecosystem Services (IPBES). (2019). *Global assessment report of the Intergovernmental Science-Policy Platform on Biodiversity and Ecosystem Services*. E. S. Brondízio, J. Settele, S. Díaz and H. T. Ngo (Eds.). Bonn: IPBES secretariat.
Keitsch, M. M., and Vermeulen, W. J. V. (Eds.) (2021). *Transdisciplinarity for sustainability: Aligning diverse practices*. London and New York: Routledge.
Kok, M. T. J., Meijer, J. R., van Zeist, W. J., et al. (under review). Assessing ambitious nature conservation strategies within a 2 degree warmer and food-secure world. www.biorxiv.org/content/10.1101/2020.08.04.236489v1.
Lawhon, M., and Murphy, J. T. (2012). Socio-technical regimes and sustainability transitions: Insights from political ecology. *Progress in Human Geography* 36, 354–378.
Leclère, D., Obersteiner, M. L., Barrett, M., et al. (2020). Bending the curve of terrestrial biodiversity needs an integrated strategy. *Nature* 585, 551–556.
Le Prestre, P. G. (Ed.) (2002). *Governing global biodiversity: The evolution and implementation of the Convention on Biological Diversity* (1st ed.). London and New York: Routledge.
Li, L., and Kampmann, M. (2017). A common vision among divergent interests: New governance strategies and tools for a sustainable urban transition. *Procedia Engineering* 198, 813–825.
Linnér, B.-O., and Wibeck, V. (2019). *Sustainability transformations: Agents and drivers across societies*. Cambridge, UK: Cambridge University Press.

Loorbach, D., Frantzeskaki, N., and Avelino, F. (2017). Sustainability transitions research: Transforming science and practice for societal change. *Annual Review of Environment and Resources* 42, 599–626.

Mace, G. M., Barrett, M., Burgess, N. D., et al. (2018). Aiming higher to bend the curve of biodiversity loss. *Nature Sustainability* 1, 448–451.

Meadowcroft, J. (2009). What about the politics? Sustainable development, transition management, and long term energy transitions. *Policy Sciences* 42, 323–340.

Mehrabi, Z., Ellis, E. C., and Ramankutty, N. (2018). The challenge of feeding the world while conserving half the planet. *Nature Sustainability* 1, 409–412.

Moser, S. C. (2016). Can science on transformation transform science? Lessons from co-design. *Current Opinion in Environmental Sustainability* 20, 106–115.

Otero, I., Farrell, K. N., Pueyo, S., et al. (2020). Biodiversity policy beyond economic growth. *Conservation Letters* 13, e12713.

Otsuki, K. (2015). *Transformative sustainable development: Participation, reflection and change.* New York: Routledge.

Patterson, J., Schulz, K., Vervoort, J., et al. (2017). Exploring the governance and politics of transformations towards sustainability. *Environmental Innovation and Societal Transitions* 24, 1–16.

Platto, S., Xue, T., and Carafoli, E. (2020). COVID19: An announced pandemic. *Cell Death & Disease* 11, 1–13.

Polanyi, K. (1944). *The great transformation: The political and economic origins of our time.* New York: Farrar and Rinehart.

Pörtner, H. O., Scholes, R. J., Agard, J., et al. (2021). Scientific outcome of the IPBES-IPCC co-sponsored workshop on biodiversity and climate change. Bonn, Germany: IPBES secretariat. DOI:10.5281/zenodo.4659158.

Razzaque, J., Visseren-Hamakers, I. J., Gautam, A. P., et al. (2019). Options for policymakers. *Global Assessment Report on Biodiversity and Ecosystem Services of the Intergovernmental Science-Policy Platform on Biodiversity and Ecosystem Services.* Bonn: IPBES Secretariat.

Schleicher, J., Zaehringer, J. G., Fastré, C., et al. (2019). Protecting half of the planet could directly affect over one billion people. *Nature Sustainability* 2, 1094–1096.

Scoones, I., Stirling, A., Abrol, D., et al. (2020). Transformations to sustainability: Combining structural, systemic and enabling approaches. *Current Opinion in Environmental Sustainability* 42, 65–75.

Termeer, C. J. A. M., Dewulf, A., and Biesbroek, G. R. (2017). Transformational change: Governance interventions for climate change adaptation from a continuous change perspective. *Journal of Environmental Planning and Management* 60, 558–576.

UN. (2015). General Assembly, Resolution 70/1, Transforming our world: The 2030 Agenda for Sustainable Development. United Nations General Assembly. A/RES/70/1.

van den Bergh, J. C. J. M., Truffer, B., and Kallis, G. (2011). Environmental innovation and societal transitions: Introduction and overview. *Environmental Innovation and Societal Transitions* 1, 1–23.

Visseren-Hamakers, I. J. (2015). Integrative environmental governance: Enhancing governance in the era of synergies. *Current Opinion in Environmental Sustainability* 14, 136–143.

Visseren-Hamakers, I. J. (2018a). A framework for analyzing and practicing integrative governance: The case of global animal and conservation governance. *Environment and Planning C: Politics and Space* 36, 1391–1414.

Visseren-Hamakers, I. J. (2018b). Integrative governance: The relationships between governance instruments taking center stage. *Environment and Planning C: Politics and Space* 38, 1341–1354.

Visseren-Hamakers, I. J. (2020). The 18th sustainable development goal. *Earth System Governance* 3, 100047.

Visseren-Hamakers, I. J., Razzaque, J., McElwee, P., et al. (2021). Transformative governance of biodiversity: Insights for sustainable development. *Current Opinion in Environmental Sustainability* 53, 20–28.

Wagner, P., and Wilhelmer, D. (2017). An integrated transformative process model for social innovation in cities. *Procedia Engineering* 198, 935–947.

WEF. (2021). Nature and net zero. Available from: https://bit.ly/3sIOLRp.

Wolfram, M. (2016). Conceptualizing urban transformative capacity: A framework for research and policy. *Cities* 51, 121–130.

WWF. (2020). *Living planet report 2020 – Bending the curve of biodiversity loss*. R. E. A. Almond, M. Grooten, and T. Petersen (Eds.). Gland, Switzerland: WWF. Available from www.worldwildlife.org/publications/living-planet-report-2020.

Part II

Unpacking Central Concepts

2
Defining Nature

HANS KEUNE, MARCO IMMOVILLI, ROGER KELLER, SIMONE MAYNARD, PAMELA MCELWEE, ZSOLT MOLNÁR, GUNILLA A. OLSSON, UNNIKRISHNAN PAYYAPPALLIMANA, ANIK SCHNEIDERS, MACHTELD SCHOOLENBERG, SUNEETHA M. SUBRAMANIAN AND WOUTER VAN REETH

2.1 Introduction

In any attempt to "rethink" biodiversity governance, we need to consider that defining nature (and related concepts such as biodiversity, ecosystems, landscapes or green infrastructure) is not merely an objective scientific exercise. In reality, context-specific, subjective, normative and dynamic worldviews and values are at play in any definition of nature, whether explicitly or implicitly. Being aware of this pluralism is essential for avoiding "objective" definitional attitudes that risk disregarding and marginalizing the plurality of values and worldviews connected to different definitions of nature. In fact, paternalistic positions can create breeding grounds for fruitless dialogues between stakeholders, and thus pluralistic approaches help open up spaces for discussion.

In the modern era, Western worldviews have emphasized the separation between culture, humans and nature, dating back to at least the era of the Old Testament. This distinction has come to be known as the nature/culture divide, a dichotomy that posits nature as a separate and discrete object that can be known, conquered and used at will for humankind's benefit, with consequences beyond theoretical and philosophical discussions (Castree, 2013). Different interpretations exist on when and how this divide came to be (Pattberg, 2007; Uggla, 2010). In her classic book *The Death of Nature: Women, Ecology and the Scientific Revolution*, Carolyn Merchant (1980) pointed out how the image of nature as a nurturing mother was gradually transformed during the sixteenth and seventeenth centuries into an image of nature as being wild, chaotic and uncontrollable, a position directly related to the dominant view on women at the time and a view that justified the domination of nature and the exploitation of its resources.

The environmental historian Donald Worster has proposed that since the Industrial Revolution, two key threads can be discerned in the way Western societies relate to nature. First, the "imperial" or Linnean tradition emerging from the development of biological classification of species and scientific exploration had the ambition to "establish, through the exercise of reason and by hard work, man's dominion over nature" (Worster, 1977: 2). At the same time, the Industrial Revolution led to a second strand that emerged as a countermovement to the idea of human domination, which Worster terms "Arcadian," and that "advocated a simple, humble life for man with the aim of restoring him to a peaceful coexistence with other organisms," given the depredations of industrial life (Worster,

1977: 2). This second strand has taken many different forms over time; for example, in the later nineteenth century, Romanticism, despite being a heterogeneous movement, challenged the idea of human domination over nature and modernity by idealizing wild nature for its beauty and purity (Uggla, 2010).

The nature/culture divide has come under criticism as a cultural construction not universally applicable to the whole of human societies (Descola, 2013), and as an invalid dichotomy for the West as well (Latour, 1991). These criticisms are not solely theoretical, as they raise the fundamental question "what is nature?" and reject a single objective answer. Thus, nature is a plural concept, and in this chapter we argue that this plurality reflecting the different values of nature will play a fundamental role in transformative biodiversity governance. Yet this does not come easily, as a plurality of values means a plurality of ontologies, epistemologies, interests and needs.

The authors do not pretend to present an exhaustive nature-definition overview in this chapter, nor to be without bias: The content of this chapter largely builds on the expertise and experience of the collaboration between them. And of course, explicitly or implicitly, certain accents or interpretations may come across more strongly than others. Nevertheless, we mainly hope to share with the reader a rich display of definition examples and elements, illustrating the core intention of this chapter: to show that nature is defined, and cannot be taken for granted as one objectifiable concept. After a brief introduction of the concept of biodiversity (Section 2.2) as a root scientific concept for conservation, we provide an overview of some of the ways nature has been defined over time and what this means for biodiversity conservation. Section 2.3 deals with wilderness, intrinsic value and how these are interlinked with protected areas. Section 2.4 addresses the concept of landscape via two lenses: ecosystem services and biocultural diversity. Instrumental and relational values of nature are also discussed. Section 2.5 takes the increasingly popular tool of conferring nature with legal rights (Rights of Nature) as demonstrating hybrid forms of biodiversity governance that attempt to merge Western and non-Western ontologies and definitions of nature. Section 2.6 discusses the importance of scenarios for nature in order to develop alternative pathways grounded on value pluralism. Section 2.7 concludes the chapter by drawing general conclusions for transformative biodiversity governance.

2.2 Nature Defined in the History of "Biodiversity"

Attention to the conservation of nature often manifests as a response to the widespread unsustainable and unethical use of nature (however defined) that stems from a view of nature from an instrumental value perspective, resulting in overlogging, overfishing, large-scale land-use change, etc. The concept of biodiversity emerged from the scientific community and, despite criticisms, represents one of the most common and recognized concepts for scientists and the general public. The term dates back to 1968, when Dasmann used it for the first time in his book *A Different Kind of Country* (Dasmann, 1968). While concepts of nature and wilderness had been commonly used previously, with this new term, global diversity that had evolved over more than 3.6 billion years was emphasized, as well as the fact that human impact extended beyond just endangered species. As the term began to circulate and become widely used, one of the first uses of the term was "biological diversity" in the United States. The United States historically played an important role in the design of conservation, where it was mentioned in the Global 2000 Report to the president, written by biologist Tom Lovejoy for President Jimmy Carter in 1980 (Lovejoy, 1980). The popularity enjoyed by the term partly lies in the increasing concern about an accelerating "extinction crisis" (Ehrlich and Ehrlich, 1981; Myers, 1979), as well as the fact that it was a useful catch-all representing the need for increased conservation for the underpinnings of life (Heywood, 1995), and the National Forum on BioDiversity in 1985 cemented the idea that the concept was fundamental for shaping conservation policy (Wilson, 1988). In other words, as biologist E. O. Wilson put it, "Biological diversity – 'biodiversity' in the new parlance – is the key to the maintenance of the world as we know it" (Wilson, 1992: 15).

Although the last decades saw a surge in the use of the concept of biodiversity in the scientific community and beyond, the term itself is not uncontested. One "formal" definition of biodiversity, adopted by the Convention on Biological Diversity (CBD) in 1992, defines it as "variability among living organisms from all sources including, inter alia, terrestrial, marine and other aquatic ecosystems and the ecological complexes of which they are part; this includes diversity within species, between species and of ecosystems" (Article 2) (CBD, 1992). Many have argued, since the emergence of the term, that it still remains vague and imprecise: "the term biodiversity is beginning to fail as a useful catch-all term for the current planetary environmental crisis ... ambiguity of meaning has, in my opinion, rendered the concept of biodiversity increasingly useless as a rallying-point by which to focus attention on the current and on-going dramatic changes to the biosphere" (Bowman, 1998: 239).

Further uncertainty emerges from the task of measuring biodiversity (Walpole et al., 2009). Early discussions about how different dimensions of biodiversity might best be measured included basic species/area ratios, which, as species diversity generally increases from the poles to the equator, led to biodiversity protection efforts centered in the tropics (Harper and Hawksworth, 1994); a focus on rarity and endemism, such as in "biodiversity hotspots" where such endemic species are under particular threat (Myers et al., 2000); or on taxonomic character differences within populations, indicating genetic richness to be conserved for the sake of future evolution (Humphries et al., 1995). In practical terms, the idea of sheer species numbers as equivalent to biodiversity has largely predominated (Takacs, 1996), although it has led some to question "whether it is adequate – or correct – to base the priorities for global biodiversity conservation simply on the quantity of biological diversity, as is often done" (Fjeldsa and Lovett, 1997: 319). More recent discussions have focused on questions of "biodiversity intactness," "biodiversity health," "species viability," and, as we note in the next section, ecological functions and services provided by biodiversity (Dinerstein et al., 2020; Mace et al., 2018; Schneiders and Müller, 2017).

As concerns over the ambiguity of the term and how to measure it allude to, there remained no clear consensus on a single standard interpretation of biodiversity for many years. The difficulty of reconciling alternative interpretations has made critical engagement with definitions of biodiversity difficult and contested when the conceptual roots of the term are questioned (see also Sarkar, 2016). At the same time, biodiversity has entered the public discourse and is commonly used by newspapers and mass media; as a term, it is gaining in popularity (Levé et al., 2019), although not (yet) as much as climate change (Legagneux et al., 2018).

Despite these debates, the concept of biodiversity has, more than any other concept in the last decades in Western ecological thinking, been a key contribution in shaping the governance of nature conservation. For example, defining the boundaries of what biodiversity is and where it can be found is required for the creation of targets to "halt biodiversity loss" and, more recently, to "bend the curve of biodiversity loss" (Mace et al., 2018). Yet, as we have noted, these targets do not "naturally" and "neutrally" emerge from agreements within the scientific community. On the contrary, they are negotiated and contested, and they lend themselves to alternative conservation strategies and practices (Bhola et al., 2021; Immovilli and Kok, 2020; Keune and Dendoncker, 2013). In the next two sections, we discuss possible ways to look at biodiversity governance and further reflect on how these approaches are grounded in different definitions of nature.

2.3 Nature Defined as Wilderness

The concept of wilderness emerged from the US context in the nineteenth century and soon gained momentum in the wider international conservation debate. As European settlers arrived in the Americas, wild nature was considered the enemy, to be replaced with traces of "modern civilization" (Nash, 1967). Later, this attitude shifted, and wild nature started to be praised as sacred havens that would spare humanity from the unstoppable expansion of modernity; for example, the well-known American writer Henry David Thoreau advocated for wild nature as a space where modern humans' excesses could be purified and limited. The cerebral and aesthetic values being praised in this context were advocated by upper-middle class and white American men, whose communing with nature conferred intellectual life, arts and letters (McDonald, 2001; Nash, 1967). In other words, wilderness, particularly in Thoreau's work, resembled an ontological claim to a different life, one not completely devoted to modernity and urbanism (McDonald, 2001; Nash, 1967).

Yellowstone National Park was established in 1872 in the United States, marking a historical moment in the movement for the protection of the wild, although as historians have subsequently pointed out, the protection of this wilderness required the eviction of Indigenous Native Americans (Spence, 1999). Yet these divisions between man and wilderness continued, eventually culminating in the passage of the Wilderness Act in 1964, where wilderness was defined as "an area where the earth and its community of life are untrammelled by man, where man himself is a visitor who does not remain."

Yet the establishment of protected areas (PAs) and the concept of wilderness itself have been harshly criticized. Many pointed out that so-called wild areas were in fact recreated and strictly administered and managed (Denevan, 1992). Furthermore, social justice concerns were raised, pointing at the violent displacement of people and the enclosing of land that followed the establishment of many parks (Cronon, 1996). Despite these criticisms, protecting the wild still drives the expansion of PAs and other area-based measures, which remain among the most common practices for conservation governance as fears over land degradation and the extinction crisis have grown (Grove, 1992). Proposals to expand protected areas continue to play a fundamental role in biodiversity governance (Locke et al., 2013).

Additionally, a strong ecocentric rhetoric has grown in academic and public discourse, underlining the intrinsic value of nature (including humans) and its inherent right to exist, live and flourish despite human pressures. Such powerful discursive material serves as conceptual – if not philosophical – ground for many political and ecological efforts (see, for instance, the recent proposal to protect half of the Earth and how it is backed by ecocentric thinking [Kopnina, 2016]). This is well captured by Wolke (2014: 204), who states that "wilderness is about setting our egos aside and doing what is best for the land."

While this definition retains the ontological claim that wilderness is a limit to human expansion – and that indirectly we can learn from it – it shifts the value of wilderness toward intrinsic (moral, spiritual and ecological) value. This should not come as a surprise when we consider the evolution of environmental concern over the last decades and the rise of biodiversity as a concept. Indeed, the concept of biodiversity itself has often been used to reinforce the narrative of wilderness (Nash, 1967; Uggla, 2010). As such, the expansion of protected areas and other area-based conservation measures is often grounded in an ecocentric rhetoric, which claims these measures to be a vital solution to achieving global biodiversity targets.

Since 1988, there has been a 400 percent increase in the number of PAs and they now cover 15 percent of the Earth's surface land. Critics point at this data and argue that, despite this surge in protection, biodiversity has neither been conserved nor restored (Butchart, 2010). This remains a point of debate, as others have argued that the achievements of PAs, despite being insufficient, are relatively positive in terms of biodiversity conservation (Butler, 2015 in Wuerthner et al., 2015), while the evidence on PAs mitigating human impacts is more mixed. Many nongovernmental organizations (NGOs) and some scientists have advocated that current levels of protection are not enough and more is needed, arguing that protection should be expanded to cover half of the Earth (Dinerstein et al., 2017; 2019; Locke, 2015; Wilson, 2016), while for others lower percentages could be enough (Visconti et al., 2019) (see also Chapters 11 and 12 for different perspectives on this conservation).

2.4 Nature Defined through Cultural and Ecosystem Services Lenses in Landscapes

In the previous section, we saw that nature has been defined as the counterpart of culture: the physical and biological world dominated by "natural" processes, not manufactured or developed by people. This resulted in the creation of wilderness and to the deployment of PAs. However, some claim that most of what we designate as "natural" areas (e.g. what are designated as Natura 2000 habitats in Europe) are in fact historical cultural landscapes with a high biodiversity value (Hermoso et al., 2018; Pechanec et al., 2018). Following this logic, "natural" ecosystems are the outcome of a coevolutionary process in which they shape, and are shaped by, new forms of social organization, knowledge, technology and value systems (Howarth and Norgaard, 1992). With this, the conceptualization of nature has shifted for some from wilderness to that of landscape, in 2000 defined by the European Landscape Convention (European Landscape Convention of the Council of Europe) as "an area, as perceived by people, whose character is the result of the action and interaction of natural and/or human factors." This definition emphasizes the dialectic and productive relationship between humans and nature and encourages a move beyond dichotomies.

Other value perspectives correspond to a definition of nature that includes culture. In 2012, the publication of what became known as the "New Conservation Manifesto" (Marvier et al., 2012) added a new set of values of nature to the discussion: instrumental value. In their article, Marvier et al. (2012) argue that conservation in the Anthropocene must move past the idea of wilderness because humans and natural systems are profoundly intertwined. Despite the increasing number of PAs, biodiversity is still in decline due to the fact that conservation cannot succeed if it does not address social issues, they claimed, such as poverty and inequality. Thus, conservation (and conservationists) must "embrace human development and the 'exploitation of nature' for human uses, like agriculture, even while they seek to 'protect' nature inside of parks" (Marvier et al., 2012). From such a perspective, nature is no longer valued (and conserved) for its intrinsic value, but because it provides humans with services and benefits (Pearson, 2016). In this, the ethical horizon of conservation has changed toward ideas of the sustainable use of nature, and in this context, the establishment of the Millennium Ecosystem Assessment and the Ecosystem Services (ES) framework are clear milestones.

2.4.1 The Ecosystem Services Lens

One of the core conclusions of the Millennium Ecosystem Assessment (MA, 2001–2005) was the fundamental dependence of human wellbeing on ecosystems through a variety of ecosystem services. Ecosystem services have been defined as the "direct or indirect contribution to sustainable human well-being" (Costanza et al., 2017), highlighting an anthropocentric and instrumental perspective on nature while acknowledging the intrinsic value of species and ecosystems. Outside of the scientific community, ES gained momentum as well, capturing the attention of the general public and private companies, and becoming firmly settled in the international policy arena (Costanza et al., 2017). The main merit of the ES framework is that it widened the policy discussion to aspects of nature that were traditionally neglected in decision-making (Schröter et al., 2014). Ecosystem services approaches have successfully shifted conservationist attention to indirect drivers of environmental change, such as socioeconomic dynamics, and attempted to reconcile ecological knowledge with economic thinking. This marked a clear difference from previous conservation efforts grounded in the idea of "conservation against development" (Gómez-Baggethun and Ruiz-Pérez, 2011). According to critics, this specific economic turn was instrumental in winning the hearts and minds of policymakers and stakeholders (Ring et al., 2010), but it narrowed down ES to a purely economic discourse, paving the way for the commodification of nature (Díaz et al., 2018; Gómez-Baggethun and Ruiz-Pérez,

2011; see also Chapter 6 of this book for a reflection on market-based approaches and their role in transformative biodiversity governance).

This shift is captured by the creation of "The Economics of Ecosystems and Biodiversity" (TEEB, 2007–2011) research program. Another example of the domination of economic approaches to ES is the increasing attention devoted to terms such as "natural capital," which aims to embed ecosystem services within the human economy in the form of stocks and assets to be accounted for (Costanza, 1991; Costanza et al., 2017). While the MA and TEEB did not introduce new definitions of nature or biodiversity, their framing and discourse have had an influence on which components of biodiversity were selected as being more or less relevant and fit for analysis (e.g. Norgaard, 2010; see also Chapter 5). Responding to these criticisms, some argued that acknowledging ES can be the basis of different types of assessment and need not lead to commodification. While monetary valuations are common, the ES framework still directs attention to the multiple benefits of nature that would otherwise be marginalized in decision-making, including ethical and sociocultural valuations, and ES can be used for nonmonetary assessment of human well-being (Costanza et al., 2009, 2017; De Groot et al., 2012; Schröter et al., 2014).

The ES framework, however, is changing. Partly out of concern for a narrow economic framing of the concept, and critiques of the domination of a Western world view embodied in ES, the Intergovernmental Science-Policy Platform on Biodiversity and Ecosystem Services (IPBES) has developed a more holistic perspective, known as Nature's Contributions to People (NCP), in which noneconomic values and non-Western worldviews receive more attention. This is an evolution of the ES concept as it considers different types of contributions, from material to nonmaterial, as a spectrum indicating the nonmutually exclusive nature of different contributions. Thus, for instance, food can be seen as not just material (provisioning), but also linked to nonmaterial values (culture and identity), in addition to other values such as options for the future (e.g. to facilitate climate adaptation). Thus, NCP concepts purport to bring in more real-life nuances to the values held by different peoples to nature (Díaz et al., 2018), as all of these values coexist, and are not equally prioritized, which could result in potential conflicts between different stakeholders (IPBES, 2017; Pascual, 2017).

2.4.2 Biocultural Diversity Lens

One reason for the development of the NCP concept was the lack of attention to nonmaterial aspects of nature. Despite the inclusion of "cultural ecosystem services" in the original ES framework, cultural services were underrepresented, lacked suitable indicators, and encountered difficulty (and reluctance) to quantify them (Satz et al., 2013). Notwithstanding these problems, studies on nature–culture relations evolved in parallel to the ES framework and gained prominence on the international agendas of organizations like the Food and Agriculture Organization (FAO) and UN Educational, Scientific and Cultural Organization (UNESCO) (Bridgewater and Rotherham, 2019), culminating in the 1988 Declaration of Belém, which found "an inextricable link between cultural and biological diversity" (Schlebusch et al., 2017: 652).

From this, the concept of biocultural diversity was coined. Agnoletti and Emanueli (2016) consider the concept of biocultural diversity to be a useful term to represent the dialectic relation between the biological and cultural diversity of a (cultural) landscape. As such, two complementary and reciprocally dependent dimensions exist within biocultural diversity: the human shaping of biodiversity and the evolution of cultural practices related to biodiversity.

Modern humans (*Homo sapiens*) developed in southern Africa some 260,000 to 350,000 years ago (Schlebusch et al., 2017), emerging from local dryland ecosystems and later found, through dispersal over the globe, in a multitude of different ecosystems. Through foraging, ancient humans shaped and impacted local ecosystems in a similar way to other animal species. Along with the development of human culture, the use of tools and implements for hunting, and later crop cultivation and the raising and maintaining of domesticated livestock, shaped distinct ecosystem patterns (Küster, 2003). The continuous harvesting of food, the hunting of animals, and the collection of medicinal and other plants influenced the composition of biological communities over time, making it impossible to distinguish "untouched" nature from human-altered ecosystems. According to Moran (2006), hardly any ecosystem on Earth has not been shaped by human action. Long before the Neolithic, our ancestors modified their environment to facilitate their quest for food. Olsson (2018) shows how the myth of untouched wilderness as a treasure for biodiversity was contested. Joint work by ecologists and anthropologists showed – through observations of tropical forests presumably untouched by humans, like large parts of the Amazon – that the habitat had in fact been used through different forms of shifting cultivation for long periods of time, thereby influencing biological diversity. This should therefore more accurately be called biocultural diversity (Gómez-Pompa and Kaus, 1992). Similar results and interpretations have been confirmed by other researchers (Padoch and Pinedo-Vasquez, 2010), such as the use of fires for hunting in shaping biodiversity (Sevink et al., 2018).

Cultural practices can also view biodiversity as a resource (Bridgewater and Rotherham, 2019). An important aspect to highlight here concerns the meaning of culture, for which Cocks' (2006) work is central in arguing that biocultural diversity has so far been linked to the cultural activities of local and Indigenous groups. In his view, this is too limited and should be extended to include non-Indigenous groups, based on observations of the variety of cultural practices regarding the use of wild plants by non-Indigenous peoples (Cocks, 2006).

This dialectic relation between nature and culture remains at the core of biocultural diversity and characterizes both rural and urban landscapes (Elands et al., 2019). Examples include seminatural vegetation, like grasslands and West-European heathlands. In seminatural grasslands in Europe, biological communities (plant species and their associated insects and other organisms) depend on continuous interference by humans, such as through fire, mowing or grazing by large herbivores like domesticated livestock. Without such activities, the seminatural grassland will return to forest and lose species richness (Babai and Molnár, 2014). Some of these grasslands existed in prehuman times and were shaped and maintained by wildfires and large wild herbivores, but the extent of seminatural vegetation from the Neolithic onward is due mainly to human interference

(Olsson, 2018; Oteros-Rozas et al., 2013). Another example relevant for agricultural systems is that of *biocultural refugia* (Barthel et al., 2013). This concept directly relates to human food provisioning, as embracing (biocultural) diversity can be seen as an agricultural strategy, and involves ensuring crop and habitat diversity as important tools for resilience in facing different disturbances and uncertainties, as well as the effects of climate change.

In Europe, traditional agricultural landscapes are often abandoned or transformed into urban or more intensively managed agricultural areas (Agnoletti, 2014; EEA, 2010; 2015; 2020). When abandoned, native shrubs, trees and invasive alien species may spread. Local farmers often perceive these changes negatively: from a landscape-in-order where "each corner had a role," reverting into a landscape-in-disorder that is "getting wild" (Babai and Molnár, 2014; Ujházy et al., 2020). This "getting wild" causes loss of cultural practices and associated biocultural diversity (Agnoletti and Rotherham, 2015), offering an interesting comparison with the interpretation of wilderness in the context of PAs given earlier. What is seen as the loss of biocultural diversity from the perspective of cultural landscapes from a traditional ecological point of view is often framed as a positive gain for biodiversity because land abandonment offers possibilities for "rewilding" (Agnoletti and Rotherham, 2015). Agnoletti (2014) acknowledges this tension and complains that many conservation approaches are too guided by the concept of wilderness when dealing with cultural landscapes, thereby neglecting biocultural diversity.

Frameworks are emerging for the conservation of landscapes that are coproduced by humans and nature, such as in the International Union for Conservation of Nature (IUCN) Category V (Protected Landscapes/Seascapes) (Schneiders and Müller, 2017; IUCN, n.d.). Furthermore, cultural aspects are included in discussions of the CBD regarding the establishment of "sustainable use" as one of the three main goals of the convention, which hints in the direction of valuing cultural landscapes (Bridgewater and Rotherham, 2019). Another noteworthy development is that of the "Other effective area-based conservation measures" (OECMs) introduced by Aichi Target 11, which allow other sustainability-related goals along with conservation objectives in management and governance (Laffoley et al., 2017).

An important step toward the protection of cultural landscapes and biocultural diversity is the increasing attention in the conservation debate to so-called relational values. Chan et al. (2016: 1462) argue that "[f]ew people make personal choices based only on how things possess inherent worth or satisfy their preferences (intrinsic and instrumental values, respectively). People also consider the appropriateness of how they relate with nature and with others, including the actions and habits conducive to a good life, both meaningful and satisfying. In philosophical terms, these are relational values." The introduction of relational values aims to capture another dimension that can support the concept of biocultural diversity by enriching understandings of human–nature interactions within the landscape.

In conclusion, the introduction of concepts like ecosystem services and biocultural diversity have broadened the horizons of biodiversity conservation in the past decades, shifting the attention from wilderness protection to also include sustainable use and cultural landscapes, from intrinsic values of nature to a plurality of other values, including instrumental and relational. These concepts have been important influences on how biodiversity governance is conceptualized and practiced, as seen in the development of numerous international policy agendas and new forms of protection. The two frameworks discussed in this section emphasize different elements and can complement each other (Bridgewater and Rotherham, 2019; Buizer et al., 2016). However, tensions exist, particularly on issues of quantification and monetization at the center of discussion within the ES framework that run the risk of objectifying and separating nature from humans.

2.5 Nature Defined as Rights of Nature

In the previous sections, we described the processes that led to the inclusion and engagement with a plurality of values and knowledge systems within mainstream conservation. This is all the more needed when one considers the importance of Indigenous Peoples and local communities (IPLC) in managing and meeting global biodiversity targets. These groups use, manage, own or occupy a quarter of the globe, including 35 percent of the formally protected land area (Garnett et al., 2018, IPBES, 2019). Despite globally-declining biodiversity trends, nature is declining less rapidly in these IPLC-managed lands (Garnett et al. 2018, IPBES, 2019).

Indigenous and local knowledge systems are mobilized by IPLC, who live within natural and rural settings and make a living through an intimate relationship with nature (UNESCO, n.d.). Examples of different conceptualizations of nature from Indigenous communities include Pachamama (Mother Earth) or Country (Australia) (McElwee et al., 2020). Across many communities, nature is considered to be reciprocal kin, such as a mother or a deity, signifying a harmonious relationship between nature and humans (Cano Pecharroman, 2018). For instance, the concept of Pachamama, despite differences across populations using the term, translates into an actual philosophy of life ("*buen vivir*" in Spanish) that permeates the daily life and practices of these communities. The formulation of *buen vivir* as an alternative to modern Western ideas of development has been embraced by numerous social mobilizations (Gudynas, 2011; Kothari, Demaria and Acosta, 2014). Once again, multiple definitions of nature and the worldviews articulated around it play a role in shaping proposals for conservation governance and, more broadly, sustainability.

Rights of Nature (RoN) is an emerging legal framework that aims at integrating IPLC knowledge with Western legal systems (also see Chapter 9). It has gained vast momentum over the last decade and confers legal rights to individual ecosystems (or the whole of nature) that are then represented in court by one or more legal representatives or guardians (Cano Pecharroman, 2018). These changes in the legal system around nature represent a fracture with previous approaches (Chapron et al., 2019), as proponents argue that the mainstream Western legal system is anthropocentric and legalizes environmental exploitation for the fulfillment of human needs (Burdon, 2011). Nature, in an ecocentric legal system, would thus be recognized a legal entity and be conferred with the status of legal subject (O'Donnell and Talbot-Jones, 2018). Starting from local ordinances in the United States, RoN have been included in the Ecuadorian Constitution in 2008, and in 2011 Bolivia passed its own Law on the Rights of Mother Earth. More recently, in 2016, the Atrato river in Colombia was given legal personhood, quickly followed by the Whanganui river in New Zealand (2017) and the Ganga and Yamuna rivers in India (2017). In 2019, Lake Erie in Ohio, United States, was granted the rights "to exist, flourish and naturally evolve" (Lake Erie Bill of Rights Charter Amendment 2018), and a proposal to confer legal rights to the Dutch Wadden Sea has recently been discussed (Lambooy et al., 2019).

The RoN framework poses an ontological quandary because it introduces nature as a subject, rather than object, not only in legal but also in moral terms (de Sousa Santos, 2015). Yet, as detailed in the previous sections, such a conceptualization of nature may perhaps be less obvious in the context of the traditional Western ontological divide between nature and culture. The challenge lies in the fact that Western national legislations and worldviews, traditionally anthropocentric, are now confronted with IPLC conceptualizations of nature and of life. Rights of Nature thus is more than a mere legal tool, as it can create encounters between different epistemologies and ontologies, as Western concepts such as "rights" and "ecosystem" meet with Indigenous worldviews and concepts such as "Pachamama" and "buen vivir" in what has been defined an "epistemic pact" (Valladares and Boelens, 2017).

The establishment of RoN presents fundamental questions concerning the way we relate to and see nature. From a conservation point of view, the narrative around nature as a subject and nature's intrinsic rights, as defined within "ecocentrism" (Washington et al., 2017), has been widely deployed for the conceptual backing of PAs expansion (Kopnina, 2016). However, ecocentric approaches are contested by critics for their lack of attention for the human dimension (Büscher et al., 2017; see also Chapter 12 on Convivial Conservation). Similarly, RoN is criticized for the risk of pitting humans against nature and neglecting human needs that are embedded in nature (Kothari and Bajpai, 2017). As such, ongoing discussions on who will represent nature and how legal representatives or guardians will play a role in trying to address these issues might offer useful examples for broader conservation debates on whether and how to integrate ecological and social concerns.

> **The Example of the Case of the Atrato River in Colombia.**
> In 2016, the Colombian Constitutional Court recognized the Atrato as subject and assigned "biocultural rights" to recognize the inextricable connection between the river and local practices and culture. These biocultural rights formed a framework wherein conservation objectives relating to the river were reconciled with the sociocultural needs of local communities (Kauffman and Martin, 2018; Roncucci, 2019). While promising, the Atrato case is relatively recent and more time is needed to draw any conclusion regarding the success (or not) of integrating environmental and sociocultural needs.

Ultimately, the integration of the Rights of Nature with the rights of people is contested, as it brings us back to the nature/culture divide and to the risk of seeing humans (or rather, some humans) as separated from and opposite to nature. Nonetheless, the inclusion of Indigenous knowledges and worldviews as exemplified by RoN frameworks is contributing to transformative biodiversity governance by proposing novel hybrid legal arrangements and by challenging dominant Western ontologies and epistemologies.

2.6 Scenarios of Nature

In this section, we deal with scenarios of nature as a way to develop future pathways that are inclusive of the plurality of definitions and values of nature encountered thus far. Scenarios of nature are qualitative and quantitative descriptions of a desirable nature future and are widely employed in environmental policymaking. Díaz et al. (2018) note that most scenarios do not take into account the complexity of human–nature relations, but in fact only consider human impacts on nature, neglecting the importance of nature in supporting human wellbeing. To remedy this and to include a plurality of values of nature into scenario exercises, a new framework is being developed by IPBES, known as the Nature Futures Framework (Pereira et al., 2020), where the three value perspectives discussed in this chapter (intrinsic, instrumental and relational) would be used to develop future visions for society and nature.

Similarly, the Nature Outlook study by PBL Netherlands Environmental Assessment Agency elaborated four perspectives based on different values of nature and explored alternative futures at the EU level (Van Zeijst et al., 2017). The result was the development of four perspectives underpinned by different value assumptions: strengthening cultural identity, allowing nature to find its way, going with the economic flow and working with nature. This exercise did not aim to identify one optimal way forward but rather to facilitate imagining alternative futures. These types of exercises are fundamental for thinking about transformative change because they allow scope for alternatives and create space for confrontation and decision-making with transparent values and inclusive practices.

A key element that is relevant for transformative biodiversity governance is that every perspective of nature comes with different sociocultural, political and economic implications for the future. At a policy level, prioritizing the intrinsic value of nature will result in adopting conservation strategies, envisioning human–nature relations or recalibrating the economic system in a very different way than if relational or instrumental values were prioritized. Moving across perspectives of nature, prioritizing one over another and referring to biodiversity instead of Mother Nature (or vice versa) imply different future worlds. This makes biodiversity governance a contested field, characterized by continual negotiation between different ontologies and epistemologies. The key to transformative biodiversity governance lies in the capacity to embrace and handle this contestation and negotiation without denying the radical value-based differences between perspectives but rather finding ways for them to coexist.

2.7 Discussion and Conclusion

This chapter introduced how different conceptions of nature have developed over time and in different geographies, as well as how different normative value perspectives shape and are reproduced by these definitions of nature. Ultimately, these conceptions and values influence strategies and targets for conserving and using nature. At the core, the nature/culture divide has been a foundational dichotomy in the way nature comes to be defined.

While this divide has been criticized both within and outside the Western context in which it was created, nonetheless, it remains essential to much of the debate around conservation.

We argue that defining nature is far from an objective and conflict-free exercise. On the contrary, defining nature is a value-laden task with theoretical and material repercussions. Choosing one definition and value of nature over another implies imagining and advocating for different worlds and nature futures. It means legitimizing one worldview over another. While this is inevitable, we must be aware of the implications for transformative biodiversity governance. Defining nature as wilderness generates conservation strategies that are not only different but possibly at odds with conservation strategies deriving from other conceptualizations of nature.

In this regard landscapes, ecosystem services and biocultural diversity are concepts that, despite differences, aim at integrating human and natural systems. Conservation strategies stemming from these concepts require a different approach to that of traditional protected areas, and much work remains to be done to understand how to integrate different strategies. It is important for transformative biodiversity governance to avoid reductionist approaches that smooth over important ontological or epistemological differences and to embrace pluralistic approaches, as well as to envision governance tools and mechanisms to navigate the political space offered by these multiple perspectives, such as legal Rights of Nature. Additionally, it will also be important to understand what pluralism materially means in terms of biodiversity governance. Does pluralism mean developing hybrid conservation strategies and targets that include multiple perspectives of nature? If so, it would be necessary to first reflect on the extent to which current strategies and targets (at both local and international levels) are receptive of this or, if not, how they favor – more or less implicitly – some perspectives over others.

Another crucial point for transformative biodiversity governance is that of transparency and clarification of choices. Many concepts and approaches are presented as "black boxes," without a clear view of the premises, rationales, norms and values included. This treats concepts and governance approaches as "truths," which is problematic for multiple reasons. Firstly, it hides (or at best marginalizes) any uncertainties, unknowns, discordant voices and ambiguity that may exist behind a concept. For example, in our discussion of the concept of "biodiversity," we noted that it did not emerge from a general consensus within the scientific community, and from the outset its usefulness was criticized.

The second problem that stems from treating concepts and approaches as truth-claims is that it makes them less open to influence by other perspectives. This is at odds with the new attention to inclusivity, plurality and justice that is emerging in biodiversity governance, and that is seen in recent multiperspective scenario exercises. In these, the objective was not to identify one single optimal vision for the future but, on the contrary, to create a space where multiple visions could come together and be realized. Truth-claims that do not acknowledge disagreement and diversity become markedly less tenable given calls for inclusivity and plurality. This requires a serious rethinking of the concepts and the practices that are employed in the name of biodiversity conservation, in order for those who deploy these concepts to become more self-reflective and aware of their own limits and of the values they hold.

References

Agnoletti, M. (2014). Rural landscape, nature conservation and culture: Some notes on research trends and management approaches from a (southern) European perspective. *Landscape and Urban Planning* 126, 66–73.

Agnoletti, M., and Emanueli, F. (Eds.). (2016). *Biocultural diversity in Europe*. Cham, Switzerland: Springer International Publishing.

Agnoletti, M., and Rotherham, I. D. (2015). Landscape and biocultural diversity. *Biodiversity and Conservation* 24, 3155–3165.

Babai, D., and Molnár, Zs. (2014). Small-scale traditional management of highly species-rich grasslands in the Carpathians. *Agriculture, Ecosystems and the Environment* 182, 123–130.

Barthel, S., Crumbley, C., and Svedin, U. (2013). Bio-cultural refugia: Safeguarding diversity of practices for food security and biodiversity. *Global Environmental Change* 23, 1142–1152.

Bhola, N., Klimmek, H., Kingston, N., et al. (2021). Perspectives on area-based conservation and what it means for the post-2020 biodiversity policy agenda. *Conservation Biology* 35, 168–178. https://doi.org/10.1111/cobi.13509.

Bowman, D. (1998) Death of biodiversity – the urgent need for global ecology. *Global Ecology and Biogeography Letters* 7, 237–240.

Bridgewater, P., and Rotherham, I. D. (2019). A critical perspective on the concept of biocultural diversity and its emerging role in nature and heritage conservation. *People and Nature* 1, 291–304. https://doi.org/10.1002/pan3.10040

Buizer, M., Elands, B., and Vierikko, K. (2016). Governing cities reflexively – The biocultural diversity concept as an alternative to ecosystem services. *Environmental Science & Policy* 62, 7–13. https://doi.org/10.1016/j.envsci.2016.03.003.

Burdon, P. (Ed.). (2011). *Exploring wild law: The philosophy of earth jurisprudence*. Adelaide: Wakefield Press.

Büscher, B., Fletcher, R., Brockington, D., et al. (2017). Half-earth or whole earth? Radical ideas for conservation, and their implications. *Oryx* 51, 407–410.

Butchart, S. (2010). Global biodiversity: Indicators of recent declines. *Science* 328(5982), 1164–1168.

Cano Pecharroman, L. (2018). Rights of nature: Rivers that can stand in court. *Resources* 7, 13. https://doi.org/10.3390/resources7010013.

Castree, N. (2013). *Making sense of nature*. New York: Routledge.

CBD. (1992). Convention on biological diversity. Available from www.cbd.int/doc/legal/cbd-en.pdf.

Chan, K. M. A., Balvanera, P., Benessaiah, K., et al. (2016). Opinion: Why protect nature? Rethinking values and the environment. *Proceedings of the National Academy of Sciences* 113, 1462–1465. DOI: 10.1073/pnas.1525002113

Chapron, G., Epstein, Y., and López-Bao, J. V. (2019). A rights revolution for nature: Introduction of legal rights for nature could protect natural systems from destruction. *Science* 363, 1392–1393.

Cocks, M. (2006). Biocultural diversity: Moving beyond the realm of "indigenous" and "local people." *Human Ecology* 34, 185–200.

Costanza, R. (1991). Ecological economics: A research agenda. *Structural Change and Economic Dynamics* 2, 335–357. https://doi.org/10.1016/S0954-349X(05)80007-4

Costanza, R., de Groot, R., Braat, L., et al. (2017). Twenty years of ecosystem services: How far have we come and how far do we still need to go? *Ecosystem Services* 28, 1–16. https://doi.org/10.1016/j.ecoser.2017.09.008

Costanza, R., Hart, M., Posner, S., and Talberth, J. (2009). Beyond GDP: The need for new measures of progress. Pardee Paper No. 4. Boston, MA: Pardee Center for the Study of the Longer-Range Future.

Cronon, W. (1996). The trouble with wilderness: Or, getting back to the wrong nature. *Environmental History* 1, 7–28.

Dasmann, R. F. (1968). *A different kind of country*. New York: Macmillan.

de Groot, R., Brander, L., van der Ploeg, S., et al. (2012). Global estimates of the value of ecosystems and their services in monetary units. *Ecosystem Services* 1, 50–61. https://doi.org/10.1016/j.ecoser.2012.07.005

Denevan, W. M. (1992). The pristine myth: The landscape of the Americas in 1492. *Annals of the Association of American Geographers* 82, 369–385.

Descola, P. (2013). *Beyond nature and culture*. Chicago, IL: University of Chicago Press.

de Sousa Santos, B. (2015). *Epistemologies of the South: Justice against epistemicide*. London; New York: Routledge.

Díaz, S., Pascual, U., Stenseke, M., et al. (2018). Assessing nature's contributions to people. Recognizing culture, and diverse sources of knowledge, can improve assessments. *Science* 359, 270–272. DOI: 10.1126/science.aap8826

Dinerstein, E., Olson, D., Joshi, A., et al. (2017). An ecoregion-based approach to protecting half the terrestrial realm. *BioScience* 67, 534–545.

Dinerstein, E., Vynne, C., Sala, E., et al. (2019). A global deal for nature: Guiding principles, milestones, and targets. *Science Advances* 5, eaaw2869.

Dinerstein, E., Joshi, A. R., Vynne, C., et al. (2020). A "global safety net" to reverse biodiversity loss and stabilize Earth's climate. *Science Advances* 6, eabb2824.

EEA. (2010). *The European environment – State and outlook 2010*. European Environment Agency. Available from https://bit.ly/3Jb7WbH.

(2015). *The European environment – State and outlook*. European Environment Agency. Available from www.eea.europa.eu/soer/2015.

(2020). *The European environment: State and outlook 2020. Knowledge for transition to a sustainable Europe*. European Environment Agency. Available from www.eea.europa.eu/soer/2020.

Ehrlich, P. and Ehrlich, A. (1981). *Extinction: The causes and consequences of the disappearance of species*. New York: Random House.

Elands, B. H. M., Vierikko, K., Andersson, E., et al. (2019). Biocultural diversity: A novel concept to assess human-nature interrelations, nature conservation and stewardship in cities. *Urban Forestry and Urban Greening* 40, 29–34.

European Landscape Convention of the Council of Europe. Available from www.coe.int/en/web/landscape/the-european-landscape-convention.

Fjeldsaa, J., and Lovett, J. (1997). Biodiversity and environmental stability. *Biodiversity and Conservation* 6, 315–323.

Garnett, S. T., Burgess, N. D., Fa, J. E. , et al. (2018). A spatial overview of the global importance of Indigenous lands for conservation. *Nature Sustainability* 1, 369–374.

Gómez-Baggethun, E., and Ruiz-Pérez, M. (2011). Economic valuation and the commodification of ecosystem services. *Progress in Physical Geography: Earth and Environment* 35, 613–628. DOI: 10.1177/0309133311421708

Gómez-Pompa, A., and Kaus, A. (1992). Taming the wilderness myth. *BioScience* 42, 271–279.

Grove, R. H. (1992). Origins of Western environmentalism. *Scientific American* 267, 42–47.

Gudynas, E. (2011). Buen Vivir: Today's tomorrow. *Development* 54, 441–447.

Harper, J. and Hawksworth, D. (1994). Biodiversity: Measurement and estimation. *Philosophical Transactions of the Royal Society B: Biological Sciences* 345, 5–12.

Hermoso, V., Morán-Ordóñez, A., and Brotons, L. (2018). Assessing the role of Natura 2000 at maintaining dynamic landscapes in Europe over the last two decades: Implications for conservation. *Landscape Ecology* 33, 1447–1460.

Himes, A., and Muraca, B. (2018). Relational values: The key to pluralistic valuation of ecosystem services. *Current Opinion in Environmental Sustainability* 35, 1–7.

Howarth, R.B., and Norgaard, R.B. (1992). Environmental valuation under sustainable development. *The American Economic Review* 82, 473–477.

Humphries, C., Williams, P. H., and Vane-Wright, R. I. (1995). Measuring biodiversity value for conservation. *Annual Review of Ecology and Systematics* 26, 93–111.

Immovilli, M., and Kok, M. T. J. (2020). *Narratives for the "half earth" and "sharing the planet" scenarios. A literature review*. The Hague: PBL Netherlands Environmental Assessment Agency.

IPBES. (2016). *The methodological assessment report on scenarios and models of biodiversity and ecosystem services*. S. Ferrier, K. N. Ninan, P. Leadley, et al. (Eds.). Bonn: Secretariat of the Intergovernmental Science-Policy Platform on Biodiversity and Ecosystem Services.

(2019). *Summary for policymakers of the global assessment report on biodiversity and ecosystem services of the Intergovernmental Science-Policy Platform on Biodiversity and Ecosystem Services*. S. Díaz, J. Settele, E. S. Brondízio, et al. (Eds.). Bonn, Germany: IPBES Secretariat.

IUCN. (n.d.). Category V: Protected Landscape/Seascape. Available from https://bit.ly/3sNq9ab.

Kauffman, C. M., and Martin, P. L. (2018). When rivers have rights: Case comparisons of New Zealand, Colombia, and India. *International Studies Association Annual Conference, San Francisco*, April 4, 2018. Available from http://files.harmonywithnatureun.org/uploads/upload585.pdf.

Keune, H., and Dendoncker, N. (2013). Negotiated complexity in ecosystem services science and policy making. In: *Ecosystem services: Global issues, local practices*. S. Jacobs, N. Dendoncker, and H. Keune (Eds.), pp. 167–180. San Diego, CA: Elsevier.

Kopnina, H. (2016). Half the earth for people (or more)? Addressing ethical questions in conservation. *Biological Conservation* 203, 176–185.

Kothari, A., and Bajpai, S. (2017). We are the river, the river is us. *Economic and Political Weekly* 52, 103–109.

Kothari, A., Demaria, F., and Acosta, A. (2014). Buen Vivir, degrowth and ecological Swaraj: Alternatives to sustainable development and the green economy. *Development* 57, 362–375.

Küster, H. (2003). *Geschichte des Waldes: von der Urzeit bis zur Gegenwart*. Munich: Beck.

Laffoley, D., Dudley, N., Jonas, H., et al. (2017). An introduction to "other effective area-based conservation measures" under Aichi Target 11 of the Convention on Biological Diversity: Origin, interpretation and emerging ocean issues. *Aquatic Conservation: Marine and Freshwater Ecosystems* 27, 130–137.

Lake Erie Bill of Rights Charter Amendment (2018). Available from https://bit.ly/341s7ti.

Lambooy, T., van de Venis, J., and Stokkermans, C. (2019). A case for granting legal personality to the Dutch part of the Wadden Sea. *Water International* 44, 786–803. DOI: 10.1080/02508060.2019.1679925

Latour, B. (1991). *We have never been modern*. Cambridge, MA: Harvard University Press.

Legagneux, P., Casajus, N., Cazelles, K., et al. (2018). Our house is burning: Discrepancy in climate change vs. biodiversity coverage in the media as compared to scientific literature. *Frontiers in Ecology and Evolution* 5. DOI: 10.3389/fevo.2017.00175

Levé, M., Colléony, A., Conversy, P., et al. (2019). Convergences and divergences in understanding the word biodiversity among citizens: A French case study. *Biological Conservation* 236, 332–339. https://doi.org/10.1016/j.biocon.2019.05.021.

Locke, H. (2015). Nature needs (at least) half: A necessary new agenda for protected areas. In: *Protecting the wild: Parks and wilderness, the foundation for conservation*. G. Wuerthner, E. Crist, and T. Butler (Eds.), 3–15. Washington, DC: Island Press.

Locke, J. M., Coates, K. A., Bilewitch, J. P., et al. (2013). Biogeography, biodiversity and connectivity of Bermuda's coral reefs. In: *Coral reefs of the United Kingdom overseas territories*. Coral Reefs of the World, vol 4., C. Sheppard (Ed.), pp. 153–172. Dordrecht: Springer. https://doi.org/10.1007/978-94-007-5965-7_12

Lovejoy, T. E. (1980). Changes in biological diversity. In: *The Global Report to the President: Entering the Twenty-First Century*, Vol. 2 (The Technical Report), G. O. Barney (Ed.). 328–331. Washington, DC: Council of Environmental Quality and the Department of State.

Mace, G. M., Barrett, M., Burgess, N. D., et al. (2018). Aiming higher to bend the curve of biodiversity loss. *Nature Sustainability*, 1, 448–451.

Marvier, M., Kareiva, P., Lalasz, R. (2012). Conservation in the Anthropocene: Beyond solitude and fragility. *Breakthrough Journal* 2. Available from https://bit.ly/3Jpr0Uu.

McDonald, B. (2001). Considering the nature of wilderness: Reflections on Roderick Nash's Wilderness and the American Mind. *Organization & Environment* 14, 188–201. doi:10.1177/1086026601142004

McElwee, P., Fernández-Llamazares, Á., Aumeeruddy-Thomas, Y., et al. (2020). Working with Indigenous and local knowledge (ILK) in large-scale ecological assessments: Reviewing the experience of the IPBES Global Assessment. *Journal of Applied Ecology* 57, 1666–1676.

Merchant, C. (1980). *The death of nature: Women, ecology and the scientific revolution*. New York: Harper and Row.

Moran, E. E. (2006). *People and nature. An introduction to human ecological relations*. Malden, MA: Blackwell.
Myers, N. (1979). *The sinking ark: A new look at the problem of disappearing species*. New York: Pergamon Press.
Myers, N., Mittermeier, R. A., Mittermeier, C. G., Da Fonseca, G. A., and Kent, J. (2000). Biodiversity hotspots for conservation priorities. *Nature* 403, 853–858.
Nash, R. (1967). *Wilderness and the American mind*. New Haven, CT: Yale University Press.
Norgaard, R. B. (2010). Ecosystem services: From eye-opening metaphor to complexity blinder. *Ecological Economics* 69, 1219–1227. https://doi.org/10.1016/j.ecolecon.2009.11.009.
O'Donnell, E. L., and Talbot-Jones, J. (2018). Creating legal rights for rivers: Lessons from Australia, New Zealand, and India. *Ecology and Society* 23, 7.
Olsson, E. G. A. (2018). The shaping of food landscapes from the Neolithic to Industrial period. Changing agro-ecosystems between three agrarian revolutions. In: *Routledge Handbook of Landscape and Food*. J. Zeunert and T. Waterman (Eds.), pp. 24–40. Oxford: Routledge.
Oteros-Rozas, E., Ontillera-Sánchez, R., Sanosa, P., et al. (2013). Traditional ecological knowledge among transhumant pastoralists in Mediterranean Spain. *Ecology and Society* 18, 33.
Padoch, C., and Pinedo-Vasquez, M. (2010). Saving slash-and-burn to save biodiversity. *Biotropica* 42, 550–552.
Pascual, U., Balvanera, P., Díaz, S., et al. (2017). Valuing nature's contributions to people: The IPBES approach. *Current Opinion in Environmental Sustainability* 26–27, 7–16. https://doi.org/10.1016/j.cosust.2016.12.006.
Pattberg, P. (2007). Conquest, domination and control: Europe's mastery of nature in historic perspective. *Journal of Political Ecology* 14, 1–9.
Pearson, R. G. (2016). Reasons to conserve nature. *Trends in Ecology and Evolution* 31(5), 366–371.
Pechanec, V., Machar, I., Pohanka, T., et al. (2018). Effectiveness of Natura 2000 system for habitat type protection: A case study from the Czech Republic. *Nature Conservation* 24, 21–41.
Pereira, L., Davies, K., den Belder, E., et al. (2020). Developing multiscale and integrative nature–people scenarios using the Nature Futures Framework. *People and Nature* 2, 1172–1195. doi:10.1002/pan3.10146
Roncucci, R. (2019). *Rights of Nature and the Pursuit of Environmental Justice in the Atrato Case*. Doctoral Thesis. Wageningen University & Research. Available from https://edepot.wur.nl/504758.
Sarkar, S. (2016). Approaches to biodiversity. In: *The Routledge handbook of philosophy of biodiversity*, J. Garson, A. Plutynski and S. Sarkar (Eds.), pp. 43–55. New York: Routledge.
Satz, D., Gould, R. K., Chan, K. M. A., et al. (2013). The challenges of incorporating cultural ecosystem services into environmental assessment. *AMBIO* 42, 675–684. https://doi.org/10.1007/s13280-013-0386-6.
Schlebusch, C. M., Malmström, H., Günther, T., et al. (2017). Southern African ancient genomes estimate modern human divergence to 350,000 to 260,000 years ago. *Science* 358, 652–655.
Schneiders, A., and Müller, F. (2017). Natural basis for ecosystem services. In: *Mapping Ecosystem Services*, B. Burkhard and J. Maes (Eds.), pp. 33–38.
Schröter, M., van der Zanden, E. H., van Oudenhoven, A. P., et al. (2014). Ecosystem services as a contested concept: A synthesis of critique and counter-arguments. *Conservation Letters* 7, 514–523. https://doi.org/10.1111/conl.12091
Sevink, J., van Geel, B., Jansen, B., and Wallinga, J. (2018). Early Holocene forest fires, drift sands, and Usselo-type paleosols in the Laarder Wasmeren area near Hilversum, the Netherlands: Implications for the history of sand landscapes and the potential role of Mesolithic land use. *Catena* 165, 286–298.
Spence, M. (1999). *Dispossessing the wilderness: Indian removal and the making of the national parks*. New York: Oxford University Press.
Takacs, D. (1996). *The idea of biodiversity: Philosophies of paradise*. Baltimore and London: The Johns Hopkins University Press.
Uggla, Y. (2010). What is this thing called "natural"? The nature-culture divide in climate change and biodiversity policy. *Journal of Political Ecology* 17, 79–91.

Ujházy, N., Molnár, Zs., Bede-Fazekas, Á., Szabó, M., and Biró, M. (2020). Do farmers and conservationists perceive landscape changes differently? *Ecology and Society* 25, 12.

UNESCO. (n.d.). Local and Indigenous knowledge systems. Available from https://bit.ly/3HuPNF3.

Valladares, C., and Boelens, R. (2017). Extractivism and the rights of nature: Governmentality, "convenient communities" and epistemic pacts in Ecuador. *Environmental Politics* 26, 1015–1034.

Van Zeijst, Prins, A.G., Dammers, E., Vonket, M., et al. (2017). European nature in the plural. Finding common ground for a next policy agenda. The Hague: PBL Netherlands Environmental Assessment Agency.

Visconti, P., Butchart, S. H., Brooks, T. M., et al. (2019). Protected area targets post-2020. *Science* 364, 239–241.

Walpole, M., Almond, R. E., Besançon, C., et al. (2009). Tracking progress toward the 2010 biodiversity target and beyond. *Science* 325, 1503–1504.

Washington, H., Taylor, B., Kopnina, H., Cryer, P., and Piccolo, J. J. (2017). Why ecocentrism is the key pathway to sustainability. *The Ecological Citizen* 1, 35–41.

Wilson, E. O. (Ed.). (1988). *Biodiversity*. Washington, DC: National Academy Press.

(1992). *The diversity of life*. Cambridge, MA: Harvard University Press.

(2016). *Half-earth: Our planet's fight for life*. New York: WW Norton and Company.

Wolke, H. (2014). Wilderness: What and why? In: *Keeping the wild*. G. Wuerthner and E. Crist (Eds.), pp. 197–204. Washington, DC: Island Press.

Worster, D. (1977). *Nature's Economy. The Roots of Ecology*. San Francisco, CA: Sierra Club Books.

Wuerthner, G., Crist, E., and Butler, T. (Eds.). (2015). *Protecting the wild: Parks and wilderness, the foundation for conservation*. Washington, DC: Island Press.

3

Global Biodiversity Governance: What Needs to Be Transformed?

JOANNA MILLER SMALLWOOD, AMANDINE ORSINI, MARCEL T. J. KOK, CHRISTIAN PRIP AND KATARZYNA NEGACZ

3.1 Introduction

The Post-2020 Global Biodiversity Framework (GBF) of the Convention on Biological Diversity (CBD) (the Post-2020 Framework) is expected to embody transformative change through the adoption of the framework's "Theory of Change" (CBD, 2020). Its implementation must recognize that the global biodiversity governance architecture needs to transform to lead the required personal and social transformations, including shifts in values, beliefs and patterns of social behaviors (Chaffin et al., 2016), necessary to successfully tackle biodiversity loss. Against this backdrop, the overarching goal of this chapter is to analyze what needs to be transformed in global biodiversity governance, including institutional structures that shape values, beliefs and behavioral change. The chapter examines obstacles and opportunities for transformation, with the indirect objective of informing implementation of the Post-2020 Framework; at the time of writing, the CBD is expected to adopt the Post-2020 GBF in 2022.

The chapter firstly introduces the key global biodiversity treaty, the 1992 UN Convention on Biological Diversity, and its principal institutional body, the Conference of the Parties (COP) (Section 3.2). The evolution of the CBD is analyzed along with its procedural mechanisms, including its decision-making and review mechanisms. Secondly, the chapter presents the other relevant international institutions in what constitutes the "regime complex" for global biodiversity governance (Section 3.3). Within this complex, biodiversity governance takes place at multiple levels, from global to local, and in different sectors, including some of those most responsible for biodiversity loss such as agriculture, trade and development. The evolution of biodiversity governance beyond the CBD is also explored by analyzing the role of private actors, including business and civil society, in global biodiversity governance. Thirdly, the implementation of global biodiversity laws and policies is examined through global and national governance processes (Section 3.4). The final section draws upon the analyses to propose ways to transform and strengthen global biodiversity governance (Section 3.5), before concluding. The chapter is mainly based on legal analyses, while also drawing on more generic biodiversity governance literature.

3.2 The Convention on Biological Diversity

3.2.1 The CBD, from Seed to Sapling

The CBD opened for signatures at the United Nations Conference on Environment and Development, known as the Earth Summit, in Rio in 1992, marking the start of the "postmodern era" of environmental regulation (Sands, 2007). The Convention, having now near universal ratification (with the major exception of the United States), marked a paradigm shift, from earlier species-specific and ecosystem-based nature conservation conventions to a holistic and development-oriented approach to biodiversity. The CBD is a framework convention that sets out basic principles, general objectives, and rather broad and qualified provisions. The three objectives are biodiversity conservation, sustainable use, and the fair and equitable sharing of benefits. Legal polycentricity, intergenerational responsibilities, and the need for inclusive and participatory processes were new concepts recognized by the treaty (Sands, 2007).

In addition, three legally binding protocols have been agreed to date under the CBD Art 28 mechanism: the 2000 Cartagena Protocol on Biosafety, the 2010 Nagoya Protocol on Access to Genetic Resources and the 2010 Kuala Lumpur Supplementary Protocol on Liability and Redress (Supplementary to the Cartagena Protocol). While these protocols cover the second and third objective of the CBD respectively, it is remarkable that no protocol has been agreed relating to the first objective of the CBD, biodiversity conservation. Thus, the first objective has been addressed by the COP only through its non-legally binding instruments like strategic plans, visions, goals and targets, decisions, guidelines and recommendations.

The design of CBD targets has improved since the first broad "2010" biodiversity target, which called state parties "to achieve a significant reduction of the current rate of biodiversity loss at the global, regional and national level by 2010 as a contribution to poverty alleviation and to the benefit of all life on earth" (CBD COP6, 2002). This target was unmet and superseded by the 2020 strategic plan and the twenty Aichi Targets (ATs), agreed at CBD COP10 in 2010 (see Chapter 1). The ATs were designed to be SMART (specific, measurable, ambitious, realistic and time-bound) and to improve the initial 2010 target (Harrop and Pritchard, 2011). However, well before the 2020 deadline it was clear that most of the ATs would not be achieved (IPBES, 2019; SCBD, 2020).

3.2.2 An Active Body: The CBD COP

The CBD COP is the governing body of the CBD, where state parties make decisions by consensus to advance implementation of the Convention. It is in a unique position to strengthen global biodiversity governance to steer change. The COP can advance the evolution and implementation of the CBD by (i) agreeing and furthering ambitions through decisions that are soft law but guide parties, and (ii) creating a space to positively encourage and promote implementation of obligations. It creates a space for the development of shared understandings of the legal regulation of biodiversity, and norms through the elaboration of guidelines on various topics. The thematic priorities of COPs (see Table 3.1) have changed

Table 3.1 *CBD COP themes*

COP	Location, year	Theme(s)
COP1	Nassau, Bahamas, 1994	–
COP2	Jakarta, Indonesia, 1995	Marine and coastal biodiversity
COP3	Buenos Aires, Argentina, 1996	Agricultural biodiversity
COP4	Bratislava, Slovakia, 1998	Inland water ecosystems
COP5	Nairobi, Kenya, 2000	Dryland, Mediterranean, arid, semi-arid, grassland and savannah ecosystems
COP6	The Hague, Netherlands, 2002	Forest ecosystems and alien species
COP7	Kuala Lumpur, Malaysia, 2004	Mountain ecosystems
COP8	Curitiba, Brazil, 2006	Island biodiversity
COP9	Bonn, Germany, 2008	One nature, one world – our future
COP10	Nagoya, Aichi Prefecture, Japan, 2010	Life in harmony into the future and the 2050 vision, focused toward developing the strategic plan
COP11	Hyderabad, India, 2012	Nature protects if she is protected
COP12	Pyeongchang, Republic of Korea, 2014	Biodiversity for sustainable development
COP13	Cancun, Mexico, 2016	Mainstreaming the conservation and sustainable use of biodiversity for well-being
COP14	Sharm El-Sheikh, Egypt, 2018	Investing in biodiversity for people and planet, and for the high-level segment: mainstreaming of biodiversity in the energy and mining; processing industry; infrastructure and health sectors
COP15	Kunming, China, scheduled for the second quarter of 2022	Ecological civilization: building a shared future for all life on Earth

from predominantly ecosystem-based themes (COP1–COP9) to addressing the main drivers of biodiversity loss (COP10–COP14). Themes of earlier COPs do not necessarily tally with their focus or substantial outcomes. For example, COP7's theme was "Mountain Ecosystems" and, while a work program on this theme was adopted, more notably a work program on protected areas and the Addis Ababa principles on sustainable use were also adopted, which received more attention and subsequently are seen as more important. Changing narratives indicate the broadening of agendas of the CBD and the themes of more recent COPs better match their outcomes.[1] COP15 follows this trend and hooks onto an important concept: "Ecological Civilization: Building a Shared Future for All Life on Earth."

Due to the broad scope and comprehensive character of the CBD COP, it is essential that there is buy-in from a very wide range of actors. The Open-Ended Working Group (OEWG)

---

[1] COP10 in 2010 adopted the "Nagoya Package," with the Nagoya Protocol, the Strategic Plan and a decision on resource mobilization, and was thus in good harmony with its broad theme, "Life in harmony into the future and the 2050 Vision." The same applies to COP13, with its overall mainstreaming theme, which resulted in various outputs to integrate biodiversity values into other sectors, including the high-level segment Cancun declaration on mainstreaming the conservation and sustainable use of biodiversity for well-being, and the CBD Business and Biodiversity Pledge.

responsible for developing the Post-2020 Framework utilizes a theory of change approach to guide the development of a nature framework for all, not just for signatories from the Ministry of Environment, but for the whole of government, multilateral institutions, Indigenous People and local communities (IPLC), nongovernmental organizations (NGOs) and business. This could be challenging. A study of the 2016 CBD COP13 in Cancun, Mexico, found a poor representation of government ministers from the economic sectors from both the global north and south, indicating the limited buy-in of biodiversity negotiations nationally, and that disadvantaged actors from the global south were unable to participate as effectively in negotiations due to the limited size of their delegations and lack of expertise to cover all agenda items (Smallwood, 2019). This unbalanced dimension creates power dynamics that are problematic in consensus decision-making and in creating obligations that rest on genuine shared understandings: not all relevant actors are present and exposed to the processes of influence and persuasion at COP meetings (Brunnée, 2002; Smallwood, 2019).

The CBD COP has a long history of engagement with stakeholders such as women, children and youth, NGOs, local authorities, trade unions, business and industry, science and technology, and farmers as observers to its meetings. IPLC have a well-established engagement and influence that is unique for the CBD compared to other intergovernmental processes (Parks, 2018). Such nongovernmental actors are central actors in international environmental regimes including the CBD (Spiro, 2007), exerting influence through: domestic political processes such as rallying voters, lobbying law makers, disseminating information, bringing legal actions and working with media and academia (Chayes and Chayes, 1995); advancement of domestic NGO agendas in the international sphere (Spiro, 2007); and agenda-setting (Arts and Mack, 2006). Nongovernmental actors also take on certain key functions within international negotiations, including supplying policy research and development to states (for instance, the 5th Global Biodiversity Outlook is a product of "collected efforts" including individuals from nongovernmental organizations and scientific networks), supplying information on compliance,[2] facilitating negotiations[3] and participating in national delegations (Smallwood, 2019).

A specificity of the CBD COP has also been its ambition to include businesses in its activities. A 2006 COP decision on business participation defines a "business and biodiversity" agenda.[4] Subsequent COP decisions aim to facilitate private sector engagement and encourage businesses to "adopt practices and strategies that contribute to achieving the goals and objectives of the Convention and the Aichi Targets" (COP12 Decision XII/10). A Global Partnership for Business and Biodiversity and a Business and Biodiversity Forum have been established, and the 2017 Business and Biodiversity Pledge has 141 signatories, including some large corporations such as Monsanto, L'Oréal and DeBeers; however, most relevant multinational corporations to biodiversity loss are not signatories. Despite these

---

[2] Among others, at CBD COP13, a coalition of NGOs produced a report on the alignment of countrys' national targets to the ATs and progress toward achievement of the ATs (RSPB et al., 2016).
[3] For example, for each CBD COP, a civil society publication known as ECO and the Earth Negotiations Bulletin provide daily reports to delegates on complex negotiation topics.
[4] CBD decision VIII/172.

decisions and initiatives on business, to date the level of business involvement has been less than aimed for by the CBD COP (van Oorschot et al., 2020).

The CBD stresses the importance of "mainstreaming," that is, the inclusion of biodiversity considerations into nonenvironmental policy areas that impact or rely on biodiversity (Young, 2011). Art 6(b) of the CBD requires Parties to integrate the conservation and sustainable use of biodiversity into sectoral and cross-sectoral activities. Subsequently, means of furthering mainstreaming have been an endeavor of the CBD COP. The first goal of the 2011–2020 CBD strategic plan, agreed at COP10, was to address the underlying causes of biodiversity loss by mainstreaming biodiversity across production sectors and society (GEF, 2016; GEF et al., 2007; SCBD, 2020).[5] In addition, COP decisions on mainstreaming have been agreed, and mainstreaming was adopted as the key theme at COP13 and COP14. So far, mainstreaming is mostly considered an issue of policy coherence that is yet to be realized at global and national levels, let alone making significant links with communities such as business to realize the whole of society approach advocated by the CBD.

The CBD has two permanent subsidiary bodies: First, Art 25 of the Convention established an open-ended intergovernmental scientific advisory board, the Subsidiary Body on Scientific, Technical and Technological Advice (SBSTTA). The SBSTTA provides advice and makes recommendations to the COP and has met twenty-four times from 1995 to 2020. Second, COP12 established a Subsidiary Body for Implementation (SBI) in 2014, whose mandate includes strengthening mechanisms to support implementation of the Convention and any strategic plans adopted under it, and identifying and developing recommendations to overcome obstacles encountered. Due to the soft law nature of most CBD decisions, the CBD has adopted a facilitative approach toward implementation by monitoring national implementation through national reporting (Art 26). Besides, a system of voluntary peer review of National Biodiversity Strategies and Action Plans (NBSAPs) and their implementation is under development. The methodology was tested in two countries (Ethiopia and India), and later three countries have been reviewed in a pilot phase (Montenegro, Sri Lanka, Uganda) (CBD, 2020).

3.3 The Biodiversity Regime Complex

3.3.1 The Intergovernmental Components of the Regime Complex

Intergovernmental biodiversity governance has also evolved beyond the CBD. Indeed, due to its comprehensive scope, the CBD has gradually become the central element of a biodiversity regime complex, consisting of five pre-existing international regimes that progressively became regime complexes as well (see Figure 3.1, based on Morin and Orsini, 2014).

The first is the environmental regime. The first objective of the CBD, biodiversity conservation, facilitated interactions between the CBD and a pre-existing cluster of

---

[5] 2011–2020 Strategic Goal A, consisting of ATs 1–4, specifically addresses mainstreaming to address the underlying causes of biodiversity loss, and ATs 6–8 call for the direct pressures on biodiversity to be reduced and to promote sustainable use in the fishery, agriculture, aquaculture and forestry sectors (see Chapter 1 for an overview of the ATs).

48 *Joanna Miller Smallwood et al.*

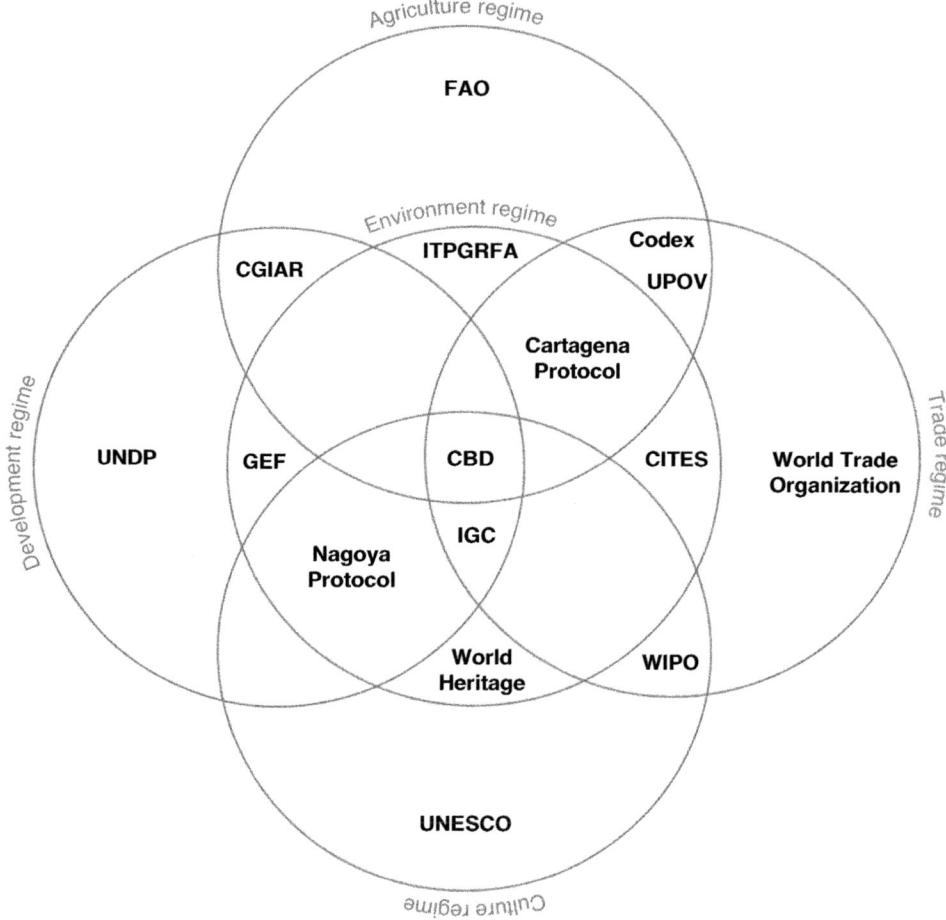

Figure 3.1 The regime complex on biodiversity (with a selection of international institutions provided as illustrations of the constituent elements)
CBD: Convention on Biological Diversity
CGIAR: Consultative Group on International Agricultural Research
CITES: Convention on International Trade in Endangered Species of Wild Fauna and Flora
FAO: Food and Agriculture Organization of the United Nations
GEF: Global Environment Facility
IGC: WIPO Intergovernmental Committee on Intellectual Property and Genetic Resources, Traditional Knowledge and Folklore
ITPGRFA: International Treaty on Plant Genetic Resources for Food and Agriculture
UNDP: United Nations Development Programme
UNESCO: United Nations Educational, Scientific and Cultural Organization
UPOV: International Union for the Protection of New Varieties of Plants
WIPO: Word Intellectual Property Organization

multilateral agreements within the environmental regime. Some of these agreements are biodiversity-related conventions such as the Ramsar Convention on Wetlands, the Convention on Migratory species (CMS) and the Convention on International Trade in Endangered Species of Wild Fauna and Flora (CITES). In 2007, these conventions started to collaborate in the framework of a broader Liaison Group of the Biodiversity-Related Conventions. The environmental conservation regime also consists of treaties that are not exclusively biodiversity-related, such as the United Nations Framework Convention on Climate Change (UNFCCC) and the UN Convention to Combat Desertification (UNCCD) (also adopted at the Rio Summit). A Joint Liaison Group of the Rio conventions has been established to enhance coordination and explore options for cooperation and synergistic action.[6]

The second is the agricultural regime. The interactions here are established on a dual basis: agriculture practices are one of the main drivers for biodiversity loss, but agricultural biodiversity is also under threat, and constitutes the basis of food security (IPBES, 2019, see also Chapter 13). How best to manage agricultural biodiversity raises several questions, as agricultural genetic resources are not only important components of biodiversity but also constitute essential food resources (Spann, 2017). In addition, the Cartagena Protocol on Biosafety to the CBD also interacts with the agricultural regime by developing rules concerning the use, especially in agriculture, of genetically modified organisms. The CBD has always considered the agricultural sector to be a priority for mainstreaming.

The third is that of trade. Natural resources, like any other type of good, are traded; and biodiversity is subject to innovation protection, through instruments of intellectual property rights such as patents under the Agreement on Trade-Related Aspects of Intellectual Property Rights (TRIPS) of the World Trade Organization (Raustiala and Victor, 2004). To counter TRIPS, the CBD stated the principle of state sovereignty over natural resources, which allows states to regulate access to biodiversity within their borders.

The fourth regime is the international development regime. Sustainable development was at the heart of the priorities of the 1992 Rio Summit, which adopted the CBD (Ademola et al., 2015). The development regime includes, among others, financial provisions through, for instance, the Global Environment Facility, to assist developing countries to achieve the objectives of the CBD.

The fifth is that of culture. Originally, the main focus of this regime was on cultural heritage through the United Nations Educational, Scientific and Cultural Organization (UNESCO) World Heritage Convention (WHC). The WHC is part of the Liaison Group of the Biodiversity-Related Conventions and is increasingly connected with biocultural diversity, alongside other international policies such as the Nagoya Protocol to the CBD, which recognizes the importance of the traditional knowledge associated with genetic resources (Morgera et al., 2014), and the positive role of IPLC in conservation and the biocultural values that they represent (IPBES, 2019).

The existence of a regime complex is both a strength and a weakness for the CBD ("*be at the table or be on the menu*"). On the one hand, it ensures biodiversity is "at the

---

[6] UNCCD-ICCD/CRIC(11)/INF.3.

table" and the various elements of the regime complex give resonance and amplify the biodiversity issue with its multiple dimensions and values (see Chapter 2). On the other hand, it is a weakness and can be seen to be "on the menu" with more powerful components of the regime deciding the fate of biodiversity. Lack of integrative governance between the different intergovernmental components of the complex, and tensions between biodiversity and the trade, agriculture and development dimensions has led to insufficient attention to biodiversity, as evidenced by poor progress on mainstreaming, and missed biodiversity targets. Policy coherence for biodiversity at the global level is an important precondition for "whole of government" approaches for biodiversity, as is being discussed in the Post-2020 Framework.

3.3.2 Governance beyond the Intergovernmental Realm

Since the 1980s, the institutional landscape of global biodiversity governance has shifted from predominantly public to more private and hybrid (public–private) forms of governance involving private actors (Kok et al., 2019; Negacz et al., 2020). The regime complex has expanded and includes new nonstate dimensions that work across state borders; this is referred to as transnational environmental governance (Bulkeley and Jordan, 2012). Neoliberalism has steered the privatization of state functions and promoted the commodification of biodiversity within global markets, thus shifting power relations (Büscher et al., 2012). For example, in agricultural commodity chains, public, private and, to a lesser extent, not-for-profit organizations play roles in global environmental governance, extending governance beyond legal and policy regimes.

The broader trend toward increased transnational governance can be seen in biodiversity policy as well as other areas, such as climate change and sustainable development (Bansard et al., 2017; Bulkeley & Newell, 2015; Jordan et al., 2015; Pattberg, 2010; Pattberg et al., 2019; van Oorschot et al., 2020; Visseren-Hamakers, 2013). An increasing number of nonstate and subnational actors (e.g., cities, regions, business and finance) participate in a plethora of national and international cooperative initiatives with the aim of addressing biodiversity loss (Pattberg et al., 2019; Visseren-Hamakers, 2013).

The increasing importance of nonstate and subnational actors, as well as their formal involvement, poses challenges to a state-based UN process like the CBD and the Post-2020 Framework and its further implementation. Collaboration with transnational actors entered a new stage in 2018 when, at COP14, COP presidencies Egypt and China, with the CBD Secretariat, launched the "Sharm El-Sheikh to Kunming Action Agenda for Nature and People" (Kok et al., 2019; Pattberg et al., 2019). The action agenda's aim is to raise public awareness about the urgent need to stem biodiversity loss and restore biodiversity for both nature and people; to inspire and implement nature-based solutions to meet key global challenges; and to catalyze nonstate and subnational initiatives in support of global biodiversity goals. The action agenda is hosted on an online platform that has received and showcased commitments and contributions to biodiversity from stakeholders across all sectors in advance of COP15. This platform enables the mapping of global biodiversity efforts and helps to identify key gaps and estimate impact. With such a platform, the CBD follows current governance trends "towards transnational environmental governance and the inclusion of non-state action in multilateral agreements" (Pattberg et al., 2019: 385). Increasing inclusivity is considered an important element of transformative biodiversity governance (see Chapter 1); this is an important development in contributing to the mainstreaming of biodiversity where it matters as part of integrative governance (Bulkeley et al., 2020; Karlsson-Vinkhuyzena et al., 2017), and is being framed as a "whole of society approach" in the Post-2020 Framework.

Within the category of nonstate actors, the important role of subnational actors, cities, regions and local authorities has been recognized in the CBD since 2010. The "Edinburgh process" allows the active participation of subnational actors in consultations, therefore shaping the Post-2020 Framework and targets. With the global growth of urban populations, Puppim de Oliveira et al. (2011) argue that, even though cities are not directly involved in negotiating environmental agreements, they can play a major role in implementation and influence biodiversity conservation (Bulkeley et al., 2012). Increasingly, large urban and regional initiatives, such as the International Council for Local Environmental Initiatives, or Covenant of Mayors, actively engage in diverse biodiversity activities and policies (see Chapter 14).

The involvement of business and the financial sector in the CBD is more contested. The first COP decision to encourage stronger business involvement was made in 1996 at COP3, but it took until 2010 for a CBD Business and Biodiversity platform to be established. Businesses within primary sectors, which exert direct pressure on biodiversity but also highly depend on it, have started to develop more biodiversity-friendly production methods, see opportunities in developing nature-based solutions and contribute to various sustainability and corporate social responsibility (CSR) goals, although pressure on biodiversity continues to grow (SCBD, 2020). Furthermore, international networks for business and biodiversity are starting to emerge: In 2019, the Business for Nature network was created with the aim of encouraging the adoption of a post-2020 biodiversity transformative agenda.

This diverse and polycentric institutional landscape of global biodiversity governance, described by Pattberg et al. (2017; 2019), is rapidly expanding. Negacz et al. (2020) and Curet and Puydarrieux (2020) identified 331 international collaborative initiatives forming a crowded and diverse governance landscape, with international collaborative initiatives transitioning from predominantly public to more hybrid forms, including state, market and civil society actors, performing a broad array of governance functions. Most initiatives focus on information sharing and networking, followed by on-the-ground activities, setting standards and certification. Their activities mostly focus on sustainable use and conservation efforts for sectors such as agriculture, forestry and fisheries, rather than solely conservation. The geographical coverage of the initiatives suggests a wide but uneven distribution of activities. The efforts of the initiatives focus on Europe and Africa, leaving areas of high biodiversity in Asia and Latin America with much less attention (Negacz et al., 2020). Most initiatives monitor their performance, and more than half report their progress annually. Yet, only one-fourth of them has a verification mechanism in place, making review of progress more challenging (Negacz et al., 2020).

These more inclusive forms of biodiversity governance that commit to action for biodiversity, by a broad coalition of nonstate and subnational actors, could facilitate transformative change for biodiversity by breaking gridlocks in current negotiations through: fostering a nature-inclusive agricultural transition; pushing governments to increase their ambition levels to create a level playing field for front runners; building new multistakeholder coalitions and finding innovative solutions to existing problems (Hale et al., 2013; Pattberg et al., 2019). Yet, business engagement also raises serious concerns with business taking a powerful role in reshaping the biodiversity regime to its own profit-making agendas (Büscher et al., 2012; Corson and MacDonald, 2012; MacDonald, 2010; Spann, 2017). Therefore, to avoid greenwashing, it is important to monitor and review progress. However, tracking the impact of international cooperative initiatives on the ground remains a challenge (Arts et al., 2017), and the impact, accountability, legitimacy and transparency of transnational biodiversity initiatives require more research (Gupta, 2008; Jones and Solomon, 2013).

3.4 Implementing Biodiversity Law and Policy

3.4.1 NBSAPs: Strengths and Limitations

National Biodiversity Strategies and Action Plans provide the foundation for national implementation of the CBD. In fact, their provision in the CBD, Article 6(a), is one of only two provisions that are unqualified and binding on Parties to the CBD whatever the circumstances; the other is Article 26 on national reporting. Its twin provision, Article 6(b), requires state parties to integrate the conservation and sustainable use of biodiversity into sectoral and cross-sectoral activities, signaling that such mainstreaming should be a key element of NBSAPs.

An upgrade of the role of NBSAPs was made in 2010 by the inclusion of AT 17, stating that "By 2015, each Party has developed, adopted as a policy instrument, and has commenced implementing, an effective, participatory and updated national biodiversity strategy and action plan."

In early 2021, 191 out of 196 CBD state parties (97%) have developed at least one NBSAP, among which 169 have been developed after the adoption of the ATs. NBSAP processes have led to a better understanding of biodiversity, its value and what is required to address its threats. However, for many first-generation NBSAPs (developed before the ATs), development processes were more technical than political and did not manage to sufficiently influence policy beyond the remit of the Ministry of Environment (or whichever ministry is directly responsible for biodiversity) (Prip et al., 2010).

Second-generation NBSAPs were therefore proposed for the post-2010 period. These include national targets to a larger extent and offer an opportunity for a diversity of actors to engage with biodiversity policies and connect relevant decision-makers within a country (Ademola et al., 2015). However, the potential to "make NBSAPs matter" (Ademola et al, 2015: 105) is challenged using national targets more oriented toward classic nature conservation than systemically oriented to address the underlying causes of biodiversity loss through mainstreaming. Such goals and targets are often expressed in general, aspirational terms, without specifications as to how they could be operationalized. Many countries seem to be at a preliminary stage in terms of mainstreaming because a necessary first step is a basic review of all policies and legislation relevant to biodiversity (Prip and Pisupati, 2018). Moreover, many first-generation NBSAPs have not been endorsed beyond the ministry directly responsible for the CBD, indicating that mainstreaming goals and targets has not always been fully coordinated at the political level. Some NBSAPs specify that this remains to be done (Prip and Pisupati, 2018).

While the post-2010 NBSAPs reveal that biodiversity mainstreaming is gaining recognition, the process is at a very early stage and a considerable amount of political and legal work still needs to be done before tangible results can be achieved on the ground. Considering the missed Aichi Targets, this work needs to be prioritized to address the biodiversity crisis in time.

3.4.2 The Implementation Gap

Effective implementation has long been a challenge for the CBD (Butchart et al., 2016). Theorists offer different explanations for poor implementation and lack of compliance, and these can be explored in the context of the CBD. International relations rationalists see power dynamics and self-interest as motivations for states to act (Goldsmith and Posner, 2005). Enforcement theorists indicate that compliance may require considerable resources in time, political engagement and financing; therefore, sanctions and other enforcement mechanisms are required to incentivize states to comply (Koskenniemi, 2011). Managerial schools understand that states will generally comply with international law because: (i) it is consent-based and therefore generally serves their interests, (ii) it is an effective cooperative problem-solving method saving costs and (iii) there is a general norm of compliance among states. Subsequently, noncompliance can be explained by ambiguity in international law and capacity limitations (Chayes and Chayes, 1993).

Positivist lawyers argue that the lack of hard law provisions in the CBD is a key factor for explaining why there are large gaps in implementation and state parties are not sufficiently achieving the CBD objectives, targets and goals (Harrop and Pritchard, 2011). As a treaty, the CBD is a hard law instrument and contains "hard" obligations, such as Art 6 relating to NBSAPs and Art 26 relating to national reporting. Otherwise, the CBD has largely developed through "soft" or qualified legal obligations, and the treaty itself uses vague and noncommittal language, such as "as appropriate," "as far as possible" and "subject to other existing international/national legislation," which essentially renders these provisions "soft" (Harrop and Pritchard, 2011: 477). Decisions, including strategic plans and targets, of the CBD COP are "soft" obligations. Significant gaps in national implementation suggest the design of targets is problematic due to their ambiguity, lack of quantifiability, complexity and redundancy (Butchart et al., 2016), and therefore they lack institutional fit at the national level (Hagerman and Pelai, 2016).

However, states can take nonbinding or "soft" international environmental legal obligations seriously.[7] If soft law can guide or influence behavior (Bodansky, 2016), then different explanations for what makes law effective must be considered. Interactive law blends law with constructivist understandings (Brunnée and Toope, 2010), and is relevant to understanding the CBD with its plethora of soft law provisions. It recognizes that law (hard or soft) can draw compliance: (i) through the fulfillment of certain internal criteria of legality; (ii) when it is based on genuine shared understandings formed by broad participation of all relevant actors in legal decision-making fora and (iii) when a practice of legality is established that reenforces and revisits the legal obligation. When applied to the CBD ATs, new explanations for implementation gaps arise:

- Clarity: Many targets are unquantifiable and complex;
- Achievability: Some ATs ask the impossible,[8] yet are still not ambitious enough to achieve the CBD's conservation objective;
- Promulgation: General lack of awareness of biodiversity issues and the biodiversity targets. The CBD COP fails to attract some relevant actors, and this influences the adopted shared understandings;
- Lack of a compliance mechanism: This poses a challenge to creating a clear practice of legality (Smallwood, 2019).

Practical challenges for implementation include: the CBD's broad scope, expanding subject-matter and failure to identify priority targets (Mace et al., 2018), thus allowing parties to cherry pick on implementation; the complexity of biodiversity as a subject-matter, coupled by lack of data, capacity and funding; power asymmetries in relation to trade-related treaties (see Section 3.3.1); lack of vertical mainstreaming to production sectors at the domestic level (Section 3.4.1); lack of coordination between ministries, state and local authorities at the national level; and a general lack of prioritization (Morgera and Tsioumani, 2010).

Another key challenge for the CBD is for state parties to effectively implement global decisions into national obligations that are relevant to the localized context in which biodiversity loss and change happens. The CBD has a system of designated national focal points (representatives of state parties) to facilitate implementation through coordination, information sharing and planning at the national level, but they lack the capacity and support needed to inspire action across sectors to achieve national contributions toward global biodiversity targets (Smith and Maltby, 2003).

Redgwell (2007) sees the top-down vertical journey toward national implementation as key to ensuring compliance with international obligations. As international obligations such as the ATs travel to the domestic level, they pass through different layers of governance and

---

[7] For example, the formal verification system of CITES was developed through resolutions and decisions of the COP (Reeve, 2001); Art 3 of the UNFCCC is an informal but influential norm laying forward key guiding principles such as sustainable development, intergenerational equality, precaution, and common but differentiated responsibilities (Toope, 2007).

[8] AT9, on invasive alien species, asks state parties to identify invasive alien species pathways, identify and eradicate priority species and take measures to prevent introduction. Identifying priority species is complex and lists at the EU level and UK level contain only some of the relevant species (Roy et al., 2014). Further, as invasive alien species are hard to control and eradication is complicated and resource-heavy, this places considerable strain on state parties, making it impossible to achieve the aims of AT9 unless political will increases and many more resources are put into such efforts (Smallwood, 2019).

are exposed to different practices that shape and reinterpret them in different contexts. These layers are important because international obligations, such as those arising from the CBD, are an ongoing challenge rather than a "fait accompli," and each stage of the journey can strengthen or weaken them (Smallwood, 2019).

Scholars argue that domestic levels of governance can also shape and influence international processes from local to global (Newell and Bumpus, 2012; Smallwood, 2019). The connections between international and regional/domestic governance are poorly understood despite their indivisible nature (Koh, 1997; 1998; Smallwood, 2019). The domestic level can strengthen global biodiversity governance during implementation without the ongoing constraints of achieving global consensus at the international level. Understandings formed at the domestic level may feed back to the CBD COP and influence and push forward shared understandings at the international level (Smallwood, 2019; 2021).

3.5 Transforming Global Biodiversity Governance

Based on the review of global biodiversity governance provided above, we identify the following four lessons learned for the transformative potential of global biodiversity governance.

3.5.1 Strengthen the Integration of International Treaties through Integrative Governance

Despite repeated attempts by the CBD COP to mainstream and attract political actors from agriculture, trade and development, it has made little progress in reaching out beyond international biodiversity-related institutions. In this respect, the Liaison Group of the Biodiversity-Related Conventions has organized several international workshops, known as the Bern I and Bern II processes, to collaborate jointly for the post-2020 biodiversity agenda.

Within the environmental regime, an integration of agendas that is also essential, yet to be realized, is between the global biodiversity and the climate change agendas. Despite many interrelated issues, the UNFCCC is largely absent from the biodiversity regime complex, with silos between climate and biodiversity responses remaining in science, international governance and civil society, thereby undermining opportunities for synergies in addressing climate change while also preserving ecosystems (Deprez et al., 2019). The focus on nature-based solutions at the 2019 UN Climate Summit marked an emerging understanding of the need for convergence between climate and biodiversity within the international political agenda. The chairs of two main science–policy international interfaces, the Intergovernmental Panel on Climate Change and the Intergovernmental Science-Policy Platform on Biodiversity and Ecosystem Services, have expressed their will to work together, and their first meeting was held in December 2020, resulting in a joint report (Pörtner et al., 2021). These efforts should be pursued and multiplied.

Besides the environmental regime, the main regime impacting biodiversity is the trade regime, due to large-scale trade in natural resources. Since its initiation, the CBD has called for integrative biodiversity governance through a comprehensive ecosystem approach, rather than focusing solely on species or genetic resource conservation (see above). However, the true realization of this comprehensive approach has been neglected due to an emphasis on profits from trade in individual species and genetic resources. Critiques of the biodiversity regime suggest that it is too much in line with trade agendas and therefore lacks the ability to achieve transformative change by implicitly supporting neoliberal globalization, especially embedded in the trade regime, as opposed to challenging it (Brand and Wissen, 2013; Brand et al., 2008; MacDonald, 2010) with broader, ecosystemic approaches.

Attempts have been made to mainstream biodiversity in the trade, agriculture, cultural and development regimes. The CBD has aimed to influence the agendas of other international initiatives and conventions within the regime complex through global targets (Harrop and Pritchard, 2011). While the strategic plan and global target for 2010 was adopted for the CBD only, the Strategic Plan for Biodiversity 2011–2020, including the ATs, was adopted as an overarching framework on biodiversity reaching out to the other biodiversity-related conventions, the entire UN system and all other partners engaged in biodiversity conservation and sustainable development policy. Although most of the ATs have not been met, the wide endorsement by these partners showed a sign of broadened recognition of the role of biodiversity conservation and sustainable use for human well-being.

This recognition was further broadened by the adoption of the 2030 Agenda for Sustainable Development by the UN General Assembly in 2015, with its seventeen Sustainable Development Goals (SDGs). Biodiversity appears as an important component of these goals: Goals 14 and 15 explicitly address life below water and on land with sub-targets consistent with the ATs (see Chapter 1). Biodiversity also plays an essential role in the achievement of most of the other SDGs, including climate action with forests as climate adaptation and mitigation options, or zero hunger with agricultural genetic resources being essential for food security (CBD Secretariat, 2017). This political upgrading of biodiversity, as expressed by the SDGs, is one important step for potentially obtaining transformative change to reverse the negative trend for biodiversity, even if the effectiveness of Agenda 2030 is yet to be shown. All in all, coordination attempts exist at the international level to mainstream biodiversity, but should be strengthened for transformative change.

3.5.2 Strengthen Inclusive Governance through the Inclusion of Nonstate Actors

Polycentric governance processes including nonstate actors are increasing in global biodiversity governance, both within the CBD and more broadly across the biodiversity regime complex (Kok et al., 2019). Inclusion of various state, market and civil society actors would empower those whose interests are not sufficiently recognized, represent transformative values and facilitate co-construction of shared understandings and social learning between actors. The question for the implementation of the CBD Post-2020 GBF is how to best involve underrepresented actors into the hierarchical and state-led process.

Stronger representation of stakeholders, such as IPLC and NGOs, that have been underrepresented so far could enable true knowledge-sharing to inform international decision-making (Tengö et al., 2017). So far, IPLC have been particularly successful in increasing their participation in the CBD and in strengthening their position. IPLC have been successful in challenging dominant discourses around biodiversity, including neoliberal valuations of nature (see Chapter 2), and in highlighting their possible contribution to the realization of the new post-2020 biodiversity targets, although this recognition at the global level is not always reflected during implementation at the domestic level.

The current role of governments in biodiversity governance may be challenged by nonstate and subnational actors to provide the stronger leadership needed to accelerate the momentum for biodiversity and to strengthen international and national policies. Civil society initiatives could scrutinize national government actions and their contributions to the realization of the goals and targets of the CBD and step up their ambition levels and increase action. Hybrid initiatives involving both public and private actors may also offer a point of leverage for transformation, although there are risks that inclusion of private business actors may preclude transformation. Analyses of international nonstate action initiatives for biodiversity show that to increase the legitimacy of their efforts, business actors usually prefer to cooperate with civil society and/or public actors rather than act alone (Negacz et al., 2020).

The development and implementation of the Sharm-el-Sheik to Kunming Action Agenda also poses challenges to the CBD (Kok et al., 2019). Solutions included in the action agenda aim to: ensure nonstate actors actively contribute to biodiversity goals; avoid overlaps and confusion in a plethora of nonstate actors and action to achieve biodiversity goals; and avoid the risk of national governments shirking established norms and responsibilities under the CBD, leaving action to nonstate and subnational actors. This would require that the CBD: provides a collaborative framework for nonstate action within the CBD and Post-2020 Global Biodiversity Framework that builds upon existing and emerging activities of nonstate action; organizes monitoring and review as part of an accountability framework of state and nonstate actors as part of the wider responsibility and transparency framework under the CBD; and provides for learning, capacity-building and follow-up action between state and nonstate actors (Chan et al., 2015; Kok and Ludwig, 2021).

3.5.3 Improve Implementation

Barriers to CBD implementation include the use of poorly designed soft law, "political" targets (as opposed to scientifically informed binding targets or protocols), reliance on NBSAPs and national reports for implementation, lack of transparent means of review, the inability of the CBD to engage economic and production sectors and business more broadly and the lack of any consequences for failure to meet targets.

Implementation is severely hindered by the lack of accountability mechanisms. The CBD Art 27 dispute mechanism has never been used, no compliance committee has been adopted and there is no compliance mechanism, whether it be through an enforcement mechanism in the form of financial or trade sanctions, such as in CITES (under which countries risk trade sanctions) or facilitative in the form of "naming and shaming," such as in the 2015 Paris Agreement on climate change (under which individual countries can make voluntary pledges, with a comparison and review of each state party's performance). Subsequently, if state parties fail to fulfill their obligations (reporting, implementation, contribution toward the ATs), there are no consequences (Le Prestre, 2017). The absence of accountability and the lack of a compliance mechanism create an obstacle to effective implementation and efficient governance, and are ultimately a result of political choice, reflecting the low priority placed on biodiversity. The CBD needs to introduce a more structured approach to implementation than practiced so far to address biodiversity loss and decline on a global level.

The CBD review mechanism could be strengthened. While most state parties submit national reports, the feedback given by the CBD on individual state party progress and their contribution to the realization of international targets lacks transparency. A strengthened review mechanism would facilitate a more structured approach to implementation, for example the provision by the CBD of basic information on who implements which provisions, and national progress toward global goals (Smallwood, 2019). NGOs have taken the lead to break down data in relation to compliance in a more meaningful way to highlight individual state party progress toward the ATs (Smallwood, 2019).

There are discussions within the CBD for adoption of a strengthened review and accountability mechanism.[9] Increased political will is needed to adopt such mechanisms, but if agreed to they would strengthen implementation. Negotiations to adopt compliance mechanisms can be quite time-consuming and burdensome (Morgera et al., 2014), but the successful agreement to create a compliance committee during the Paris Agreement climate negotiations (Bodansky, 2016) shows that this may not be beyond the reach of the CBD. Agreement on strong means of compliance may be politically difficult, but increased transparency and introducing a system of accountability (including a compliance committee) through a "pledge, review and ratchet" mechanism would help facilitate CBD compliance (Kok et al., 2019).

Another approach could be through the adoption of a "naming but not shaming" approach, which, rather than punish noncompliance, aims to support state parties struggling to reach their goals through increased financial support and capacity-building. This could be achieved through the development of the NBSAP peer review mechanism (Smallwood, 2019). Learning and accountability approaches may also be combined to further strengthen implementation.

---

[9] S18 of draft 1.0 of the Post-2020 Global Biodiversity Framework recognizes the importance of responsibility and transparency; SBI3 draft recommendations to COP include the adoption of an enhanced multidimensional approach to planning, monitoring, reporting and review with a view to enhancing implementation of the CBD and the Post-2020 Global Biodiversity Framework (CBD/SBI/5/CRP.5).

Focus should also be given to strengthening multilevel governance processes to improve implementation. International obligations can be strengthened or weakened through inclusive and integrative practices during implementation; therefore, careful attention must be paid to their dynamics at all levels of governance. If resourced properly, the CBD national focal points and other relevant actors could play a greater role in implementation, and better catalyze action across sectors to achieve national contributions toward global biodiversity objectives, targets and goals. Failure to engage all relevant actors at the national level is largely because implementation of biodiversity policies falls upon conservation sectors with limited or no buy-in from production sectors. Strengthened integrative processes at the national level are essential to engage production sectors to address biodiversity loss.

3.5.4 Increase Anticipatory Adaptive Capacities

In some respects, the CBD has shown its ability to learn and adapt to the ongoing challenge of nature conservation, sustainable use and benefit sharing. It has gradually developed more defined strategic plans with targets, as well as specific work programs and guidance for state parties. While these efforts should not be underestimated, a key challenge for the CBD is to evolve more rapidly and counter the escalating rates of biodiversity loss.

The preparation of the Post-2020 Framework has been an important moment of reflection, deliberation and joint learning as a basis for changing course guided by the OEWG. Quite extensive regional and thematic consultations have been held in-person before the second meeting of the OEWG, and online thereafter, that have fed into the negotiations. They have highlighted important elements of the Convention, including mainstreaming, finance and capacity-building in further implementing the Post-2020 Framework. The results of the IPBES assessments and especially the Global Assessment (IPBES, 2019), and to a lesser extent also the CBD Global Biodiversity Outlook (SCBD, 2020) and the two Local Biodiversity Outlooks (Forest Peoples Programme et al., 2020), have played an important role in the process by informing the negotiations and strengthening the science–policy interface, including through its emphasis on the co-construction of transdisciplinary knowledge (Díaz et al., 2015).

Improved transparency of efforts of state parties and nonstate actors, and identification of ambition and implementation gaps, are key to strengthening the adaptive capacity of the CBD. Improved monitoring of implementation attributed to specific state parties (which has up to 2020 not been the case), stocktaking, review and possible follow-up in terms of a "ratchet" mechanism in the Post-2020 Framework (as discussed above) would allow for more timely course corrections and create a basis for joint learning between state parties, and between state parties and nonstate actors.

A further underlying limitation of transformative governance by the CBD is its UN context, which requires consensus from all state parties on CBD COP decisions, thus allowing little room for adaptive governance through experimentation and reflexivity or anticipatory governance due to lack of political will. One actor of change could be the CBD Secretariat. CBD parties have indeed traditionally given a rather large leeway to the CBD Secretariat (Siebenhüner, 2007), although perhaps not in comparison to other biodiversity conventions such as Ramsar (Bowman, 2002) and CITES.

Does the secretariat of the CBD provide institutional memory that lends itself well to the adaptability needed to achieve transformative governance? The secretariat interacts with informal expert and liaison groups to advise the COP, drafts background documents and agendas, and facilitates negotiations, and is thereby able to play a key role in the adaptability of the CBD. Yet the creation of the OEWG to develop the Post-2020 Framework marked a change to the freedom given to the secretariat, as the OEWG process is mostly managed by cochairs, representing state parties. The emphasis on the OEWG process to inform the Post-2020 Framework, led by state parties, suggests that the secretariat's contribution to adaptability within governance processes has lessened. While the secretariat still has significance in intergovernmental cooperative processes (Biermann and Siebenhüner, 2009), its roles as an emerging political actor and a "norm entrepreneur" (Jinnah, 2008; 2011; 2012) have been toned down and this may signify a challenge to the pace of adaptability within the CBD, unless political will for transformative change is deepened among state parties.

Reconfiguring how the CBD operates is complex and lengthy due to the restraints of the institutional mechanisms in place, such as gaining multilateral consensus and the adoption of protocols. However, procedurally it is possible and under the Convention there is a process for actors (state and nonstate) to identify new and emerging issues for future work programs relating to the conservation and sustainable use of biodiversity and the fair and equitable sharing of benefits arising from the use of genetic resources (Siebenhüner, 2007). This mechanism offers potential to advance and adapt governance processes at the CBD (Le Prestre, 2017). Ambitious, anticipatory and innovative proposals can be introduced to the CBD as "new and emerging issues" with the potential to form future work programs (see Chapter 7). The agreement by state parties on the criteria for the adoption of new and emerging issues by the COP is an essential step forward to make this procedural mechanism workable, and their application has proved to be challenging in practice.

Another important change in how governance takes place through the CBD could be through initiating change in the scales of governance, for example by breaking down the "global" scale of the CBD and achieving agreement on the adoption of differentiated approaches according to regions, priority ecosystems, countries, sectors or themes, following the example of the Convention on Migratory Species. This would change the dynamics of agreement and operation and would be a step toward more meaningful large-scale action on biodiversity at a subglobal level, while still in a unified global framework.

3.6 Conclusion

Currently, global biodiversity governance fails to address the indirect drivers of biodiversity loss, and is unable to confront the economic, political and social paradigms that drive the destruction of biodiversity globally. This chapter has presented the current state of global biodiversity governance and suggested how it could be improved, thus transforming biodiversity governance. We conclude with Table 3.2, which summarizes the strengths, weaknesses and transformative potential of global biodiversity governance.

Table 3.2 *Strengths, weaknesses and transformative potential of global biodiversity governance*

Strengths	Weaknesses	Lessons learned and transformative potential
International institutions and architecture		
The global biodiversity regime and its different elements amplify the theme of biodiversity. There are commitments across biodiversity conventions and SDGs to global biodiversity targets.	There is little engagement with the trade or climate regime; integration with the agricultural, development and cultural regimes must be strengthened.	Biodiversity governance needs active support from a range of other international agreements, including those related to trade, climate, agriculture, development and culture.
Engagement with nonstate actors		
Polycentric governance processes including nonstate actors around biodiversity are increasing. The CBD COP attracts a wide range of sectors and stakeholders.	The involvement of nonstate actors comes with several risks, such as risks of commodification of the biodiversity agenda and lowering of ambition due to actors' interests.	Inclusive governance must be strategic and purposeful, with an aim of focusing on the indirect drivers of biodiversity loss and empowering those who represent transformative values. Means of accountability for nonstate actors such as businesses would facilitate transformation.
Implementation		
During implementation, processes of multilevel governance can strengthen CBD obligations (e.g. domestic levels have integrated global obligations into laws or more concrete policies, host more inclusive decision-making processes, have better accountability mechanisms, etc.) and these interactions feed back into global governance processes (negotiation process, national reports, peer review, etc.).	Generally weak implementation of CBD obligations due to lack of political will and societal understanding, poorly worded targets, lack of accountability and pragmatic challenges.	Multilevel governance processes can offer leverage points for transformation. Objectives could include: Strengthen the focus of implementation on addressing the indirect drivers. Better designed obligations including protocols or "harder" obligations that will facilitate national implementation. Strengthened compliance mechanisms through more transparency in reporting back on progress of individual state parties. Strengthen peer review mechanisms.

Table 3.2 (*cont.*)

Strengths	Weaknesses	Lessons learned and transformative potential
Adaptation potential		
Well-established procedures through decision-making at the CBD COP have enabled institutional evolution through the adoption of protocols, strategic plans and targets, reviews of national reports, tracking of NBSAP implementation and development of this process.	Adaptability is not sufficient compared to the rate of biodiversity loss.	Strengthen the role of the CBD Secretariat. Better use the "new and emerging issues" identification process. Diversify the global scale of governance and adopt differentiated approaches (e.g. regional, priority ecosystems, themes, etc.). Biodiversity is in essence local, and global decisions should be better linked to local/regional specificities.

References

Ademola, A., Casey, S., and Bridgewater, P. (2015). Global conservation and management of biodiversity in developing countries: An opportunity for a new approach. *Environmental Science & Policy* 45, 104–108.

Arts, B. J. M., Buijs, A. E., Gevers, H., et al. (2017). *The impact of international cooperative initiatives on biodiversity (ICIBs)*. Wageningen: Forest and Nature Conservation Policy Group, Wageningen University.

Arts, B., and Mack, S. (2006). NGO strategies and influence in the biosafety arena, 1992–2005. In: *The international politics of genetically modified food.* R. Falkner (Ed.), pp. 48–64. London: Palgrave Macmillan.

Bansard, J. S., Pattberg, P. H., and Widerberg, O. (2017). Cities to the rescue? Assessing the performance of transnational municipal networks in global climate governance. *International Environmental Agreements: Politics, Law and Economics* 17, 229–246.

Biermann, F., and Siebenhüner, B. (2009). *Managers of global change: The influence of international environmental bureaucracies.* Cambridge, MA: MIT Press.

Bodansky, D. (2016). The Paris Climate Change Agreement: A new hope? *American Journal International Law* 110, 288–319.

Bowman, M. (2002). The Ramsar convention on wetlands: Has it made a difference? In: *Yearbook of international co-operation on environment and development.* O.S. Stokke and Ø. B. Thommessen (Eds.), pp. 61–68. London: Earthscan.

Brand, U., Görg, C., Hirsch, J., and Wissen, M. (2008). *Conflicts in environmental regulation and the internationalisation of the state contested terrains.* London; New York: Taylor and Francis.

Brand, U., and Wissen, M. (2013). Crisis and continuity of capitalist society-nature relationships: The imperial mode of living and the limits to environmental governance. *Review of International Political Economy* 20, 687–711, DOI: 10.1080/09692290.2012.691077

Brunnée, J. (2002). COPing with consent: Law-making under multilateral environmental agreements. *Leiden Journal of International Law* 15, 1–52.

Brunnée, J., and Toope, S. (2010). *Legitimacy and legality in international law: An interactional account*. Cambridge; New York: Cambridge University Press.

Bulkeley, H., and Jordan, A. (2012). Guest editorial. *Environment and planning. C, Government and Policy* 30, 556–570.

Bulkeley, H., Andonova, L., Bäckstrand, K., et al. (2012). Governing climate change transnationally: Assessing the evidence from a database of sixty initiatives. *Environment and Planning. C: Government & Policy* 30, 591–612.

Bulkeley, H., Kok, M., van Dijk, J., et al. (2020). *Harnessing the potential of the post-2020 global biodiversity framework. Report prepared by an Eklipse Expert Working Group*. Wallingford, United Kingdom: UK Centre for Ecology & Hydrology.

Bulkeley, H., and Newell, P. (2015). *Governing climate change*. Abingdon; New York: Routledge.

Büscher, B., Sian, S., Neves, K., Igoe, J., and Brockington, D. (2012). Towards a synthesized critique of neoliberal biodiversity conservation. *Capitalism, Nature, Socialism* 23, 4–30. DOI: 10.1080/10455752.2012.674149

Butchart, S. H., Di Marco, M., and Watson, J. E. (2016). Formulating smart commitments on biodiversity: Lessons from the Aichi Targets. *Conservation Letters* 9, 457–468.

CBD. (2020). *Post 2020 thematic consultation on transparent implementation, monitoring, reporting and review*. Available from https://bit.ly/3o24Q1F.

CBD COP6. (2002). CBD decision VI/26. CBD Secretariat. Available from www.cbd.int/decisions/cop/6/26/2.

CBD OEWG. (2020). Update of the zero draft of the post-2020 global biodiversity framework. Available from https://bit.ly/3GdKvMV.

CBD Secretariat. (2017). *Biodiversity and the 2030 agenda for sustainable development*. Available from https://bit.ly/3s6oavO.

Chaffin, B. C., Garmestani, A. S., Gunderson, L. H., et al. (2016). Transformative environmental governance. *Annual Review of Environment and Resources* 41, 399–423.

Chan, S., van Asselt, H., Hale, T., et al. (2015). Reinvigorating international climate policy: A comprehensive framework for effective nonstate action. *Global Policy* 6, 466–473.

Chayes, A., and Chayes, A. H. (1993). On compliance. *International Organization* 47, 175–205.

Chayes, A., and Chayes, A. H. (1995). *The new sovereignty: Compliance with international regulatory agreements*. Cambridge, MA: Harvard University Press.

Corson C., and MacDonald, K. I. (2012). Enclosing the global commons: The convention on biological diversity and green grabbing. *The Journal of Peasant Studies* 39(2), 263–283, DOI: 10.1080/03066150.2012.664138

Curet, F., and Puydarrieux, P. (2020). *Catalyzing state and non-state actors for nature: Mapping coalitions and their potential contribution to reduce pressures on biodiversity*. Gland, Switzerland: IUCN.

Deprez, A., Vallejo, L., and Rankovic, A. (2019). Towards a climate change ambition that (better) integrates biodiversity and land use. *IDDRI, Study* N°08/19. Available from https://bit.ly/3G3I13K.

Díaz, S., Demissew, S., Carabias, J., et al. (2015). The IPBES conceptual framework: Connecting nature and people. *Current Opinion in Environmental Sustainability* 14, 1–16.

Forest Peoples Programme et al. (2020). *Local biodiversity outlooks 2. The contributions of indigenous peoples and local communities to the implementation of the Strategic Plan for Biodiversity 2011–2020 and to renewing nature and cultures*. A complement to the fifth edition of Global Biodiversity Outlook. Moreton-in-Marsh, England: Forest Peoples Programme. Available from www.localbiodiversityoutlooks.net.

GEF. (2016). *Biodiversity mainstreaming in practice: A review of GEF experience*. Available from https://bit.ly/3rb7PHa.

GEF, UNEP and CBD. (2007). Mainstreaming biodiversity into sectoral and cross-sectoral strategies, plans and programmes. Module B-3 Version 1. Available from https://bit.ly/3GiMaAU.

Goldsmith, J., and Posner, E. (2005). *The limits of international law*. Oxford; New York: Oxford University Press.

Gupta, A. (2008). Transparency under scrutiny: Information disclosure in global environmental governance. *Global Environmental Politics* 8, 1–7. DOI: 10.1162/glep.2008.8.2.1

Hagerman, S. M., and Pelai, R. (2016). "As far as possible and as appropriate": Implementing the Aichi biodiversity targets. *Conservation Letters* 9, 469–478.

Hale, T., Held, D., and Young, K. (2013). *Gridlock: Why global cooperation is failing when we need it most.* Cambridge: Polity.

Harrop, S., and Pritchard, D. (2011). A hard instrument goes soft: The implications of the Convention on Biological Diversity's current trajectory. *Global Environmental Change* 21, 474–480.

IPBES. (2019). *Summary for policymakers of the global assessment report on biodiversity and ecosystem services of the Intergovernmental Science-Policy Platform on Biodiversity and Ecosystem Services.* Bonn: IPBES secretariat. https://doi.org/10.5281/zenodo.3553579.

Jinnah, S. (2008). Who's in charge? International bureaucracies and the management of global governance. Dissertation, University of California, Berkley.

(2011). Marketing linkages: Secretariat governance of the climate biodiversity interface. *Global Environmental Politics* 11, 23–43.

(2012). Singing the unsung: The key role of secretariats in global environmental politics. In: *The roads from Rio: Lessons learned from twenty years of multilateral environmental negotiations.* P. Chasek and L. Wagner (Eds.), pp. 107–126. London: Routledge.

Jones, M. J., and Solomon, J. F. (2013). Problematising accounting for biodiversity. *Accounting, Auditing and Accountability Journal* 26, 668–687. doi/10.1108/AAAJ-03-2013-1255

Jordan, A. J., Huitema, D., Hildén, M., et al. (2015). Emergence of polycentric climate governance and its future prospects. *Nature Climate Change* 5, 977–982. doi: 10.1038/nclimate2725

Karlsson-Vinkhuyzena, S., Kok, M. T. J., Visseren-Hamakers, I. J., and Termeera, C. J. A. M. (2017). Mainstreaming biodiversity in economic sectors: An analytical framework. *Biological Conservation* 210: 145–156.

Koh, H. (1997). Why do nations obey international law? *Yale Law Journal* 106, 2599–2659.

(1998). The 1998 Frankel Lecture: Bringing international law home. *Houston International Law Journal* 35, 626–681.

Kok, M. T. J., and Ludwig, K. (2021). Understanding international non-state and sub-national action for biodiversity and their possible contribution to the post-2020 CBD global biodiversity framework: Insights from six international cooperative initiatives. *International Environmental Agreements: Politics, Law and Economics.* 1–25. https://doi.org/10.1007/s10784-021-09547-2.

Kok, M., Widerberg, O., Negacz, K., Bliss, C., and Pattberg, P. (2019). *Opportunities for the action agenda for nature and people.* The Hague: PBL.

Koskenniemi, M. (2011). *The politics of international law.* Portland, OR: Hart.

Le Prestre, P. G. (Ed.). (2017). *Governing global biodiversity: The evolution and implementation of the convention on biological diversity.* London; New York: Routledge.

MacDonald, K. (2010). The devil is in the (bio)diversity: Private sector "engagement" and the restructuring of biodiversity conservation. *Antipode* 42, 513–550.

Mace, G. M., Barrett, M., Burgess, N. D., et al. (2018). Aiming higher to bend the curve of biodiversity loss. *Nature Sustainability* 1, 448–451.

Morgera, E., and Tsioumani, E. (2010). Yesterday, today, and tomorrow: Looking afresh at the convention on biological diversity. *Yearbook of International Environmental Law* 21, 3–40.

Morgera, E., Tsioumani, E., and Buck, M. (2014). *Unravelling the Nagoya Protocol: A commentary on the Nagoya Protocol on access and benefit-sharing to the Convention on Biological Diversity.* Leiden, the Netherlands: Brill.

Morin, J.-F., and Orsini, A. (2014). Policy coherency and regime complexes: The case of genetic resources. *Review of International Studies* 40, 303–324.

Negacz, K., Widerberg, O. E., Kok, M., and Pattberg, P. H. (2020). *BioSTAR: Landscape of international and transnational cooperative initiatives for biodiversity: Mapping international and transnational cooperative initiatives for biodiversity.* Amsterdam: IVM.

Newell, P., and Bumpus, A. (2012). The global political ecology of the clean development mechanism. *Global Environmental Politics* 12, 49–67.

Parks, L. (2018). Spaces for local voices? A discourse analysis of the decisions of the Convention on Biological Diversity. *Journal of Human Rights and the Environment* 9, 141–170.

Pattberg, P. (2010). Public–private partnerships in global climate governance. *WIREs Climate Change* 1, 279–287. DOI: 10.1002/wcc.38

Pattberg, P., Kristensen, K., and Widerberg, O. (2017). *Beyond the CBD. Exploring the institutional landscape of governing for biodiversity*. Amsterdam: IVM Institute for Environmental Studies.

Pattberg, P., Widerberg, O., and Kok, M. T. J. (2019). Towards a global biodiversity action agenda. *Global Policy* 10, 385–390. DOI: 10.1111/1758-5899.12669

Pörtner, H., Scholes, B., Agard, J., et al. (2021). *Scientific outcome of the IPBES-IPCC co-sponsored workshop on biodiversity and climate change*; Bonn, Germany: IPBES secretariat. DOI:10.5281/zenodo.4659158.

Prip C., and Pisupati, B. (2018). *Assessment of post-2010 national biodiversity strategies and action plans*. Nairobi, Kenya: UNEP.

Prip, C., Gross, T., Johnston, S., and Vierros, M. (2010). *Biodiversity planning: An assessment of national biodiversity strategies and action plans*. Yokohama, Japan: United Nations University Institute of Advanced Studies.

Puppim de Oliveira, J. A., Balaban, O., Doll, C. N. H., et al. (2011). Cities and biodiversity: Perspectives and governance challenges for implementing the convention on biological diversity (CBD) at the city level. *Biological Conservation* 144, 1302–1313. DOI: 10.1016/j.biocon.2010.12.007

Raustiala, K., and Victor, D. G. (2004). The regime complex for plant genetic resources. *International Organization* 58, 277–309. DOI: 10.1017/S0020818304582036

Redgwell, C. (2007). National implementation. In: *The Oxford handbook of international environmental law*. D. Bodansky, J. Brunnée and E. Hey (Eds.), pp. 922–946. Oxford: Oxford University Press.

Reeve, R. (2001). Verification mechanisms in CITES. *Verification Yearbook*, 137–156. The Verification Research, Training and Information Centre (Vertic). Available from: https://bit.ly/3r6JxxS.

Roy, H. E., Preston, C., Harrower, C. A., et al. (2014). GB non-native species information portal: Documenting the arrival of non-native species in Britain. *Biological Invasions* 16, 2495–2505.

RSPB et al. (2016). Progress and alignment of national targets to the Aichi biodiversity targets, World Maps. Available from https://bit.ly/3Gizj1J.

Sands, P. (2007). Evolution of international environmental law. In: *The Oxford handbook of international environmental law*. D. Bodansky, J. Brunnée and E. Hey (Eds.), pp. 29–43. Oxford: Oxford University Press.

SCBD. (2020). Global biodiversity outlook 5. Available from www.cbd.int/gbo5/.

Siebenhüner, B. (2007). Administrator of global biodiversity: The secretariat of the Convention on Biological Diversity. *Biodiversity Conservation* 16, 259–274.

Smallwood, J. (2019). *The Convention on Biological Diversity's objectives include conservation of biological diversity at a global level, but has it become another victim of extinction as a result of its text and strategic plan?* PhD Thesis. University of Sussex.

(2021). Whose utopia?: The complexity of incorporating diverse ethical views within nature governance frameworks. In: *Reconsidering extinction in terms of the history of global bioethics*. S. Booth and C. Mounsey (Eds.), pp. 184–204. New York: Routledge.

Smith, R., and Maltby, E. (2003). *Using the ecosystem approach to implement the convention on biological diversity: Key issues and case studies* (no. 2). Gland, Switzerland; Cambridge, UK: IUCN.

Spann, M. (2017). Politics of poverty: The post-2015 sustainable development goals and the business of agriculture. *Globalizations* 14, 360–378.

Spiro, P. (2007). Non-governmental organizations and civil society. In: *The Oxford handbook of international environmental law*. D. Bodansky, J. Brunnée and E. Hey (Eds.), pp. 770–790. Oxford: Oxford University Press.

Tengö, M., Hill, R., Malmer, P., et al. (2017). Weaving knowledge systems in IPBES, CBD and beyond – Lessons learned for sustainability. *Current Opinion in Environmental Sustainability* 26–27, 17–25. DOI: 10.1016/j.cosust.2016.12.005

Toope, S. (2007). Formality and Informality. In: *The Oxford handbook of international environmental law*. D. Bodansky, J. Brunnée and E. Hey (Eds.), pp. 107–124. Oxford: Oxford University Press.

van Oorschot, M., van Tulder, R., and Kok, M. (2020). *Business for biodiversity: Mobilising business towards net positive impact*. The Hague: PBL.

Visseren-Hamakers, I. J. (2013). Partnerships and sustainable development: The lessons learned from international biodiversity governance. *Environmental Policy and Governance* 23, 145–160. DOI: 10.1002/eet.1612

Young, O. R. (2011). Effectiveness of international environmental regimes: Existing knowledge, cutting-edge themes, and research strategies. *Proceedings of the National Academy of Sciences of the United States of America* 108, 19853–19860.

4

How to Save a Million Species? Transformative Governance through Prioritization

INGRID J. VISSEREN-HAMAKERS, BENJAMIN CASHORE, DERK LOORBACH, MARCEL T. J. KOK, SUSAN DE KONING, PIETER VULLERS AND ANNE VAN VEEN

4.1 Introduction

Around one million species of animals and plants are threatened with extinction. It is increasingly clear that this tragedy can only be avoided through transformative change (IPBES, 2019). This chapter aims to understand why the current state of biodiversity is so fragile, despite over half a century of global conservation efforts, and develop insights for more effective ways forward. We argue that past efforts have failed in part because they are based on an "ill-fit for purpose" problem analysis, and that reconfiguring problem conceptions shows promising directions for identifying novel strategies for triggering transformative change.

The chapter develops this argument by: (a) bringing together literatures on how to govern transformative change, transformations and transitions; (b) distinguishing their insights against a problem typology that identifies different perspectives on how to conceive of, and address, sustainability challenges and, as a result, (c) providing new insights for transformative governance.

The chapter is organized as follows. In the next section, we discuss and integrate different contributions to the literatures on transformative change, transformations, transitions and their governance, in order to better understand and govern transformative change. We then apply the four problem conceptions that Cashore (2019) has developed with colleagues (Cashore and Bernstein, 2022; Cashore et al., 2019; Humphreys et al., 2017) to assess how different schools of applied sustainability scholarship have shaped how to conceive of, and address, environmental challenges. Sections 4.4 and 4.5 then discuss the implications for transformative governance, including the need for much greater thinking about the contribution of scientific knowledge. Finally, we identify key conclusions that, together, offer a novel contribution to the academic and practitioner debates on transformative change and governance.

4.2 Transformations and Transitions: Integration and Reflection

It is clear that the dominant sustainability strategies to date have failed to "bend the curve" (Mace et al., 2018) of biodiversity loss. A consensus is now emerging that a fundamentally different approach to how governance and science address the biodiversity challenge – through a focus on transformative change – is needed. Such fundamental change is called for since current structures often inhibit sustainable development and actually represent the underlying societal causes of biodiversity loss. To accomplish such transformative change, attention must not only be placed on the apparent direct drivers of ecological degradation (the physical causes of biodiversity loss, including land-use change, climate change, overfishing and pollution) that have guided so much of environmental and biodiversity policy analysis, design and implementation to date (IPBES, 2019; also see Chapter 1), but especially on the underlying societal causes, or indirect drivers, of biodiversity loss. But what exactly do these concepts of (governing) transformative change, transformations and transitions entail, and how do they relate to one another?

Over the past decades, new governance approaches have been developed under the headers of transformation and transition. Coming from different scientific disciplines and methodological traditions, these approaches share a recognition of the need for fundamental change, as well as a focus on the complexity, patterns and dynamics of structural and systemic change and the broader societal agency and governance that do, or do not, accelerate and guide such change. However, there is a distinction. The differentiation by Linnér and Wibeck (2019) is useful here, with macrotransformations referring to transformations that have spanned across entire civilizations, while particular transformations (or transitions) refer to transformations within subsystems of society, such as parts of specific socioecological systems (e.g. the food, mobility or energy transition).

We here provide a brief overview of the literatures on transformative change, transformations, transformative governance, transitions, and transition management and governance, which have all contributed to the thinking on fundamental societal change. We focus on governance, governance instruments and mixes of governance instruments (instead of governmental policy only) in order to recognize the role of different societal actors, including governments, market actors, civil society and researchers, in transformative change.

4.2.1 *Transformations, Transformative Change and Their Governance*

Linnér and Wibeck (2019: 4) define transformations as "profound and enduring non-linear systemic changes, typically involving social, cultural, technological, political and/or environmental processes." Approaches that deal with problems on a global socioecological scale, such as approaches in resilience thinking (Olsson et al., 2014; Westley et al., 2013) and transformative adaptation (O'Brien, 2012), use the notion of "transformation" to refer to the essential and rudimentary shifts in nature–culture interactions and feedbacks. According to O'Brien and Sygna (2013), transformations consist of three spheres, the practical, political and personal sphere, which all need to be addressed to enable societal transformations. Based on the IPBES Global Assessment (GA), Chapter 1 defines transformative change in a similar manner, namely as "a fundamental, society-wide reorganization across technological, economic and social factors and structures, including paradigms, goals and values."

The GA operationalizes transformative change in terms of pathways, and levers and leverage points (IPBES, 2019). Because of the transformative change required, existing unsustainable development pathways and vested interests and existing structures should make space for new and more sustainable pathways (Loorbach et al., 2017; Sharpe et al., 2016). Part of this departure may occur by deepening and accelerating existing processes of change. The IPBES GA suggests that these outcomes can be achieved through complementary top-down and bottom-up action on eight key points of intervention, or "leverage points" (Abson et al., 2017; Meadows, 2008), and five types of "levers," or management or governance interventions to effect the transformative change.

Visseren-Hamakers et al. (2021: 400) have defined transformative governance as "the formal and informal (public and private) rules, rule-making systems and actor-networks at all levels of human society (from the local to global) that enable transformative change, in our case, toward biodiversity conservation and sustainable development more broadly." Building on the IPBES GA and Visseren-Hamakers et al. (2021), Chapter 1 of this volume further operationalizes the concept of transformative governance as including five approaches (integrative, inclusive, adaptive, transdisciplinary and anticipatory) which should be: (a) focused on addressing indirect drivers underlying sustainability issues; (b) implemented in conjunction and (c) operationalized in specific manners.

Similarly, Linnér and Wibeck (2019) stress the importance of integrative and inclusive governance through developing smart governance mixes, involving nonstate actors and the general public, and developing transformative capacity to be adaptive, creative and innovative, and to be able to deal with uncertainty. The authors highlight the need for transformative governance to aim at achieving different sustainability goals in an integrative manner instead of focusing on particular transitions.

An alternative approach to governing transformations is to think in terms of principles that might provide guidance to realize transformative change (Bulkeley et al., 2020). The process of transformation itself is then one through which new solutions are generated, thus requiring a pragmatic and adaptive approach.

4.2.2 Conceptualizing Transitions and Their Governance

According to Hölscher (2018), a societal "transition" refers to a fundamental, systemic shift in the structure, culture and practices of sociotechnical, socioeconomic or socioinstitutional processes. Basic concepts in sustainability transitions research include regimes, landscapes and niches, with regime referring to an ecosystem, sector, technological system, area or organization that develops toward an optimum by gradually reducing diversity and optimizing efficiency (see e.g. Geels, 2002). The societal context (or landscape), however, changes autonomously (the climate, demographic change, or political, economic or technological developments). From a certain point in time, adapting the regime to this changing context becomes harder and tensions start to build. At the same time, alternatives (niches) start to develop (new technologies, practices or models), which can become more competitive over time, especially when the regime is disrupted (through e.g. economic crisis, technological breakthroughs, forest fires or social revolution). In most disciplines the concept of transitions is used analytically (e.g. in ecology and literature on resilience) or descriptively (historical transition studies). However, transition governance uses this idea prescriptively: If persistent sustainability problems are rooted in existing regimes then existing knowledge frames and political strategies that deal with them are inherently part of perpetuating a development pathway that causes the "symptoms" of unsustainability. The transition premise is that this pathway will inevitably be disrupted by external pressures, internal crises and emerging alternatives. Transition management literature thus conceptualizes systemic change as a nonlinear process that takes us from one dynamic equilibrium to another as a result of destabilization of the status quo and breakthrough of alternatives (Grin et al., 2010).

Over time, the dynamics of transitions evolve, together with the types of agency that drive it. To initiate transitions and go against a very stable societal regime typically requires strong vision, radical voices, experimentation and leadership. As more people become aware of the need for transitions, alternatives become more attractive and mainstream. New combinations and collaborations between niche-actors and regime-actors can start to develop. Contrary to these bottom-up changes, spaces for rapid institutional change occur typically in a more top-down manner. Transition governance is then the strategy that combines this actor perspective and the dynamics of transitions with action-oriented instruments (see de Haan and Rotmans, 2018).

By necessity, transition governance is multi-actor, multilevel and multidomain in its analysis and selective when it comes to participation by only involving actors already committed to transformative change to achieve common goals. It is also by definition based on co-construction, backcasting and reflexivity, as it acknowledges structural uncertainties while trying to use the mechanisms of social construction and social learning. Experimentation is also an important aspect in transition management, based on learning-by-doing. These principles have been translated in a number of instruments and tools, such as transition arenas, scenarios and experiments, with the idea of bringing transformative thinking – critical toward the status quo in order to improve it, assuming disruptive systemic change ahead and assuming positive futures are already emerging somewhere – into contexts and networks where people implicitly or actively work on sustainable alternatives to the regime.

4.2.3 Integrating Transformations and Transitions through Transformative Governance

The literatures on sustainability transformations and transitions share many similarities. They both recognize the need for fundamental change and the roles of different actors in governing such change, and they share a normative starting point, aiming to contribute to transforming our societies to become sustainable, equitable and just.

Interestingly, they emphasize different aspects of fundamental change, with transformations by definition focused on changing societal structures, or the underlying societal causes of unsustainable practices, and transition approaches often zooming in on change in specific systems or regimes (while recognizing the interrelationship between these regimes and broader societal structures).

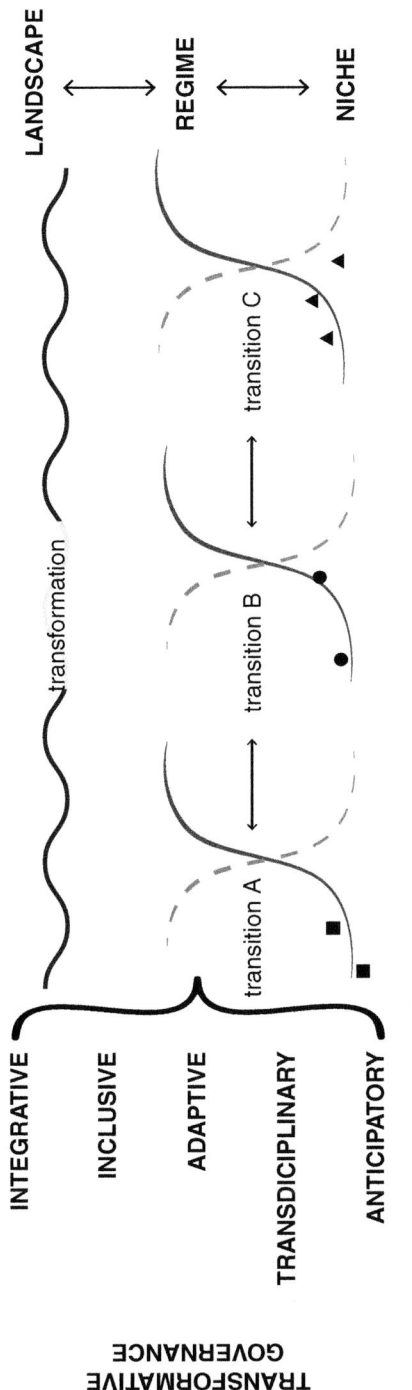

Figure 4.1 Integrating transformations and transitions through transformative governance Transformative governance enables transformative change through governance mixes that include instruments focused on niches, transitions (and their interactions) and transformations. Transformative change encompasses both transformations and transitions, and is thereby focused on both the generic societal underlying causes and those specific to certain regimes.

We propose here that transformative change encompasses both transformations and transitions, and is thereby focused on both the generic societal underlying causes and those specific to certain regimes. So transformative change includes a focus on enabling change in what is referred to in the transitions literature as the "landscape." It also (implicitly) assumes more agency to actually directly enable change in these societal structures, instead of only through niches and regime change, for example by promoting alternatives to paradigms of globalization, neoliberalism, economic growth or current discourses on relationships between humans and nonhumans.

The transformation and transition literatures can be integrated by positioning transitions in a broader societal context of transformations: from the transitions perspective seeing transformation as a "family of transitions" (Loorbach, 2014), or from the transformation perspective approaching transformative change to include multiple specific transitions (e.g. the transitions on energy, mobility, animal-free innovation, food), that also influence one another. Some of the change takes place in specific regimes or sectors, and some of the change is inherent in multiple regimes. More importantly, some of the societal causes underlying our current inherently unsustainable societies are generic (e.g. values, paradigms and goals; economic structures; generic institutions; ways of governing), and thus influence all specific transitions. Together, the stronger focus on generic societal change of the transformations literature, combined with the detailed focus on specific transitions, represents an important new avenue for understanding transformative change and its governance. With this, transformative governance entails agency at the niche, regime *and* landscape level, and governance mixes need to include instruments meant to enable transformative change both within specific regimes, among regimes and in society more broadly (Figure 4.1).

While both literatures highlight the need for adaptive, anticipatory and transdisciplinary governance, the transformation literature is more explicit about the need for integrative governance. Also, some authors from both literatures agree on the need to strategically think about participatory processes, highlighting the crucial role of those actors with transformative ambitions and the danger of including actors with vested interests in the old regime too early on in the process. However, many authors, especially from the transformative change literature, see inclusive governance in terms of its representativeness of different views, and promote pluralist approaches. We here follow the former, more strategic approach, also in light of the "problem-solving through prioritization" approach we are proposing, as elaborated below.

4.3 Four Sustainability Problem Conceptions, Not One

The role of cognitive frames in shaping policy and governance in general (Douglas and Wildavsky, 1982; Stone, 1997) and on the environment in particular (Bernstein, 2001) has long been recognized by a range of scholars within public policy, transnational governance and global environmental politics (e.g. Haas, 2002). Cashore and colleagues contributed to this literature by reflecting on the types of problems that confronted environmental and sustainability challenges. Doing so led to three observations. First,

Table 4.1 *The four problem conceptions (adapted from Cashore and Bernstein, 2022)*

			Rationale	
			Do economic or utility rationales dominate the underlying moral philosophy?	
			Yes	No
Problem orientation	Analysis justified based on features of a specific kind of problem?	Yes	Type 1: Commons	Type 4: Prioritization
		No	Type 2: Economic optimization	Type 3: Compromise

practitioners and applied scholars were involved, often unwittingly, in a narrowing of attention to environmental problems to those that, when solved, created "win-win" outcomes with economic goals. Second, the championing of "evidence-based" science often narrowed data collection that reinforced, rather than confronted, this bias (Cashore, 2019). Third, widespread emphasis among the private sector and international agencies on sustainable development tended to drift toward ameliorating economic sustainability challenges that, ironically, contributed to environmental degradation (Cashore and Bernstein, 2020). Overcoming this drift required consciously identifying a "learning protocol" among scientists and stakeholders through which four different types of sustainability problem conceptions, and corresponding evidence, would be rendered explicit (Cashore et al., 2019). Such exercises, they argued, can lead to innovating insights for ameliorating environmental and social problems (Humphreys et al., 2017) rather than "drifting" away from them (Cashore and Bernstein, 2022).

This quest to help ameliorate the environmental (and social) problems that were usually caused by championing economic goals led Cashore and colleagues to offer a three-part framework that is relevant to, and helps frame, the literature on transformative governance.

First, they identified two ways to disentangle four types of approaching sustainability issues: those that champion economic utility as the goal versus those that do not; and those that justify their approach to applied policy analysis based on the particular features of a problem in question versus those that offer universalistic frameworks (Cashore et al., 2019). The corresponding four types (Table 4.1) are innovative in that they simultaneously capture (subjective) constructed notions of particular problems but also point researchers to collect (seemingly objective) empirical evidence that narrows "lessons learned" to those that reinforce particular problem conceptions over others (Cashore and Bernstein, 2022).

Second, they found that four different sustainability schools tended to reinforce each type.

The Type 1 reinforcing commons school captures those sustainability scholars who target overuse of resources (Araral, 2014; Ostrom, 1990) commonly referred to as "tragedies of the commons." This orientation, which dominates schools of resource and agricultural economics, leads experts to focus on developing policies and institutions that limit the extraction of any resource to the same level as they reproduce. This approach also shows up in biodiversity cases when viewing them as a global tragedy of the commons that stems from a failure or absence of collective action that produces suboptimal economic results.

The Type 2 reinforcing economic optimization school shares Type 1 conceptions advancing overall economic utility or welfare. However, it is guided by a moral philosophy that evaluates solutions to any problem on whether they enhance economic welfare in society as a whole. It finds economically optimal solutions through cost–benefit analyses in which a range of environmental, social and economic outcomes are all granted some type of utility decreasing or increasing value, which then allows comparison across all outcomes (Arrow et al., 1996). Environmental goals are often converted into economic values through willingness to pay by consumers. Only those solutions that are deemed to enhance, rather than reduce, economic utility are considered rationally appropriate (Sinden et al., 2009). The economic optimization school has dominated the vast majority of environmental governance over the last thirty years (Hepburn and Stern, 2008; Nordhaus, 2019). It explains why Nordhaus (2019) has found that limiting carbon emissions to a 3.1 degree world is the rational approach, even though environmental scientists have found that maintaining 1.5 degrees is required to avert catastrophic ecological outcomes.

The Type 3 reinforcing compromise school emerged out of a critique of the economic optimization school and advances a moral philosophy championed by many applied political scientists and sociologists who seek balance and compromise across different values. Also disconnected from problem structure, it advances multistakeholderism and "multigoal" policy analysis as the appropriate and legitimate way to understand and manage trade-offs that seek some type of balance among competing perspectives (Eckersley, 2019; Weimer and Vining, 1999). This school has dominated many global processes over the past thirty years, including the United Nations Sustainable Development Goals (SDGs) (Bebbington and Unerman, 2018). This school and its Type 3 reinforcing approach also tends to dominate high-level global reports on sustainability challenges (Cashore and Nathan, 2020; IPBES, 2019; IPCC, 2019). Moreover, the formal goals of the Convention on Biological Diversity (CBD) actually include three main pillars, namely conservation, sustainable use and the equitable sharing of the benefits of the use. Such a problem conception can also be considered a Type 3 typology.

In contrast, *Type 4 reinforcing prioritization conceptions* identify those problems that, for either moral or scientific reasons, cannot, by definition, be ameliorated by subjecting them to Type 3, 2 or 1 schools. Cashore and Bernstein refer to antislavery as an undisputed example of a moral argument for prioritization. Adjudicating whether society should be against allowing humans to own other humans based on optimality or compromise calculations to permit some types of slavery is considered abhorrent and absurd by almost every country and citizen across the world (although modern slavery still exists). Since the nineteenth century, antislavery is considered a universal norm, which means that it cannot be addressed by a universal framework meant to apply to any class of problems.

A second kind of Type 4 conception emerges from scientific evidence about the problem at hand, for example about what type of conservation efforts must be in place to ensure addressing an irreversible problem like extinction. Disciplines that tend to treat problems as Type 4 include scientists who study biodiversity loss, as well as philosophers and social scientists who focus on ways in which universally shared norms emerge and permeate societal attitudes. Their general agreement is based on science: The rate of biodiversity loss is real, alarming and caused by human activity.

This Type 4 school, for instrumental reasons, turns to "lexical" or sequential policy analysis in which policy solutions are adjudicated against a particular problem at hand, and then, once resolved, it turns to second and third order challenges – but only in ways that do not undermine the higher level problems. Long ago, Cashore and Bernstein (2022) point out, Tribe made this point when referring to species extinction (Tribe, 1972). Put succinctly, he posited that since extinctions are irreversible and often caused by championing economic utility, the only way to address them is to grant them lexical status. The point here is that the underlying moral philosophy of the universalism of the compromise school or economic optimization school usually works against solving Type 4 problems, when, tragically, in today's world they are often offered as transformative solutions for doing so. While Type 4 conceptions were prevalent in global and domestic environmental governance in the 1970s (Bernstein, 2001; Yaffee, 1994), this thinking has been marginalized owing to the dominance of Type 2 and 3 frames. Recently, however, Type 4 conceptions are again gaining increasing salience (Geels, 2020; Lockwood et al., 2017).

Third, they offered that "fit for purpose" governance requires explicit and continuous attention to problem conception, instead of applying "ill-fit for purpose" policy analyses and solutions. This contributes to the literature on transformative governance as it reinforces the need to be very clear about what actual problems, and corresponding outcomes, are being advocated when the literature makes conclusions about how to foster transformations. Put another way, proposed solutions that seek to value the environment through its economic values and that pose no threat to economic growth will look fundamentally different to those that champion the environment and justice. We therefore argue that if governments and scholars seek to address the environment then they must begin, and end, with attention to the problem at hand, rather than narrowing it to those cases that appear synergistic with other problems.

The question then becomes how we can accelerate such a norm shift from Type 2 or 3 conceptions to Type 4 for biodiversity conservation, as part of transformative change in terms of goals, values and paradigms (see Section 4.2 for the definition of transformative change used in this volume). Cashore and Bernstein (2022) argue that doing so requires greater interrogation of disciplines and literatures that have tended to maintain Type 4 conceptions in the midst of so much drift over the last thirty years to Types 1, 2 and 3. These tend to include critical and discursive political scientists, legal scholars and some strands of philosophy – the very disciplines that have been undermined in the shift toward a "data driven," "evidence-based" and artificial intelligence (AI) world – while their general agreement is based on science: The rate of biodiversity loss is real, alarming and caused by human activity.

However, one of the complications of academic debates on biodiversity loss is not only that scholars do not conceptualize biodiversity-related issues as Type 4 problems, but that different scholars actually prioritize different biodiversity-related issues, and therefore also propose different solutions, as shown in the different chapters in this volume (see

Chapter 2 for an overview of different perspectives). A first group (e.g. Dinerstein et al., 2019 and Chapter 11) places biodiversity conservation at the top of the lexical ordering, and, as a result, proposes to protect large areas of land and ocean to halt biodiversity loss. A second group prioritizes improving the lives and livelihoods of local communities living in biodiversity-rich landscapes, which often leads them to oppose formal protection. Yet another, third, group prioritizes moving away from the human–nature dichotomy, and promotes addressing the indirect drivers of biodiversity loss and integrating multiple land uses, and thereby are also often against formal protection (see e.g. Chapter 12). A fourth prioritizes rights of nature, animal rights, antispeciesism, or posthumanism, thereby also moving beyond the human–nature dichotomy but in a different manner, criticizing positioning human wellbeing as more important than that of animals or nature (see Chapter 9). And even when prioritizing biodiversity conservation, scientists often disagree on what types of biodiversity can be best conserved and how, for example ecosystem approaches, focused approaches for specific species, or ex-situ approaches (Cashore and Bernstein, 2020).

So, while the scientific evidence for the fragile state of global biodiversity is clear, academic conceptualizations of the problem and solutions that should be prioritized differ among different groups of scholars. Many scholars would therefore actually disagree with framing the problem as "how to save a million species" – the title of this chapter. Obviously, these groups overlap, as the boundaries are not set in stone, and views evolve over time. Also, different arguments are used by different groups for the prioritization, with the first and fourth groups mainly recognizing the intrinsic value and rights of nature and animals, the second group mainly arguing for biodiversity conservation because humans depend on it, based on instrumental values, and the third group mostly representing relational values. Interestingly, academics representing the different schools of prioritization often collaborate without being explicit about these problem conceptions (see Pascual et al., 2021). So not only in policymaking in general, but also within Type 4 problem analysis, more explicit attention to problem conception is needed.

Integrating explicit attention to problem typologies in biodiversity governance requires that actors first ask how they conceptualize the problem at hand. If they have determined that they conceive of the problem as akin to antislavery norms, or in line with scientific knowledge of ecological tragedies, then they also need to be careful not to inadvertently undertake policy options in ways that are based on or strengthen Type 1, 2 or 3 rationales. Following Cashore and Bernstein, we argue that only Type 4 is "fit for purpose" to ameliorate the problem of global biodiversity loss that threatens one million species with extinction, since it's the only one that addresses the problem as an ecological catastrophe or moral obligation. This does not mean that governance instruments identified by other schools have become obsolete, but that they need to be converted in service of ameliorating Type 4 problems, as elaborated below.

4.4 Implications for Transformative Biodiversity Governance

4.4.1 Prioritizing Biodiversity

What does all of this mean for halting biodiversity loss, or in other words, saving one million species? As shown, among others, in Chapter 6, most biodiversity policies have recently been based on Type 2 and Type 3 thinking, with local initiatives sometimes based on Type 1. Perhaps some protected areas (PAs) could be considered as fitting a Type 4 conception, although the trend in PAs is moving from strict protection to combining land uses, so moving toward Type 3 thinking. Also, PA policy, including deciding where to realize PAs, is often based on Type 2 or 3 thinking. Perhaps the emerging rights-based approaches (see Chapters 2 and 9) could be considered as representing Type 4 thinking. But overall, we have to conclude that most biodiversity policies and initiatives have not been based on Type 4 thinking – biodiversity loss is not treated as a priority in biodiversity governance.

When integrating problem-type analysis into the debate on transformative change and governance, we can conclude that defining biodiversity loss as a Type 4 problem in essence represents an integral part of transformative change: a change in terms of values, goals and paradigms. This would mean transforming biodiversity governance – this volume's title – would mean *prioritizing* biodiversity concerns.

Interestingly, the transformation and transition literatures are not explicit about how they conceptualize sustainability problems. In general, sustainability transitions research (Loorbach et al., 2017; Rotmans et al., 2001) acknowledges the importance of problem framings and implicitly makes the case for transition governance that supports the shift from Type 2 and 3 thinking to Type 4. Also, by highlighting the need for fundamental change, the transformative change literature implicitly tries to address the fact that existing institutions and governance systems do not prioritize biodiversity or sustainability concerns, so could be seen as Type 4 thinking. However, the dominance of pluralist approaches in the transformative change literature and (science) policy debates, as discussed in the above, reflect Type 3 typologies.

Incorporating the focus on Type 4 problems thus provides a goal to transformative change, for example the goal of saving one million species. So, while we agree with the often-heard argument that different actors have different perspectives on the envisioned goal of transformative change and the ways to achieve these, we suggest another way forward. Instead of trying to accommodate all of these different views in the proposed solutions (which in essence reflects Type 2 or 3 thinking), we propose to explicitly discuss these different perspectives in order to come to a clearer understanding of what the problem is that needs to be prioritized and what types of solutions would be appropriate. Being aware of the differentiation between the four problem types thus makes governance more problem-focused.

In other words, explicitly prioritizing biodiversity conservation, and transforming to a truly sustainable society in order to avoid biodiversity loss, has consequences for the types of governance instruments that are required – and perhaps more importantly, those that are less relevant. Taking such a starting point would thereby radically change the way governance would be implemented, since the prioritization would be the basis for strategies and interventions, as discussed in more detail below.

4.4.2 Toward Ecocentric, Compassionate and Just Sustainable Development

Integrating problem-type thinking into transformative governance has consequences for the latter concept, defined in Chapter 1 as "the formal and informal (public and private) rules, rule-making systems and actor-networks at all levels of human society (from local to global) that enable transformative change, in our case, toward biodiversity conservation and sustainable development more broadly." Especially the reference to the concept of sustainable development, currently operationalized around the world through the SDGs, needs further thought (see Chapter 1 for an overview of the SDGs). With the currently dominant Type 2 and 3 thinking, implementing the SDGs quickly becomes a matter of optimizing, or compromising between, the different goals.

Instead, approaches such as Raworth's doughnut economy prioritize the ecological and social SDGs to inform how to operationalize the economic ones to create a "safe and just space for humanity" (Raworth, 2017: 218). However, Raworth's doughnut mainly focuses on human justice, since the planetary boundaries are based on an instrumental perspective, and not necessarily on the intrinsic value of nature. Two important omissions of the doughnut include: (a) attention to the interests of the individual animal – it does not address speciesism, and (b) the intrinsic value and rights of nature. Therefore, we propose to include nonhuman animals and nature in the consideration of the safe and just space – so an ecocentric, compassionate (Bekoff, 2013) and just doughnut economy (see Burgerboerderijen, 2021 and The Vegan Society, 2021) (Figure 4.2). In line with the

Figure 4.2 The ecocentric, compassionate and just doughnut economy (adapted from Raworth, 2017).

proposal by Visseren-Hamakers (2020) for an eighteenth SDG on animal health, welfare and rights, this would represent a transformation of the definition of sustainable development, from "meeting the needs of current generations without compromising the ability of future [human] generations to meet their own needs" (Brundtland et al., 1987), a rather anthropocentric definition, to a definition that includes more ecocentric approaches: "meeting the needs of humans and nonhumans, while respecting the constraints of the planetary boundaries and the intrinsic value of nature." This implies a prioritization of People and Planet over Profit, instead of regarding the three Ps as equal, while also recognizing animal interests (see Chapter 9). So, integrating Type 4 thinking into the definition of transformative governance changes the interpretation of the concept of sustainable development. Redefining sustainable development thus also represents an integral part of transformative change – a change in terms of values, goals and paradigms. This would mean transforming biodiversity governance would not only mean prioritizing biodiversity concerns, but prioritizing ecological, justice and equity concerns over economic ones more broadly, with a view to enabling ecocentric, compassionate and just sustainable development (Elder and Olsen, 2019; Gericke, 2021; Stockholm Resilience Centre, 2016; United Nations Environment Programme, 2021).

4.4.3 Further Operationalizing Transformative Governance

So how can governance support and accelerate this change in problem definition from optimization or compromise to prioritization? As previously stated, transformative governance, as operationalized in Chapter 1 (focused on the indirect drivers, and operationalizing integrative, inclusive, adaptive, transdisciplinary and anticipatory governance in a specific manner), implicitly already starts from Type 4 problem-solving. However, the concept can be further specified to enable prioritization approaches in the following manners.

This focus of transformative governance on the indirect drivers should include addressing those institutions, modes of governance and characteristics of our economic structures that do *not* prioritize ecocentric, compassionate and just sustainable development, since these actually represent an integral part of the *indirect drivers* (or underlying societal causes) of biodiversity loss. With this, addressing the indirect drivers becomes focused on enabling the prioritization of ecological and social societal goals.

The definitions of *integrative, inclusive* and *anticipatory* governance already implicitly reflect Type 4 thinking, with integrative governance (working through governance mixes) basically aimed at ensuring that biodiversity conservation (and ecocentric, compassionate and just sustainable development more broadly) is a priority across sectors, issues, levels of governance and places, and inclusive governance, operationalized in a manner that emancipates those stakeholders who prioritize biodiversity conservation (and ecocentric, compassionate and just sustainable development). With this, transformative inclusive governance could strengthen, support, emancipate and empower those parts of society and the economy where biodiversity loss and its associated negative impacts are already perceived and treated as a Type 4 problem. Anticipatory governance ensures prioritization in contexts of uncertainty by applying the precautionary approach. *Adaptive governance* then becomes focused on reflecting on whether governance still reflects Type 4 thinking, or whether the process is "drifting" (Cashore and Bernstein, 2022) toward optimization or compromise approaches. Stakeholders can together reflect on the extent to which governance is becoming and remains transformative. When integrating priority type thinking, *transdisciplinary governance* becomes focused on ensuring the needed types of knowledge are available and applied, as elaborated in Section 4.5. Through the iterative process of governance that combines these five approaches in this manner, over time, governance becomes increasingly transformative and thereby able to address indirect drivers (see Visseren-Hamakers et al., 2021).

Type 4 problem-solving thereby has significant consequences for governance mixes (combinations of public, private and hybrid governance instruments): as they become more transformative over time, they will increasingly include Type 4 solutions, with the aim of becoming fully focused on the prioritized objective. The question then becomes what types of governance instruments enable Type 4 solutions. Clear examples include prohibiting biodiversity-unfriendly practices, or conservation on the ground through well-placed, strictly protected and effectively managed PAs or other conservation measures. During the evolution of governance becoming increasingly transformative, Type 1 self-governing, Type 2 market-based, cost–benefit solutions and Type 3 deliberative or synergies-oriented approaches can play a role in the governance mix, applied in ways that contribute to Type 4 problem-solving and with this mix changing over time.

Diercks et al. (2020) discuss four governance roles and four processes in transitions. We here apply these in reflecting on transformative governance, as operationalized in the above to include both transitions and transformations. The *four governance roles* include:

- Regulating,
- Collaborating,
- Stimulating and
- Facilitating.

The *four processes*, which take place in parallel, include:

- Emergence (developing new ways of thinking, working and organizing),
- Changing (changing existing elements for new applications or a new context),
- Institutionalization (becoming the norm),
- Phasing out (of ways of thinking, working and organizing).

When combining these governance roles and processes with the four problem conceptions and the main governance instruments based on their logics, the following contributions to transformative governance emerge (see Table 4.2).

Type 1 self-governing solutions have a role to play throughout the transformation, in specific contexts in which local communities informally regulate natural resource use in a collaborative manner. These processes, however, need to be aligned with generic societal priorities. Type 2 market-based and financial solutions (e.g. subsidies, taxation, certifications schemes) can support actors (companies, consumers) during the transformation toward a fully sustainable economy by making sustainable options more competitive.

Table 4.2 Problem conceptions and transformative governance

Problem conception	Main governance roles	Main processes	Main governance instruments	Contributions to transformative governance
Type 1 self-govern	Collaborating, regulating	Institutionalization	Informal local rules	- Role throughout transformation in specific contexts - Needs to reflect broader societal priorities
Type 2 optimize	Stimulating	Phasing out, changing, institutionalization	Market-based, financial instruments	- Decreasing role as transformation evolves
Type 3 compromise	Collaborating, facilitating	Changing, emergence, institutionalization	Deliberative, synergies-oriented instruments	- Role throughout transformation: deliberation remains important to discuss priorities - Synergies within ecocentric, compassionate and just sustainable development remain important
Type 4 prioritize	Regulating	Phasing out, institutionalization	Formal rules	- Increasing role as norms change

They can especially play a role in phasing out, changing and institutionalization processes, and represent stimulating governance roles. Type 3 deliberative, synergies-oriented solutions (multistakeholder processes, partnerships) can facilitate discussing the perspectives of different stakeholders on what priorities should be. They can especially play a role in changing, emergence and institutionalization processes, and represent collaborating and facilitating governance roles. They have a role to play throughout the transformation to avoid "drifting" to nonprioritizing solutions, and to find synergies among different Type 4 problems within the context of ecocentric, compassionate and just sustainable development. Type 4 solutions, including formal rules that enable prioritization, have a regulating role and mainly play a role in phasing out nonsustainable practices and the institutionalization of sustainable ones.

Transformative governance thus evolves over time. As the indirect drivers become increasingly addressed over time, the governance mix can become more focused on Type 4 solutions, since economic structures and institutions, and societal values, paradigms and goals, are evolving to become more sustainable, making Type 4 solutions more feasible. Also, as a Type 4 understanding of the issue of biodiversity loss (and ecocentric, compassionate and just sustainable development more generally) gains prominence in society, Type 1, 2 and 3 policy approaches can be revisited in the light of the emerging transformations. Type 2 policy analysis starts to change, as can be seen with the Stern review and the Dasgupta review (Dasgupta, 2021; Stern, 2007), and there will be gradually growing attention for concerns beyond gross domestic product (GDP) and post-growth approaches. In order to accelerate the process, different actors can reflect on the most appropriate governance mix in different phases of the transformation, through the transformative governance approach discussed in the above.

Some interesting questions remain. What does Type 4 thinking mean for trade-offs between different sustainability or societal concerns, for example climate change mitigation and biodiversity conservation, or biodiversity conservation and local livelihoods, or biodiversity conservation and animal rights? In other words – what should be done if two Type 4 problems meet? In essence, most transformative solutions address multiple sustainability concerns simultaneously, since the same societal structures cause various sustainability issues, as discussed above, so in theory Type 4 governance mixes to address biodiversity loss would simultaneously help mitigate climate change, and vice versa. However, sometimes trade-offs are unavoidable, for example in the case of Invasive Alien Species (IAS). We could have avoided, and still can prevent, IAS through preventative measures (less trade and travel), but the damage in some cases has already been done. The rights of which animal then has priority in a situation where they cannot coexist – the one considered local or the one considered invasive? In such cases, the only way forward would be for actors to explicitly discuss what the priority should be.

4.5 Implications for the Role of Science in Transformative Governance

What is the role of science in transformative governance focused on prioritizing biodiversity conservation? It is important to realize that knowledge, science and the scientific community can be considered part of the problem, or perhaps more gently, not part of the solution; (parts of) our knowledge systems may be part of the indirect drivers of biodiversity loss, including our perception of the problem, how we relate to nature and how we understand what nature is (Stengers, 2011).

As discussed in Section 4.3, parts of the scientific community represent Type 4 thinking but prioritize different biodiversity-related problems, while other parts of the scientific community represent other problem types. The main social scientific theories also represent different problem conceptions. While recognizing many possible exceptions, one could say that rational choice scholars mostly represent Types 1 and 2; different institutionalist approaches cover Types 1–3; discursive theories are mainly aimed at understanding different perspectives, thereby best matching Type 3; and critical theory is clearly focused on a Type 4 problem conception.

Moreover, there are significant epistemological differences between the natural sciences promoting prioritizing biodiversity conservation and those social sciences and humanities also representing Type 4 problem conceptions. So, while their problem conceptions converge, their scientific practices differ to the extent that collaboration becomes difficult. Instead, and as a result, ecologists tend to gravitate toward Type 2 environmental economists, with whom they share similar methods, but who reinforce moral philosophies representing "rational" approaches to addressing ecological catastrophes.

The consequence is that the message in science–policy interfaces is diffuse. While there is academic consensus that biodiversity loss is a problem, scientists characterize the problem and its solutions in many different ways. Moreover, because most current policy processes actually represent Type 2 and 3 conceptions, Type 4 messages on prioritizing biodiversity conservation do not match policy practices and are not integrated in governance efforts. What can we learn from different scientific schools of thought in addressing these dilemmas?

Research on uncertainties (van Asselt et al., 1996) postulates the idea that reductionist and logical empiricist or positivist knowledge approaches are not able to effectively address the most wicked or unstructured problems. In these approaches, "scientific evidence" is used as a basis for policymaking aimed at tackling the complexity of sustainability problems. However, this evidence is never neutral, as is also stressed in literature about political epistemology: Its nuances and uncertainties will be used to misinterpret, modify or motivate interventions in line with powerful interests or dominant perspectives. The objective position of the research(ers) related to policy and, in general, the science–policy interface has already been the subject of debate for decades (e.g. Hoppe and Hisschemoller, 1996; Wildavsky, 1979), but has been revived in the context of sustainable development and biodiversity loss.

While the unstructured nature of complexity points at a need for the involvement of diverse knowledge systems and sources in science for policy, given the inherent uncertainties and values in policy-related science, we need to critically reflect on the "contributions" of academia to noneffective approaches or reinforcing certain typologies, including the synergies norm of Type 3 thinking.

Examples include sustainability science (Clark and Dickson, 2003) and integrated assessment (Rotmans and Van Asselt, 1996), developed as integrated sciences to deal with unstructured "sustainability problems." The core idea, for example, is that the future effects of biodiversity loss are unknown and will also be interpreted and perceived in different ways depending on context. Integrated assessment, transdisciplinarity, cocreation and participatory research engage different types of scientific and practitioner knowledge to create shared analyses and consensus about complex problems as a basis for solutions. However, while we agree that such processes of sense-making and problem-structuring (Rosenhead, 2006) are critical in order to explore why persistent and unstructured problems are seemingly unsolvable, the danger of "drifting" to Type 2 and 3 solutions is tremendous. So, transformative change and governance need a realist ontology: Problems are real but our way of understanding them differs. Therefore, regular deliberation on what exactly are the priorities is vital.

Disciplinary knowledge remains important. Political theory, for example, showcases how vested interests may be reinforced within current regimes, by analyzing processes through which dominant regime-actors (within policy and markets) are able to influence innovation, thereby maintaining their influential position. In other words, these dominant regime-actors make sure that their interests flow into the mainstream debate and policy discourse. This helps them to improve their position and work against potential emerging disruptors (Sterling, 2001). This tendency is also elaborated in institutional theory, which points at the inertia and incremental nature of policymaking and change, and also addresses how powerful actors seek to reinforce and maintain their position. More broadly, institutional theory addresses how organizational structures keep cultural norms and behavioral routines intact in order to stabilize societal systems.

In order for science–policy interfaces to be able to contribute to transformative governance, stakeholders and academics can together codesign governance approaches focused on Type 4 problem-solving. The role of researchers within a Type 4 conception also changes: they are not simply knowledge providers or "experts that resolve needs" (Illich, 1977: 11), but they also act as change agents to establish the much-needed modes of thinking, participation and dialogue for the purpose of transformative change (Fazey et al., 2018; Wittmayer and Schäpke, 2014).

Natural science can continue providing scientific evidence for biodiversity loss, through which biophysical nature – one millions species – gains a voice through the scientists' activities and their instruments (Latour, 2020). In biodiversity governance, this marks the role of the natural sciences: They provide species with a voice, and as biodiversity declines, this voice also increasingly demands political representation. Social sciences that are especially needed in transformative governance include knowledge on institutional change and stability, path dependency, economics that moves beyond the economic growth paradigm, and governance focused on changing values, paradigms and goals. Cashore (2019) proposes an emphasis on qualitative disciplines in history, philosophy, law, historical sociology, political science, sociology and some strands of geography in order to address the nature of Type 4 problems properly.

Transformative governance perhaps also includes a more fundamental reflection of the institutional structures of academia, and in our case the science–policy interfaces around biodiversity. These are in many ways intimately linked to the dominant discourses in science (disciplinary, descriptive, objective) and policy (solution-oriented, formal, power-based) rather than around the transformative governance principles (integrative, inclusive, transdisciplinary, adaptive and anticipatory). If we take these as design principles for transformative science–policy interfaces, it would mean a completely different way of bringing together knowledge perspectives and societal governance. It would mean facilitating communities of stakeholders that work on transformative change in practice, and working with them to identify the institutional principles and conditions needed to mainstream their practices (e.g. regenerative agriculture, biodiversity conservation, cooperative models, de-growth economies, circular economic models and social enterprises). In other words, such a new institutional design would provide mechanisms for transforming biodiversity governance by actually prioritizing the new practices of governance that prioritize biodiversity governance. Together, practitioners and academics could reflect on the main bottlenecks in the transformation, and address them together, whether they be at the landscape, regime or niche level, and whether they would be relevant for only one transition or for sustainability transformations more generally.

4.6 Conclusions

In this chapter we have combined various literatures in order to provide answers to the question of how to save one million species. We have combined the literatures on transformative change, transformations, transitions, transformative governance and problem typologies, which has allowed us to develop the following unique insights.

Bringing together the literature on (governing) transformations and transitions combines the strengths of both bodies of knowledge. The combined perspective allows more focused attention to the generic societal underlying causes of sustainability issues than the transition literature has done so far. These indirect drivers are now better represented as not only influencing transitions in regimes, but also as objects to be changed through transformative governance. The renewed perspective also allows sustainability transformations scholars to operationalize the transformation to – in essence – sustainable societies as "a family of transitions," thereby enabling integrative governance (Visseren-Hamakers, 2015; 2018) of transitions, focused on the interrelationships between different transitions and the underlying causes they have in common. It's perhaps through this enhanced attention to the underlying causes of sustainability problems in multiple transitions that both the transitions and the transformations they are embedded in can be accelerated.

Integrating problem-type thinking (Cashore and Bernstein, 2020) into the transformative change and governance literature has contributed to furthering the conceptualization and operationalization of the concept of transformative governance in the following ways.

First, through the development of this chapter, we have come to realize that most biodiversity policies and initiatives have (purposefully or inadvertently) not been based on Type 4 thinking: Biodiversity loss is not considered as a priority, but instead often regarded as part of problems of optimization or compromise. Based on this analysis, we conclude that defining biodiversity loss as a Type 4 problem in essence represents an integral part of transformative change: a change in terms of values, goals and paradigms. Transforming biodiversity governance would then mean *prioritizing* biodiversity concerns. Incorporating the focus on Type 4 problems thus provides a goal to transformative change, in our case the goal of saving one million species.

Integrating problem-type thinking also has consequences for the reference to the concept of sustainable development in the definition of transformative governance, as introduced in Chapter 1. Transforming biodiversity governance would then mean prioritizing ecological, justice and equity concerns over economic ones to come to mean *ecocentric, compassionate and just sustainable development*, which can be defined as meeting the needs of humans and nonhumans, while respecting the constraints of the planetary boundaries and the intrinsic value of nature.

Transformative governance then becomes focused on the role of current institutions, modes of governance or characteristics of our economic structures that do *not* prioritize ecocentric, compassionate and just sustainable development as part of addressing the *indirect drivers* (or underlying societal causes) of biodiversity loss.

Type 4 problem-solving also radically changes governance. Governance mixes will need to increasingly include Type 4 solutions with the aim of becoming fully focused on prioritization. During the evolution of governance becoming increasingly transformative, Type 1 self-governing, Type 2 market-based, cost–benefit solutions and Type 3 deliberative or synergies-oriented approaches can play a role in the governance mix, adjusted and applied in such ways that they contribute to Type 4 problem-solving, and with this mix changing over time. Through adaptive governance, actors can reflect on whether governance mixes are focused enough on Type 4 problem-solving, or whether implemented solutions are "drifting" toward optimization or compromise solutions. Only if we treat the threat of losing one million species as a priority will we succeed in avoiding this potentially historic loss of life.

References

Abson, D. J., Fischer, J., Leventon, J., et al. (2017). Leverage points for sustainability transformation. *Ambio* 46, 30–39.

Araral, E. (2014). Ostrom, Hardin and the commons: A critical appreciation and a revisionist view. *Environmental Science and Policy* 36, 11–23. https://doi.org/10.1016/j.envsci.2013.07.011

Abson, D. J., Fischer, J., Leventon, J., Newig, J., Schomerus, T., Vilsmaier, U., von Wehrden, H., Abernethy, P., Ives, C. D., Jager, N. W., Lang, D. J., (2017). Leverage points for sustainability transformation. *Ambio* 46, 30–39. https://doi.org/10.1007/s13280-016-0800-y

Arrow, K. J., Cropper, M. L., Eads, G. C., et al. (1996). Is there a role for benefit-cost analysis in environmental, health, and safety regulation? *Science* 272, 221–222.

Bacchi, C., (2012). Why study problematizations? Making politics visible. *Open Journal of Political Science* 2 (1), 1–8. http://dx.doi.org/10.4236/ojps.2012.21001

Bebbington, J., and Unerman, J. (2018). Achieving the United Nations sustainable development goals. *Accounting, Auditing & Accountability Journal* 31, 2–24.

Bekoff, M. (2013). *Ignoring nature no more: The case for compassionate conservation*. Chicago, IL: University of Chicago Press.

Bernstein, S. (2001). *The compromise of liberal environmentalism.* New York: Columbia University Press.

Brundtland, G. H., Khalid, M., Agnelli, S., Al-Athel, S., and Chidzero, B. J. N. Y. (1987). *Report of the World Commission on environment and development: Our common future (A/42/427).* New York: United Nations.

Bulkeley, H., Kok, M., van Dijk, J. J., et al. (2020). *Moving towards transformative change for biodiversity: Harnessing the potential of the post-2020 global biodiversity framework.* Report prepared by an Eklipse Expert Working Group. Wallingford: UK Centre for Ecology & Hydrology.

Burgerboerderijen. (2021). MGV donut. Available from https://burgerboerderijen.nl/mgv-donut/.

Cashore, B. (2019). A growing disconnect between environmental problems and solutions. *Distilled* (spring), 6–7.

Cashore, B., and Bernstein, S. (2022). Bringing the environment back in: Overcoming the tragedy of the diffusion of the commons metaphor. *Perspectives on Politics.* Available from: https://scholarbank.nus.edu.sg/handle/10635/167880. https://doi.org/10.1017/S1537592721002553

Cashore, B., Bernstein, S., Humphreys, D., Visseren-Hamakers, I., and Rietig, K. (2019). Designing stakeholder learning dialogues for effective global governance. *Policy and Society* 38, 118–147.

Cashore, B., and Nathan, I. (2020). Can finance and market driven (FMD) interventions make "weak" states stronger? Lessons from the good governance norm complex in Cambodia. *Ecological Economics* 177, 106689.

Clark, W. C., and Dickson, N. M. (2003). Sustainability science: The emerging research program. *Proceedings of the National Academy of Sciences* 100, 8059–8061. https://doi.org/10.1073/pnas.1231333100

Dasgupta, P. (2021). *The economics of biodiversity: The Dasgupta review.* London: HM Treasury.

De Haan, F. J., and Rotmans, J. (2018). A proposed theoretical framework for actors in transformative change. *Technological Forecasting and Social Change* 128, 275–286.

Diercks, G., Loorbach, D., Van der Steen, M., et al. (2020). *Sturing in transities; een raamwerk voor strategiebepaling.* Rotterdam, The Netherlands:DRIFT and NSOB.

Dinerstein, E., Vynne, C., Sala, E., et al. (2019). A global deal for nature: Guiding principles, milestones, and targets. *Science Advances* 5, eaaw2869. http://dx.doi.org/10.1126/sciadv.aaw2869ESA

Douglas, M., and Wildavsky, A. (1982). How can we know the risks we face? Why risk selection is a social process. *Risk Analysis* 2, 49–58.

Eckersley, R. (2019). *Just transitions and great green transformations: The green state revisited.* Spetses, Greece: Environmental Politics Workshop.

Elder, M., and Olsen, S. H. (2019). The design of environmental priorities in the SDGs. *Global Policy* 10, 70–82.

Fazey, I., Schäpke, N., Caniglia, G., et al. (2018). Ten essentials for action-oriented and second order energy transitions, transformations and climate change research. *Energy Research & Social Science* 40, 54–70. http://dx.doi.org/10.1016/j

Geels, F. W. (2002). Technological transitions as evolutionary reconfiguration processes: A multi-level perspective and a case-study. *Research Policy* 31, 1257–1274.

(2020). Micro-foundations of the multi-level perspective on socio-technical transitions: Developing a multi-dimensional model of agency through crossovers between social constructivism, evolutionary economics and neo-institutional theory. *Technological Forecasting and Social Change* 152, 119894.

Gericke, M. (2021). The doughnut model, SDGs, ESG factors and SDG entry points, a mapping experiment. Available from: www.scrypt.media/2021/02/17/doughnut-model-sdg-esg-mapping/.

Grin J., Rotmans J., Schot J., Geels F., and Loorbach D. (2010). *Transitions to sustainable development: New directions in the study of long term transformative change.* New York: Routledge.

Haas, P. (2002). UN conferences and constructivist governance of the environment. *Global Governance* 8, 73–79.

Hepburn, C. and Stern, N. (2008). A new global deal on climate change. *Oxford Review of Economic Policy* 24, 259–279.

Hölscher, K., Wittmayer, J. M., and Loorbach, D. (2018). Transition versus transformation: What's the difference? *Environmental Innovation and Societal Transitions* 27, 1–3.

Hoppe, R., and Hisschemoller, M. (1996). Coping with intractable controversies: The case for problem structuring in policy design and analysis. *Knowledge for Policy*, 8, 40–60.

Humphreys, D., Cashore, B., Visseren-Hamakers, I. J., et al. (2017). Towards durable multistakeholder-generated solutions: The pilot application of a problem-oriented policy learning protocol to legality verification and community rights in Peru. *International Forestry Review* 19, 278–293.

Illich, I. (1977). *Disabling professions*. London; New York: Marion Boyars.

IPBES (2019). Summary for policymakers of the global assessment report on biodiversity and ecosystem services of the Intergovernmental Science-Policy Platform on Biodiversity and Ecosystem Services. Bonn: IPBES secretariat. Available from https://zenodo.org/record/3553579#.YdXHYlnLeM8.

IPCC (2019). *AR6 synthesis report: Climate change 2022*. New York: United Nations.

Latour, B. (2020). Het Parlement van de Dingen: Over Gaia en de Representatie van Niet-Mensen. Boom filosofie. Amsterdam: Boom.

Linnér, B.-O., and Wibeck, V. (2019). *Sustainability transformations: Agents and drivers across societies*. Cambridge: Cambridge University Press.

Lockwood, M., Kuzemko, C., Mitchell, C., and Hoggett, R. (2017). Historical institutionalism and the politics of sustainable energy transitions: A research agenda. *Environment and Planning C: Politics and Space* 35, 312–333.

Loorbach, D. (2014). To transition! Governance penarchy in the new transformation. Inaugural address of Prof. Dr. Derk Loorbach. Erasmus University Rotterdam, the Netherlands. Available from: https://bit.ly/3KUIR6E.

Loorbach, D., Frantzeskaki, N., and Avelino, F. (2017). Sustainability transitions research: Transforming science and practice for societal change. *Annual Review of Environment and Resources* 42, 599–626.

Mace, G. M., Barrett, M., Burgess, N. D., et al. (2018). Aiming higher to bend the curve of biodiversity loss. *Nature Sustainability* 1, 448–451.

Meadows, D. (2008). *Thinking in systems: A primer*. Hartford, Vermont: Chelsea Green Publishing.

Nordhaus, W. (2019). International carbon pricing: The role of carbon clubs. In: *A Better Planet: 40 Big Ideas for a Sustainable Future*. D. C. Esty (Ed.), pp. 274–278. New Haven, CT: Yale University Press.

O'Brien, K. (2012). Global environmental change II: From adaptation to deliberate transformation. *Progress in Human Geography* 36, 667–676.

O'Brien, K., and Sygna, L. 2013). Responding to climate change: The three spheres of transformation. *Proceedings of Transformation in a Changing Climate, 19–21 June 2013, Oslo, Norway*, pp.16–23. Oslo: University of Oslo.

Olsson, P., Galaz, V., and Boonstra, W. J. (2014). Sustainability transformations: A resilience perspective. *Ecology and Society* 19.

Ostrom, E. (1990). *Governing the commons: The evolution of institutions for collective action*. Cambridge: Cambridge University Press.

Pascual, U., Adams, W. M., Díaz, S., et al. (2021). Biodiversity and the challenge of pluralism. *Nature Sustainability* 4, 567–572. https://doi.org/10.1038/s41893-021-00694-7

Raworth, K. (2017). Why it's time for doughnut economics. *IPPR Progressive Review* 24, 216–222.

Rosenhead, J. (2006). Past, present and future of problem structuring methods. *Journal of the Operational Research Society* 57, 759–765. https://doi.org/10.1057/palgrave.jors.2602206

Rotmans, J., Kemp, R., and Van Asselt, M. (2001). More evolution than revolution: Transition management in public policy. *Foresight* 3, 15–31.

Rotmans, J., and Van Asselt, M. (1996). Integrated assessment: A growing child on its way to maturity: An editorial essay. *Climatic Change* 34, 327–336. https://doi.org/10.1007/BF00139296

Sharpe, B., Hodgson, A., Leicester, G., Lyon, A., and Fazey, I. (2016). Three horizons: A pathways practice for transformation. *Ecology and Society* 21, 47.
Sinden, A., Kysar, D. A., and Driesen, D. (2009). Cost-benefit analysis: New foundations on shifting sand. *Regulation & Governance* 3, 48–71. https://doi.org/10.1111/j.1748-5991.2009.01044.x
Stengers, I. (2011). *Thinking with Whitehead*. Cambridge, MA: Harvard University Press.
Sterling, S. (2001). *Sustainable education: Revisioning learning and change*. Bristol: Schumacher Briefings.
Stern, N. (2007). *The economics of climate change: The Stern review*. Cambridge; New York: Cambridge University Press.
Stockholm Resilience Centre. (2016). *How food connects all the SDGs*. Available from: https://bit.ly/3JJxXjv.
Stone, D. A. (1997). *Policy paradox: The art of political decision making (Vol. 13)*. New York: W.W. Norton.
The Vegan Society (2021). *Planting value in the food system. Part 2: The research*. Available from: www.plantingvalueinfood.org.
Tribe, L. H. (1972). Policy science: Analysis or Ideology? *Philosophy & Public Affairs* 2, 66–110.
United Nations Environment Programme (2021). *Adapt to survive: Business transformation in a time of uncertainty*. Nairobi:UNEP.
Van Asselt, M. B. A., Beusen, A. H. W., and Hilderink, H. B. M. (1996). Uncertainty in integrated assessment: A social scientific perspective. *Environmental Modeling & Assessment* 1, 71–90. https://doi.org/10.1007/BF01874848
Visseren-Hamakers, I. J. (2015). Integrative environmental governance: Enhancing governance in the era of synergies. *Current Opinion in Environmental Sustainability* 14, 136–143.
  (2018). Integrative governance: The relationships between governance instruments taking center stage. *Environment and Planning C: Politics and Space* 38, 1341–1354.
  (2020). The 18th sustainable development goal. *Earth System Governance* 3, 100047.
Visseren-Hamakers, I. J., Razzaque, J., McElwee, P., et al. (2021). Transformative governance of biodiversity: Insights for sustainable development. *Current Opinion in Environmental Sustainability* 53, 20–28.
Weimer, D. L., and Vining, A. R. (1999). *Policy analysis: Concepts and practice* (2nd ed.). Englewood Cliffs, NJ: Prentice-Hall.
Westley, F. R., Tjornbo, O., Schultz, L., et al. (2013). A theory of transformative agency in linked social-ecological systems. *Ecology and Society* 18, 27.
Wildavsky, A. (1979). *Speaking truth to power: The art and craft of policy analysis*. Piscataway, NJ: Transaction Publishers.
Wittmayer, J. M., and Schäpke, N. (2014). Action, research and participation: Roles of researchers in sustainability transitions. *Sustainability Science* 9, 483–496. http://dx.doi.org/10.1007/s11625-014-0258-4
Yaffee, S. L. (1994). *The wisdom of the spotted owl: Policy lessons for a new century*. Covelo, CA: Island Press.

Part III

Cross-Cutting Issues Central to Transformative Biodiversity Governance

5
One Health and Biodiversity

HANS KEUNE, UNNIKRISHNAN PAYYAPPALLIMANA, SERGE MORAND
AND SIMON R. RÜEGG

5.1 Introduction

The main aim of this chapter is to discuss linkages between nature and generic health from a One Health as well as transformative biodiversity governance perspective. Due to the COVID-19 pandemic, the interest in the linkages between nature and human health has increased drastically, in general but also in the biodiversity realm. The origin of the virus is still under investigation, but Haider et al. (2020) propose classifying COVID-19 as an "emerging infectious disease of probable animal origin." The tens of millions of human COVID-19 infections reported internationally appear to have primarily emerged through human-to-human transmission. Thus, amidst the pandemic, the potential animal origin is of secondary interest for further containment of the disease. Still, in the public and international governance debate for example in the Intergovernmental Science-Policy Platform on Biodiversity and Ecosystem Services (IPBES, 2020), a link is clearly made between zoonotic infectious diseases and the effects of human pressures on ecosystems. The dissemination of the virus, facilitated by intense global travel and high local connectivity, should also cause us to question our understanding of the fragilities of human health in a globalized world.

Early foundational steps regarding nature–human health linkages were present in the World Health Organization's (WHO's) contribution to the Millennium Ecosystem Assessment (WHO, 2005) and the State of Knowledge Review that was jointly produced by the Convention on Biological Diversity (CBD) and WHO (WHO-CBD, 2015). Until recently, however, for many in the biodiversity domain, linkages with human health were little known or taken into account in science, policy and practice. The concept of One Health is now often mentioned as a "silver bullet" solution to challenges like the COVID-19 pandemic (e.g. IPBES, 2020). More or less in the background, One Health has been around for quite some time, including in the WHO-CBD knowledge review (2015), where it was proposed as an overarching concept for biodiversity and health governance. The concept was supported by the CBD member states in the final declaration of the Conference of the Parties in 2018, which "Invites Parties and other Governments to consider integrating One Health policies, plans or projects, and other holistic approaches in their national biodiversity strategies and action plans, and, as appropriate, national health plans" (CBD, 2018). But what does One Health

---

Part of this chapter builds on, and is very grateful to work conducted in the frame of, the European Cooperation on Science and Technology (COST) 582 Action TD 1404 "Network for Evaluation of One Health."

entail, or rather, what can it entail, as we can question whether the beauty of One Health is the same in the eyes of many beholders? We do not have the ambition to present an exhaustive overview of nature–human health linkages or of One Health. We aim to discuss key aspects and challenges of One Health, highlight definitional diversity, and in doing so hope to give inspiration for transformative biodiversity governance.

5.2 Understanding the Concept of One Health

5.2.1 Biodiversity and Health

From the perspective of nature's contributions to people (see Chapter 2 for more details on definitions of nature), it may seem that human health is only one of many elements of the ways in which nature and biodiversity can contribute to human well-being. This is illustrated by the fact that in modern scientific literature on the conceptual and operational development of the concept of ecosystem services, health is often "only" considered to be a subsection of cultural values (Bryce et al., 2016; Bullock et al., 2018), or is even absent (Cheng et al., 2019). An explanation is that the concept emerged in the realm of biological sciences, with biologists trying to link the importance of "their world" to societal relevance, with as a main first step economic valuation (Ring et al., 2010). This is the same the other way around: Until recently the word "ecology" in the health sector often had limited reference to nature, but rather to the social or societal environment of a patient (Hoffmann et al., 2019; White, 1997), and nature was only considered to a limited extent in, for example, primary health care (Lauwers et al., 2020), and even the concept of "green prescription" initially had few linkages with nature, but mainly referred to environmental pollution and climate change challenges, lifestyle and nonmedicinal prescriptions (Anderson et al., 2015; Patel et al., 2011; Swinburn et al., 1997). A prominent exception is the WHO Ottawa Charter on Health Promotion (WHO, 1986: 1), which has highlighted the importance of a stable ecosystem: "The fundamental conditions and resources for health are peace, shelter, education, food, income, a stable ecosystem, sustainable resources, social justice and equity. Improvement in health requires a secure foundation in these basic prerequisites." Apart from this example, the (more tangible) negative drivers relating to environment, like pollution, have dominated. There was relatively little discussion on the positive and negative contributions of ecosystems and biodiversity.

The mechanisms linking nature and biodiversity on the one hand and human health on the other are complex and intertwined, and can result in human health benefits and risks (IPBES, 2018a; WHO-CBD, 2015). Figure 5.1 (Marselle et al., 2021) shows how biodiversity and human health and well-being are related through diverse pathways and a wide array of moderating factors.

Biodiversity supports the ecosystem services that mitigate heat, noise and air pollution, which all mediate the positive health effects of green spaces (see Chapter 14). In the topical domain of medicinal plants, significant work has been done regarding biodiversity and health, including a vast body of Indigenous traditional knowledge (IPBES, 2018b; WHO-CBD, 2015). In more mainstream contemporary environmental health science, direct health

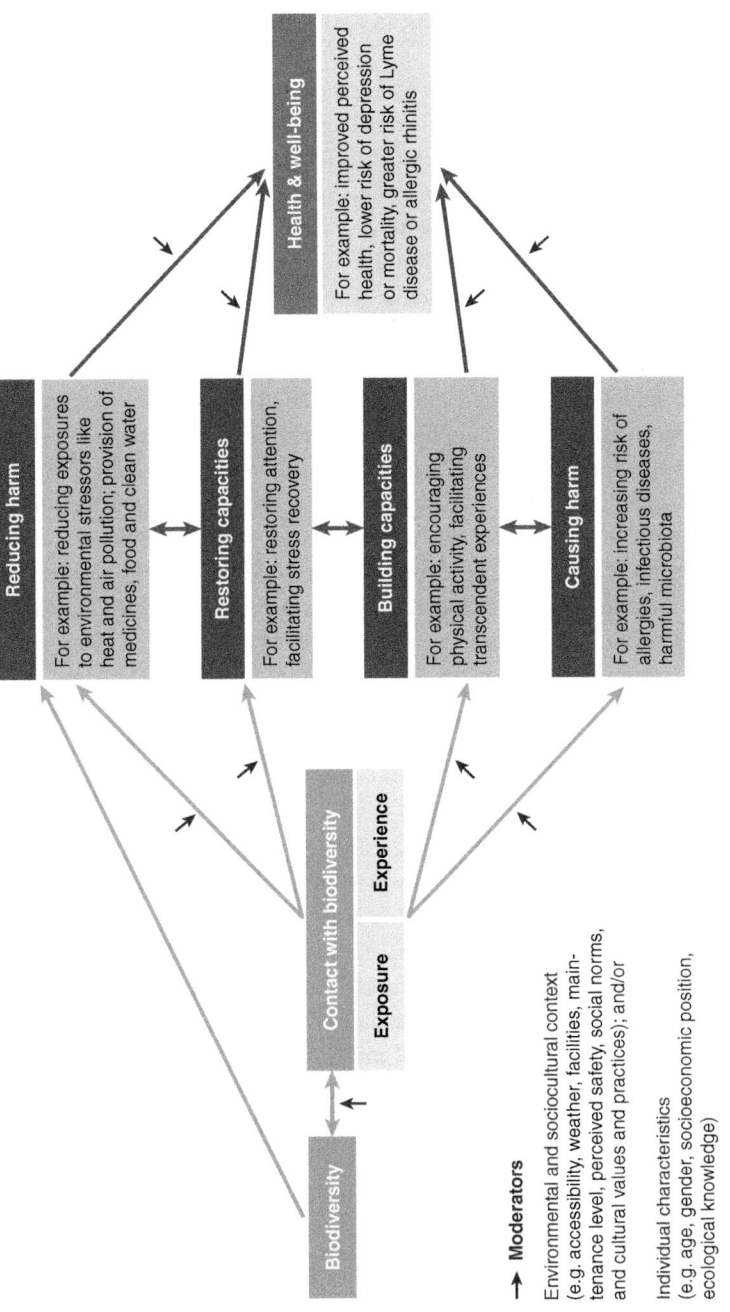

Figure 5.1 Pathways linking biodiversity to human health (Marselle et al., 2021)

outcomes of biodiversity have been understudied and underverified so far. There is evidence for positive associations between species and ecosystem diversity, and psychological and physical well-being and immune system regulation. There is more evidence for self-reported psychological well-being than for well-defined clinical outcomes. High biodiversity has been associated with both reduced and increased vector-borne disease risk (Aerts et al., 2018).

Ecosystem change is recognized as a risk factor for disease emergence and spread, but a specific role for biodiversity is not always clear. Biodiversity may reduce disease risk by what is called the *dilution effect*. The dilution effect hypothesis proposes that high vertebrate species richness reduces the risk of infectious diseases among humans because pathogens are "diluted" among a high number of animal reservoir species that differ in their capacity to infect invertebrate vector species (Schmidt and Ostfeld, 2001). Under the dilution effect hypothesis, the transmission and burden of infectious diseases are expected to be lower in animal species-rich, natural environments through lower infection prevalence in vectors (Johnson et al., 2015; Ostfeld and Keesing, 2017), even when higher species richness also implies higher pathogen richness (Dunn et al., 2010). However, factors such as species composition, persistence of contacts between reservoirs and vectors, and the various ways in which reservoirs and nonreservoirs are affected by environmental change may all affect the dilution mechanism. The *amplification effect*, in which the infection prevalence in vectors increases following an environmental change affecting biodiversity, has also been observed (Faust et al., 2017). The conditions in which dilution or amplification will be observed are still the object of research (Johnson et al., 2015; Kilpatrick et al., 2017; Morand, 2018). However, it has been established that the risk of disease spread appears higher in human-dominated and simplified habitats (Morand, 2018). Habitat fragmentation affects both pathogen diversity and pathogen prevalence. The *perturbation hypothesis* holds that if a habitat is fragmented, the sum of fragments will not be able to sustain the same diversity and prevalence of pathogenic species (but also reservoirs and vectors) as the original habitat (Murray and Daszak, 2013). However, fragmentation also leads to a longer boundary between the habitat(s) and those of other communities. This in turn increases the chance of encounters between communities of hosts and vectors. The *pathogen pool diversity hypothesis* thus assumes that this intensified interaction raises the transmission of pathogens between habitats and species, and within populations. Hence, ongoing habitat fragmentation may both decrease and increase disease transmission risk. Beyond fragmentation, the ongoing "Anthropocene defaunation" leads to almost empty tropical forests (Dirzo et al., 2014). The sharp decline of many animal populations has dramatic implications for zoonotic diseases, by both decreasing and increasing transmission risks. As the diversity of host populations decreases, so will the diversity of the microbes (including pathogens) they harbor. Decreasing host diversity means the loss of important interspecific regulations provided by predation or competition. The remaining pathogens hosted by more abundant but less diverse hosts or vectors released from competition or predation show enhanced transmission. This is particularly evident for pathogens able to switch host species easily and those living in synanthropic species such as rodents or some mosquito vectors. The recent study by Gibb et al. (2020) demonstrates how global land-use changes favor zoonotic reservoirs and increase the risks of zoonotic diseases, and more specifically in Southeast Asian environments with critical ongoing defaunation (Morand, 2018).

5.2.2 Integrative Concepts

Integrative approaches to health have quite a long history. The WHO Constitution in 1946 envisioned a comprehensive view of health: "health is a state of complete physical, mental and social well-being and not merely the absence of disease or infirmity" (WHO, 2006: 1). In the WHO meeting in Alma-Ata (today Almaty, Kazakhstan) in 1978, a holistic and intersectoral conceptualization of health assumed importance: "[health] involves, in addition to the health sector, all related sectors and aspects of national and community development, in particular agriculture, animal husbandry, food, industry, education, housing, public works, communications and other sectors; and demands the coordinated efforts of all those sectors" (WHO, 1978: 2). As mentioned above, in 1986, the WHO Ottawa Charter for Health Promotion highlighted the need for a stable ecosystem as a basis for good health (WHO, 1986). In 2002, the World Summit on Sustainable Development initiated the foundation for an inclusive framework: WEHAB (Water, Energy, Health, Agriculture [food, nutrition] and Biodiversity and Ecosystems) (United Nations, 2002). In 2005, the Millennium Ecosystem Assessment identified key connections between biodiversity, ecosystems and human well-being (WHO, 2005), and in 2006 the Finnish presidency of the European Union presented the concept of "Health in All Policies" as a main health theme (Puska, 2006). In the Finnish opinion, the core of "Health in All Policies" was to focus on health determinants mainly controlled by policies of sectors other than health. The wish was to address policies in the context of policy-making at all levels of governance. The idea in fact dates back even further: In 1978, at the WHO International Conference on Primary Health Care, the Alma-Ata Declaration emphasized the role of sectors other than health in the creation of public health: "the highest possible level of health is a most important worldwide social goal whose realization requires the action of many other social and economic sectors in addition to the health sector" (cited in Ståhl, 2018: 38). Health as overarching generic principle raises the question: Can One Health follow in these footsteps as an overarching governance integrator, while also being more inclusive by incorporating animal, plant and ecosystem health?

Several integrative governance perspectives regarding challenges with environmental (natural and built) determinants of health are gaining traction today, even if some of these concepts already have some history. This is driven by concern for emerging infectious diseases, rapid increases of noncommunicable diseases, rising morbidity due to ecosystem and climatic changes, and increased awareness of challenges of chemical use in human living environments and in livestock farming, including antibiotics, fertilizers and pesticides in agroecological systems and so on (WHO, 2012). One Health, EcoHealth, planetary health, global health, conservation medicine, biodiversity and health, agrihealth and health pluralism are examples of these broader frameworks, which aim for an integrated perspective on health and the living environment (Assmuth et al., 2019).

EcoHealth encompasses ecosystem approaches to health, covering the biological, physical, social and economic environments and their relation to human health (Lebel, 2003). The concept One Health originated at the interface of animal and human health (Woods and Bresalier, 2014) with the aim of covering a larger diversity of expertise than health and veterinary sciences, and over time broadened its perspective to the environment (Rüegg et al., 2017). Zinsstag et al. (2011) proposed One Health as an approach aimed at tackling complex patterns of global change, in which the inextricable interconnection of humans, pets, livestock and wildlife, along with their social and ecological environments, is evident and requires integrated approaches to human and animal health and their respective social and environmental contexts. The WHO and CBD State of Knowledge Review on biodiversity and health (2015) proposed One Health as an overarching framework for integrated efforts, while also recognizing and relating to other relevant approaches, such as EcoHealth. Earlier, a tripartite collaboration among the Food and Agriculture Organization (FAO), the World Organisation for Animal Health (OIE) and WHO (2010) proposed a similar integrated effort, also called One Health. A related concept is One Welfare, which aims to relate animal, human and environmental welfare under one umbrella (Bourque, 2017; also see Chapter 9). Similarly, the Lancet Commission on planetary health (Whitmee et al., 2015) highlights the integrated nature of human and planetary health.

In a different vein, there has been fresh thinking on alternative worldviews and perspectives provided by diverse knowledge systems on health and well-being for tackling sustainability challenges. The idea of holistic health traditions has existed for centuries, but recently there have been new frames of reference that allow mainstreaming of such holistic approaches. According to some health cultures, optimal health is "To be established in one's self or own natural state" (Payyappallimana, 2013: 105). To achieve this, one must have a balance of physical, mental, spiritual, social and ecological dimensions of existence. Based on this philosophy, there are distinct epistemological principles and practices for the prevention of disease and promotion of health and health care in several Indigenous and Local Knowledge cultures. Shared explanatory frameworks, healing practices including rituals, physical healing environments and so on become central in such a context. Sacredness is attributed to trees, grains, animals, hills, forests, streams, mountains and caves that are worshiped through rituals, ceremonies, festivals and fairs. Such knowledge, belief systems and worldviews find expression in agroecological traditions, art, songs and other symbolic representations and practices linked to well-being. For instance, in a study among communities of coastal Tamil Nadu, Sujatha (2007: 178) states, "the body is seen as being constituted by food which is the vehicle by which the external ecology is internalized."

A shared perspective across Indigenous and local communities in the Indian subcontinent is the inherent relationship between the "outside" and "inside" worlds. In Āyurveda and other traditional knowledge systems of medicine in the subcontinent, this is known in terms of "loka" (macrocosm) and "puruṣa" (microcosm). Similar traces of this principle form an underlying basis for all Indigenous and Local Knowledge traditions. Health in Āyurveda is understood as a positive state and is based on the outcomes of adaptive feedback that each person establishes with the environment and determined by the ability of a person to adapt and self-manage (Morandi et al., 2011). Similarly, in other cultures the biopsychosocial model of health (Engel, 1977) brings the concept of health from a purely biological realm into, as the name suggests, the psychological and social realms of health. The concept has gained popularity with health professionals, making them consider the broader factors impacting on the health and well-being of individuals and communities, indicating that health care alone does not provide health. Likewise, the concept of "salutogenesis," coined by Aaron Antonovsky (1979), depicts an approach that focuses on the drivers of health and well-being rather than focusing on morbidities or pathogenesis.

Though seemingly quite similar in holistic and integrative ambition, these overarching concepts do not necessarily result in identical definitions of nature and linkages with human health, nor in common framing of challenges and remedies (Keune and Assmuth, 2018). Different expert groups may identify themselves differently with the concept of One Health. On the one hand, there is a community of expertise and practice focusing mainly on nature-related health benefits, and on the other one concerned mainly with its risks (Keune and Assmuth, 2018). While the former community advocates for nature-based solutions as a path to a better future, some prominent virologists representing the latter community label nature as an extreme threat to human health. Some of the latter group even state, "nature is the biggest bioterrorist," from which yet unknown threats should be avoided: one must "intervene in the conditions of emergence of the future, before one may be besieged by nature's own act of emergence" (Mutsaers, 2015: 128). This biosecurity framing has led to the development of vaccines, but also brought forward preventative culling of wildlife and domestic animals, resulting in a strategy with questionable ethics. Clearly, a balancing of perspectives is needed to escape such paradigmatic deadlocks. An approach coined Structural One Health (Wallace et al., 2015) extended the concept of One Health to include the socioeconomic perspective more clearly. It criticized the prior iteration of One Health for failing to address the fundamental structural, political and economic causes underlying collapsing health ecologies, similar to ideas of transformative change. Figure 5.2 illustrates Structural One Health compared to other approaches, highlighting different characteristics of different health approaches and interventions.

5.2.3 Dilemmas in Nature-Based Approaches to Health

Horwitz et al. (2012) and Roiko et al. (2019) summarize the complex character of nature–health linkages with reference to the paradox of the health imperative, and the opposite of the environmentalist's paradox: Where, from an ecosystem services point of view, one would expect a clear relation between a healthy ecosystem and human health, the environmentalist's paradox points at the fact that degradation of an ecosystem, for example by using DDT for malaria control, can in fact be beneficial in the short-term for human health. The health imperative exemplifies cases where a healthy ecosystem can, in fact, pose human health threats, for example the presence of mosquitoes in urban nature conservation areas, which may support spreading infectious diseases under specific conditions.

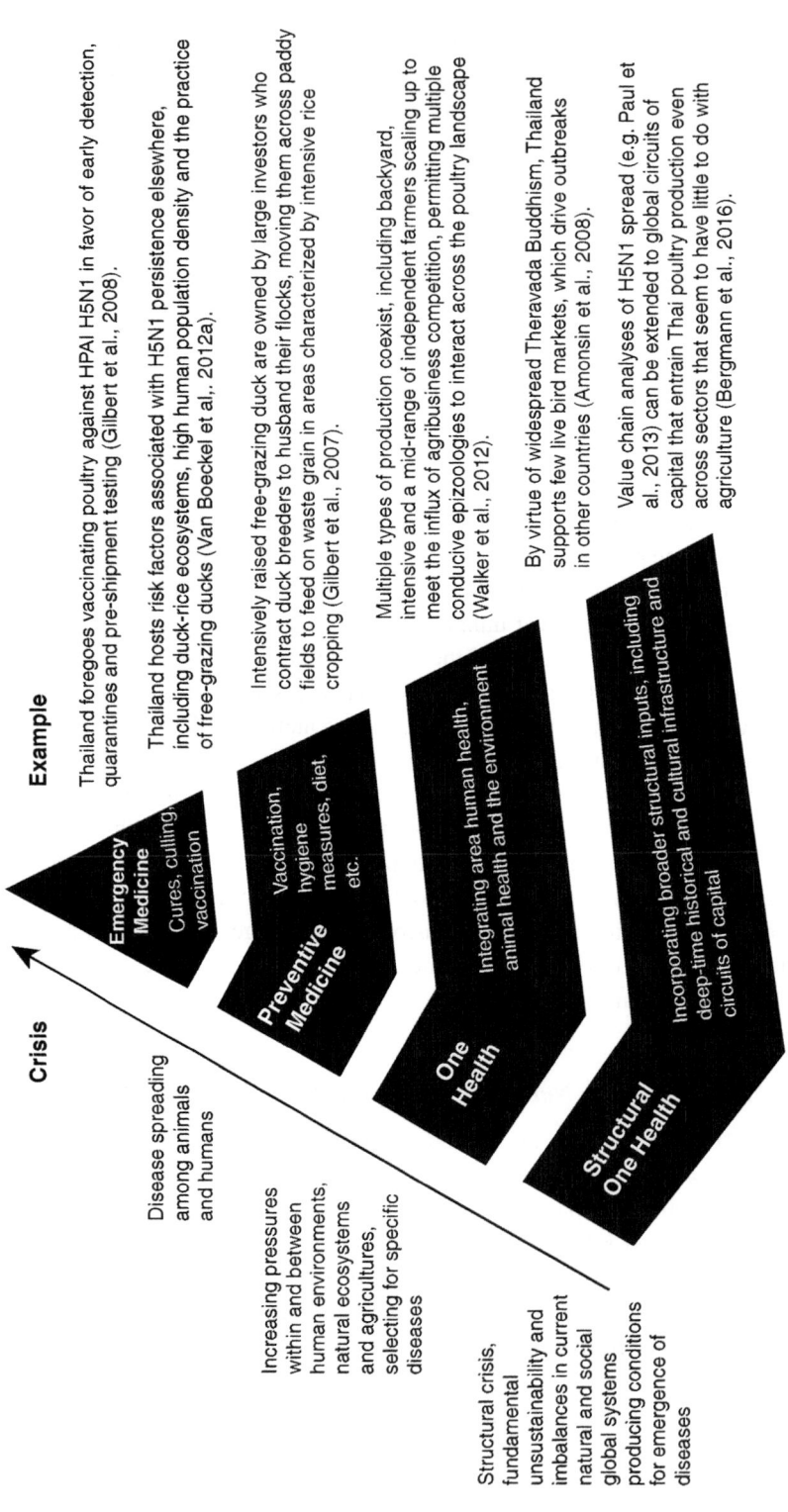

Figure 5.2 Structural One Health

"Structural One Health investigates the broader context of a disease, including out beyond the local, more proximate mechanisms of emergence on which more episodic One Health focuses. Preventive and emergency medicine are deployed in response to threats on the health of specific populations and individuals. For all mechanisms that promote disease (under 'crisis'), the proximity in space, time and causal origin to any given outbreak increases up the pyramid. The relative importance of each point along the scale is dependent on the collective interplay between all parts of the pyramid. An array of inputs and outcomes for highly pathogenic avian influenza H5N1 in Thailand is shown across the schematic" (Wallace et al., 2015: 5).

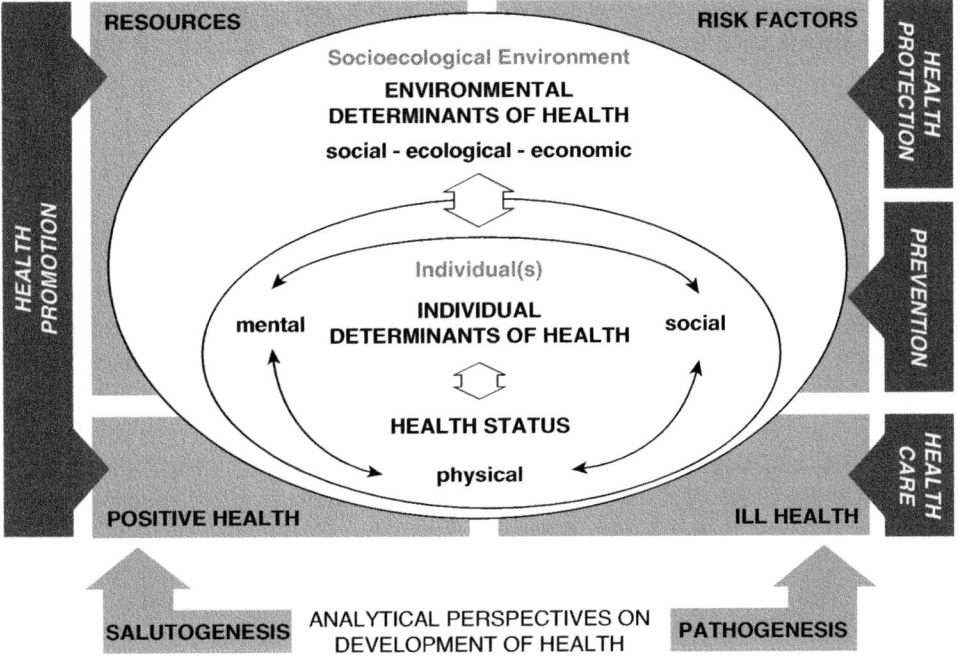

Figure 5.3 Salutogenis and pathogenis (Bauer et al., 2006)

Bauer et al. (2006: 156) illustrate this dilemma by comparing the focus on pathogenesis to that on salutogenesis (Figure 5.3). With pathogenesis, the focus is mainly on health risk factors for individuals in their living environment, leading to "disease, disorders, subjective sickness, malfunctioning and impairment." With salutogenesis, the focus is mainly on human health-supporting resources, including "fitness, subjective wellbeing, optimal functioning, meaningful life and positive quality of life." Both concepts should be considered to be complementary and interacting throughout life.

Balancing these two perspectives in relation to nature is also a clear challenge in primary health care (Lauwers et al., 2020). In the fast-growing body of scientific literature on nature–human health linkages, a role for primary health care is still only marginally present. Also, specific uptake tools for practical consideration of these linkages in primary health care seem lacking. Besides, the need for a primary One Health care approach has been highlighted (Lauwers et al., 2020).

Further scientific challenges on nature–human health linkages remain. One recent review on "types and characteristics of urban and peri-urban green spaces having an impact on human mental health and wellbeing" (Beute et al., 2020) illustrates this for an important subdomain of nature-related health benefits in the urban context (see Chapter 14). Clearly, the extensive review could not find a gold standard for a particular green space type or characteristic working best for everyone, everywhere and at every time. This heterogeneity may be explained in terms of differences in exposure duration and differences in experiences, and there are different effects for different target groups. This would lead to recommendations for a variety of green space types to capture all potential users, their different needs and their activities.

5.3 Challenges in One Health Governance

These apparent contradictions and dilemmas at the conceptual and practical level form the challenging landscape in which One Health governance should intervene. Currently, there is no clear agreement on, or understanding of, what is best practice regarding One Health knowledge (Rüegg et al., 2018). A key challenge is knowledge integration (incorporating a diversity of knowledge related to different disciplines, topical areas and practices) and learning by doing. Clearly this takes time and effort: More mature initiatives become more holistic as they evolve in a trial and error process (Buttigieg et al., 2018; Fonseca et al., 2018; Hanin et al., 2018; Paternoster et al., 2017). In addition, the importance of knowledge integration and particularly the sharing of data is well-recognized but is often hampered by political boundaries. A phenomenon that has been reported for the governance of the Sustainable Development Goals (SDGs) (Nilsson et al., 2018) can also be observed in One Health (Hanin et al., 2018). The evaluation of an international effort for infectious disease surveillance showed that national as well as institutional borders are challenging for the sharing of data (Hanin et al., 2018). Whether this has structurally changed during the COVID-19 pandemic remains to be seen.

Another important One Health challenge is interdisciplinary and transdisciplinary approaches, which appear to be one of the most demanding practices in the academic context (Léger et al., 2018; Muñoz-Prieto et al., 2018). There seems to be a disconnect between the ambition to work across disciplines and the cultural practice in science of evaluating achievements based on scientific, preferably high-impact, publications. The prevailing competitive mentality in academia is a serious obstacle to the trusted collaboration required for interdisciplinary progress. An explicit mandate to reach beyond academia and connect to practitioners can result in a surprisingly good alignment with the One Health concept (Radeski et al., 2018). Partnerships spanning collaborators from government, academia and practitioner circles may generate more holistic solutions.

5.3.1 The Scission between Human Health Benefits and Threats from Nature

As already mentioned above, although One Health acts as an integrating umbrella for talking about health, there appear to be two main opposing narratives around nature–human health linkages, under the same heading of "One Health" (Keune et al., 2013). But even without explanatory causal links, a comprehensive conversation about the management of our environment requires a constructive dialogue between those two communities. To move from a struggle for prerogative of interpretation to a co-construction of understanding, it will be necessary to have more direct interaction and discourse between the different viewpoints and groups, through transdisciplinary governance.

5.3.2 *Which Ethics?*

As emphasized by Morand and Lajaunie (2019) and Lainé and Morand (2020), ethical reflection in the field of health and biodiversity would require examining the relevant scientific domains (i.e. biology, ecology, evolution, human medicine, animal medicine, political science, environmental studies, anthropology and law), their epistemology and, for some, deep roots in the colonial sciences based on a paternalistic perspective, dominated by the lens of the Western worldview on reality. Consequently, several ethical responses to public health crises have been proposed with "One Bioethics," "One Health ethics," "Global Health ethics" and, more recently, "Planetary Health ethics," with no consensus among bioethicists. The need to recognize scientific pluralism appears essential for interdisciplinarity, but it requires acknowledging the values and practices of each scientific domain. It requires also a decolonized (less Western paternalistic) and a more-than-human (respecting also nonhuman health) One Health approach (Lainé and Morand, 2020). Further, it needs to be stressed that even though perspectives like One Health are more encompassing, they are to be implemented in a context of highly linear positivist science and a practice structure of current health systems that have limited capacities related to human resources, knowledge and so forth.

While "Global Health ethics" is essential in underlining the importance of justice and equity, a "One Health ethics" or a "Planetary Health ethics" could refer more to a metaethics regarding the ecological crisis and its implications for the study of nature or biodiversity. The question is, then: Is nature reducible to a simple mechanism such as the dilution effect, or is it a complex adaptive system of physical and sensible interactions between various life forms including humans? Considering what kind of nature is at play in a health crisis has profound consequences for the attitudes toward nature and people and for health policy responses. COVID-19, as well as previous pandemics, shows that crises are often systemic, which calls for the development of systemic actions with better nature stewardship, and resonates well with the ideas of transformative change.

5.3.3 *Balancing Top-Down and Bottom-Up Health Norms and Challenges*

Contemporary medical practice relies heavily on norms and reference values. A strong deviation from a mean is commonly considered as pathology, implying that regularity (i.e. the mean) is a healthy objective. Consequently, decision matrixes are often positivist, objective and deterministic, with the aim of reestablishing normalcy. Similarly, in public health, veterinary health and food safety, solutions are often prescribed top-down, implying singular linear pathways in isolated aspects of health. There are obvious advantages of this approach when it comes to health management at scale, such as decision-making for resource allocation in a national health service. However, current health management is in stark contrast to the observation that complex systems show fractal behavior, in a coherent variation and diversity (West, 2012). A complex adaptive systems approach in medicine would require moving away from preestablished medical problems with expected solutions, and working with people toward defining the medical goal itself. Such an approach requires, of course, an acceptance of unpredictability, uncertainty and ambiguity (Strand et al., 2004) – something most health care systems are not set up to deal with. At the onset of the COVID-19 pandemic, many aspects were unknown. Nevertheless, in order to prevent its spread, swift action was needed. It proved more successful to take some generic assumptions to contain highly infectious diseases and to implement a crude strategy in Mongolia and Taiwan, for example, than to delay action waiting for sufficient detailed knowledge. The ambiguity of the evidence and the unclear relation to the situation in the field kept fueling public debates about the way to deal with the pandemic in many other democratic states, while lives were lost to the disease.

There is no doubt that norms and reference values have an important place in daily practice, but there is a risk that such norms may obliterate other potential pathways to health. In the context of One Health, the question arises: To what degree are such norms universal and time independent, and to what degree would they require contextualization? While a strong focus on individual choice in health care has the advantage of more tailor-made health strategies, the right to individualism stands in contrast with the needs of communities or societies. Especially in developed countries, when people make unhealthy choices, the health costs either result in a loss of solidarity because the community does not want to cover the consequences of individual behavior, or in rising health expenses for the community. Another example is the individual choice of vaccination, where people who may choose to abstain from vaccination contribute to lowering community or herd immunity and thereby increase disease risks. Here again, cohesion appears to be an important concept, that is, solidarity needs to be reciprocal: While individuals consider the resilience of the community in their acts, the community can offer solidarity in return. Importantly, at various levels of socioeconomic status, health should be discussed and co-produced. This shows that One Health is more than an integrated approach to emerging infectious diseases, but a way to address many health concerns, from malnutrition to traffic accidents, in an integrative and inclusive governance process. The impacts are considerable as they affect legislation and require, and represent, transformative change. Some possible approaches have been proposed such as social prescription (Jani and Gray, 2019; Jani et al., 2019), positive health dialogue (Huber et al., 2016), quintuple helix innovation (Carayannis et al., 2012), critical complexity (Cilliers, 2005; Keune, 2012), participatory action research (Kincheloe, 2009) and salutogenesis (Lindström and Eriksson, 2005; 2006).

5.3.4 What Are the Values Associated with Health?

In the search for generic validity of concepts and frameworks, it goes unnoticed that we know very little about the lives of those who experience the complex entanglements between humans, animals and ecosystems on a daily basis, and whose stewardship is decisive for change to occur. Although there are studies on more general values (World Values Survey Association, n.d.) – particularly the comparative value of health for oneself – people, animals and ecosystems have not been explored. While currently, with few exceptions, justice is an anthropocentric notion, the aim of achieving interspecies health equity as an outcome of One Health suggests that there may be a need to develop a framework for biocentric social justice (see Chapters 8 and 9).

5.3.5 *The Need for a Scalable Definition of Health*

Such a framework would need to be grounded in a generic understanding of health. Exchange across different disciplines and sectors in charge of different scales of life, from microorganisms to national and global economies, reveals a large variety of definitions of health. At the level of ecosystems, the concept of health is controversial (Rapport, 1998). But also at an individual level, our concepts of personal health are diverse. Health can be regarded as a dynamic, adaptive process rather than a static state. A potential framing would be health as resilience at the individual level, with well-being and welfare as emerging properties of a functional co-adaptation between an individual and their direct environment. In some Indigenous cultures, an individual is also seen as a constantly changing substratum and thus health as interaction between two dynamic (in some contexts deteriorating) systems. The concept of resilience can be evaluated at multiple levels of social-ecological systems. Metrics for resilience are different at different scales, primarily because change occurs at much slower rates at larger scales and is faster at smaller scales, thus preventing the same relative time resolution at all scales. Nonetheless, the principal idea can be transferred across all scales and can also accommodate for cultural differences. Consequently, One Health approaches would need to foster resilience at all scales, and as a minimal requirement not reduce resilience at any scale in a social-ecological system. This would allow humans and nonhumans to live together and allow adaption to various challenges in the short and long term.

5.3.6 *Will Egoism Define the Boundaries?*

Inclusive governance – as used in the field of sustainable development – may help to make use of One Health opportunities and to promote dialogue and solutions for intergenerational health if there is propensity among participants to engage, connect, reflect and change. It is expected that economic activities that promote human well-being, sustainability and justice will need to be coupled with a steady-state or degrowth economy respecting planetary boundaries. This is essentially the premise of ecological economics. The future will show whether people are willing to rethink today's concept of prosperity driven by continuous increase in economic growth. Data show that the link between income and life-satisfaction is only linear up to a certain point (Clark et al., 2018). Given that the paradigm from which a system arises has a high leverage on the system outcomes, it appears intuitive that there are important drivers of well-being, health and disease rooted in our current shared values (Meadows, 2008). It may be time for health professionals to engage in a broader conversation about transformative change.

5.4 Methodological Gaps

So far, many participatory methods rely on workshops and group facilitation. In order to operationalize participation at a larger scale, scalable tools must be developed. While these are available for example for smart cities, in the field of One Health this has not been developed. Furthermore, the call for transdisciplinarity would require multiple perspectives and the facilitation of interactions across many social boundaries.

While the skill set usually associated with public health, veterinary health or conservation relies strongly on natural science, it appears much more important to be equipped with skills unusual in these fields, such as nonviolent communication, philosophy of science, history of science, macroeconomics, systems thinking, designing thinking, dealing with scales, and (nonequilibrium) social sciences. Also, the importance of self-reflection can be stressed: dealing with ambiguity and uncertainty, and critiquing our own and others' paradigms.

While we have discussed the concerns about the prescriptive nature of legislation previously, market mechanisms (see Chapter 6) are also failing to provide public health, animal health and welfare, and environmental protection, as the latter are not restricted to tangible entities and not tradable. Impaired health and reduced resilience at all scales is often a result of cumulative behavior. The current socio-ecological context does not seem to provide the appropriate feedback and incentives for sustainable behavior. In the light of modern neuroscience and nonequilibrium social sciences, it appears to be an achievable target to reflect on the processes and features needed in a social-ecological system for all life to thrive. Solutions may be found in ecological economics, where concepts of degrowth, green growth and similar are discussed to provide alternatives to the prevailing increasing economic growth theory. Daly (2003) observed that beyond a certain point, growth is uneconomic and that multiple forms of ill health and the costs thereof can increase faster than wealth. Consequently, novel conceptualizations of growth and their measurement tools provide an opportunity for different narratives, research and strategies, and relate well with, and are an integral part of, ideas of transformative change and governance.

5.5 Early One Health Lessons from COVID-19

The COVID-19 pandemic, a singular disruptive event in recent human history, has required rapid, innovative, coordinated and collaborative approaches to manage and ameliorate its worst impacts. However, the threat remains, and learning from initial efforts may benefit the response management in the future. One Health approaches to managing health challenges through multistakeholder engagement need an enabling environment, for example in terms of available budgets or the instigation of integrative and inclusive processes. Häsler et al. (2020) described three case studies from state (New South Wales, Australia), national (Ireland) and international (sub-Saharan Africa) scales that illustrate different aspects of One Health in action in response to the COVID-19 pandemic. In Ireland, a One Health team was assembled to help design complex mathematical and resource models. In New South Wales, state authorities engaged collaboratively with veterinarians and epidemiologists to leverage disease outbreak knowledge, expertise and technical and support structures for application to the COVID-19 emergency. The African One Health University Network linked members from health institutions and universities from eight countries to provide a virtual platform for knowledge exchange on COVID-19 to support the response. Themes common to successful experiences included a shared resource base, interdisciplinary engagement, communication network strategies and a global perspective for addressing local needs.

The authors concluded that the COVID-19 pandemic showed the need for improvement of emerging infectious disease (EID) preparedness, early warning and prevention. The cost of unpreparedness is high, leading to high mortality rates and draconic measures like lockdowns. Early warning systems in support of more targeted and rapid responses need to be strengthened. Better/broader understanding of the consequences of human–environment interactions is also needed. Several key drivers for EID clearly came to the foreground: 1. Human population density, with degrading natural ecosystems associated with increased disease transmission risk. 2. Global travel and trade. 3. Excessive consumption: resulting in the aforementioned environmental degradation, which is a defining factor for facilitating pandemics and exacerbating the effects. Barriers for overcoming these challenges are largely structural in character, both institutional (governance) and socioeconomic (see Chapter 4).

Next to direct COVID-19 / One Health related challenges, some generic challenges are relevant to One Health operationalization. The need for better interdisciplinary and transdisciplinary collaborative arrangements is one of the core ambitions of One Health. Structural barriers for collaboration remain, including a lack of mutual understanding regarding the expertise of others, meaning that work continues in silos within rigid structures. Also, attitudinal barriers remain, such as lack of openness toward collaboration. "Old" governance challenges appear even more prominent: well-coordinated multilevel, integrative governance at local, regional, national and global levels remains a crisis management challenge. Current governance structures clearly showed deficiencies in adequate crisis management, including a general lack of preparedness and lack of coordination. A better balance between relevant governance issues is needed, including social issues.

Enhanced scientific capacity is needed; there is currently insufficient long-lasting research capacity in all sectors: animal health, human health, plant health and ecosystem health. This warrants increased mutual understanding and overcoming silos: There is lack of sufficient knowledge of the expertise of the others. We need open science: sharing instead of competing on crucial knowledge. The connection between science and policy is problematic: The science-policy interface was already struggling at the beginning of the COVID-19 crisis, when early warnings from scientists were not taken seriously.

Systemic health challenges, like COVID-19, need a systemic approach, such as Structural One Health. This requires an integrative perspective, overcoming barriers between disciplines, sectors and topical foci. This also requires a One Health funding framework, in order to provide sufficient resources. The COVID-19 crisis clearly revealed some systemic weaknesses, and may offer momentum for change. Finally, we notice the positive role and importance of nature for health during the COVID-19 pandemic and resulting confinement measures. The lockdown policies adopted in several countries, encouraging outdoor physical activity, highlighted the role of nature recreation facilities in the urban context for human health, and the challenge of accessibility for many urban households. In situations where visits to natural surroundings were still possible, an increase in visits was observed, as shown, for example, by a public survey during the first COVID-19 wave in Belgium (Lenaerts et al., 2021). People also reported a positive effect on human health and well-being. In situations where such visits were restricted, people looked forward to using parks and other natural areas, resulting in an increase in visits when allowed under lockdown restrictions. This highlights the need to account for social differences in options for contact with nature. The least deprived often live in single family dwellings with gardens and thus enjoy natural surroundings, even when confined to their homes. In preparation for future pandemics, policies should plan for socially equal access to natural surroundings (Slater et al., 2020), including for human health care workers, who during a pandemic have to perform their tasks under severe pressure. In return, the increased visiting intensity of natural spaces in high density areas also poses a threat to those very spaces, and the related health benefits, and requires attention in a sustainable governance context.

5.6 Conclusions

We see many opportunities for applying One Health to transformative biodiversity governance. The transformative governance ambitions (see Chapter 1) resonate quite well with the One Health ambitions and challenges presented in this chapter. A synthesis is presented in Table 5.1.

Table 5.1 *One Health transformative biodiversity governance potential*

Generic transformative governance challenges	One Health challenges			
	1. Practical implementation of One Health	2. Integration of animal, human, plant and ecosystem health	3. Integrated view on nature-related health risks and benefits	4. Integration of structural societal One Health drivers
A. Integrative	Combining different relevant ecosystem and health issues, sectors, and structural systemic drivers and outcomes			
B. Inclusive	Choosing how to deal with system complexity is inherently normative, which warrants the inclusion of societal deliberation next to scientific analysis			
C. Transdisciplinary	Combining different relevant forms of knowledge, stemming both from different scientific disciplines and different societal perspectives			
D. Adaptive	We cannot wait for perfect understanding or consensus; we need to take One Health to iterative implementation: learning by doing			
E. Anticipatory	Complexity, ongoing normative debate and development of insight need to be incorporated in analytical–deliberative transformative processes			

We discuss the specific elements of the table and how they are linked in further detail. The specific One Health aspects concern the following challenges: (1) *Practical implementation of One Health*. This still is considered a challenge, especially when taking into account the other aspects (expectations, demands) mentioned below. Initially, (2) *Integration of animal, human, plant and ecosystem health* was mainly considered as the core aim of One Health. As described in the chapter, there are still challenges in that respect. This very much relates to (3) *An integrated view on nature-related health risks and benefits*: traditionally One Health was mainly focused on health risks, taking potential health benefits of nature contact far less into account. Finally, (4) *Integration of structural societal One Health drivers*, or Structural One Health, which can be seen as a more critical, fundamental and preventative turn in the One Health debate, taking it beyond the development of vaccines and culling of "dangerous" animals.

One Health, like transformative change, deals with systemic challenges. Taking into account and structuring complexity and decision-making, and dealing with inherent uncertainties, unknowns and ambiguities, is therefore at the core. The process of how to deal with complexity, also from the scientific perspective, can be perceived as a social and normative process in itself. Complexity can never be fully grasped and should encourage us to choose what has to be taken into account for understanding and action. These choices have an important framing effect and are normative in nature, requiring a combined scientific and deliberative effort (Cilliers, 2005; Keune, 2012). In order not to stand still, we need to act wisely and deliberatively, in an adaptive learning-by-doing approach.

Collaboration is key to One Health to overcome silos. The implementation of One Health can benefit from transdisciplinary and iterative processes between policy, science and practice, and will enhance practical relevance of these collaborations (Hitziger et al., 2019). This also requires a collaborative attitude (soft skills) and a sharing attitude (open data, data sharing, integrated data base management).

In support of the above-mentioned One Health challenges, several elements of an enabling environment are to be considered. An important element is a dedicated network for professionals, practitioners and stakeholders. When the ambition of integration leads to the creation of large One Health institutions, this runs the risk of building fences rather than creating openness to (new) collaborations. This may be overcome by focusing on open, collaborative networks like Communities of Practice, which are less (institutionally) bound and more flexible, and are open to newcomers and new ideas and approaches (Keune et al., 2017). Such networks should not be limited to scientific experts, but also need to include policy experts, local knowledge holders, practitioners, grassroots organizations and all relevant stakeholders. The Network for EcoHealth and One Health (NEOH), the European chapter of EcoHealth International, is a good example, and so are other similar nature–health initiatives (Keune et al., 2019).

One Health approaches aim to overcome ad hoc reactive actions responding to emerging health challenges. It is better to develop proactive anticipatory governance capacity and preparedness, to allow us to better foresee health risks. The introduction of One Health concepts in primary, secondary and tertiary education, with the aim to raise awareness and create a natural understanding of systems and their interlinked nature, is important. Finally, the availability of sufficient financial and other resources for One Health science, policy and practice remains another crucial challenge. Current investment practices then have to put less focus on a purely economic rationale, and focus more on other rationales for society at large. A One Health funding framework could be supportive in allocation of funding, both in science, policy and practice.

References

Aerts, R., Honnay, O., and Van Nieuwenhuyse, A. (2018). Biodiversity and human health: Mechanisms and evidence of the positive health effects of diversity in nature and green spaces. *British Medical Bulletin* 127, 5–22.

Amonsin, A., Choatrakol, C., Lapkuntod, J., et al. (2008). Influenza virus (H5N1) in live bird markets and food markets, Thailand. *Emerging Infectious Diseases* 14, 1739e1742.

Anderson, Y., Taylor, G., Grant, C., Fulton, R., and Hofman, P. (2015). The green prescription active families programme in Taranaki, New Zealand 2007–2009: Did it reach children in need? *Journal of Primary Health Care* 7, 192–197. https://doi.org/10.1071/HC15192

Antonovsky, A. (1979). *Health, stress and coping*. San Francisco, CA: Jossey-Bass.

Assmuth, T., Chen, X., Degeling, C., et al. (2019). Integrative concepts and practices of health in transdisciplinary social ecology. *Socio-Ecological Practice Research* 2, 71–90. DOI: 10.1007/s42532-019–00038-y

Bauer, G., Kenneth, J., and Pelikan, J. (2006). The EUHPID health development model for the classification of public health indicators. *Health Promotion International* 21, 153–159. DOI: 10.1093/heapro/dak002

Bergmann, L., and Holmberg, M. (2016). Land in motion. *Annals of the American Association of Geographers* 106, 932–956. DOI: 10.1080/24694452.2016.1145537

Beute, F., Andreucci, M. B., Lammel, A., et al. (2020). Types and characteristics of urban and peri-urban green spaces having an impact on human mental health and wellbeing. EKLIPSE report. Available from https://bit.ly/3G1rnCP.

Bourque, T. (2017). One welfare. *The Canadian veterinary journal = La revue veterinaire canadienne* 58, 217–218.

Bryce, R., Irvine, K. N., Church, A., et al. (2016). Subjective well-being indicators for large-scale assessment of cultural ecosystem services. *Ecosystem Services* 21, 258–269. https://doi.org/10.1016/j.ecoser.2016.07.015

Bullock, C., Joyce, D., and Collier, M. (2018). An exploration of the relationships between cultural ecosystem services, socio-cultural values and well-being. *Ecosystem Services* 31, 142–152. https://doi.org/10.1016/j.ecoser.2018.02.020

Buttigieg, S. C., Savic, S., Cauchi, D., et al. (2018). Brucellosis control in Malta and Serbia: A One Health evaluation. *Frontiers in Veterinary Science* 5, 1–15. https://doi.org/10.3389/fvets.2018.00147

Carayannis, E. G., Barth, T. D., and Campbell, D. F. (2012). The quintuple helix innovation model: Global warming as a challenge and driver for innovation. *Journal of Innovation and Entrepreneurship* 1, 1–12. https://doi.org/10.1186/2192-5372-1-2

Cheng, X., Van Damme, S., Li, L., and Uyttenhove, P. (2019). Evaluation of cultural ecosystem services: A review of methods. *Ecosystem Services* 37, 1–10. https://doi.org/10.1016/j.ecoser.2019.100925

Cilliers, P. (2005). Complexity, deconstruction and relativism. *Theory, Culture and Society* 22, 255–267.

Clark, A. E., Flèche, S., Layard, R., Powdthavee, N., and Ward, G. (2018). *The origins of happiness: The science of well-being over the life cycle*. London: Princeton University Press.

Convention on Biological Diversity (CBD). (2018). Decision adopted by the Conference of the Parties to the Convention on Biological Diversity – 14/4. Health and biodiversity, CBD/COP/DEC/14/4, 30 November 2018, available from www.cbd.int/doc/decisions/cop-14/cop-14-dec-04-en.pdf.

Daly, H. E. (2003). The illth of nations and the fecklessness of policy: An ecological economist's perspective. *Post-Autistic Economics Review* 30, 1–23.

Dirzo, R., Young, H. S., Galetti, M., et al. (2014). Defaunation in the anthropocene. *Science* 345: 401–406.

Dunn, R. R., Davies, T. J., Harris, N. C., and Gavin, M. C. (2010). Global drivers of human pathogen richness and prevalence. *Proceedings of the Royal Society B: Biological Sciences* 277, 2587–2595. http://doi.org/10.1098/rspb.2010.0340

Engel, G. L. (1977). The need for a new medical model: A challenge for biomedicine. *Science* 196, 129–136.

Faust, C. L., Dobson, A. P., Gottdenker, N., et al. (2017). Null expectations for disease dynamics in shrinking habitat: Dilution or amplification? *Philosophical Transactions of the Royal Society B: Biological Sciences* 372, 20160173.

Fonseca, A. G., Torgal, J., de Meneghi, D., et al. (2018). One Health-ness Evaluation of Cysticercosis Surveillance Design in Portugal. *Frontiers in Public Health* 6, 1–10. https://doi.org/10.3389/fpubh.2018.00074

Gibb, R., Redding, D. W., Chin, K. Q., et al. (2020). Zoonotic host diversity increases in human-dominated ecosystems. *Nature* 584, 398–402.

Gilbert, M., Xiao, X., Chaitaweesub, P., et al. (2007). Avian influenza, domestic ducks and rice agriculture in Thailand. *Agriculture, Ecosystems & Environment* 119, 409e415.

Gilbert, M., Xiao, X., Pfeiffer, D. U., et al. (2008). Mapping H5N1 highly pathogenic avian influenza risk in Southeast Asia. *Proceedings of the National Academy of Sciences of the United States of America* 105, 4769e4774.

Haider, N., Rothman-Ostrow, P., Osman, A. Y., et al. (2020). COVID-19 – Zoonosis or emerging infectious disease? *Frontiers in Public Health* 8, 596944. DOI: 10.3389/fpubh.2020.596944

Hanin, M. C. E., Queenan, K., Savic, S., Rüegg, S. R., and Häsler, B. (2018). A One Health evaluation of the Southern African Centre for Infectious Disease Surveillance. *Frontiers in Veterinary Science* 5, 1–16. https://doi.org/10.3389/fvets.2018.00033

Häsler, B., Bazeyo, W., Byrne, A. W., et al. (2020). Reflecting on One Health in action during the COVID-19 response. *Frontiers in Veterinary Science* 7, 578649. DOI: 10.3389/fvets.2020.578649

Hitziger, M., Aragrande, M., Berezowski, J. A., et al. (2019). EVOLvINC: EValuating knOwLedge INtegration Capacity in multi-stakeholder governance. *Ecology and Society* 24. Available from www.ecologyandsociety.org/vol24/iss2/art36/.

Hoffmann, K., Ristl, R., George, A., Maier, M., and Pichlhöfer, O. (2019). The ecology of medical care: Access points to the health care system in Austria and other developed countries. *Scandinavian Journal of Primary Health Care* 37, 409–417. DOI: 10.1080/02813432.2019.1663593

Horwitz, P., Finlayson, M., and Weinstein, P. (2012). Healthy wetlands, healthy people: A review of wetlands and human health interactions. Ramsar Technical Report No. 6. Secretariat of the Ramsar Convention on Wetlands, Gland, Switzerland, and The World Health Organization, Geneva, Switzerland.

Huber, M., van Vliet, M., Giezenberg, M., et al. (2016). Towards a "patient-centred" operationalisation of the new dynamic concept of health: A mixed methods study. *BMJ Open* 6, e010091. https://doi.org/10.1136/bmjopen-2015-010091

IPBES (2018a), Appendix 2.8: Contributions to physical, mental and social dimensions of health section. Appendix to IPBES (2018): *The IPBES regional assessment report on biodiversity and ecosystem services for Europe and Central Asia*. Rounsevell, M., Fischer, M., Torre-Marin Rando, A. and Mader, A. (eds.). Secretariat of the Intergovernmental Science-Policy Platform on Biodiversity and Ecosystem Services, Bonn, Germany, https://www.ipbes.net/system/tdf/eca_ch_2_appendix_2.8_assessment_of_health.pdf?file=1&type=node&id=16593

IPBES. (2018b). Appendix 2.5: Provision of medicinal plant resources in Europe and Central Asia. Appendix to *The IPBES regional assessment report on biodiversity and ecosystem services for Europe and Central Asia*. M. Rounsevell, M. Fischer, A. Torre-Marin Rando and A. Mader (Eds.). Bonn, Germany: Secretariat of the Intergovernmental Science-Policy Platform on Biodiversity and Ecosystem Services. Available from https://bit.ly/3pWxdj3.

IPBES. (2020). Workshop report on biodiversity and pandemics of the Intergovern-mental Platform on Biodiversity and Ecosystem Services. Daszak, P., das Neves, C., Amuasi, J., et al. (Eds.). Bonn: IPBES Secretariat.
Jani, A., and Gray, M. (2019). Making social prescriptions mainstream. *Journal of the Royal Society of Medicine* 112, 459–461. https://doi.org/10.1177/0141076819848304
Jani, A., Pitini, E., Jungmann, S., Adamo, G., Conibear, J., Mistry, P., 2019. A social prescriptions formulary: bringing social prescribing on par with pharmaceutical prescribing. *Journal of the Royal Society of Medicine* 112, 498–502. https://doi.org/10.1177/0141076819877555
Johnson, C. K., Hitchens, P. L., Smiley Evans, T., et al. (2015). Spillover and pandemic properties of zoonotic viruses with high host plasticity. *Scientific Reports* 5, 14830. DOI: 10.1038/srep14830
Keune, H. (2012). Critical complexity in environmental health practice: Simplify and complexify. *Environmental Health* 11, 1–10.
Keune, H., and Assmuth, T. (2018). *Framing complexity in environmental and human health*. Oxford research encyclopedia of environmental science. Oxford University Press.
Keune, H., Flandroy, L., Thys, S., et al. (2017). The need for European OneHealth/EcoHealth networks. Archives of Public Health 75, 64.
Keune H., Friesenbichler K., Häsler B., et al. (2019). European nature and health network initiatives. In: Biodiversity and health in the face of climate change. M. Marselle, J. Stadler, H. Korn, et al. (Eds.), pp. 329–362. Cham, Springer. Available from https://bit.ly/34Ewta6.
Keune, H., Kretsch, C., De Blust, G., et al. (2013). Science–policy challenges for biodiversity, public health and urbanization: Examples, from Belgium. *Environmental Research Letters* 8, 025015.
Kilpatrick, A. M., Dobson, A. D. M., Levi, T., et al. (2017). Lyme disease ecology in a changing world: Consensus, uncertainty and critical gaps for improving control. *Philosophical Transactions of the Royal Society B: Biological Sciences* 372, 20160117.
Kincheloe, J. L. (2009). Critical complexity and participatory action research: Decolonizing "democratic" knowledge production. In: *Education, participatory action research, and social change: International perspectives*. D. Kapoor and S. Jordan (Eds.), pp. 107–121. New York: Palgrave Macmillan. https://doi.org/10.1057/9780230100640_8
Lainé, N., and Morand, S. (2020), Linking humans, their animals, and the environment again: A decolonized and more-than-human approach to "One Health." *Parasite* 27, 55. https://doi.org/10.1051/parasite/2020055
Lauwers, L., Bastiaens, H., Remmen, R., and Keune, H. (2020). Nature's contributions to human health: A missing link to primary health care? A scoping review of international overview reports and scientific evidence. *Frontiers in Public Health* 8, 52. doi: 10.3389/fpubh.2020.00052
Lebel, J. (2003). *Health: An ecosystem approach*. International Development Research Centre Canada. Available from https://bit.ly/34fYoO1.
Léger, A. L., Stärk, K., Rushton, J., and Nielsen, L. R. (2018). A One Health evaluation of the University of Copenhagen Research Centre for Control of Antibiotic Resistance. *Frontiers in Veterinary Science* 5, 1–14. https://doi.org/10.3389/FVETS.2018.00194
Lenaerts, A., Heyman, S., De Decker, A., et al. (2021). Vitamin nature: How coronavirus disease 2019 has highlighted factors contributing to the frequency of nature visits in Flanders, Belgium. *Frontiers in Public Health* 9, 646568. https://doi.org/10.3389/fpubh.2021.646568
Lindström, B., and Eriksson, M. (2005). Salutogenesis. *Journal of Epidemiology and Community Health* 59, 440–442. https://doi.org/10.1136/jech.2005.034777
Lindström, B., and Eriksson, M. (2006). Contextualizing salutogenesis and Antonovsky in public health development. *Health Promotion International* 21, 238–244. https://doi.org/10.1093/heapro/dal016
Marselle, M. R., Hartig, T., Cox, D. T. C., et al. (2021). Pathways linking biodiversity to human health: A conceptual framework. *Environment International* 150, 106420.
Meadows, D. H. (2008). *Thinking in systems – A primer*. White River Junction, VT: Chelsea Green Publishing Co.
Morand, S. (2018). Biodiversity and disease transmission. In: *The connections between ecology and infectious disease. Advances in environmental microbiology*, vol 5. C. Hurst (Ed.), pp. 39–56. Cham: Springer. https://doi.org/10.1007/978-3-319-92373-4_2

Morand, S., and Lajaunie, C. (2019). Linking biodiversity with health and wellbeing: Consequences of scientific pluralism for ethics, values and responsibilities. *Asian Bioethics Review* 11, 153–168. https://doi.org/10.1007/s41649-019-00076-4

Morandi, A., Tosto, C., Roberti di Sarsina, P., and Dalla, L. D. (2011). Salutogenesis and Ayurveda: Indications for public health management. *EPMA Journal* 2, 459–465.

Muñoz-Prieto, A., Nielsen, L. R., Martinez-Subiela, S., et al. (2018). Application of the NEOH framework for self-evaluation of One Health elements of a case-study on obesity in European dogs and dog-owners. *Frontiers in Veterinary Science* 5, 1–9.

Murray, K. A., and Daszak, P. (2013). Human ecology in pathogenic landscapes: Two hypotheses on how land use change drives viral emergence. *Current Opinion in Virology* 3, 79–83. https://doi.org/10.1016/j.coviro.2013.01.006

Mutsaers, I. (2015). One-health approach as counter-measure against "autoimmune" responses in biosecurity. *Social Science and Medicine* 129, 123–130.

Nilsson, M., Chisholm, E., Griggs, D., et al. (2018). Mapping interactions between the sustainable development goals: Lessons learned and ways forward. *Sustainability Science* 13, 1489–1503. https://doi.org/10.1007/s11625-018-0604-z

Ostfeld, R. S., and Keesing, F. (2017). Is biodiversity bad for your health? *Ecosphere* 8, e01676. DOI: 10.1002/ecs2.1676

Patel, A., Schofield, G. M., Kolt, G.S., et al. (2011). General practitioners' views and experiences of counselling for physical activity through the New Zealand Green Prescription program. *BMC Family Practice* 12, 119. https://doi.org/10.1186/1471-2296-12-119

Paternoster, G., Tomassone, L., Tamba, M., et al. (2017). The degree of One Health implementation in the West Nile virus integrated surveillance in Northern Italy, 2016. *Frontiers in Public Health* 5, 1–10. https://doi.org/10.3389/fpubh.2017.00236

Paul, M., Baritaux, V., Wongnarkpet, S., et al. (2013). Practices associated with Highly Pathogenic Avian Influenza spread in traditional poultry marketing chains: Social and economic perspectives. *Acta Tropica* 126, 43e53.

Payyappallimana, U. (2013). Health and well-being in Indian local health traditions. In: *An integrated view of health and well-being*. Cross-cultural advancements in positive psychology, vol 5. A. Morandi and A. Nambi (Eds.), pp. 99–112. Dordrecht: Springer.

Puska, P. (2006). Health in all policies. *European Journal of Public Health* 17, https://doi.org/10.1093/eurpub/ckm048

Radeski, M., O'Shea, H., De Meneghi, D., and Ilieski, V. (2018). Positioning animal welfare in the One health concept through evaluation of an animal welfare center in Skopje, Macedonia. *Frontiers in Veterinary Science* 4, 1–11. https://doi.org/10.3389/fvets.2017.00238

Rapport, D. (1998). Assessing ecosystem health. *Trends in Ecology & Evolution* 13, 397–402. https://doi.org/10.1016/S0169-5347(98)01449-9

Ring, I., Hansjürgens, B., Elmqvist, T., Wittmer, H., and Sukhdev, P. (2010). Challenges in framing the economics of ecosystems and biodiversity: the TEEB initiative. *Current Opinion in Environmental Sustainability* 2, 15–26. DOI: 10.1016/j.cosust.2010.03.005

Roiko, A., Kozak, S., Cleary, A., and Murray, Z. (2019). Managing the public health paradox: Benefits and risks associated with waterway use. In: *Moreton Bay Quandamooka and Catchment: Past, present, and future*. I. R. Tibbetts, P. C. Rothlisberg, D. T. Neil, et al. (Eds.). Brisbane: The Moreton Bay Foundation.

Rüegg, S. R., Häsler, B., and Zinsstag, J. (Eds.). (2018). *Integrated approaches to health: A handbook for the evaluation of One Health*. Wageningen: Wageningen Academic Publishers. Available from www.wageningenacademic.com/doi/book/10.3920/978-90-8686-875-9.

Rüegg, S. R., McMahon, B. J., Häsler, B., et al. (2017). A blueprint to evaluate One Health. *Frontiers in Public Health* 5, 20. DOI: 10.3389/fpubh.2017.00020

Schmidt, K. A., and Ostfeld, R. S. (2001). Biodiversity and the dilution effect in disease ecology. *Ecology* 82, 609–619.

Slater, S. J., Christiana, R. W., and Gustat, J. (2020). Recommendations for keeping parks and green space accessible for mental and physical health during COVID-19 and other pandemics. *Preventing Chronic Disease* 17, 200204. https://doi.org/10.5888/pcd17.200204

Ståhl, T. (2018). Health in all policies: From rhetoric to implementation and evaluation – the Finnish experience. *Scandinavian Journal of Public Health* 46, 38–46. https://doi.org/10.1177/1403494817743895

Strand, R., Rortveit, G., and Schei, E. (2004). Complex systems and human complexity in medicine. *Complexus* 2, 2–6. https://doi.org/10.1159/000087849

Sujatha, V. (2007). Pluralism in Indian medicine: Medical lore as a genre of medical knowledge. *Contributions to Indian Sociology* 41, 169–202. DOI: 10.1177/006996670704100202

Swinburn, B. A., Walter, L. G., Arroll, B., Tilyard, M. W., and Russell, D. G. (1997). Green prescriptions: Attitudes and perceptions of general practitioners towards prescribing exercise. *British Journal of General Practice* 47, 567–569.

United Nations. (2002). World Summit on Sustainable Development. A framework for action on health and the environment. WEHAB Working Group. Available from https://sustainabledevelopment.un.org/milesstones/wssd.

Van Boeckel, T. P., Thanapongtharm, W., Robinson, T., D'Aietti, L., and Gilbert, M. (2012). Predicting the distribution of intensive poultry farming in Thailand. *Agriculture, Ecosystems & Environment* 149, 144e153.

Walker, P., Cauchemez, S., Hartemink, N., Tiensin, T., and Ghani, A. C. (2012). Outbreaks of H5N1 in poultry in Thailand: The relative role of poultry production types in sustaining transmission and the impact of active surveillance in control. *Journal of the Royal Society Interface* 9. http://dx.doi.org/10.1098/rsif.2012.0022

Wallace, R. G., Bergmann, L., Kock, R., et al. (2015). The dawn of structural One Health: A new science tracking disease emergence along circuits of capital. *Social Science and Medicine* 129, 68–77.

West, G. B. (2012). The importance of quantitative systemic thinking in medicine. *Lancet* 379, 1551–1559. https://doi.org/10.1016/S0140-6736(12)60281-5

White, K. (1997). The ecology of medical care: Origins and implications for population-based healthcare research. Health Services Research 32, 11–21.

Whitmee, S., Mace, G. M., Haines, A., et al. (2015). Safeguarding human health in the Anthropocene epoch: report of The Rockefeller Foundation–Lancet Commission on planetary health. *The Lancet* 386, 1973–2028. https://doi.org/10.1016/S0140-6736(15)60901-1.

Woods, A., and Bresalier, M. (2014). One health, many histories. *Veterinary Record* 174, 650–654. https://doi.org/10.1136/vr.g3678

World Health Organization (WHO). (1978). Primary health care: Report of the International Conference on Primary Health Care, Alma-Ata, USSR, 6–12 September 1978 / jointly sponsored by the World Health Organization and the United Nations Children's Fund. Available from https://apps.who.int/iris/handle/10665/39228.

(1986). Ottawa charter for health promotion. Available from https://bit.ly/3GHuhMA.

(2005). *Millennium Ecosystem Assessment, 2005*. Ecosystems and Human Well-being: Synthesis. Washington, DC: Island Press. Available from www.millenniumassessment.org/documents/document.356.aspx.pdf.

(2006). Constitution of the World Health Organization. Available from www.who.int/governance/eb/who_constitution_en.pdf.

(2012). WHO Executive Board EB132/14. 132nd session, November 23, 2012. Provisional agenda item 7.3. Available from www.who.int/social_determinants/B_132_14-en.pdf?ua=1.

World Health Organization and Secretariat of the Convention on Biological Diversity (WHO-CBD). (2015). *Connecting global priorities: Biodiversity and human health – A state of knowledge review*. Geneva, Switzerland: World Health Organization.

World Values Survey Association. (n.d.). World Values Survey. Available from www.worldvaluessurvey.org/wvs.jsp.

Zinsstag, J., Schelling, E., Waltner-Toews, D., and Tanner, M. (2011). From "one medicine" to "one health" and systemic approaches to health and well-being. *Preventive Veterinary Medicine* 101, 148–156.

6

Biodiversity Finance and Transformative Governance: The Limitations of Innovative Financial Instruments

RICHARD VAN DER HOFF AND NOWELLA ANYANGO-VAN ZWIETEN

6.1 Introduction

The urgency to halt and reverse the alarming rates of biodiversity loss is grounded in the most comprehensive and up-to-date evidence (e.g. Dasgupta, 2021; Díaz et al., 2019) and has been translated into a forward-looking governance agenda for stimulating biodiversity conservation (CBD, 2020a; see Chapter 1 for a more detailed overview). Preparations for this Post-2020 Global Biodiversity Framework have centralized the issue of raising the financial resources necessary for promoting this agenda. This outlook has spurred a wealth of new publications in recent years that address the financial challenges for the foreseeable future (OECD, 2019; 2020; Tobin-de la Puente and Mitchell, 2021; Turnhout et al., 2021; UNDP, 2018; 2020). Although the new challenges raised by the COVID-19 pandemic have postponed the development of the Post-2020 framework (see Chapter 1), they have also kindled debates on a reconfiguration of the global economic system through a "green recovery" that potentially benefits biodiversity conservation (McElwee et al. 2020; Sandbrook et al. 2020). These developments underline that now is the right time for critically reflecting on how to maintain and enhance a biodiverse world.

Building primarily on a critical review of literature on biodiversity finance instruments, in this chapter we aim to take these reflections a step further by assessing the role of finance from the transformative biodiversity governance perspective adopted in this book. This perspective emphasizes the necessity of a transformative change to address the underlying drivers of biodiversity loss. To realize this change, this book argues that governance approaches must be integrative, inclusive, adaptive, transdisciplinary and anticipatory (see Chapter 1). We start by defining biodiversity finance, classifying the diversity of instruments that it encompasses and exploring the challenges that it seeks to address. This sets the stage for a critique of the fundamental premises of what we refer to as "innovative financial instruments" (see below) based on four interrelated questions that capture the five dimensions of transformative governance.

1. *How comprehensive is "financeable" biodiversity?* Biodiversity finance conceptualizes nature from an anthropocentric, mechanical and managerial perspective;
2. *Who values "financeable" biodiversity (and how)?* Although transformative governance requires a recognition of value pluralism, biodiversity finance instruments inherently transpose monetary values;

3. *How does biodiversity finance deal with uncertainty?* Biodiversity finance instruments frame biodiversity loss as a (manageable) material risk;
4. *How profound are the transformative changes fostered by biodiversity finance?* There are many ways in which biodiversity finance can foster integrative governance, but it does not challenge the systemic drivers of biodiversity loss.

Our critical reflection on biodiversity finance instruments and their role in a broader governance setting points to the strengths and weaknesses of these instruments, which are presented and discussed in the concluding section.

6.2 Key Developments in Biodiversity Finance

In this section, we provide our understanding of biodiversity finance, which serves as the basis for critique in the subsequent section. We start by arguing that despite the broad range of instruments, most biodiversity finance instruments have common roots in a "nature-as-natural-capital" view (see Sullivan, 2018). Subsequently, we discuss three interrelated arguments found in the literature that reflect the core challenges for biodiversity finance (see Anyango-van Zwieten, 2021). First, it is generally asserted that there is a "funding gap" for biodiversity conservation, which leads to the argument that financial instruments need upscaling. Second, one of the primary candidates for this upscaling is a greater involvement of the private sector and market-based instruments, as most biodiversity finance still comes from public sources. Third, key to leveraging or "unlocking" private finance for conservation are financial instruments built on the view of biodiversity loss as material risk (Dempsey, 2016). These three combined arguments are the primary target of our critical assessment in Section 6.3.

6.2.1 The Diversity of Biodiversity Finance

Biodiversity finance encompasses a diversity of instruments. A widely used definition provided by UNDP (2018: 6) describes biodiversity finance as "the practice of raising and managing capital and using financial and economic mechanisms to support sustainable biodiversity management" (see Tobin-de la Puente and Mitchell, 2021). Alternatively, the Organisation for Economic Co-operation and Development (OECD, 2020: 7) refers to biodiversity finance as any "expenditure that contributes – or intends to contribute – to the conservation, sustainable use and restoration of biodiversity." These definitions suggest a breadth of possibilities and require some sorting out. The lexicon offered by Pirard (2012) offers some clarity. It states, firstly, that not all economic instruments are markets, pointing to regulatory price signals (e.g. eco-taxes) or voluntary price signals (e.g. certification, labels, norms) that intervene in existing markets to correct for market failures. There is also the establishment and regulation of "direct markets" for products and services directly derived from biodiverse ecosystems, such as ecotourism, forest and fisheries products, and others. Finally, we group together three remaining categories – Pirard (2012) refers to these as "tradable permits" (e.g. carbon credits or fishing quotas), "reverse auctions" (e.g. payments for ecosystem services – PES) and "coasean-type agreements" (e.g. conservation easements or concessions) – that demand innovative ways of addressing biodiversity loss through processes of agreements, auctions or trade. Moreover, these categories encompass instruments that are highly heterogeneous with respect to the type of exchange and the involvement of public and/or private organizations (Koh et al., 2019; Pirard and Lapayre, 2014). This chapter primarily addresses this third heterogenous conglomerate of categories, also referred to as "innovative financial mechanisms" (Anyango-van Zwieten, 2021), which is distinct from other instruments that are premised on the stimulation or correction of existing social relations (i.e. direct markets and regulatory and voluntary price signals). They are innovative in the way in which they materialize specifically for biodiversity conservation in new hybrid forms of governance arrangements and represent new products and services, including through modifications to traditional mechanisms.

Although quite comprehensive, Pirard's (2012) lexicon does not encompass all biodiversity finance, as the role of the financial sector is becoming increasingly recognized in biodiversity conservation debates. Direct involvement of this sector was still incipient in the early 2010s. Early gray literature had already begun advocating for the pivotal role that the financial sector could play in stimulating biodiversity conservation (e.g. Huwyler et. al., 2014; IUCN, 2012), but estimates of the contribution by such instruments were still absent from key biodiversity finance publications (e.g. Parker et al., 2012). Fast-forward a decade and the financial sector becomes increasingly important for its potential to "unlock" private capital for biodiversity conservation (UNDP, 2020). According to Deutz et al. (2020), for example, green financial products like green bonds, green loans, equity funds and others account for US$3.8–6.3 billion (Table 6.1; see also Tobin-de la Puente and Mitchell, 2021). Green (or blue) bonds, of which biodiversity is a small share of the total green bonds market, offer the possibility of raising financial resources for green development projects and natural assets (e.g. marine protected areas and sustainable fisheries management in Seychelles) in exchange for a return to the investor after the contract period ends (Tobin-de la Puente and Mitchell, 2021). We distinguish between these biodiversity-related green financial products and other approaches that redirect existing investment flows without a clear link to biodiversity, such as "divesting," environmental, social and governance (ESG) criteria, positive and negative screening, or other norms and standards that guide investment portfolios away from unsustainable practices and sectors (e.g. the oil industry) and toward sustainable ones (Deutz et al., 2020).

Despite myriad differences, most gray literature produced in recent years indicates that the overarching purpose of these innovative financial instruments is to redirect socioeconomic practices through value or price signals in a way that benefits biodiversity conservation. The UNDP (2018: 6) states that biodiversity finance "is about leveraging and effectively managing economic incentives, policies, and capital to achieve the long-term well-being of nature and our society" (see also Tobin-de la Puente and Mitchell, 2021). Alternatively, Dasgupta (2021) suggests that "finance is an enabling asset that facilitates investments in capital assets [... and ...] plays a role in determining both the stock of natural capital and the extent of human demands on the biosphere" (p. 467). This means that a core function of finance is to "confer value to the three classes of capital goods [produced capital, human capital, natural capital] by facilitating their use" (p. 325). Moreover, Dasgupta argues that "the value of biodiversity is embedded in the accounting prices of natural capital" (p. 43). These conceptualizations suggest that the contribution of finance to biodiversity conservation is to value or price natural capital. This is the case even in the financial sector, where biodiversity loss may be viewed as a calculable material risk in terms of physical flows (Dempsey, 2016), corporate reputation or broader impacts (e.g. Deutz et al., 2020; DNP and PBL, 2020; see also Section 6.3.3). We therefore argue that the view of "nature-as-natural-capital" (Sullivan, 2018) forms the foundation for most innovative biodiversity finance mechanisms and, therefore, the critiques presented in this chapter are directly targeted at this view.

6.2.2 Principal Challenges for "Unlocking" Biodiversity Finance

Much biodiversity finance literature often proceeds from a compelling argument that, on the one hand, biodiversity conservation is economically important as many sectors rely on it, but, on the other hand, effective implementation of biodiversity conservation is costlier than is currently provided by financial instruments. The implementation of the CBD Strategic Plan for Biodiversity (2011–2020), for example, would incur annual costs of US$150–440 billion (UNDP, 2018). More recently, Deutz et al. (2020) have reported an annual funding need of US$722–967 billion by 2030 for the sustainable management of protected areas, landscapes and seascapes, and urban environments (see also Tobin-de la Puente and Mitchell, 2021). Such estimates have been used as the basis for estimating what is called the "funding gap."

Many studies that estimate the funding gap compare the funding needs discussed above with the financial resources spent on biodiversity conservation (see Table 6.1). Although an accurate comparison of these results needs to account for differences in definitions, methodologies, assumptions and epistemologies, they illustrate the general trends over time in emphasizing the funding gap. At the global level, for example, Parker et al. (2012) have estimated biodiversity finance resources to be US$50.8–52.7 billion in 2010, while Deutz et al. (2020) estimated this to be US$123.6–142.9 billion in 2019. More important than the apparent growth of available biodiversity finance over time, both studies report a funding gap of US$99.2–387.3 billion and US$598.4–824.1 billion, respectively. This funding gap problem plays out at lower levels of governance as well, particularly with respect to protected areas. The European Union Natura 2000 network of protected areas, for example, requires a total investment of €5.8 billion per year for its maintenance and ecological improvement (Kettunen et al., 2014), but the EU's advance budgetary allocation between 2007 and 2013 was only €0.6–1.2 billion per year (Kettunen et al., 2011). Likewise, lion conservation in protected areas in Africa receives US$0.4 billion annually despite indicating a need for US$1.2–2.4 billion (Lindsey et al., 2018), while the Brazilian protected areas had a funding deficit of nearly US$360 million for their management costs in 2016 (Silva et al., 2021). Notwithstanding the estimate variation or the scale of governance, the central argument remains the same: finance needs upscaling to address the funding gap.

Table 6.1 *Overview of global biodiversity finance sources and needs. Amounts are in billion US$ (categories are based on Deutz et al., 2020)*

	Category	Parker et al., 2012; UNDP, 2018	OECD, 2020	Tobin-de la Puente and Mitchell, 2021; Deutz et al., 2020
Reference year		2010	2015–2017	2019
Natural infrastructure[1]	Public	Unspecified	Unspecified	26.9
Domestic budgets and tax policy	Public	33.4	67.7	74.6–77.7
Official development aid	Public	6.3	3.9–9.1	4.0–9.7
Other public finance flows	Public		<0.1–0.9	Unspecified
Total public finance		**39.7**	**71.6–77.0**	**US$ 105.5–114.3**
Biodiversity offsets	Public-Private	2.5–4.1	2.6–7.3	6.3–9.2
Green financial products	Public-Private	Unspecified	Unspecified	3.8–6.3
Nature-based solutions and carbon markets	Public-Private	Unspecified	<0.1–0.1	0.8–1.4
Sustainable supply chains and commodities	Private	6.6	2.3–2.8	5.5–8.2
Philanthropy, conservation NGOs	Private	1.4–1.7	1.4–2.7	1.7–3.5
Other private finance flows	Private	0.4–0.5	0.2–0.9	Unspecified
Total private and hybrid finance		**10.9–12.9**	**6.6–13.6**	**18.1–28.6**
Total biodiversity finance		**50.8–52.7**	**78.2–90.6**	**123.6–142.9**
Total financing needs		150–440	Unspecified	722–967
Finance gap		**99.2–387.3**		**598.4–824.1**

[1] According to Deutz et al. (2020: 121), natural infrastructure involves "networks of land and water bodies that provide ecosystem services for human populations, which produce similar outcomes to implemented gray infrastructure."

In addition to the identification of a funding gap, the studies reported here identify another feature of biodiversity finance, which is that the bulk of this finance still comes from public sources. The comparisons in Table 6.1 demonstrate this clearly for global biodiversity finance, where contributions from public sources currently vary between 73.8 percent and 92.5 percent (percentages were based on the estimates reported by Deutz et al., 2020). Moreover, public finance for biodiversity conservation competes with other important goals. For instance, international funding through conservation NGOs is less than 1 percent of official development assistance (ODA) to Africa (Brockington and

Scholfield, 2010). While public finance alone is unlikely to be sufficient for closing the funding gap (Huwyler et. al., 2014), private finance has been slow in directing financial resources to biodiversity conservation. Between 2004 and 2015, most private investments were made in (more) sustainable food and fiber production (US$6.5 billion), so outside the innovative financial instruments that we are focusing on here. Investments in habitat conservation (US$1.3 billion) and water quality and quantity (US$0.4 billion) were much lower, although the latter was still backed by substantial public investments (US$21.5 billion between 2009 and 2015) (Hamrick, 2016).

To address this gap, most studies argue for "unlocking" private finance (e.g. UNDP, 2020). In this respect, many innovative financial mechanisms are targeted at enhancing private sector funding, increasing involvement of private capital and implementing market-based instruments (Anyango-van-Zwieten, 2021; Clark et al., 2018; EC, 2011; Gutman and Davidson, 2007; Miles, 2005; Pirard, 2012; Thiele and Gerber, 2017; UNDP, 2020). Similarly, stakeholders have started to build the "business case" for biodiversity conservation to attract private sector involvement by pointing out cost reduction, return-on-investment and risk mitigation motives, among others (IUCN, 2012; OECD, 2019). The UNDP and the European Commission, for example, launched the Biodiversity Finance Initiative (BIOFIN) in 2012 to seek new methodologies for "optimal" and "evidence-based" biodiversity finance plans and solutions (UNDP, 2018; 2020). The European Commission also launched its own EU Business @ Biodiversity Platform (B@B) in 2007. Arguably, the most promoted instruments for leveraging financial resources are deemed to be market-based, meaning that "biodiversity conservation [is] financed through and undertaken with the aim of generating profitable returns for their investors" (Dempsey and Suarez, 2016: 654). At the same time, such for-profit instruments still face challenges, including lack of scale (often the projects are too small), lack of financial track record, lack of so-called angel investors at the risky early-stage phase and poor project design without "investable, simple and understandable conservation asset classes" (Anyango-van Zwieten, 2020; Huwyler et al., 2014: 27). The task ahead, these publications assert, is to address these challenges and scale up private finance to close the funding gap.

6.2.3 Toward a Critical Assessment of Biodiversity Finance

Unlocking private finance has a broader and more important role in mainstreaming biodiversity in all socioeconomic sectors by closing a "different gap" between the current state-of-affairs and a transformative change thereof. In practice, this requires catalyzing more structural transformations of economic and financial systems because "all economic sectors need to contribute to conserving biodiversity and ecosystems and their sustainable management" (CBD, 2020b; Díaz et. al., 2019; UNDP, 2020: 12). In this context, the CBD's Post-2020 Global Biodiversity Framework was, at the time of writing this chapter, expected to incite new and additional financial resources, stimulate corporate sector accountability and establish more rigorous safeguards for private sector engagement (Ching and Lin, 2019). Greening finance, then, involves a broader transition of biodiversity governance into a "whole-of-society approach" (Van Oorschot et al., 2020) where existing biodiversity finance instruments catalyze this transition rather than merely addressing the "funding gap" for biodiversity conservation. The establishment of the Network for Greening the Financial System (NGFS) in 2017, for example, aims to "mobilize mainstream finance to support the transition toward a sustainable economy" (NGFS, 2020), promote the adoption of sustainable and responsible investment principles and address the environmental and societal impacts of the policy portfolios of central banks across the world (NGFS, 2019; see Section 6.3.3. for an example from Brazil). At the same time, this approach still faces substantial challenges, such as reshaping entrenched investment norms, risk definitions and investment practices in the financial sector (Crona et al., 2021).

Recognizing that the whole-of-society approach advocated by the Post-2020 Framework was still in the initial stages of development, the critical assessment of biodiversity finance presented in the remainder of this chapter focuses on the innovative financial instruments that aim to catalyze this approach. For purposes of clarity, we understand such instruments to encompass not only "tradable permits," "reverse auctions" and "coasean-type agreements," in Pirard's (2012) lexicon, but also new financial products like nature derivatives and weather insurances that mitigate the material risks of biodiversity loss (Anyango-van Zwieten, 2021). Our analysis thereby excludes price signals (e.g. US$274–542 of harmful subsidies, see Deutz et al., 2020), although we acknowledge their importance within the broader context of biodiversity finance. Furthermore, we acknowledge the intense controversies around the extent to which instruments like biodiversity offsetting, PES or nature derivatives are market-based, economic or financial, but at the same time argue that this variety of instruments share common ontological and epistemological foundations. Focusing on innovative financial instruments is therefore our attempt to capture this common ground.

6.3 Deconstructing Biodiversity Finance for Transformative Change

This section addresses the four central questions that are in line with the core purposes of this book, as presented in the introduction. It also critically discusses innovative financial instruments in light of the five dimensions of transformative governance (i.e. integrative, inclusive, adaptive, transdisciplinary and anticipatory; see Chapter 1 for full definitions). Based on this framework, we first deconstruct discussions in the literature and then summarize each subsection with our critique.

6.3.1 How Comprehensive Is "Financeable" Biodiversity?

All innovative finance instruments have a material basis for making transactions possible. Many instruments tie financial resources to objects like credits, rights, quotas, offsets and permits that in many ways give access to natural capital (e.g. Koh et al., 2019; May et al., 2015; van der Hoff and Rajão, 2020). This access to natural capital should be understood as its utilization either as a source of natural resources (e.g. permits to extract fish from

Antarctic waters) or as a sink for the wasteful byproducts of economic activity (e.g. credits for greenhouse gas emissions or Tradable Development Rights). Nonmarket instruments like results-based payments require a clear definition of the "results" or "performance" (e.g. emissions reductions) in relation to conservation objectives (Van der Hoff et al., 2019). In the financial sector, we encounter bonds, derivatives, securities, swaps, futures and insurances, among others, that facilitate investments in conservation (e.g. green bonds) or hedge against the risk of biodiversity loss (e.g. weather derivatives) (Bracking, 2012; Little et al., 2014; Ouma et al., 2018; Sullivan, 2018). For purposes of argumentation, we will refer to this material basis as "financeable objects."

Following Callon and Muniesa (2005: 1233–1234), these financeable objects are the outcome of processes of "objectification" and "singularization" of (parts of) biodiversity and by which financial transactions become possible. Objectification emphasizes the materiality of this object, which means that they have tangible and objective properties that characterize them as a "good" (e.g. rubber), "service" (e.g. pollination) or more abstract (financial) products like derivatives. These objects become financeable through "singularization," which "consists in a gradual definition of the properties of the product [or object], shaped in such a way that it can enter into the consumer's world and become attached to it." This means that the object can be assigned a value (see below) and appropriated by others. Take biodiversity offsets as an example (Koh et al., 2019): In most schemes, the biodiversity in areas with natural vegetation is assessed based on indicators of habitat type, species, threat level, richness, rarity, diversity and connectivity, among others. These indicators are then used to classify these areas and establish biodiversity offset credits. The number of credit types range from only one (e.g. the Rio Tinto QIT Madagascar Minerals [RTQMM] offsets) or two (e.g. species and ecosystem credits in the New South Wales Biodiversity Conservation Trust), to up to eight (wetland mitigation banking in the United States). These credits are the financeable objects of biodiversity offsetting that can be acquired by developers to compensate for their impact on nature. Even in cases where such exchange does not take place (say, results-based payments for REDD+), one may argue that financing parties may obtain other gains from the "investment," like satisfying domestic political constituencies (e.g. Angelsen [2017] calls this "political offsets").

The translation of biodiversity into "financeable objects" poses several challenges to transformative biodiversity governance because it denotes a very managerial approach to nature conservation. Sullivan (2017, 2018) calls this approach a "nature-as-natural-capital" view that is enacted through processes of commensuration (i.e. enhancing the comparability of nature), aggregation (i.e. a preference of total quantities over qualitative specificity) and capitalization (i.e. producing natural assets or, in this chapter, financeable nature). It embodies an ontological understanding of nature as mechanically composed of "gears and bolts" (Worster, 1994) or "rivets" (Dempsey, 2016) that, epistemologically, can be fully known and, more importantly, used and managed to meet human needs and preferences, thereby representing instrumental values (see Chapter 2). Although this ontological and epistemological view is enormously powerful (think about the ecosystem services concept), the downside is that it excludes a vast array of alternative ways of knowing and interacting with nature, which precludes possibilities for transdisciplinary governance. Although ecologists and economists have been working closely together on nature conservation issues since the 1980s, Dempsey (2016) argues that this collaboration leans more toward economic than ecological pragmatics. Many studies have lamented the ecologically reductionist conceptualizations of nature hidden in the "nature commodification" of PES schemes (e.g. Kosoy and Corbera, 2010; Wilson, 2013), the metrics of biodiversity offsetting (Marshall et al., 2020) and the methodology of biodiversity valuation (Farnsworth et al., 2015). Finally, such objects exclude alternative sources of intrinsic, spiritual and other forms of meaning (Laband, 2013) in order to only reflect the measurable and delineable properties of the financeable object.

Another problem with financeable objects is that they need to be rigid in order to become operational, which allows little space for adaptation. The market for Tradable Development Rights (TDRs) in Brazil, also called Environmental Reserve Quota (or *Cota de Reserva Ambiental* – CRA), is a case in point. Rural landowners in Brazil are obliged by law to conserve native vegetation on their properties (up to 80 percent in the Amazon), demanding restoration in case of a deficit and allowing deforestation in case of surplus. The CRA market offers an alternative option: Landowners with a surplus may issue and sell CRAs rather than deforest, while those with a deficit may acquire CRAs instead of restoring native vegetation (May et al., 2015). For over two decades of political development, this market has been subject to substantial expansions, one of which involves the geographical boundaries of trade (i.e. from trade within watershed to trade within biome and across states) (van der Hoff and Rajão, 2020). These expanded trade boundaries, the outcome of political pressure from the rural caucus, were challenged by a supreme court ruling that demanded a proof of similar "ecological identity" of properties engaged in a CRA exchange. Although this ruling is considered positive from a biodiversity conservation standpoint, it also poses significant challenges to ecologists to establish a workable indicator and thus slows down the operationalization of the market (Rajão et al., 2021).

Our critical assessment of the nature of financeable objects denotes an argument against the role of finance in transformative biodiversity governance. Such objects necessarily build on an economic conceptualization of nature that emphasizes its measurability, its manageability, its anthropocentrism and its instrumentalism. More importantly, this economism can potentially drown out other approaches to nature conservation, such as arguments for conserving pristine nature (Dempsey, 2016) or a harmonious relationship with nature that embeds local livelihoods (e.g. buen vivir, see Chapters 2, 8 and 9), which attests to poor inclusive governance. The difficulty (if not impossibility) of other ontologies and epistemologies to shape this financeable object also preclude the manifestation of a truly transdisciplinary governance. Moreover, this constrained transdisciplinarity limits possibilities for adaptive governance, as the CRA trade in Brazil exemplifies.

6.3.2 Whose Values Does "Financeable" Biodiversity Represent (and Whose Are Excluded)?

The process of singularization does not stop at defining the financeable object. According to Callon and Muniesa (2005: 1233), "the thing that 'holds together' [the financeable object] is a good if and only if its properties represent a value for the buyer." Applied to biodiversity finance, it suggests that financing biodiversity conservation occurs only if the destination (i.e. the financeable object) of these resources is considered to be valuable. Biodiversity indicators by themselves do not immediately prompt a mobilization of financial resources, but once they are packaged in, say, development rights or biodiversity offsets, they become valuable to potential financers. This value perception is fundamental. Results-based payments to the Brazilian Amazon Fund, for example, were based on demonstrated deforestation reductions in the Amazon region,[1] but its financers (mainly the Norwegian government) had slightly different criteria for "valuable" results than Brazil. Brazil held the belief that it deserved to be rewarded for past achievements (deforestation fell from nearly 30,000 km^2 in 2004 to less than 5,000 km^2 in 2012) and therefore maintained that annual results accumulate over time. By contrast, financers retained the preference for financing only the most recent results (e.g. Norway's payments in 2017 referred to results obtained in 2016). As deforestation rates went up in the 2010s, annual "results" significantly declined and financers were compelled to stop payments due to lack of "valuable results" (van der Hoff et al., 2018). In other words, the financeable object – be it an offset, a bond or a permit – needs to be perceived as valuable by the financer, otherwise financing is unlikely to take place.

Innovative financial instruments communicate the value to financiers in monetary terms. Section 6.2 already noted Dasgupta's (2021) conceptualization of biodiversity finance as a conveyor of biodiversity value through natural capital accounting prices. Economists claim that the previous inexistence of such prices was (and still is) the underlying problem of biodiversity loss. Pearce, Markandya and Barbier (1989: 5), for example, argued that when "something is provided at a zero price, more of it will be demanded than if there was a positive price." For landowners in the Brazilian Amazon, for example, standing forests have little value and legislation obliging them to conserve forests is perceived as an obstruction to land development (e.g. agriculture) and thus incurs high opportunity costs (Metzger et al., 2019; Stickler et al., 2013). Putting a price on these forests could change these perceptions. One of the main ideas behind the CRA market in Brazil, for example, was to allow landowners with vegetation beyond legal requirements to sell quota to those with deficits (in the final regularization, this was expanded to include PES as well) instead of legally clearing the land for, say, agricultural development (van der Hoff and Rajão, 2020; see also Section 6.2.1). Other finance instruments raise the costs of development projects (Koh et al., 2019) or risks related to biodiversity loss (Little et al., 2014). The value of

---

[1] Actual deforestation rates each year were compared to a ten-year average (baseline) that would actualize every five years. For instance, actual deforestation rates between 2011 and 2015 were compared to the baseline of 2001–2010. The difference between the baseline and actual deforestation rates would represent the "result" for which Brazil could receive REDD+ payments.

biodiversity reflected in these prices transposes the idea that using (or destroying) nature is no longer for free, but involves foregone opportunities or additional costs.

Prices, however, muddle the value of biodiversity in two ways. Firstly, the anthropocentrism implied in the type of biodiversity knowledge that forms the foundation of financeable objects (see Section 6.3.1), to which economists assign a "use value" and, subsequently, an exchange value. The ecosystem services concept is a notable reflection of these use values of biodiversity and there is currently a wealth of different tools to inform decision-makers (Grêt-Regamey et al., 2017; Martinez-Harms et al., 2015). According to critical scholars, however, this use value of biodiversity overemphasizes those aspects of nature that instrumentally benefit humankind, but downplays, excludes or even fails to perceive others that may be otherwise valuable. Economists have come a long way in identifying future use or non-use values (e.g. option, bequest and existence values; [see Tietenberg and Lewis, 2018]), but other uses of ecosystems that reflect cultural, aesthetic, spiritual and intrinsic values are extremely hard to express numerically (Small et al., 2017; see also Chapters 2, 8 and 9). Recognition of such value pluralism is not new, but has been advocated in predominantly noneconomist disciplines like anthropology (e.g. Graeber, 2001) and environmental ethics (e.g. Hourdequin, 2015) and has become an important theme in the critical discipline of ecological economics (Spash, 2017). Even Costanza et al. (2017), who famously and controversially valued the world's ecosystem services at US$16–54 trillion per year, acknowledge that the economic definition of value is too narrow as individuals are unable to appreciate or even perceive how some ecosystem services are valuable to them. The prevalence of use values in biodiversity finance (see Dempsey, 2016) is a far cry from this value pluralism, which attests to its constrained ability to promote transdisciplinary governance.

The second layer of problems with the prices of financeable objects refers to the repercussions of translating nature into use values and exchange values. Firstly, prices exacerbate the commensurability of inherently distinct dimensions of nature that are reflected in nonmonetary numeric assessments of biodiversity (Sullivan, 2017). Monetary valuation reduces "the problem of scarcity [of nature] into a problem of scarcity of capital, considered as an abstract category expressible in homogeneous monetary units" (Naredo, 2003: 250). Commensurate nature can thus be considered on a par with economically or technologically alternative actions. For instance, Brazilian landowners can choose their preferred course of action depending on their situation. Those with conservation deficits can choose between restoring degraded land or acquiring CRA, while those with vegetation beyond legal requirements can choose to legally clear it or sell CRA credits (May et al., 2015). Secondly, an emphasis on prices widens the gap between what innovative financial instruments define as valuable and the local perceptions and values of peoples on the ground. For example, the Brazilian Amazon Fund disburses financial resources to a myriad of projects that contribute to regional sustainability despite unclear contributions to emissions reductions (Correa et al., 2019), which become prejudiced as Brazil's basis for receiving donations is eroded (see above; Van der Hoff et al., 2018). Conversely, the introduction of monetary values for biodiversity through, say, PES initiatives may risk "crowding out" the intrinsic motivations of local people to conserve nature (Akers and

Yasué, 2019). Nonmonetary values thus become sidelined, while "valuable" development and conservation projects prevail (see also Laschefski and Zhouri, 2019; Villén-Pérez et al., 2018). These problems pose significant challenges for integrative and inclusive governance.

6.3.3 How Does Biodiversity Finance Deal with Uncertainty?

There are many similarities between the "nature-as-natural-capital" view and what Dempsey (2016) calls the "biodiversity loss as material risk" perspective. The central tenet is that biodiversity loss is a financial and economic risk that has (or will have) an impact on the bottom line. This is a fast-developing awareness: in 2010 biodiversity loss featured inconspicuously as "less prominent" in the World Economic Forum's (WEF) Global Risks Landscape report but dominated its global risks reports in 2021 (WEF, 2010; 2021). Two key responses to this growing awareness are that biodiversity loss needs to be managed as a business risk as well as treated as an opportunity for profit-making. The management and commodification of biodiversity risks have translated into new financial products including green bonds, rainforest bonds and climate bonds, biodiversity and nature derivatives, weather derivatives, catastrophe bonds and commodity index funds (Ouma et al., 2018; Sullivan, 2018). This calculative management of biodiversity risks is different from a precautionary approach that acknowledges the difficulty or impossibility of such calculations, preferring not to seek out the threshold of the "critical rivet" (Paul and Anne Ehrlich, cited in Dempsey, 2016). The agricultural sector, for example, may insure itself against unpredictable climate patterns like low precipitation, severe drought and destructive storms (e.g. Souza and Assunção, 2020), but cannot account for the full complexity of impending ecosystem "tipping points" to irreversibly transition to unfavorable landscapes (e.g. Lovejoy and Nobre, 2019). The calculative, managerial approach to uncertainty adopted by the financial sector, therefore, does not correspond with the precautionary definition of anticipatory governance.

In terms of inclusive and transdisciplinary governance, risk management instruments such as biodiversity derivatives, bonds and futures are designed to give preeminence to financial actors, their expertise and knowledge (Bracking, 2012). Though "spark[ing] the interest and imagination of investors" (Brockington, 2014: 123), these instruments are severed from actual conservation (Büscher, 2013). Take regional precipitation patterns as an example. Strand et al. (2018) estimate that a decreased capacity of Amazonian forests to provide this climate regulation service reduces rents and productivity for the soybean, beef and hydroelectricity sectors, incurring an average cost of US$1.81, US$5.43 and US$0.32 per hectare per year, respectively. Although understanding how these sectors negatively impact their own business through land clearing has the potential to raise awareness about the "real costs" of biodiversity loss, the challenge is to make these costs felt at the individual company level (see Dempsey, 2016). Rode et al. (2019: 7) found that the identification and valuation of ecosystem services does not readily attract investments, but "require[s] specific stakeholder processes and verification procedures" for this information to become part of these stakeholders' worlds (see also Callon and Muniesa, 2005).

Using the concepts of Sullivan (2018), investable nature requires not only its understanding as capital (qualification) in numeric or monetary terms (quantification), but also its subsequent "fabrication" into a "leverageable" asset class (materialization). Some risks become financeable objects (e.g. bonds, futures and other derivatives), while others become quantitative indicators that inform decision-making.

Not all uncertainties can readily become "calculated" risks and require substantial initial investment to catalyze private sector interest. In this respect, according to Christiansen (2021: 96), blended finance emphasizes the role of public finance "to pursue so-called 'crowding-in' of investments by either lowering [real or perceived] risks or increasing [anticipated] returns for private investments," especially during the initial "seed-stages" of conservation projects. Blended finance is the use of public and philanthropic funds to leverage private finance. Evaluating the Unlocking Forest Finance (UFF) project in Brazil and Peru, Rode et al. (2019: 7) emphasize that investor expectations and requirements do not "reflect the realities of the current scale, return and risk structures of sustainable landscape investments on the ground." These challenges, they argue, could be mitigated through the mobilization of blended finance that includes philanthropy to ensure direct conservation benefits or impact monitoring, NGOs to offer technical support for implementation, and governments to reduce risk of investment. Blended finance, then, may offer a "proof of concept" to build investor confidence in making sustainable investments (Christiansen, 2021). It is in these initial stages that learning – or adaptive governance – is most likely to take place (Rode et al., 2019). At the same time, the investor requirements related to financial returns and risk exposure tend to drown out other criteria for assembling the investment portfolio, at least in the case of sustainable agriculture. In catering to these requirements, blended finance adheres to the predominant investor milieu and thereby risks relinquishing aspects of inclusive (not all projects are financed) and transdisciplinary (not all criteria are weighed equally) governance.

In practice, businesses, farmers, investors and corporations perceive biodiversity losses as reputational or regulatory risks (Dempsey, 2016). With respect to the latter, for example, introducing sustainability performance as a condition for granting rural credit has great potential to prompt the immediate behavioral change of rural producers (e.g. Rode et al., 2019). In Brazil, the introduction of such sustainability criteria in 2008 by the Central Bank has had significant repercussions for its agricultural sector and contributed to the declining deforestation rates in the Brazilian Amazon at the time (Assunção et al., 2019). In this case, biodiversity loss comes at a price: restricted access to finance. This example underscores that consideration of biodiversity loss as a material risk by private sector organizations still requires strong encouragement through blended finance initiatives and strong governmental institutions. Moreover, it signals that economic efficiency continues to prevail even in the "triple bottom-line" over environmental protection and social equality (Christiansen, 2021). Despite its contribution to internalizing externalities, the "biodiversity loss as material risk" perspective still denotes a limited contribution to transformative governance.

6.3.4 How Profound Are the Transformative Changes Fostered by Biodiversity Finance?

Innovative financing instruments for biodiversity conservation commonly involve multi-actor networks. Firstly, they establish connections between the "users" and "providers" of biodiversity. Examples abound: the CRA market links landowners with vegetation beyond legal requirements to landowners with legal deficits (May et al., 2015; van der Hoff and Rajão, 2020); biodiversity offsetting ties potentially harmful development projects to conservation efforts (Koh et al., 2019); responsible investors can buy green bonds from organizations or governments that develop sustainable economic activities or strengthen conservation (for examples, see Deutz et al., 2020); and polluting countries make results-based payments to forested countries (Angelsen, 2017; van der Hoff et al., 2018). Secondly, the actor networks of innovative finance instruments often extend beyond "users" and "providers." Koh et al. (2019) make this abundantly clear with respect to biodiversity offsetting. In Germany, for example, municipal governments are responsible for matching the supply side (i.e. buying or leasing land for conservation) and the demand side (i.e. reviewing assessments of biodiversity losses at impact sites) of development impact compensation. Alternatively, wetland mitigation banking in the United States is a mandatory market arrangement under the Clean Water Act (1980) that potentially harmful development projects must adhere to. Koh et al. (2019) also argue that many biodiversity offsetting schemes include conservation NGOs (e.g. England, South Africa, Madagascar), consultancies (nearly all schemes evaluated), trust funds (e.g. Australia), and brokers (England, Australia, United States). Barton et al. (2017) have taken this argument a step further by describing Costa Rica's PES program as a policy mix that combines different actor types in different roles following specific rules ("rules-in-use") in order to attain conservation objectives (see also Ring and Barton, 2015). These examples suggest a potential of some biodiversity finance instruments to foster coordination among different actors toward biodiversity conservation objectives.

Some finance instruments also link conservation actions across governance levels. In the case of the Amazon Fund, the financial resources are passed on by the recipient (i.e. Amazon Fund) to projects that correspond with core categories of Brazilian environmental policies, most notably (1) monitoring and control, (2) land tenure and regularization and (3) sustainable economic activities. More importantly, the Amazon Fund, mediated by the Brazilian Development Bank, acts more like a mediator than a recipient. The transaction of financial resources from investors (e.g. the Norwegian government) to the Amazon Fund is not the final objective, since these resources are passed on to a plethora of other stakeholders across Brazil that comply with specific access requirements (e.g. project documentation). For example, this allowed the Amazon Fund to strengthen and empower protected areas with an investment of over US$66 million (Correa et al., 2019; Van der Hoff et al., 2018). Such an arrangement of transactions enacts what some REDD+ scholars have called a "nested approach," where individual projects are embedded in broader national and international governance networks (Angelsen et al., 2008). More recent efforts at integration aim to build an architecture for REDD+ transactions (ART) that demand upscaling efforts to national levels and subsuming lower-level performance (e.g. biome or states) within national accounting (see ART, 2021).

Despite the potential of innovative financial instruments to contribute to integrative governance through coordination (e.g. a "nested approach" to REDD+) and combination (e.g. PES policy mix) (see Chapter 1), some nuancing is appropriate here. Firstly, the very rules-in-use that enable such integration to take place also constrain the finance instruments that apply them. For instance, the Brazilian Amazon Fund distributes financial resources based on criteria that include organizational capacity to comply with its strict reporting demands, making it harder for finance to flow to smaller (but no less important) projects (Correa et al., 2019; van der Hoff et al., 2018). It must further be noted that these rules are politically negotiated. In Brazil's CRA market, smallholders may supply credits that represent all vegetation on their properties (even when they have a legal deficit), while uncompensated properties located inside protected areas (already protected by law) may supply credits representative of their legal surpluses (van der Hoff and Rajão, 2020).[2] The degree to which biodiversity finance instruments are inclusive depends to a large extent on how these rules-in-use are defined.

Another limitation, closely related to the former, is that there are limits to the degree of integration that innovative finance instruments can foster. Outcomes of PES programs, for example, challenge the characterization as a policy mix (see above) evidenced by contextual factors that are unaccounted for and that (positively or negatively) affect their performance. The Costa Rican government actively portrays its PES program as a market instrument, whereas in practice the program has been accepted by recipient farmers as a recognition of their stewardship, more than the prospect of being rewarded, which enhances the likelihood of positive outcomes (Chapman et al., 2020), and the PES program in Chiapas, Mexico, has faced substantial social conflict that threatens its continuity (Corbera et al., 2020). Mixed outcomes were also found for biodiversity offsets (Bidaud et al., 2017). Alternatively, deforestation rates in the Brazilian Amazon have been rising since 2012 despite increased disbursements from the Amazon Fund, which denotes that such instruments rarely operate in isolation and that conservation outcomes are just as much the result of the synergetic effects of factors like a hostile political climate (e.g. the Amazon Fund was extinguished in 2019) and broader commodity market developments. These examples illustrate that the outcomes of innovative financial instruments are affected by contextual factors that cannot be fully accounted for, which suggests that they themselves need to be integrated into a broader policy or governance mix.

Finally, and most importantly, biodiversity finance does not challenge the foundations of the capitalist system that is often argued to reinforce many of the known drivers of biodiversity loss (Díaz et al., 2019), because it reproduces the existing (skewed) power relations that this system builds on. The adoption of the CRA market in Brazil, for example, does not challenge the notion that, by federal constitution, private land needs to be used "productively" and could not prevent the "flexibilization" of nature conservation

---

[2] The problem with these latter supplier groups is that CRA credits will not add to the protection of its vegetation, because these lands are already legally prohibited from clearing this vegetation.

requirements via a new Forest Code in 2012 that mostly benefits dominant agribusiness interests (Rajão et al., 2021; Van der Hoff and Rajão, 2020). In addition, blended finance exacerbates global economic imbalances by giving preferential treatment to donors' own private sector firms and focusing on middle income countries (Pereira, 2017). These instruments typically aim to influence decision-making processes at the individual level (for example institutional investors) but do not challenge systemic or structural drivers of biodiversity loss. These perennial issues jeopardize the inclusive dimension of transformative governance. By insufficiently challenging the indirect drivers of biodiversity loss, moreover, they cannot be considered transformative as they do not correspond with the definition of transformative governance in Chapter 1, which states that addressing these indirect drivers is fundamental.

6.4 Conclusions and Ways Forward

The challenges for innovative financial instruments to support transformative biodiversity governance are substantial as they pose multiple limitations for transformative governance both in terms of its five dimensions and with respect to addressing the drivers of biodiversity loss. Starting with the dimensions (see Table 6.2), our analysis shows that while these instruments may foster integrative governance to some extent (see Section 6.3.4), they exacerbate the marginalization of local communities and values. In addition, the emphasis on financeable objects and monetary values promotes the biodiversity-as-natural-capital and biodiversity-loss-as-material-risk views that underpin the mobilization of financial resources. At the same time, these traits advance an ontological and epistemological understanding of biodiversity that is inherently narrow in terms of both its substance and its value, which undermines the inclusive and transdisciplinary dimensions of transformative governance. Other dimensions contain mixed considerations. With respect to adaptive governance, evidence in the reviewed literature indicates processes of learning taking place, although these mostly tend to occur in the initial stages of instrument development (see Section 6.3.3). In addition, the incorporation of biodiversity-related uncertainties into financial decisions, although in itself positive, follows a managerial and calculative approach that translates these into material risks. In terms of the five dimensions of transformative governance, therefore, innovative financial instruments must be approached cautiously and critically.

Some scholars have pointed to interesting measures for moving toward transformative governance. Kenter (2016) suggests that deliberative and participatory approaches to valuation could be an appropriate format for supplementing monetary approaches to valuing ecosystem services, which would improve the inclusive governance dimension. Participation and deliberation may also counterbalance the emphasis on anthropocentric, mechanistic and managerial approaches to nature conservation, building toward transdisciplinary governance. With respect to anticipatory governance, innovative financial instruments (and biodiversity finance in general) may consider what Chenet et al. (2021) refer to as "precautionary financial policy" to better deal with uncertainties that escape biodiversity risk assessments, thereby improving the anticipatory governance dimension. The limits to

Table 6.2 *Assessment summary for innovative financial instruments. Symbols refer to positive (+), negative (−) and mixed or neutral (*) assessments and reflect author interpretations*

Governance	Assessment	Evidence	Potential ways forward
Integrative	Mixed	(+) Potential for multiactor and multilevel governance (−) Capitalist foundations remain unchallenged	• "Whole-of-society approach" (Van Oorschot et al., 2020)
Inclusive	Negative	(−) Does not foster value pluralism (*) Rules-in-use govern and restrict participation	• Participation and deliberative valuation (Kenter, 2016)
Adaptive	Mixed	(*) Responsive to political pressure (−) Slow to adapt to new knowledge (*) Lessons learned during initial/pilot stages	• Biodiversity Finance Initiative (BIOFIN) (UNDP, 2018; 2020)
Transdisciplinary	Negative	(*) Anthropocentric ontology of nature (−) Mechanic epistemology of nature (*) Emphasis on capital and risk management	• Participation and deliberation
Anticipatory	Mixed	(+) Biodiversity risks mobilize financial resources (−) Uncertainties as manageable calculated risks	• Precautionary financial policy (Chenet et al., 2021)

strengthening integrative governance through innovative financial instruments underscores the importance of developing a "whole-of-society approach" (Van Oorschot et al., 2020). For improvements in the adaptive governance dimension, one may look to the BIOFIN as a platform for learning and feedback (UNDP, 2018; 2020).

It is doubtful, however, that such developments can shape up innovative financial instruments to manifest the transformative governance envisioned in this book. As this chapter has made abundantly clear, the prevailing logics of innovative financial instruments often fall short of the five dimensions discussed above. One may even argue that their proper functioning depends on clear definitions of "financeable objects," their monetary values and the rules-in-use that govern financial transactions. Moreover, they fail to address the deeper (capitalist) structures that indirectly drive biodiversity loss. In this respect, the new Forest Code in 2012 marked a turning point in Brazilian environmental politics that prompted rising deforestation rates, expanding agricultural production and exports, and dismantling of environmental political structures, among others, that neither the CRA market, the Amazon Fund, REDD+ or PES schemes were able to avoid (Rajão et al., 2021). To borrow loosely from IPBES' list of key indirect drivers of transformation (Balvanera et al., 2019), this underscores that we need to rethink the ways in which we conceive of and value nature; how we live, learn, move and appreciate one another; how we produce, consume and trade; and how we govern and confer rights and obligations. It calls for wider structural and systemic changes to our economies, societies and cultures where finance is a component of a broader system of transformative governance (see Chapter 4 on governance mixes). Biodiversity finance, even if optimally funded, is an iota in the world of global finance and trade that drive biodiversity loss, which means that a serious consideration of the ideas proposed throughout this book is warranted.

References

Akers, J., and Yasué, M. (2019). Motivational crowding in payments for ecosystem service schemes: A global systematic review. *Conservation & Society* 17, 377–389.

Angelsen, A. (2017). REDD+ as result-based aid: General lessons and bilateral agreements of Norway. *Review of Development Economics* 21, 237–264.

Angelsen, A., Streck, C., Peskett, L., Brown, J., and Luttrell, C. (2008). What is the right scale for REDD? The implications of national, subnational and nested approaches. *CIFOR Infobrief*. https://doi.org/10.17528/cifor/002595

Anyango-van Zwieten, N. (2020). Networks and flows of conservation finance: The case of World Wide Fund for Nature (WWF). Doctoral dissertation, Wageningen University.

(2021). Topical themes in biodiversity financing. *Journal of Integrative Environmental Sciences* 18, 19–35.

ART – Architecture for REDD+ Transactions (2021). Available from www.artredd.org.

Assunção, J., Gandour, C., Rocha, R., and Rocha, R. (2019). The effect of rural credit on deforestation: Evidence from the Brazilian Amazon. *The Economic Journal* 130, 290–330.

Balvanera, P., Pfaff, A., Vina, A., et al. (2019). Chapter 2.1. Status and trends – Drivers of change. In: *Global assessment report of the Intergovernmental Science-Policy Platform on Biodiversity and Ecosystem Services*. E. S.Brondízio, J. Settele, S. Díaz and H. T. Ngo (Eds.), pp. 49–200. Bonn: IPBES secretariat. DOI: 10.5281/zenodo.3831881

Barton, D. N., Benavides, K., Chacon-Cascante, A., et al. (2017). Payments for ecosystem services as a policy mix: Demonstrating the institutional analysis and development framework on conservation policy instruments. *Environmental Policy and Governance* 27, 404–421. DOI: 10.1002/eet.1769

Bidaud, C., Schreckenberg, K., Rabeharison, M., et al. (2017). The sweet and the bitter: Intertwined positive and negative social impacts of a biodiversity offset. *Conservation and Society* 15, 1–13.

Bracking, S. (2012). How do investors value environmental harm/care? Private equity funds, development finance institutions and the partial financialization of nature-based industries. *Development and Change* 43, 271–293.

Brockington, D. (2014). Celebrity spectacle, post-democratic politics, and Nature™ Inc. In: *Nature™ Inc: Environmental conservation in the neoliberal age*. B. Büscher, W. Dressler and R. Fletcher (Eds.), pp. 108–126. Tempe: University of Arizona Press.

Brockington, D., and Scholfield, K. (2010). The work of conservation organisations in sub-Saharan Africa. *The Journal of Modern African Studies* 48, 1–33.

Büscher, B. (2013). *Transforming the frontier: Peace parks and the politics of neoliberal conservation in Southern Africa*. Durham, NC: Duke University Press.

Callon, M., and Muniesa, F. (2005). Peripheral vision: Economic markets as calculative devices. *Organizaton Studies* 28, 1229–1250. DOI: 10.1177/0170840605056393

CBD. (2020a). *Global Biodiversity Outlook 5*. Montreal:Secretariat of the Convention on Biological Diversity.

(2020b). *Report on the Thematic Workshop on Resource Mobilization for the Post-2020 Global Biodiversity Framework*. Workshop held in Berlin, 14–16 January 2019. CBD/POST2020/WS/2020/3/3, February 12, 2020.

Chapman, M., Satterfield, T., Wittman, H., and Chan, K. M. A. (2020). A payment by any other name: Is Costa Rica's PES a payment for services or a support for stewards? *World Development* 129, 104900. https://doi.org/10.1016/j.worlddev.2020.104900

Chenet, H., Ryan-Collins, J., and van Lerven, F. (2021). Finance, climate-change and radical uncertainty: Towards a precautionary approach to financial policy. *Ecological Economics* 183, 106957.

Ching, L. L., and Lin, L. L. (2019). Cornerstones of the post-2020 biodiversity framework. In: *Spotlight on sustainable development 2019:Reshaping governance for sustainability: Transforming institutions – shifting power – strengthening rights*. B. Adams, C. A. Billorou, R. Bissio, et al. (Eds.), pp. 174–179. Beirut; Bonn; Ferney-Voltaire; Montevideo; New York; Penang; Rome and Suva: Social Watch, Global Policy Forum, Development Alternatives with Women for a New Era, Public Services International, Third World Network, Arab NGO Network for Development, Society for International Development, Center for Economic and Social Rights. Available from https://bit.ly/3GseocN

Christiansen, J. (2021). Fixing fictions through blended finance: The entrepreneurial ensemble and risk interpretation in the Blue Economy. *Geoforum* 120, 93–102.

Clark, R., Reed, J., and Sunderland, T. (2018). Bridging funding gaps for climate and sustainable development: Pitfalls, progress and potential of private finance. *Land Use Policy* 71, 335–346.

Corbera, E., Costedoat, S., Ezzine-de-Blas, D., and Van Hecken, G. (2020). Troubled encounters: Payments for ecosystem services in Chiapas, Mexico. *Development and Change* 51, 167–195. DOI:10.1111/dech.12540

Correa, J., van der Hoff, R., and Rajão, R. (2019). Amazon fund 10 years later: Lessons from the world's largest REDD+ program. *Forests* 10, 272. https://doi.org/10.3390/f10030272

Costanza, R., de Groot, R., Braat, L., et al. (2017). Twenty years of ecosystem services: How far have we come and how far do we still need to go? *Ecosystem Services* 28, 1–16. DOI: https://doi.org/10.1016/j.ecoser.2017.09.008

Crona, B., Eriksson, K., Lerpold, L., et al. (2021). Transforming toward sustainability through financial markets: Four challenges and how to turn them into opportunities. *One Earth* 4, 599–601. DOI: 10.1016/j.oneear.2021.04.021

Dasgupta, P. (2021). *The economics of biodiversity: The Dasgupta review*. London: HM Treasury.

Dempsey J. (2016). *Enterprising nature: Economics, markets, and finance in global biodiversity politics*. Chichester: John Wiley & Sons.

Dempsey, J., and Suarez, D. C. (2016). Arrested development? The promises and paradoxes of "selling nature to save it." *Annals of the American Association of Geographers* 106, 653–671. DOI: 10.1080/24694452.2016.1140018

Deutz, A., Heal, G. M., Niu, R., et al. (2020). *Financing nature: Closing the global biodiversity financing gap*. The Paulson Institute, The Nature Conservancy, and the Cornell Atkinson Center for Sustainability. Available from https://bit.ly/3r1rWpJ.

Díaz, S., Settele, J., Brondízio, E., et al. (2019). *Summary for policymakers of the global assessment report on biodiversity and ecosystem services of the Intergovernmental Science-Policy Platform on Biodiversity and Ecosystem Services*. Bonn: IPBES Secretariat.

DNP and PBL. (2020). *Indebted to nature: Exploring biodiversity risks for the Dutch financial sector*. June 2020 Report of De Nederlandsche Bank and Planbureau voor de Leefomgeving. Available from www.dnb.nl/media/4c3fqawd/indebted-to-nature.pdf

EC. (2011). Our life insurance, our natural capital: An EU biodiversity strategy to 2020. COM (2011) 244. Brussels: European Commission. Available from: https://bit.ly/33cmYhI.

Farnsworth, K. D., Adenuga, A. H., and de Groot, R. S. (2015). The complexity of biodiversity: A biological perspective on economic valuation. *Ecological Economics* 120, 350–354. https://doi.org/10.1016/j.ecolecon.2015.10.003

Graeber, D. (2001). *Toward an anthropological theory of value: The false coin of our own dreams.* New York: Palgrave.

Grêt-Regamey, A., Sirén, E., Brunner, S. H., and Weibel, B. (2017). Review of decision support tools to operationalize the ecosystem services concept. *Ecosystem Services* 26, 306–315.

Gutman, P., and Davidson, S. (2007). *A review of innovative international financial mechanisms for biodiversity conservation with a special focus on the international financing of developing countries' protected areas.* Washington, DC:WWF-MPO.

Hamrick, K. (2016). *State of private investment in conservation 2016: A landscape assessment of an emerging market.* Ecosystem Marketplace. Available from https://bit.ly/3s0JFOJ

Hourdequin, M. (2015). *Environmental ethics: From theory to practice.* London: Bloomsbury Academic.

Huwyler, F., Käppeli, J., Serafimova, K., Swanson, E., and Tobin, J. (2014). *Conservation finance: Moving beyond donor funding toward an investor-driven approach.* Zurich:Credit Suisse, WWF, McKinsey & Company.

IUCN. (2012). Business engagement strategy. IUCN, April 2012. Available from https://bit.ly/3f7ENkB.

Kenter, J. O. (2016). Integrating deliberative monetary valuation, systems modelling and participatory mapping to assess shared values of ecosystem services. *Ecosystem Services* 21, 291–307. https://doi.org/10.1016/j.ecoser.2016.06.010

Kettunen, M., Baldock, D., Gantioler, S., et al. (2011). *Assessment of the Natura 2000 co-financing arrangements of the EU financing instrument, final report.* Brussels: Institute for European Environmental Policy (IEEP).

Kettunen, M., Torkler, P., and Rayment, M. (2014). *Financing Natura 2000 Guidance Handbook. Part I – EU funding opportunities in 2014–2020.* Luxembourg: European Commission DG Environment.

Koh, N. S., Hahn, T., and Boonstra, W. J. (2019). How much of a market is involved in a biodiversity offset? A typology of biodiversity offset policies. *Journal of Environmental Management* 232, 679–691. https://doi.org/10.1016/j.jenvman.2018.11.080

Kosoy, N., and Corbera, E. (2010). Payments for ecosystem services as commodity fetishism. *Ecological Economics* 69, 1228–1236. http://dx.doi.org/10.1016/j.ecolecon.2009.11.002

Laband, D. N. (2013). The neglected stepchildren of forest-based ecosystem services: Cultural, spiritual, and aesthetic values. *Forest Policy and Economics* 35, 39–44.

Laschefski, K., and Zhouri, A. (2019). Indigenous peoples, traditional communities and the environment: The "territorial question" under the new developmentalist agenda in Brazil. In *The Brazilian left in the 21st century: Conflict and conciliation in peripheral capitalism.* V. Puzone and L. F. Miguel (Eds.), pp. 205–236. Cham: Springer International Publishing.

Lindsey, P. A., Miller, J. R. B., Petracca, L. S., et al. (2018). More than $1 billion needed annually to secure Africa's protected areas with lions. *Proceedings of the National Academy of Sciences* 115, E10788–E10796. DOI: 10.1073/pnas.1805048115

Little, L. R., Parslow, J., Fay, G., et al. (2014). Environmental derivatives, risk analysis, and conservation management. *Conservation Letters* 7, 196–207.

Lovejoy, T. E., and Nobre, C. (2019). Amazon tipping point: Last chance for action. *Science Advances* 5, eaba2949.

Marshall, E., Wintle, B. A., Southwell, D., and Kujala, H. (2020). What are we measuring? A review of metrics used to describe biodiversity in offsets exchanges. *Biological Conservation* 241, 108250. https://doi.org/10.1016/j.biocon.2019.108250

Martinez-Harms, M. J., Bryan, B. A., Balvanera, P., et al. (2015). Making decisions for managing ecosystem services. *Biological Conservation* 184, 229–238.

May, P., Bernasconi, P., Wunder, S., and Lubowski, R. (2015). *Environmental reserve quotas in Brazil's new forest legislation: An ex ante appraisal.* Bogor, Indonesia: CIFOR.

McElwee, P., Turnout, E., Chiroleu-Assouline, M., et al. (2020). Ensuring a post-COVID economic agenda tackles global biodiversity loss. *One Earth* 3, 448–461.

Metzger, J. P., Bustamante, M. M. C., Ferreira, J., et al. (2019). Why Brazil needs its legal reserves. *Perspectives in Ecology and Conservation* 17, 91–103. https://doi.org/10.1016/j.pecon.2019.07.002

Miles, K. (2005). Innovative financing: Filling in the gaps on the road to sustainable environmental funding. *Review of European Community & International Environmental Law* 14, 202–211.

Naredo, J. M. (2003). *La economía en evolución: historia y perspectivas de las categorías básicas del pensamiento económico*. Madrid:Siglo XXI de España.

NGFS. (2019). *A sustainable and responsible investment guide for central banks' portfolio management*. Technical document, October 2019. Network for Greening the Financial System, NGFS Secretariat/Banque de France. Available from https://bit.ly/3AIMwzP

(2020). Website of the Network for Greening the Financial System. Available from www.ngfs.net/en.

OECD. (2019). Biodiversity: Finance and the economic and business case for action. Report prepared for the G7 Environment Ministers' Meeting, 5–6 May 2019. Available from https://bit.ly/3KXvxys

(2020). *A comprehensive overview of global biodiversity finance*. Final report, April, 2020. Organisation for Economic Cooperation and Development: OECD Publishing. Available from https://bit.ly/3rf72Vs

Ouma, S., Johnson, L., and Bigger, P. (2018). Rethinking the financialization of "nature." *Environment and Planning A: Economy and Space*. 50, 500–511.

Parker, C., Cranford, M., Oakes, N., and Leggett, M. (2012). *The little biodiversity finance book. A guide to proactive investment in natural capital (PINC)*. Third Edition. Oxford: The Global Canopy Programme. Available from https://bit.ly/3ujCDrb

Pearce, D., Markandya, A., and Barbier, E. (1989). *Blueprint for a green economy*. London: Earthscan Publications Limited.

Pereira, J. (2017). *Blended finance: What it is, how it works and how it is used*. The Hague: Oxfam Novib.

Pirard, R. (2012). Market-based instruments for biodiversity and ecosystem services: A lexicon. *Environmental Science & Policy* 19, 59–68.

Pirard, R., and Lapeyre, R. (2014). Classifying market-based instruments for ecosystem services: A guide to the literature jungle. *Ecosystem Services* 9, 106–114.

Rajão, R., Del Giudice, R., van der Hoff, R., and de Carvalho, E. B. (2021). *Uma Breve História da Legislação Florestal Brasileira*. Rio de Janeiro: OCF.

Ring, I., and Barton, D. N. (2015). Economic instruments in policy mixes for biodiversity conservation and ecosystem governance. In: *Handbook of Ecological Economics*. J. Martínez-Alier and R. Muradian (Eds.), pp. 413–449. Northampton: Edward Elgar Publishing.

Rode, J., Pinzon, A., Stabile, M. C., et al. (2019). Why "blended finance" could help transitions to sustainable landscapes: Lessons from the Unlocking Forest Finance project. *Ecosystem Services* 37, 100917. https://doi.org/10.1016/j.ecoser.2019.100917

Sandbrook, C., Gómez-Baggethun, E., and Adams, W. M. (2020). Biodiversity conservation in a post-COVID-19 economy. *Oryx* 1–7. DOI: 10.1017/S0030605320001039

Silva, J. M. C. da, Dias, T. C. A. de Castro, Cunha, A. C. da, and Cunha, H. F. A. (2021). Funding deficits of protected areas in Brazil. *Land Use Policy* 100, 104926. https://doi.org/10.1016/j.landusepol.2020.104926

Small, N., Munday, M., and Durance, I. (2017). The challenge of valuing ecosystem services that have no material benefits. *Global Environmental Change* 44, 57–67. https://doi.org/10.1016/j.gloenvcha.2017.03.005

Souza, P., and Assunção, J. (2020). *Risk management in Brazilian agriculture: Instruments, public policy, and perspectives*. Rio de Janeiro: Climate Policy Initiative.

Spash, C. L. (Ed.). (2017). *Routledge handbook of ecological economics: Nature and society*. New York: Routledge.

Stickler, C. M., Nepstad, D. C., Azevedo, A. A., and McGrath, D. G. (2013). Defending public interests in private lands: Compliance, costs and potential environmental consequences of the Brazilian Forest Code in Mato Grosso. *Philosophical Transactions of the Royal Society B: Biological Sciences* 368, 20120160. DOI: 10.1098/rstb.2012.0160

Strand, J., Soares-Filho, B., Costa, M. H., et al. (2018). Spatially explicit valuation of the Brazilian Amazon Forest's Ecosystem Services. *Nature Sustainability* 1, 657–664. DOI: 10.1038/s41893-018-0175-0

Sullivan, S. (2017). Noting some effects of fabricating "nature" as "natural capital." *The Ecological Citizen* 1, 65–73.

——— (2018). Making nature investable: From legibility to leverageability in fabricating "nature" as "natural capital." *Science and Technology Studies* 31, 47–76.

Thiele, T., and Gerber, L. R. (2017). Innovative financing for the high seas. *Aquatic Conservation: Marine and Freshwater Ecosystems* 27, 89–99.

Tietenberg, T., and lewis, L. (2018). *Environmental and natural resource economics*. Boston, MA: Pearson Education Inc.

Tobin-de la Puente, J., and Mitchell, A. W. (Eds.). (2021). *The little book of investing in nature*. Oxford: Global Canopy.

Turnhout, E., McElwee, P., Chiroleu-Assouline, M., et al. (2021). Enabling transformative economic change in the post-2020 biodiversity agenda. *Conservation Letters* 14, e12805.

UNDP. (2018). *The BIOFIN workbook 2018: Finance for nature*. New York: The Biodiversity Finance Initiative. United Nations Development Programme.

——— (2020). *Moving mountains: Unlocking private capital for biodiversity and ecosystems*. New York: United Nations Development Programme.

Van der Hoff, R., and Rajão, R. (2020). The politics of environmental market instruments: Coalition building and knowledge filtering in the regulation of forest certificates trading in Brazil. *Land Use Policy* 96, 104666. https://doi.org/10.1016/j.landusepol.2020.104666

Van der Hoff, R., Rajão, R., and Leroy, P. (2018). Clashing interpretations of REDD+ "results" in the Amazon Fund. *Climatic Change* 150, 433–445. DOI: 10.1007/s10584-018-2288-x

Van der Hoff, R., Rajão, R., and Leroy, P. (2019). Can REDD+ still become a market? Ruptured dependencies and market logics for emission reductions in Brazil. *Ecological Economics* 161, 121–129. https://doi.org/10.1016/j.ecolecon.2019.03.011

Van Oorschot, M. M. P., Kok, M. T. J., and Van Tulder, R. (2020). *Business for biodiversity. Mobilising business towards net positive impact*. The Hague: PBL Netherlands Environmental Assessment Agency.

Villén-Pérez, S., Mendes, P., Nóbrega, C., Gomes Córtes, L., and De Marco, P. (2018). Mining code changes undermine biodiversity conservation in Brazil. *Environmental Conservation* 45, 96–99. DOI: 10.1017/S0376892917000376

WEF. (2010). *Global risks 2010*. Cologny; Geneva: World Economic Forum Report.

——— (2021). *The global risks report 2021*. 16th Edition. Cologny; Geneva: World Economic Forum Report.

Wilson, M. (2013). The green economy: The dangerous path of nature commoditization. *Consilience* 10, 85–98. DOI: 10.2307/26476140

Worster, D. (1994). *Nature's economy: A history of ecological ideas*. Cambridge: Cambridge University Press.

7

Emerging Technologies in Biodiversity Governance: Gaps and Opportunities for Transformative Governance

FLORIAN RABITZ, JESSE L. REYNOLDS AND ELSA TSIOUMANI

7.1 Introduction

Emerging technologies potentially have far-reaching impacts on the conservation, as well as the sustainable and equitable use, of biodiversity. Simultaneously, biodiversity itself increasingly serves as an input or source material for novel technological applications. In this chapter, we assess the relationship between the regime of the Convention on Biological Diversity (CBD, or "the Convention") and the governance of three sets of emerging technologies: geoengineering, synthetic biology and gene drives, as well as bioinformatics. The linkages between biodiversity and technology go beyond these cases, with, for example, geographic information systems, satellite imagery or possibly even blockchain technology playing potentially important roles for implementing the CBD's objectives. Here, however, we focus on technologies that have been subject to extensive debate and rulemaking activity under the CBD.

First, geoengineering, that is, the "deliberate intervention[s] in the planetary environment of a nature and scale intended to counteract anthropogenic climate change and its impacts" (Williamson and Bodle, 2016: 8), includes both carbon dioxide removal and solar radiation management (or modification) techniques. Geoengineering techniques could mitigate climate change and its impacts on biodiversity but could also cause harmful effects. Assessing these benefits and risks is complicated by great uncertainty as well as normative and political contestation. Second, synthetic biology applications, including so-called gene drives, fall within the scope of biotechnology as defined by the CBD: "any technological application that uses biological systems, living organisms, or derivatives thereof, to make or modify products or processes for specific use" (CBD, Art. 2). Such applications may have positive impacts on the conservation and sustainable use of biodiversity (and, possibly, the fair and equitable sharing of benefits arising out the utilization of genetic resources); yet they also imply diverse and potentially severe biosafety risks, as well as possibly problematic socioeconomic impacts (SCBD, 2015: 39–40). Third, bioinformatics allows for the extraction of digital sequence information (DSI), that is, the genetic information that is

---

Florian Rabitz's contribution has been supported by Lithuanian Research Council grant no. P-MIP-19–513, "Institutional Adaptation to Technological Change (ADAPT)".

Jesse Reynolds thanks Open Philanthropy for its support of this work.

Elsa Tsioumani has received funding from the European Union's Horizon 2020 research and innovation program under the Marie Sklodowska-Curie grant agreement no. 101029634 (SynBioGov).

derived from genetic resources. DSI is increasingly used in basic and applied research, replacing the need for access to "physical" genetic resources. While DSI has the potential to facilitate research on genetic resources, its use poses challenges with regard to the CBD's objective of fair and equitable benefit-sharing (Tsioumani, 2020: 24).

The Convention facilitates political, technical and scientific deliberation on biodiversity-related technologies and partially provides for their regulation. This takes place through technical guidance, legally binding international rules under the Convention and its protocols, as well as different layers of governing body decisions. These two general functions are essential to implementing the CBD's objectives. Regarding facilitating deliberation and cooperation, the Convention created a standing Subsidiary Body on Scientific, Technical and Technological Advice (SBSTTA) to assist the Conference of the Parties (COP). The Convention also provides for access to and transfer of technology (Art. 16), exchange of information including research results (Art. 17) and scientific and technical cooperation (Art. 18) as means toward bridging capacity asymmetries in achieving its objectives. Aichi Target 19 under the Strategic Plan for Biodiversity 2011–2020 holds that by 2020, "technologies relating to biodiversity, its values, functioning, status and trends, and the consequences of its loss, are improved, widely shared and transferred, and applied." With respect to regulation, the preambular text of the CBD, the Cartagena Protocol on Biosafety and a host of COP decisions refer to the precautionary approach, thus acknowledging its applicability in regard to relevant technological issues. The customary rule of transboundary environmental harm, enshrined in CBD Article 3, applies to technologies and activities in general that may "cause damage to the environment of other States or of areas beyond the limits of national jurisdiction." Environmental impact assessment, mandated under Article 14, bears relevance for technological projects "that are likely to have significant adverse impacts" on biodiversity.

The CBD regime has responded relatively quickly to specific emerging technological opportunities and challenges: hrough publication of technical reports, deliberations at COP and SBSTTA meetings and the creation of various consultation processes and ad hoc technical expert groups (AHTEGs). This has led to diverse COP decisions on a broad range of technological issues, as well as the adoption of a series of guidelines on both methodological and substantive aspects of governing technological change. In addition, rules have been put in place for the systematic monitoring of technological developments relating to biodiversity conservation and sustainable use, with SBSTTA being mandated to "[i]dentify new and emerging issues relating to the conservation and sustainable use of biodiversity" (Decision VIII/10). However, none of the technologies we discuss in this chapter has been classified as such as of yet.

The following three sections map the rules, institutional responses and regulatory gaps with regard to climate-related geoengineering; synthetic biology, including gene drives; and bioinformatics and DSI. In the conclusions, we assess the extent to which governance of those technologies under the CBD regime can support transformative change in order to address indirect drivers of biodiversity loss (see Chapter 1). While the CBD seems reasonably effective and appropriate in most of those regards, we point out that adaptation is limited to soft-law governing body decisions as well as technical guidance, limiting its efficacy for mitigating risks or capturing potential benefits associated with technological change. This raises questions regarding the effectiveness and stringency of technology regulation within the context of the CBD's Post-2020 Global Biodiversity Framework, which, at the time of writing, contracting parties are expected to adopt in 2022.

7.2 Climate-Related Geoengineering

Anthropogenic climate change is closely related to the CBD's goals, especially the conservation of biological diversity (Bellard et al., 2012). The Intergovernmental Science-Policy Platform on Biodiversity and Ecosystem Services estimates that climate change is the third most impactful direct driver of biodiversity loss (IPBES, 2019), and deleterious effects are expected to increase as the climate further changes. However, it is not only climate change that could have impacts on biodiversity but also our responses to mitigate it, including through two sets of technology that are often collectively referred to as "geoengineering." In recent years, it has become increasingly evident that greenhouse gas emissions reductions in line with the relevant international agreements will likely be insufficient for limiting global warming to 2°C above preindustrial levels. Decision-makers, climate modelers and other scientists began to turn to anthropogenic activities and technologies that would remove carbon dioxide from the atmosphere and durably sequester it for long timescales. Such carbon dioxide removal (CDR) techniques are diverse, and some hold the potential to significantly reduce net emissions and atmospheric concentrations of CO_2 (The Royal Society, Royal Academy of Engineering, 2018). Proposed CDR techniques include: (1) bioenergy with carbon capture and sequestration (BECCS), in which plants are grown and burnt to produce energy, with the resulting CO_2 captured and stored; (2) direct air capture (DAC), in which CO_2 is captured from ambient air, and stored; (3) enhanced weathering, in which minerals are processed to accelerate natural chemical CO_2 sequestration; and (4) ocean fertilization, in which nutrients are added to accelerate natural marine biological CO_2 sequestration. CDR could make ambitious climate change targets more achievable, could later compensate for initially exceeding emissions limits, and appears essential to meeting internationally agreed-upon climate change goals. Indeed, the favorable scenarios of the Intergovernmental Panel on Climate Change (IPCC) assume very large-scale BECCS (IPCC, 2018). The 2015 Paris Agreement implicitly endorses this technique (Articles 4.1, 5). Likewise, some states have implicitly committed to them through "net zero" emissions targets (Darby, 2019). At the same time, these techniques pose environmental risks and social challenges. Furthermore, CDR techniques affect atmospheric concentrations only slowly, are relatively expensive and are unlikely to be available at scale in the short term.

In addition to CDR, the other form of geoengineering is a set of technological responses to climate change referred to as solar radiation modification (SRM), which would intentionally modify the Earth's shortwave radiative budget with the aim of reducing climate change (IPCC, 2018: 558). Models indicate that at least some approaches could reduce climate change effectively, rapidly, reversibly and at low direct financial cost (National

Research Council, 2015). The leading proposal would replicate volcanoes' natural cooling effect by injecting aerosols into the stratosphere. Another proposal is to spray seawater as a fine mist, the droplets of which would, after evaporation, brighten low-lying marine clouds. Like CDR, SRM could reduce climate change but poses environmental risks and social challenges. As it is presently understood, SRM is necessarily global, which points to issues of international decision-making that are further complicated by its low resource requirements which, in principle, might allow for its deployment by smaller clubs or even single countries. Among the social challenges are a need for long maintenance and only gradual phase-down, displacing emissions cuts, claims of blame and demands for compensation for harm, and biasing future decision-making through sociotechnical lock-in (Reynolds, 2019).

Although geoengineering is typically envisioned as a means to reduce global climate change, it could be done in ways that have local effects. This is particularly salient with respect to biodiversity, which is unevenly distributed and mostly concentrated in hotspots. These might constitute priority areas for local deployment. Consider coral reefs, which are among the most biodiverse and threatened ecosystems. Coral reefs face the double threat of warmer marine waters and ocean acidification due to dissolved CO_2, both of which result in coral bleaching. Ocean alkalinization, a marine CDR method akin to enhanced weathering, may be able to locally prevent and reduce ocean acidification (Feng [冯玉铭] et al., 2016). Local SRM through marine cloud brightening or biodegradable ocean surface films could protect corals by locally limiting warming during heat waves (McDonald et al., 2019).

Geoengineering's effects are uncertain. At a gross level, if a technology were to reduce climate change, then it would also reduce climatic impacts on biodiversity. This general claim is subject to a number of qualifications. First, geoengineering would have secondary effects, some of which would be negative. For CDR, these are relatively local, whereas the benefits of reduced atmospheric CO_2 would be global. In order to substantially reduce atmospheric CO_2 concentrations, BECCS would require vast amounts of arable land, which could reduce natural habitat, especially in (sub)tropical regions (Stoy et al., 2018). BECCS and DAC need storage, which could leak, posing risks to species and ecosystems. Enhanced weathering involves large-scale excavation, transportation and processing, and could adversely affect ocean chemistry. Ocean fertilization alters marine ecosystems in uncertain ways (Joint Group of Experts on the Scientific Aspects of Marine Environmental Protection, 2019). For SRM, impacts would be geographically distant or global. It would compensate changes to temperature and precipitation differently, imperfectly and heterogeneously. Stratospheric aerosol injection could slow the recovery of the protective stratospheric ozone layer. Other environmental risks remain unknown. A second qualification is that geoengineering's positive and negative impacts on biodiversity would be socially mediated. Although it could be used rationally to reduce climate change, it – especially SRM – might be poorly implemented. In that case, it could be deployed too rapidly or at too high of an intensity, or it could be stopped too suddenly (but see Rabitz, 2019a; Trisos et al., 2018). Similarly, BECCS could be scaled-up carefully, with relatively little biodiversity impact, or haphazardly. Third and finally, much remains unknown. Research to date has been limited, especially on SRM and on biodiversity impacts (McCormack et al., 2016).

Given the CBD's broad scope and geoengineering's potential to help conserve or potentially harm biodiversity, it is unsurprising that the Convention's bodies have engaged with the governance of geoengineering. However, the path that it took there has been somewhat reactive and arguably suboptimal. The catalyst for action was commercial firms' plans to undertake ocean fertilization, which at the time seemed to some observers to have substantial potential to remove CO_2. In response to agitation by some nongovernmental organizations and "in accordance with the precautionary approach," in 2008 the COP requested that states not allow ocean fertilization activities until there is "adequate scientific basis on which to justify such activities ... and a global, transparent and effective control and regulatory mechanism," and even then, only if they are noncommercial, scientific, subject to prior environmental impact assessment and "strictly controlled" (Decision IX/16.C). Although, as a COP decision, this statement is necessarily nonbinding, it appears to have contributed to the subsequent halt of legitimate, noncommercial ocean fertilization research, which had been occurring for about a decade (Williamson et al., 2012). The Parties to the London Convention and London Protocol, which regulate marine dumping, issued similar decisions on ocean fertilization in 2008 and 2010 (Resolutions LC-LP.1 and LC-LP.2). Parties to the latter agreement also approved an amendment that, when and if it comes into effect, would regulate marine geoengineering more broadly, although low ratification numbers indicate that this is unlikely to happen in the short term (Resolution LP.4[8]).

Since then, the CBD COPs have adopted three decisions regarding geoengineering. The first of these, in 2010, expanded the ocean fertilization decision to apply to geoengineering more broadly (Decision X/33.8[w]). In this, the COP invited Parties and other governments to consider not allowing any "climate-related geo-engineering activities that may affect biodiversity unless three criteria are met: a) 'science based, global, transparent and effective control and regulatory mechanisms'; b) an 'adequate scientific basis'; and c) 'appropriate consideration of the associated risks for the environment and biodiversity and associated social, economic and cultural impacts'." This decision has received significant attention. Some journalists and activists call it a moratorium or even a ban (e.g. Tollefson, 2010). However, that is an incorrect description (Reynolds et al., 2016). The COP does not have the authority to issue rules that are binding under international law. The text here uses particularly qualified language, in which it merely "invites" states to "consider the guidance." Both CBD reports on the topic call the decision "a comprehensive non-binding normative framework" (SCBD, 2012: 106; Williamson and Bodle, 2016: 144). Finally, its reference to being "in accordance with [...] Article 14" suggests that the decision is further limited to climate-related geoengineering activities that are likely to have *significant adverse* effects on biological diversity. In the absence of threshold criteria, it remains unclear beyond which point an activity would be classified as causing such effects.

In 2012, the Parties issued a decision on climate-related geoengineering. This, however, added little substance, only noting that no single geoengineering approach "meets basic criteria for effectiveness, safety and affordability," that significant knowledge gaps remain, and "the lack of science-based, global, transparent and effective control and regulatory mechanisms for climate-related geoengineering" (Decision XI/20). Somewhat more substantive was Decision XIII/14 of 2016, which "notes that more transdisciplinary research and sharing of knowledge ... is needed in order to better understand the impacts of climate-related geoengineering on biodiversity and ecosystem functions and services, socioeconomic, cultural and ethical issues and regulatory options." Finally, the Secretariat of the CBD has commissioned and published two major reports on geoengineering with respect to the Convention (SCBD, 2012; Williamson and Bodle, 2016).

These COP decisions are important to the global governance of geoengineering, as they remain the only explicit statements from the international community regarding geoengineering in general (notably, the UN Environment Assembly was unable to reach a consensus in a 2019 discussion). Although the Parties to the London Convention and London Protocol, as well as the International Maritime Organization, have since 2008 largely assumed the international governance of ocean fertilization, the CBD's 2010 and 2016 decisions offer significant guidance in a domain that arguably lacks it. They express caution, calling on states to ensure that geoengineering activities beyond a certain expected magnitude of impact do not take place until particular criteria are satisfied. At the same time, important ambiguities persist. Are "small scale scientific research studies that would be conducted in a controlled setting" limited to indoor activities, or could they include low-risk and/or well-contained outdoor experiments? And given that geoengineering could reduce dangerous climate change, that it poses its own threats of significant reduction or loss of biological diversity and that full scientific certainty is lacking, what are the implications of anticipatory governance for decision-making under uncertainty? Furthermore, the 2016 COP decision and report have important implications for the global governance of biodiversity: that large-scale interventions in natural systems, such as climate geoengineering, have the potential to help conserve biodiversity and that more research is consequently needed. Furthermore, the COP decisions push the boundary of the CBD's scope, engendering real and potential conflict with other international legal institutions such as the London Convention and London Protocol, and the UN Framework Convention on Climate Change (UNFCCC) (see van Asselt, 2014).

Geoengineering activities, including those that may affect biodiversity, are governed by several legal and nonlegal mechanisms beyond the CBD, including the UNFCCC, the UN Convention on the Law of the Sea, the Convention on the Prohibition of Military or Any Other Hostile Use of Environmental Modification Techniques, and the London Convention and London Protocol (Reynolds, 2019). However, almost all of these were developed without geoengineering in mind and do not explicitly reference geoengineering and/or biodiversity. Exceptions in both regards are the above-noted resolutions on ocean fertilization and amendment on marine geoengineering that the Parties to the London Convention and London Protocol have approved. The frameworks under the 2010 resolution and 2013 amendment include assessing potential impacts on marine ecosystems, and the resolution explicitly refers to biodiversity.

7.3 Synthetic Biology and Gene Drives

Synthetic biology comprises a broad variety of technologies that are at different stages of the research and development pipeline and that differ widely in terms of their practicability as well as potential benefits and risks for biodiversity. Work under the Convention is guided, for the time being, by a 2016 operational definition developed by the AHTEG on synthetic biology but not endorsed by the COP, which defines synthetic biology as "a further development and new dimension of modern biotechnology that combines science, technology and engineering to facilitate and accelerate the understanding, design, redesign, manufacture and/or modification of genetic materials, living organisms and biological systems" (Decision XIII/17; Keiper and Atanassova, 2020). How this differs from "traditional" biotechnology, such as defined under CBD Article 2, is not clear. Regardless, this includes, for instance, approaches for the computer-based design of genomes, the synthesis of DNA nucleobases that do not exist in the known universe and the deliberate engineering of metabolic pathways within cells (SCBD, 2015). Current and near-term commercial and industrial applications of synthetic biology aim mainly at creating microorganisms that synthesize products for fuels, pharmaceuticals, chemicals, flavorings and fragrances (El Karoui et al., 2019). Potential positive impacts may include pollution control through microorganisms designed for bioremediation and reduction of overharvesting of threatened wild species through development of synthesized products (SCBD, 2015). Synthetic biology may also serve a role in enhancing the resilience of agricultural systems by developing crops with improved resistance to environmental stress, chemical pollution, pesticides and fertilizers. One – currently hypothetical – application of synthetic biology of relevance to biodiversity conservation is de-extinction: the cloning of extinct species by grafting ancestor DNA onto the genome of existing species with a similar genetic profile (Church and Regis, 2014). As the history of agricultural biotechnology suggests a pattern of overpromising and underdelivering on the supposed environmental benefits of genetic engineering, many of these claims may warrant skepticism. What sets the case of synthetic biology and gene drives apart from the debate on agricultural biotechnology during the 1990s is that, at least for the time being, a significant amount of research and development is being carried out in the public and philanthropic sectors rather than in the for-profit private sector. Patent activity remains relatively limited (Oldham and Hall, 2018). In addition, as synthetic biology technologies become less expensive and more widely accessible, several small-scale, publicly accessible community laboratories, do-it-yourself and open science collaborations are emerging that may lead to a democratization of science (Laird and Wynberg, 2018).

However, the release (including from small-scale, "do-it-yourself biology") of organisms created via synthetic biology may raise environmental concerns in regard to biosafety, as well security, socioeconomic and ethical issues. Biosafety issues include, for example, the potential for survival, persistence and transfer of genetic material to other microorganisms, possible negative effects on nontarget organisms and transfer of genetic material to wild populations. Indirect negative impacts could arise from the increase in the utilization of biomass required for synthetic biology applications. Security considerations arise from the potential malicious or accidental use of synthetic biology applications. Socioeconomic considerations relate to potential impacts on community livelihoods in developing countries where traditional crops and other natural resources are replaced. Ethical concerns relate to the socially accepted level of uncertainty and predictability of its impacts and the threshold between the modification of existing organisms and the creation of new ones (SCBD, 2015). More fundamentally, transformative change may also entail deeper ethical concerns regarding the very creation of artificial life or the genetic modification of entire species.

As a specific set of emerging technologies, gene drives are conceptually easier to pin down. These are often understood as "systems of biased inheritance in which the ability of a genetic element to pass from a parent to its offspring through sexual reproduction is enhanced" (National Academies of Sciences, Engineering, and Medicine, 2016: 1). Within the CBD process, gene drives have generally been considered part of the broader issue of synthetic biology. From a technical perspective, however, gene drives are based on techniques for genome editing, such as CRISPR/Cas9, that are already firmly established in the contemporary life sciences and, while falling within the broad definition of "biotechnology" in the Convention's Article 2, do not necessarily fall within the operational definition of "synthetic biology" (Esvelt et al., 2014). By increasing the probability with which genetic traits are passed on to later generations, gene drives offer the possibility of rapidly and efficiently modifying the genetic profile of entire target populations (meaning the interbreeding members of a species that typically live in a geographic place) of sexually reproducing organisms with short gestation cycles (Esvelt et al., 2014). A major motivation for the development of gene drives is the control of disease vectors such as mosquitoes. However, they are also under discussion as a tool for combating invasive alien species, which is a crosscutting issue under the CBD (Leitschuh et al., 2018). Examples of such species include rats and other rodents, as well as organisms such as certain mussels, jellyfish and sea stars that have been introduced into vulnerable marine ecosystems through ballast water tanks. At the same time, the rapid environmental diffusion of gene drives, the potential of unforeseen effects on target species and ecosystems, the possibility for the introduction of new diseases through the replacement of the population of the original disease vector by another vector species, unpredicted mutations in the drive or unintended off-target effects raise serious biosafety questions (SCBD, 2015). Thus, while synthetic biology and gene drives could potentially contribute to the CBD's objectives of conservation and sustainable use by protecting or restoring ecosystems, or by reducing anthropogenic pressures from agricultural practices, they also pose novel and unpredictable risks and regulatory challenges.

The CBD COP started addressing synthetic biology and gene drives as a recurring agenda item in 2014. Yet by 2010, COP decision X/37 on biofuels and biodiversity urges Parties and non-Parties to apply precaution regarding "the field release of synthetic life, cell or genome into the environment." Decision XII/24 of 2014, which addresses synthetic biology in general but does not cover gene drives, urges Parties to take a precautionary approach, including by having "effective risk assessment and management procedures" or other types of regulation in place prior to any deliberate release. That decision also installed an AHTEG for collecting and synthesizing different stakeholder perspectives, for identifying existing regulatory gaps and for elaborating the operational definition of synthetic biology quoted above. Decision XIII/17 of 2016 notes the future need for developing new approaches to assessing the risks associated with synthetic biology; notes that some organisms produced through synthetic biology may fall outside the functional scope of the CBD and the Cartagena Protocol; and invites Parties to engage in further stakeholder consultations, research and knowledge synthesis for identifying potential biodiversity-related risks and benefits of synthetic biology. In that decision, the COP for the first time engages with gene drives, noting that they may fall within the category of synthetic biology, and thus may partially fall within the scope of the earlier decision XII/24. In 2018, the COP finally agreed on the need for systematic monitoring and horizon-scanning for technological developments in synthetic biology, under decision XIV/19. This decision for the first time provided more specific guidance in regard to gene drives, calling upon Parties and non-Parties to require "[s]cientifically-sound case-by-case risk assessment" as well as adequate risk management procedures prior to a deliberate release.

The primary barrier to the effective governance of synthetic biology and gene drives under the CBD framework is the stark contrast in perceptions of the Parties of the associated risks and benefits, as well as their distribution. Reminiscent of CBD debates in the 1990s with regard to modern biotechnology and LMOs, the highly politicized deliberations reflect different understandings of technology, perceptions of environmental risk and precaution, expectations regarding benefits (including commercial ones), and scientific and regulatory capacities to assess associated risks (Reynolds, 2020). At the same time, an important difference between past biotechnology debates and the current ones regarding gene drives is that, while private firms were developing and advocating for the former, they are absent from the latter, presumably due to insufficient commercialization perspectives (Mitchell et al., 2018). While there is general consensus among Parties that the use of those technologies should be subject to the precautionary approach (see CBD preamble, recital 9), how exactly precautions would be *operationalized* is a matter of ongoing dispute. Bracketed text in SBSTTA recommendation 22/3 of July 2018 – later rejected by the COP – illustrates this divergence of views: Whereas some Parties prefer precaution regarding the *extent and timeframe of the release* of gene drives, others, such as Bolivia at the time, interpret precaution as implying *refraining* from such releases (ENB, 2018a). To some extent, the debate revolves around questions of regulation of synthetic biology as an inherently risky new and emerging technology versus case-by-case assessment of its products and applications, or even prohibition of environmental releases until further knowledge is available.

Regardless of the merits of any of these approaches, nonuniversal participation in the CBD and, particularly, the Cartagena Protocol poses additional challenges and creates the risk of jurisdiction-shopping. Notably, the USA is neither a party to the Convention nor to the Cartagena Protocol, and some of the countries with strong biotechnology industries, such as Argentina, Australia and Canada, are not parties to the Protocol. Addressing this issue under both the Convention and the Protocol thus poses challenges for effective decision-making because of their different memberships. Regulating or even prohibiting environmental releases of gene drives and organisms produced via synthetic biology may generate incentives for operators to carry out such releases in jurisdictions where regulatory standards are less restrictive. Especially regarding initial, small-scale field testing that might only entail limited transboundary effects, the insufficient geographic coverage of the CBD regime severely limits the scope for effective international regulation (Rabitz, 2019b).

Beyond the CBD regime, a range of other international institutions potentially bear relevance for the governance of synthetic biology and gene drives. The WHO has developed a Guidance Framework for Testing of Genetically Modified Mosquitoes, incorporating cost–benefit analysis and precaution. The Review Conferences of the Biological Weapons and Toxins Convention have, in recent years, started considering the biosecurity implications of both synthetic biology and gene drives. Other institutions may be relevant without necessarily addressing either technology directly. International patent law might matter to the extent that the patent protection of first-generation gene drive organisms might extend to their progeny. The use of synthetic biology in the food sector would likely create a role for the World Trade Organization's Agreement on Sanitary and Phytosanitary Measures as well as the Codex Alimentarius Commission. Yet in all those cases, the governance implications of synthetic biology and gene drives are even less clear than they are for the CBD regime.

7.4 Bioinformatics and Digital Sequence Information

Synthetic biology applications have largely become possible due to advances in bioinformatics, an interdisciplinary field of knowledge that develops and uses methods and software tools to extract knowledge from biological material. It includes the collection, storage, retrieval, manipulation and modelling of data from biological resources for analysis, visualization or prediction through the development of algorithms and software. Bioinformatics tools allow for generating and analyzing large quantities of genotypic, phenotypic and environmental data. Techniques for high-efficiency genomic sequencing have been followed by methods for measuring the current molecular state of cells and organisms, for predicting classical phenotypes in an automated manner and even for reengineering the content and function of living systems. These technologies have led to the rapid generation of large amounts of data describing biological systems, and the analysis and interpretation of these data using statistical and computational expertise (Can, 2014; Diniz and Canduri, 2017).

Developments in bioinformatics pose challenges for access and benefit-sharing (ABS) frameworks. This includes the CBD and its Nagoya Protocol on ABS, which aim to ensure that users of genetic resources share (commercial and other) benefits that arise from utilization. They result in what is described as the "dematerialization" of genetic resources, suggesting that "the information and knowledge content of genetic material [could increasingly be] extracted, processed and exchanged in its own right, detached from the physical exchange of the ... genetic material" (Secretariat of the International Treaty on Plant Genetic Resources for Food and Agriculture, 2013).

Within the CBD, the term DSI is understood to refer to nucleic acid sequence reads and the associated data, and information on the sequence assembly, its annotation and genetic mapping, describing whole genomes, individual genes or fragments thereof, barcodes, information on gene expression, and behavioral data, among others (Convention on Biological Diversity, 2018). The origin of debates on DSI can be traced to the report of the 2015 meeting of the AHTEG on synthetic biology. Participating experts identified potential adverse effects of synthetic biology for the CBD objective of fair and equitable benefit-sharing, including inappropriate access without benefit-sharing due to the use of DSI, and a "shift in the understanding of what constitutes a genetic resource" (Convention on Biological Diversity, 2015: 10). As explored below, such a shift in understanding lies at the heart of the highly polarized debate on DSI (see also Keiper and Atanassova, 2020).

The issue of regulation of DSI-use has also arisen in ABS-related processes beyond the CBD and the Nagoya Protocol, including the International Treaty on Plant Genetic Resources for Food and Agriculture (ITPGRFA), the Pandemic Influenza Preparedness Framework for access to vaccines and other benefits (PIP Framework) under the WHO, and the ongoing negotiations under the UN Convention on the Law of the Sea on marine biodiversity beyond the limits of national jurisdiction (BBNJ), albeit with differing terminologies and varying political progress. While significant advances in deliberations have been made under the PIP Framework, DSI turned out to be a deal-breaker for efforts at reforming the ITPGRFA's Multilateral System, leading to the collapse of six years of negotiations at the end of 2019 (ENB, 2019; Tsioumani, 2020).

The availability and easy exchange of large amounts of sequence data have the potential to facilitate research on genetic resources, especially for actors in developed countries who have the capacities to analyse and use such data. At the same time, it poses two main regulatory issues: the possibility of appropriation of genetic sequence data, including data placed in the public domain, through intellectual property rights (IPRs), in particular patents; and the question of value generation from the use of such data, and related benefit-sharing obligations (Laird and Wynberg, 2018; Welch et al., 2017). Opinions diverge in particular as to whether and how its utilization should give rise to benefit-sharing obligations supporting the CBD's objective of fair and equitable benefit-sharing, which is intended to incentivize nature conservation, provide the financial and other means for doing so, and inject fairness and equity in bio-based research and development (Morgera, 2016; Tsioumani, 2018). The latter question further involves a series of legal interpretation issues concerning the scope of the CBD and the Nagoya Protocol, and implementation concerns involving the identification of users and monitoring/tracking of uses of such data. These issues will be briefly addressed below, in turn. Additional normative questions arise with regard to benefit-sharing from the utilization of human genetic resources which, however, fall outside the scope of the CBD and thus this chapter.

As evidenced from several open-access registries and projects, the synthetic biology community – which brings together most DSI users – has a strong open source sharing ethos and encourages the release of genomic and other datasets as public goods (Tsioumani et al., 2016). At the same time, as in all technological fields, researchers tend to patent research tools and sequences strategically, with clear commercial applications (Welch et al., 2017). As patent law is territorial in nature, and legal debates on social and moral concerns regarding patent eligibility of genetic sequences continue to rage in several jurisdictions, the patent landscape varies around the globe (Nuffield Council on Bioethics, 2002). In the United States, the 2013 Supreme Court decision in Association for Molecular Pathology v. Myriad Genetics held that DNA segments and the information they encode are not patent-eligible simply because they have been isolated from surrounding genetic material, thus reversing years of prior jurisprudence and confirming a shift in the broad scope of the patentability of genetic sequences. Under the EU's Biotechnology Directive (98/44/EC), biological material that is isolated from its natural environment or produced by means of a technical process may be the subject of an invention, even if it previously occurred in nature. The European Court of Justice subsequently clarified, in Monsanto Technology v. Cefetra BV, that, in order to meet the requirements for patent eligibility, the "functionality" of the genetic sequence must be disclosed in the patent application. Developing countries have also sought to set their own standards. Brazil, for instance, excludes living beings or biological materials found in nature from patentability, even if isolated, and this includes the genome or germplasm of any living being (Correa, 2014). Navigating the patent landscape is further complicated by the uncertainty generated by those patent applications that are still pending, resulting in an inability to locate the ownership of patents, as well as by the fees usually required for searching patent databases (Hope, 2004). Moreover, while ownership of a patent is usually a matter of public record, ownership of the rights transferred through licenses is not. Most jurisdictions do not impose a responsibility on licensees to disclose, making it almost impossible for a researcher to assemble all the licenses needed to proceed with their research (Jefferson, 2006). This complexity has devastating consequences for public sector researchers, particularly in developing countries. Adding the specificities of ABS legislation to the mix can only increase the degree of complexity and legal uncertainty, further restricting access to DSI.

Unrestricted access to DSI, in the form of public and open-access databases, can be considered an important form of nonmonetary benefit-sharing, as long as it is accompanied by capacity-building measures to ensure its fair and equitable use by actors in developed and developing countries alike. Nonmonetary benefit-sharing, via information exchange, capacity-building and technology transfer, may allow for an increase of endogenous research capacities for genetic resource utilization and thus assist in bridging the gap between developed and developing countries. However, in view of the increasing use of DSI in bio-based research and development, alongside potential restriction of its availability through IPRs, biodiversity-rich developing countries have been calling for the application of monetary benefit-sharing requirements to the use of DSI arising from genetic resources, according to the provisions of the CBD and the Nagoya Protocol. Debates have centered mainly around the interpretation of the scope of the CBD and the Protocol. At the time of writing, most developed countries oppose any benefit-sharing from DSI and argue that the CBD and the Nagoya Protocol have been developed to address exchanges of "material" resources. Their legal argumentation points to the definition of "genetic resources," as genetic "material" that contains "functional units of heredity" (CBD Art. 2 and Nagoya Protocol Art. 2). Therefore, exchanges of "immaterial" information such as DSI would fall outside the scope of the two instruments. In contrast, developing countries argue that letting DSI-use escape benefit-sharing obligations would make the Nagoya Protocol obsolete, and thus negate any progress toward the redistribution of benefits from countries that have the capacity to use genetic resources toward those that have stewarded them. In addition, developing countries hold that the use of DSI qualifies as "utilization" of genetic resources (Nagoya Protocol Art. 2), thus giving rise to benefit-sharing obligations. The issue attracted more attention than any other item under negotiation at the 2018 meeting of the COP in Egypt and is expected to be central at the negotiations for a Post-2020 Global Biodiversity Framework. In fact, several countries from the global South declared that there will be no agreement on a Post-2020 Global Biodiversity Framework unless benefit-sharing from DSI-use is ensured (ENB, 2018b; 2019).

The CBD and Nagoya Protocol objective of fair and equitable benefit-sharing has opened new ground in environmental agreements with regard to the distribution of benefits of scientific progress. However, its implementation in the bilateral system of exchanges between providers and users of genetic resources envisaged by these instruments poses challenges, particularly with regard to the determination of the value of the genetic resource under consideration, the determination of benefits, the development of mutually agreed terms for benefit-sharing and their application in the context of an interlinked web of national laws and policies, and ensuring compliance by users (Morgera et al., 2014). These challenges are exacerbated in the case of DSI. Implementation concerns involve in particular the identification of the value of DSI, its origin and its user, as well as ensuring compliance by monitoring its use (Laird and Wynberg, 2018). Digitalization raises fundamental questions regarding the long-term viability of the bilateral approach to benefit-sharing under the CBD and the Nagoya Protocol. That said, a number of CBD Parties have already enacted benefit-sharing obligations from DSI-use as part of their domestic ABS measures, including, among others, Brazil, Malaysia and South Africa.

Despite the intense political controversies, COP decision 14/20 of 2018 established a science and policy-based process that is expected to shed light on many of the regulatory challenges related to DSI. The COP invited submission of views aiming to clarify the concept, including relevant terminology and scope, as well as submission of domestic ABS measures and benefit-sharing arrangements considering DSI. It further called for submission of information on capacity-building needs, and commissioned a series of peer-reviewed studies focused on some of the more technical issues explored above, including: the concept and scope of DSI; traceability; databases; and domestic ABS measures addressing benefit-sharing arising from DSI commercial and noncommercial use. In anticipation of deliberations in the CBD subsidiary bodies and the Working Group on the Post-2020 Framework, these studies informed the debates of the AHTEG established to address the issue. The AHTEG offered clarifications on the scope of DSI; options on terminology regarding categories of information that could be considered DSI; implications concerning traceability, use, exchange of information and ABS measures; and key areas for capacity-building (Convention on Biological Diversity, 2020).

7.5 Toward the Transformative Governance of Emerging Technologies

While our cases address different issues, all highlight the challenges the CBD regime faces in governing biodiversity-related technologies. In general, the CBD regime is relatively quick to pick up novel technological issues and to process them in an inclusive manner, based on high-quality scientific and technical expert advice. In the output dimension, rulemaking has been limited to nonbinding (and frequently heavily qualified) COP decisions and assorted technical guidance. The rapid identification and addressing of governance gaps associated with novel technologies thus does not necessarily translate into strengthened international regulation. This appears linked to the Convention's broad scope and objectives, complex overlaps with other intergovernmental organizations, system of consensual and participatory decision-making, lack of compliance and enforcement mechanisms, and, crucially, frequently stark divergences in the regulatory preferences of its contracting parties.

To assess the extent to which the CBD can support transformative governance of biodiversity with respect to emerging technologies, we follow the criteria introduced in Chapter 1. The capacity of the CBD regime to *integrate* governance activities varies across our cases. For geoengineering, we witness an institutional division of labor with the London Convention / London Protocol (see Reynolds, 2018). On DSI, the parallel processes under the CBD, the WHO and the ITPGRFA are characterized by polycentric cross-institutional linkages, although debates focus more on the differences between them with regard to mandate, scope and objectives, rather than the need to address such implications in a systematic manner across sectors and processes. For synthetic biology and gene drives, the lack of rulemaking activities outside the CBD regime limits the scope for integration from the outset. At the same time, the CBD possesses a high degree of *inclusiveness*, illustrated by the establishment of an open-ended online forum on synthetic biology and stakeholder participation regarding DSI, including by representatives of Indigenous peoples and local communities, civil society, academia and research, and the private sector, as well as relevant international bodies. The CBD processes on DSI, as well as synthetic biology and gene drives, are also characterized by relatively strong *transdisciplinarity*, drawing on natural sciences, law and social sciences, as well as the knowledge of Indigenous peoples. In contrast, information uptake with regard to deliberations on geoengineering is less structured and arguably weak, with relevant COP decisions having been criticized as poorly informed (Sugiyama and Sugiyama, 2010). Regarding *adaptiveness*, all our cases are characterized by COP decisions that are vague, use heavily qualified language and fail to clarify important operational criteria. However, institutional adaptation to emerging technologies is a frequent challenge that is not necessarily specific to the CBD (Marchant et al., 2013). Finally, *anticipation* requires addressing the Collingridge dilemma, in which developing governance faces few barriers early on but too little is then known, while later on there is greater knowledge, but interests have arisen and legislation has ossified (Collingridge, 1980). From this perspective, governance responses under the CBD have indeed been anticipatory. This is most evident in the SBSTTA's mandate to identify "new and emerging issues." Also, in all three cases considered here, the CBD initiated governance processes in the very early stages of technological development. This may be a consequence of the relatively prominent position given to precaution in the CBD and in the COP's interpretation thereof. If anything, there is a reasonable argument that the CBD has engaged too early in these areas, before sufficient knowledge of potential technological impacts, limits and risks became available.

To conclude, it is important to keep in mind that the three technologies discussed above not only pose potential threats, but also offer potential benefits for the objectives of the CBD. DSI may either undermine effective benefit-sharing (by allowing users to shirk their obligations) or enhance utilization of genetic resources (by obviating the need for physical specimens), thus improving research on environmentally useful innovations as well as increasing the overall size of the "pie" from which benefits may subsequently be shared. Some proposals for geoengineering could arguably have adverse effects on biodiversity but equally have an important function for its conservation. Synthetic biology and gene drives create novel biosafety risks and could cause significant harm for species and ecosystems, yet may also contribute to the conservation objective by allowing for greater biological control of invasive alien species, pests and diseases.

Such technological solutions to environmental challenges are frequently critically referred to as "techno-fixes." On one hand, they may enable overreliance on unproven, ineffective or unsafe technologies while displacing regulatory or socioeconomic solutions that could address root causes of biodiversity loss, such as habitat loss and alteration, pollution and overexploitation of species. Faith in technological solutions further can ignore the complexity of biological diversity and interdependence of living systems, which, coupled with lack of data and knowledge, can translate into uncertainties and even ignorance. On the other hand, the history of biodiversity governance demonstrates the limited efficacy of conventional solutions and the lack of sufficiently powerful political coalitions to address the root causes of biodiversity loss. History also suggests that technological evolution is, to a certain degree, inevitable and often faster than regulation. In addition, technologies can catalyze structural social, political and economic change, often in surprising ways. The emerging synthetic biology community, for instance, could be a source of great risk, although it may in the future also produce valuable social and institutional advancements in how the CBD and other bodies govern emerging biotechnologies, including through their open data and sharing ethos.

However, within the context of the CBD, interest constellations reflect differences in socioeconomic development and innovative capacity, as well as normative disputes over the role of technology in environmental governance. Shifting toward inclusive, effective and outcome-oriented technology regulation in the post-2020 era, together with the fair distribution of costs, risks and benefits of the technologies involved, is likely to be one of the main challenges of the CBD deliberations for the years to come. In this context, given the divergences in Parties' priorities and interests and the realities of intergovernmental decision-making, it is doubtful that transformative governance of technology will originate in the realm of the CBD, or any other intergovernmental process; it will rather reflect and follow deep socioeconomic and behavioral changes.

References

Bellard, C., Bertelsmeier, C., Leadley, P., Thuiller, W., and Courchamp, F. (2012). Impacts of climate change on the future of biodiversity. *Ecology Letters* 15, 365–377.

Can, T. (2014). Introduction to bioinformatics. In: *miRNomics: MicroRNA biology and computational analysis*. M. Yousef and J. Allmer (Eds.), pp. 51–71. Totowa, NJ: Humana Press.

Church, G. M., and Regis E. (2014). *Regenesis: How synthetic biology will reinvent nature and ourselves*. New York: Basic Books.

Collingridge, D. (1980). *The social control of technology*. London: Pinter.

Convention on Biological Diversity. (n.d.) 2019–2020 Inter-sessional period. Studies on digital sequence information on genetic resources. Available from www.cbd.int/dsi-gr/2019-2021/studies.

(2015). Report of the Ad Hoc Technical Expert Group on synthetic biology, UN Doc. CBD/SYNBIO/AHTEG/2015/1/3.

(2018). Report of the Ad Hoc Technical Working Group on digital sequence information on genetic resources. UN Doc. CBD/DSI/AHTEG/2018/1/4.

(2020). Report of the Ad Hoc Technical Expert Group on digital sequence information on genetic resources, Montreal, Canada, March 17–20, 2020, UN Doc. CBD/DSI/AHTEG/2020/1/7.

Correa, C. M. (2014). Patent protection for plants: Legal options for developing countries. Geneva: South Centre. Available from https://bit.ly/33iKBVH.

Darby, M. (2019). Which countries have a net zero carbon goal? Climate Home News. Available from https://bit.ly/3JUE6cx.

Diniz, W., and Canduri, F. (2017). Bioinformatics: An overview and its applications. *Genetics and Molecular Research* 16, gmr16019645.

ENB. (2018a). Summary of the 22nd meeting of the Subsidiary Body on Scientific, Technical and Technological Advice and 2nd meeting of the Subsidiary Body on Implementation of the Convention on Biological Diversity. *Earth Negotiations Bulletin* 9, 710.

(2018b). Summary of the UN Biodiversity Conference. *Earth Negotiations Bulletin* 9, 725.

(2019). Summary of the Eighth Session of the Governing Body of the International Treaty on Plant Genetic Resources for Food and Agriculture. *Earth Negotiations Bulletin* 9, 740.

El Karoui, M., Hoyos-Flight, M., and Fletcher, L. (2019). Future trends in synthetic biology: A report. *Frontiers in Bioengineering and Biotechnology* 7, 175.

Esvelt, K. M., Smidler, A. L., Catteruccia, F., and Church, G. M. (2014). Emerging technology: Concerning RNA-guided gene drives for the alteration of wild populations. *eLife* 3, e03401.

Feng (冯玉铭), E. Y., Keller, D. P., Koeve, W., and Oschlies, A. (2016). Could artificial ocean alkalinization protect tropical coral ecosystems from ocean acidification? *Environmental Research Letters* 11, 074008.

Hope, J. E. (2004). Open source biotechnology. Available from https://ssrn.com/abstract=755244.

IPBES. (2019). *Summary for policymakers of the Global Assessment Report on Biodiversity and Ecosystem Services*. Bonn:Intergovernmental Science-Policy Platform on Biodiversity and Ecosystem Services Secretariat.

IPCC. (2018). *Special report on global warming of 1.5°C*. Geneva: Intergovernmental Panel on Climate Change.

Jefferson, R. (2006). Science as social enterprise: The CAMBIA biOS initiative. *Innovations: Technology, Governance, Globalization* 1, 13–44.

Joint Group of Experts on the Scientific Aspects of Marine Environmental Protection. (2019). High level review of a wide range of proposed marine geoengineering techniques. London: International Maritime Organization.

Keiper, F., and Atanassova, A. (2020). Regulation of synthetic biology: Developments under the Convention on Biological Diversity and its protocols. *Frontiers in Bioengineering and Biotechnology* 8, 310.

Laird, S., and Wynberg, R. (2018). A fact-finding and scoping study on digital sequence information on genetic resources in the context of the Convention on Biological Diversity and the Nagoya Protocol. UN Doc. CBD/DSI/AHTEG/2018/1/3.

Leitschuh, C. M., Kanavy, D., Backus, G. A., et al. (2018). Developing gene drive technologies to eradicate invasive rodents from islands. *Journal of Responsible Innovation* 5, S121–S138.

Marchant, G. E., Abbot, K. W., and Allenby, B. (2013). *Innovative governance models for emerging technologies*. Cheltenham: Edward Elgar.

McCormack, C. G., Born, W., Irvine, P. J., et al. (2016). Key impacts of climate engineering on biodiversity and ecosystems, with priorities for future research. *Journal of Integrative Environmental Sciences* 13, 103–128.

McDonald, J., McGee J., Brent, K., and Burns, W. (2019). Governing geoengineering research for the Great Barrier Reef. *Climate Policy* 19, 801–811.

Mitchell, P. D., Brown, Z., and McRoberts, N. (2018). Economic issues to consider for gene drives. *Journal of Responsible Innovation* 5, S180–S202.

Morgera, E. (2016). The need for an international legal concept of fair and equitable benefit sharing. *European Journal of International Law* 27, 353–383.

Morgera, E., Tsioumani, E., and Buck, M. (2014). *Unraveling the Nagoya Protocol: A commentary on the Nagoya Protocol on Access and Benefit-sharing to the Convention on Biological Diversity.* Leiden and Boston: Brill.

National Academies of Sciences, Engineering, and Medicine. (2016). Gene drives on the horizon: Advancing science, navigating uncertainty, and aligning research with public values. Washington, DC:National Academies Press.

National Research Council. (2015). Climate intervention: Reflecting sunlight to cool earth. Washington, DC: National Academies Press.

Nuffield Council on Bioethics. (2002). The ethics of patenting DNA: A discussion paper. London: Nuffield Council on Bioethics.

Oldham, P., and Hall, S. (2018). Synthetic biology: Mapping the patent landscape. Available from www.biorxiv.org/content/10.1101/483826v1.

Rabitz, F. (2019a). Governing the termination problem in solar radiation management. *Environmental Politics* 28, 502–522.

(2019b). Gene drives and the international biodiversity regime. *Review of European, Comparative & International Environmental Law* 28, 339–348.

Reynolds, J. L. (2018). Governing experimental responses: Negative emissions technologies and solar climate engineering. In: *Governing Climate Change: Polycentricity in Action?* A. Jordan, D. Huitema, H. van Asselt and J. Forster (Eds.), pp. 285–302. Cambridge: Cambridge University Press.

(2019). *The governance of solar geoengineering: Managing climate change in the Anthropocene.* Cambridge:Cambridge University Press.

(2020). Governing new biotechnologies for biodiversity conservation: Gene drives, international law and emerging politics. *Global Environmental Politics* 20, 28–48. https://doi.org/10.1162/glep_a_00567

Reynolds, J. L., Parker, A., and Irvine, P. (2016). Five solar geoengineering tropes that have outstayed their welcome. *Earth's Future* 4, 562–568.

The Royal Society, Royal Academy of Engineering. (2018). Greenhouse gas removal. Available from https://bit.ly/3JOx7C9.

SCBD. (2012). *Geoengineering in relation to the Convention on Biological Diversity: Technical and regulatory matters.* Montreal: Secretariat of the Convention on Biological Diversity.

(2015). *Synthetic biology.* Montreal: Secretariat of the Convention on Biological Diversity.

Secretariat of the International Treaty on Plant Genetic Resources for Food and Agriculture. (2013). Report of the Secretary. FAO Doc. IT/GB-5/13/4. Available from www.fao.org/3/a-be587e.pdf.

Stoy, P. C., Ahmed, S., Jarchow, M., et al. (2018). Opportunities and trade-offs among BECCS and the food, water, energy, biodiversity and social systems nexus at regional scales. *BioScience* 68, 100–111.

Sugiyama, M., and Sugiyama, T. (2010). Interpretation of CBD COP10 Decision on Geoengineering. Socio-economic Research Center, Central Research Institute of Electric Power Industry. Available from https://bit.ly/3GfgMEg.

Tollefson, J. (2010). Geoengineering faces ban. *Nature* 468, 13–14.

Trisos, C. H., Amatulli, G., Gurevitch, J., et al. (2018). Potentially dangerous consequences for biodiversity of solar geoengineering implementation and termination. *Nature Ecology & Evolution* 2, 475–482.

Tsioumani, E. (2018). Beyond access and benefit-sharing: Lessons from the law and governance of agricultural biodiversity. *The Journal of World Intellectual Property* 21, 106–122.

(2020). *Fair and Equitable Benefit-Sharing in Agriculture: Reinventing Agrarian Justice*. London: Routledge.

Tsioumani, E., Muzurakis, M., Ieropoulos, Y., and Tsioumanis, A. (2016). Following the open source trail outside the digital world: The case of open-source seeds. *tripleC: Communication, Capitalism & Critique* 14, 145–162.

van Asselt, H. (2014). *The fragmentation of global climate governance: Consequences and management of regime interactions*. Cheltenham: Edward Elgar.

Welch, E., Bagley, M. A., Kuiken, T., and Louafi, S. (2017). Potential implications of new synthetic biology and genomic research trajectories on the International Treaty for Plant Genetic Resources for Food and Agriculture. Emory Legal Studies Research Paper. Available from https://ssrn.com/abstract=3173781.

Williamson, P., and Bodle, R. (2016). *Update on climate geoengineering in relation to the Convention on Biological Diversity: Potential impacts and regulatory framework*. Montreal: Secretariat of the Convention on Biological Diversit,.

Williamson, P., Wallace, D. W., Law, C. S., et al. (2012). Ocean fertilization for geoengineering: A review of effectiveness, environmental impacts and emerging governance. *Process Safety and Environmental Protection* 90, 475–488.

8

Rethinking and Upholding Justice and Equity in Transformative Biodiversity Governance

JONATHAN PICKERING, BRENDAN COOLSAET, NEIL DAWSON,
KIMBERLY MARION SUISEEYA, CRISTINA Y. A. INOUE
AND MICHELLE LIM

8.1 Introduction

Justice and equity are fundamental to the complex choices that societies need to make to achieve transformative change (Bennett et al., 2019; IPBES, 2019; Leach et al., 2018; Martin, 2017). Evidence that more socioeconomically unequal societies tend to experience higher rates of biodiversity loss (Holland et al., 2009; IPBES, 2019) suggests that injustice and threats to biodiversity are closely intertwined. Injustice can function as an underlying cause of biodiversity loss, such as where colonial expropriation of Indigenous peoples' land paves the way for its exploitation (Martinez-Alier, 2002). Similarly, biodiversity loss can create new injustices or exacerbate existing ones, for example where the destruction of ecosystems accelerates risks such as climate change or pandemics that disproportionately affect the poor (Kashwan et al., 2020). Alleviating unjust conditions could provide a catalyst for environmentally sustainable governance (and vice versa), as where respecting and securing the land rights of marginalized groups enhances the ecological integrity of biologically diverse areas (IPBES, 2019). However, a major challenge for achieving transformative governance in practice is that measures to address biodiversity loss or social injustice can give rise to trade-offs between these goals. Accordingly, efforts to pursue transformative biodiversity governance need to acknowledge social-ecological complexity, expose existing conditions of injustice and embrace opportunities to overcome them.

In the context of this chapter, we understand justice and equity as crucial features of both the means and the ends of transformative biodiversity governance: they are important not only for their instrumental role in addressing biodiversity loss, but also because they are among the core social values that transformative governance aims to rethink and pursue (throughout the chapter, we generally use the term "justice" as shorthand for "justice and equity" unless otherwise specified; Section 8.2 notes different usages of the two terms). Accounts of transformative governance – such as the one that informs this collection – often see *inclusive* governance as an integral feature of the concept (Chapter 1; IPBES, 2019). Including different groups with diverse worldviews, experiences, knowledge systems and

---

An earlier version of this chapter was presented at the Earth System Governance conference in Oaxaca in 2019. We are grateful to the editors and to an anonymous reviewer for very helpful comments on previous drafts. Research for this chapter was supported by the following bodies: the Australian Research Council (grant number FL140100154) [JP]; the "Just Conservation" project funded by the Centre for the Synthesis and Analysis of Biodiversity (CESAB) of the French Foundation for Research on Biodiversity (FRB) [ND and BC]; Brazil's National Council for Scientific and Technological Development (CNPq) and Programa de Excelència Acadèmica (PROEX)-CAPES [CYAI].

values requires respect, trust, mutual understanding and dialogue, and can be seen as a key requirement of procedural justice. The idea of inclusive governance provides an important conceptual entry point for recognizing justice as a core element of transformative governance. However, as we will show, inclusion is only one among several principles of justice that transformative governance needs to take into account. More broadly, the pursuit of justice speaks to another key feature of transformative governance, which is that it must be *integrative* in seeking synergies and minimizing incoherence not only across sectors, institutions and policy instruments, but also across societal goals, including justice and sustainability (Chapter 1; IPBES, 2019).

The question of what justice involves is complex, contested and often overlooked in policy-making. Despite considerable advances in theorizing social and environmental justice and applying these theories to biodiversity governance, there has been little exploration to date of whether and how justice could strengthen the transformative potential of biodiversity governance. This gives rise to the overall question that this chapter addresses: *How should principles of justice and equity be interpreted and upheld in efforts to pursue transformative biodiversity governance?*

To address this question, we begin in Section 8.2 with an overview of evolving theories and norms of justice and equity in biodiversity governance. In Section 8.3 we illustrate how the need for transformative change demands a rethink about what justice entails and requires in the context of biodiversity governance. Then in Sections 8.4–8.6 we address justice in three key stages of transformative governance to address the direct and indirect drivers of biodiversity loss: How should decision-making processes be structured (Section 8.4)? How should financial resources for achieving transformative change be mobilized and allocated (Section 8.5)? And how should transformative biodiversity initiatives be designed and implemented (Section 8.6)? These three areas offer a framework for discussing several important areas of debate about justice in biodiversity governance, including the roles of Indigenous peoples and local communities (IPLC) (Section 8.4), relations between the Global South and North (Section 8.5) and the social impacts of protected area expansion and biodiversity mainstreaming (Section 8.6). While our review does not exhaustively cover all aspects of justice in transformative biodiversity governance, it is complemented by other chapters in this collection, including on emerging technologies (Chapter 7), animals (Chapter 9), and access and benefit-sharing (Chapters 10 and 15). Section 8.7 sets out policy recommendations emerging from the preceding sections, and Section 8.8 concludes.

Throughout the chapter we conduct an integrative review (Snyder, 2019) that critically assesses key theoretical and empirical literature (mainly spanning the period 2010–2020) on justice and equity in biodiversity governance, while also drawing parallels with related areas of environmental governance. Our review is supplemented by the analysis of documents produced by the UN Convention on Biological Diversity (CBD) and the UN Framework Convention on Climate Change (UNFCCC), as presented in Figure 8.1. While our primary focus is on governance at the global scale – in particular the CBD – we also discuss how concerns of justice and equity arise in local and national governance, given that these concerns are linked across multiple scales.

A core set of claims advanced in the chapter is that the depth, scale and urgency of transformative change: (a) demand heightened attention to justice in biodiversity governance; (b) reinforce the need for understandings of justice that are multidimensional (encompassing just processes and recognition as well as distributively just outcomes); and (c) underscore the importance of ensuring justice for the most vulnerable and marginalized groups in processes of transformative change. These claims converge on the idea that transformative biodiversity governance entails a "just transformation" toward a more sustainable planet.

8.2 Theories and Norms of Justice and Equity in Biodiversity Governance

Why are justice and equity so important for biodiversity governance? A first rationale rests on the idea that justice is of intrinsic moral importance. As an essential foundation for sustaining human and nonhuman wellbeing, biodiversity could be seen as a prerequisite for achieving justice (Human Rights Council, 2017). Yet, societies have strong incentives – often but not always grounded in concerns for their own wellbeing – to exploit biodiversity rather than conserve it. Whatever combination of exploitation and conservation is pursued, its impacts are unevenly distributed across human and nonhuman communities, spaces and generations (Blythe et al., 2018; Howe et al., 2014; McShane et al., 2011). This recurrent imbalanced distribution of costs and benefits poses fundamental moral questions about what a just state of affairs is and who should be responsible for envisioning and achieving it.

A second rationale relies on the instrumental importance of justice for biodiversity governance, as in the claim that injustice is an indirect driver of biodiversity loss (IPBES, 2019). According to this view, if governance is just (or at least widely perceived to be so) it will produce better ecological outcomes (Martin et al., 2020). Evaluating both of these rationales requires clarifying how the terms "justice" and "equity" are used in theory and practice.

8.2.1 Theories of Justice, Equity and Biodiversity: A Brief Overview

The meanings of justice and equity are necessarily plural and contested (see Rawls, 1999; Sen, 2009; Shelton, 2007). In the literature reviewed in this chapter, justice, equity and fairness are frequently considered to be synonymous or interchangeable, and our analysis does not rely on drawing a clear-cut distinctions between these terms. However, theorists often see justice as a more stringent set of moral (and sometimes legal) responsibilities that social institutions owe to humans (and sometimes also to nonhumans) as a matter of right, whereas equity may refer to a wider notion of fair, proportionate or nonarbitrary treatment (see e.g. Armstrong, 2019). As outlined in later sections, applied definitions frequently depart from the theoretical foundations of these terms, and the term "equity" tends to be invoked in policy contexts and at project level more than "justice."

A range of theories and conceptions of justice have emerged that relate to biodiversity. These include environmental and ecological justice (Kopnina, 2016; Schlosberg, 2007), social-ecological justice (Gunnarsson-Östling and Svenfelt, 2018), multispecies justice (Celermajer et al., 2021), just conservation (Gavin et al., 2015; Martin, 2017), just sustainabilities (Agyeman et al., 2003), equitable sustainability (Leach et al., 2018) and planetary justice (Dryzek and Pickering, 2019; Kashwan et al., 2020). One could also refer to the idea of "biodiversity justice" (Godden and O'Connell, 2015) or "just biodiversity governance" (Adam, 2014). Each of these conceptualizations of justice varies in several respects.

First, theories vary depending on who or what are the subjects of justice or rights-holders (Martin et al., 2016). These are commonly disaggregated to include gender, socioeconomic, racial, ethnic or cultural differences, while taking account of intersectionality across these characteristics (Schlosberg and Carruthers, 2010). Conventional accounts of environmental justice tend to be anthropocentric, while ecological and social-ecological accounts recognize nonhumans (e.g. animals, plants or ecosystems) as subjects of justice (Schlosberg, 2007; Chapter 9). Second, the theories operate over different spatial, temporal and sectoral scales. Some see the state as the primary site of justice, while others foreground a global perspective or underscore the agency of local communities and institutions (Sikor and Newell, 2014). Some theories focus on duties toward those living now, while others emphasize intergenerational justice (Dryzek and Pickering, 2019). A range of theories – particularly those that call for the explicit adoption of critical, decolonial, feminist and other lenses – situate questions of justice and biodiversity within broader processes that continue to perpetuate injustice, such as colonial exploitation and gender inequality (Alvarez and Coolsaet, 2020; Elmhirst, 2011; Pellow, 2017).

Three core dimensions have gained prominence in environmental justice scholarship over the last two decades: distribution, procedure and recognition (Schlosberg, 2007, building on Fraser, 1995). Distributive justice is the most widely researched and commonly recognized dimension. It encompasses who receives the benefits and opportunities versus who bears the costs and risks of social cooperation (Walker, 2012). Theories vary considerably as to what kinds of principles should determine a just distribution, such as equality, need or aggregate social utility/wellbeing (Kaswan, 2020). Procedural justice engages with the processes by which decisions are made (Davoudi and Brooks, 2014; Dawson et al., 2018a). Recognition pertains to the status afforded to multiple social groups, worldviews and cultural values and identities, and to issues of self-respect and self-esteem (Martin et al., 2016; Whyte, 2011; 2018) Examples of how each dimension of justice applies to biodiversity governance are outlined in Table 8.1. A final aspect of justice that is not always explicit in this tripartite categorization is corrective or remedial justice, which involves measures to correct or remedy unjust actions or omissions, such as sanctions for "ecocide" or violence against environmental defenders (Gonzalez, 2012; Whyte, 2011). Space constraints preclude a detailed discussion of this aspect, but related issues are discussed under distributive and procedural justice.

This chapter does not advocate any one of the conceptions of justice outlined above, but instead takes elements from each to adopt a pluralist approach spanning both social and ecological aspects, and all three dimensions of justice across multiple temporal, spatial and sectoral scales.

Table 8.1 *Dimensions of justice in biodiversity governance*

Dimension of justice	Examples in biodiversity governance
Procedural justice	• Inclusion and representation in formal processes (e.g. CBD negotiations or government policy-making) or informal/customary institutions and interactions (e.g. meetings of IPLC) • Access to information and justice (e.g. judicial review of environmental decisions)
Recognition	• Acknowledgment of and respect for Indigenous and local knowledge, diverse worldviews and ways of valuing nature • Recognition of customary land rights
Distributive justice	• Measures to address distributional impacts of biodiversity loss or of biodiversity policies (e.g. through area-based measures or mainstreaming) • International finance for conservation and sustainable use • Equitable sharing of benefits from use of genetic resources

8.2.2 Norms of Justice and Equity in Global Biodiversity Governance

Debates about justice and equity – particularly between the Global South and North – have pervaded the politics of global biodiversity governance since its emergence (Broggiato et al., 2015; Swanson, 1999). Discussions on global environmental governance since the 1970s prompted the Global South to develop a set of common demands on environmental issues (Williams, 1993), including on what Christopher Stone (1996) called the "most difficult moral question" regarding the Convention: the distribution of costs associated with conserving biodiversity. Most of the world's biodiversity is located in nonindustrialized countries, which generally have more limited capacity to pay for conservation than industrialized countries (see also Section 8.5). As a result, conservation has increasingly shifted toward more "people-friendly" and decentralized interventions such as "integrated conservation and development projects," driven by the belief that poverty was the main cause of environmental degradation (Roe, 2008).

Against this political backdrop, norms of equity, rights and justice have gained traction in key documents and practices of global biodiversity governance.[1] The CBD and the UNFCCC – both of which were adopted at the 1992 Earth Summit – were among the first multilateral environmental agreements to explicitly integrate equity. The CBD's third objective is "the *fair* and *equitable* sharing of the benefits arising out of the utilization of genetic resources" (UN, 1992, Article 1; emphasis added). While *inter*generational equity (i.e. equity between generations) was foundational to the narrative of sustainable development in the 1987 Brundtland Report, the CBD and the UNFCCC raised the profile of *intra*generational equity (i.e. equity among groups within a single generation) on the international environmental agenda (Okereke, 2006). A comparison of official documents associated with each treaty body shows how references to equity in the CBD are far more common than references to equity in the UNFCCC or to justice in either treaty (see Figure 8.1).

---

[1] We define norms as "shared expectations about appropriate behavior held by a community of actors" (Finnemore, 1996: 22).

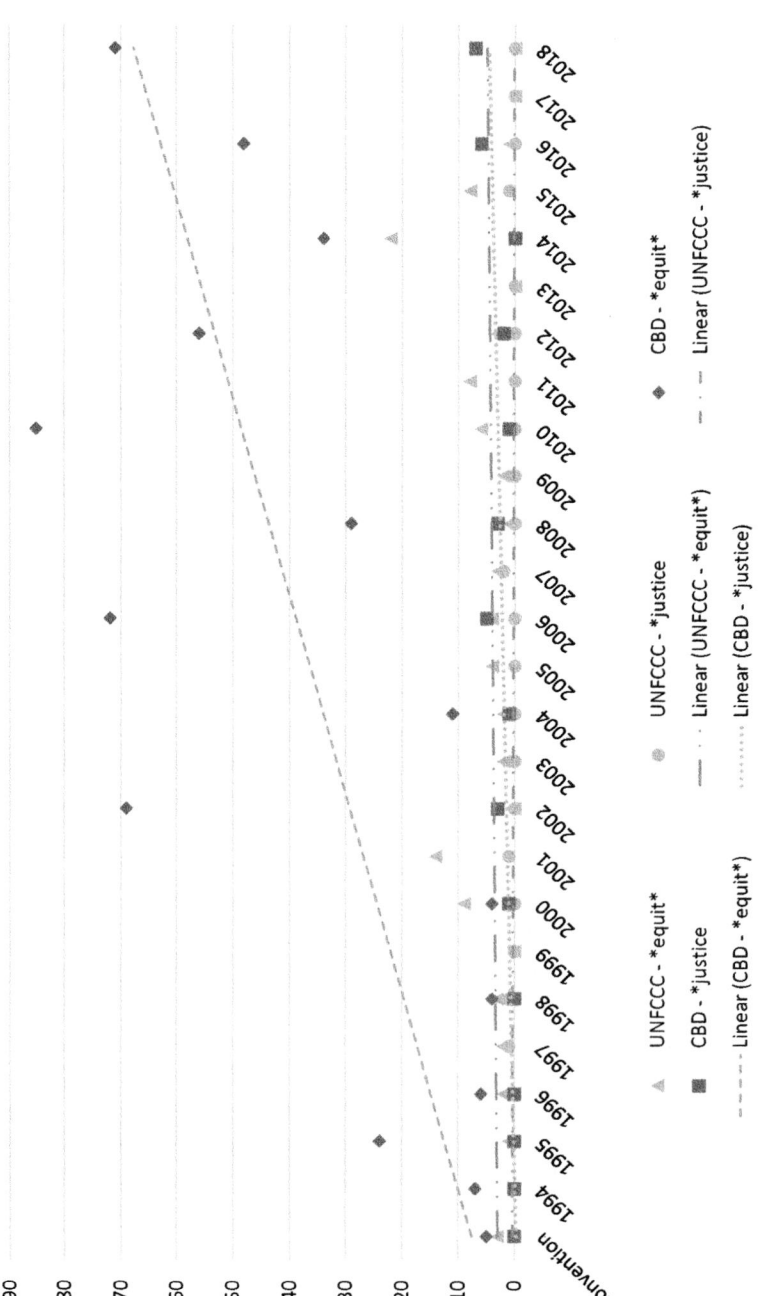

Figure 8.1 Frequency of references to equity and (in)justice in CBD and UNFCCC documents

Source: CBD and UNFCCC conventions and COP reports compiled for this chapter, excluding equity financing and names of organizations containing *equit* and/or *justice. The CBD COPs take place every other year. Peaks in CBD equity data generally coincide with heightened attention to equity in access and benefit-sharing, notably the Nagoya COP in 2010

This evidence reflects a broader observed tendency that equity is more commonly invoked than justice in international agreements (see also Okereke, 2008). In policy contexts, equity is often applied to specific policies or with a focus on a single dimension (most frequently distribution), allowing more politically sensitive issues such as historic land rights to be sidelined (Coolsaet et al., 2020).

Today, international policy norms on conservation cover most of the dimensions of justice introduced above (Dawson et al., 2018a; FAO, 2001, Article 1.1; Marion Suiseeya, 2017). In 2010, the CBD extended equity-related wording, which for a long time had been limited to access and benefit-sharing (ABS), to conservation efforts: Aichi Target 11 called for the conservation of biodiversity to take place through "effectively and equitably managed" protected areas (Zafra-Calvo et al., 2017).

Global norms entrenched in other international frameworks – especially ideas of rights – have played an increasingly important role in shaping debates about justice and equity in biodiversity governance (Coolsaet et al., 2020). Indigenous Peoples, for example, steward 85 percent of the world's remaining biodiversity, yet their ability to do so is threatened by weak and/or infringed political, economic and social rights (IPBES, 2019; Tauli-Corpuz, 2016). In recent years, the global Indigenous movement has worked to secure references to the UN Declaration on the Rights of Indigenous Peoples (UNDRIP) in texts negotiated at the CBD and the UNFCCC as ways to recognize their rights to self-determination, but also to protect their ability to steward lands and forests critical for biodiversity conservation (Marion Suiseeya and Zanotti, 2019). Linking biodiversity to the human rights to life and health, adequate standards of living and nondiscrimination in the enjoyment of rights, the UN Special Rapporteur on Human Rights and the Environment recognized that "the full enjoyment of human rights ... depends on biodiversity, and the degradation and loss of biodiversity undermine ... human rights" (Human Rights Council, 2017: 3). These developments have been complemented by the institutionalization of procedural environmental rights – particularly in regional agreements such as the Aarhus Convention and the Escazú Agreement – such as the right to participate in environmental decision-making and the recognition of rights to a healthy environment in many national constitutions (Gellers, 2017).

Despite these advances, biodiversity continues to decline at unprecedented rates, giving rise to calls to transform existing governance systems (see e.g. IPBES, 2019). The section that follows highlights justice and equity considerations that need to be taken into account specifically when moving toward transformative governance of biodiversity.

8.3 Rethinking Justice and Equity in the Context of Transformative Governance: Toward Just Transformation

What does transformation mean for justice and equity in biodiversity governance? Adopting the definition of Chapter 1, transformative governance embraces the multiple enabling processes that facilitate "fundamental system-wide reorganisation" (IPBES, 2019).

Transformative governance "seeks to achieve desired societal values" (Chaffin et al., 2016: 408; see also Chapters 1 and 4). However, determining what is desirable – including whether transformation is desirable at all – and how to achieve a desired transformation involves contestation over values, interests and worldviews. Indeed, rethinking core societal values can be seen as a constitutive feature of transformative governance (Chapters 1 and 4; IPBES, 2019). Questions about who should be involved in this contestation, how values should be rethought and who has the authority to make decisions underscore the political character of transformation (Blythe et al., 2018; Patterson et al., 2017), hence posing concerns of justice. Nevertheless, existing accounts of sustainability transformations have been criticized for their lack of attention to justice (Martin et al., 2020; Patterson et al., 2017). In contrast to more conventional or incremental approaches to biodiversity governance, the depth, scale and urgency of change associated with transformative biodiversity governance demand reflecting on its association with social and environmental justice.

First, transformative change requires *deep shifts* in existing patterns of production and consumption, disrupting inequalities of power that drive and arise from these patterns. Not only could misguided attempts at transformation result in an unjust redistribution of resources, but powerful vested interests may also resist transformative change and defend an unjust status quo. While transformative governance is often portrayed as universally beneficial, transformations inevitably produce winners and losers (Blythe et al., 2018; Morrison et al., 2017; Patterson et al., 2017). Even if the normative ideal of transformative governance entails justice (as stipulated in the Introduction), the implications of different policy options can be difficult to predict, and some forms of social transformation may in practice yield injustice, e.g. if the creation of protected areas deprives Indigenous peoples and local communities of access to their traditional lands (Chapters 2, 11, 12). Moreover, policy-makers and other powerful actors may manipulate discourses of transformation for unjust ends, for example to justify business as usual or to shift responsibility for behavioral change away from themselves and onto consumers or citizens (Blythe et al., 2018).

Second, the *geographic and temporal scale* of transformative change magnifies the justice challenges for transformative governance. Transformative change will require attention to the drivers of biodiversity loss emanating in one part of the world while affecting another (Liu et al., 2013; Chapters 1 and 4), e.g. where demand for beef or soy in Europe drives land clearing in the Amazon rainforest. Moreover, addressing transformative change over large geographic regions will inevitably need to deal with a tremendous diversity of meanings and claims of justice. Yet conventional understandings of social justice often center on relationships among participants in a domestic social contract and struggle to conceptualize relations of justice at a global level (Sikor and Newell, 2014). With regard to temporal scale, a strong argument for transformative biodiversity governance is that the continued loss of biodiversity, even if equitably distributed for present generations, will inevitably disadvantage future generations (Alvarez and Coolsaet, 2020). However, the costs of initiating transformative change rest initially on the present generation, raising questions of intergenerational equity (Martin et al., 2013).

Third, the *urgency* of transformative governance intensifies questions about the feasibility of pursuing justice. Invoking an ecological or climate emergency risks circumventing democratic safeguards and resulting in unjust reforms (Niemeyer, 2014). However, while halting biodiversity loss is long overdue, the urgency of the task does not make it impossible to consider the justice implications of critical decisions. Indeed, if hasty action results in further injustice, this is likely to damage public support for transformative governance and ultimately be counterproductive (Dryzek and Pickering, 2019).

The remaining sections explore in more depth how questions of justice and equity can be addressed in specific areas of transformative governance. Our discussion builds on ideas of a "just transition" to more sustainable societies. While the term has become prominent in climate policy – underscoring that the transition to renewable energy should not disproportionately affect groups such as coal miners or low-income electricity consumers (Stevis, 2020) – scholars and activists have applied the term to environmental justice more broadly (Ciplet and Harrison, 2020). Thus, one could think of a just transition (Newell and Mulvaney, 2013; Swilling, 2019) or even a "just transformation" (Bennett et al., 2019; Schlosberg et al., 2017) of biodiversity governance.[2] The idea of just transformation speaks to the notion that transformative governance must be integrative and inclusive (Chapter 1), and calls attention to the interests of disadvantaged or marginalized groups in the context of transformation, including nonhuman species and ecosystems. One might object that, since the ideal of transformative governance necessarily entails justice, the idea of "just transformation" is tautologous. We believe, however, that processes of *transformation* (as distinct from *transformative change*) may be just or unjust (see also Bennett et al., 2019). Moreover, considerations of justice can easily be overshadowed by the pursuit of transformations toward environmental sustainability; hence the need to foreground a *just* transformation (Martin et al., 2020).

8.4 How Should Decision-Making Processes Be Structured?

Transformative change demands a fundamental reordering and rescaling of how problems are defined, solutions are deliberated and decisions are reached. One of the five key ingredients of transformative governance set out in Chapter 1 is *inclusive* governance ("governance approaches through stakeholder engagement, including Indigenous Peoples and Local Communities, in decision-making processes." IPBES, 2019: 894). Transformative governance needs to be inclusive in order "to empower ... those whose interests are currently not being met and who represent values that constitute transformative change toward sustainability" (Chapter 1). Similarly, Chapter 1 stipulates that transformative governance needs to be *transdisciplinary*, "in ways that recognize different knowledge systems." Attention to inclusive and informed governance highlights the importance of procedural justice and recognition. At the same time, a just transformation further demands greater attention to the underlying forces that structure and constitute decision-making landscapes.

---

[2] Bennett et al. (2019: 5) define just transformations as "radical shifts in social–ecological system configurations through forced, emergent or deliberate processes that produce balanced and beneficial outcomes for both social justice and environmental sustainability." On the distinction between transition and transformation, see Chapters 1 and 4.

Relative to other environmental problems, the CBD is generally considered to be a rather inclusive arena (Coolsaet and Pitseys, 2015; Cordonier Segger and Phillips, 2015), even though debates on these questions are ongoing (Reimerson, 2013). At a local level, however, biodiversity governance most commonly remains in the control of external actors, both public and private, through management regimes that seek to amend local practices and override customary institutions (Coolsaet et al., 2020). Biodiversity conservation initiatives that fail to include affected communities in decision-making often fail to achieve their conservation objectives (Bell and Carrick, 2017; Bennett and Satterfield, 2018; Dawson et al., 2018b). Unjust decision-making processes can spark new conflicts (Paavola, 2004), compound injustices (Sikor, 2013), foment distrust of the decision-making process and its proponents (Brechin et al., 2003; Hotes and Opgenoorth, 2014), and undermine broader biodiversity governance objectives (Martin, 2017).

Drawing on a growing body of literature examining concepts and practices for ensuring broad representation and inclusive decision-making (see e.g. Walker, 2012), we direct attention to three key questions: Who should be included in decision-making processes? On what terms should decision-making processes take place? At what point do requirements of recognition and procedural justice begin?

8.4.1 Who Should Be Included in Decision-Making Processes?

At a minimum, procedural justice requires the inclusion and representation of affected parties in decision-making processes (Schlosberg, 2007). The authority to decide who should be included typically rests with powerful actors (e.g. governments or intergovernmental organizations) who may misuse this authority to entrench existing inequalities of power. However, that authority can be subjected to scrutiny and challenge by social movements or other actors. The task of assessing who counts as affected – and determining what sorts of processes justice requires – becomes even more complex in the context of transformative biodiversity governance, which may both extend and amplify the effects of ecological and policy change across different social groups.

Scholars and practitioners broadly agree that affected parties include those groups who are vulnerable to biodiversity loss and/or who might be adversely impacted by conservation policies (Martin et al., 2013). These groups include IPLC and other marginalized groups with land-, water- or sea-based identities and lifeways. Attending to how demographic features, such as gender, age, race, class and ethnicity, shape different groups' experiences with biodiversity governance is critical for understanding who affected parties are and how they are differentially affected (IPBES, 2019; Malin and Ryder, 2018; Marion Suiseeya and Zanotti, 2019). Efforts to address distributive injustice or lack of recognition may be undermined when those most affected are not part of decision-making processes (Marion Suiseeya, 2016). More contentious is how other actors affected by conservation policy – such as corporations whose practices contribute to biodiversity loss – should be included in decision-making processes in ways that do not reinforce or exacerbate asymmetries of power (Dempsey, 2016).

8.4.2 On What Terms Should Decision-Making Processes Take Place?

Procedural justice requires attention to at least four characteristics of affected actors' roles in decision-making processes: (1) physical *presence* of affected actors or their representatives in decision-making settings; (2) *access*, meaning the authority to be an active participant in decision-making processes rather than only an observer; (3) capacity to *leverage* access to exercise agency (e.g. the ability to initiate a proposal or make a statement without being first invited to do so) and (4) capacity to *influence* decision-making processes (Marion Suiseeya and Zanotti, 2019; Witter et al., 2015). Numerous studies have shown that presence and access alone are insufficient for procedural justice (Cooke and Kothari, 2001; Holland, 2017).

The CBD has been a leader among multilateral treaty bodies in the inclusion of IPLC in its decision-making processes (Jones-Walters and Çil, 2011). Unlike the UNFCCC, which severely limits how nonstate actors can directly engage in their proceedings, the CBD moves beyond presence as a measure of inclusion. For example, representatives from the IPLC constituency colead negotiations on issues that have direct implications for the wellbeing and lifeways of Indigenous peoples, such as the Convention's Working Group on Article 8(j) (which deals with traditional knowledge, innovations and practices) and related provisions. Indigenous peoples have similarly forged new ground in intergovernmental scientific bodies such as IPBES by securing formal mechanisms for integrating diverse knowledge and value systems into its processes (Tengö et al., 2017).

Procedural justice also requires attending to power inequalities and political representation. Uneven power relations – such as states' control over multilateral governance processes or the privileged access of some stakeholders to the ear of government – affect the ability of actors to contribute to decision-making processes (Schroeder, 2010). Tools such as Free, Prior and Informed Consent (FPIC) show how institutions can help to address power imbalances and dismantle barriers to direct engagement. If fully implemented, FPIC creates a formalized channel for marginalized groups to leverage their power by requiring that affected parties give consent to receiving biodiversity governance initiatives in their communities (Colchester and Ferrari, 2007).

In practice it is not feasible for everyone affected to directly engage in decision-making processes; all the more so in deliberation at a global level that affects billions of people. Not all actors may have the financial, linguistic, physical or social capabilities to participate directly (Reimerson, 2013). Where feasible, actors who cannot participate directly should be able to select their own representatives. In the case of nonhuman subjects (e.g. animals, plants and ecosystems), which cannot select humans to represent them, options include legally appointed custodians, or nongovernmental organizations or experts working on conservation or animal welfare and rights. Similarly, custodians may be formally appointed to represent future generations (Dryzek and Pickering, 2019; Schlosberg, 2007).

8.4.3 When Do Requirements of Recognition and Procedural Justice Begin?

Although most studies of procedural justice focus on collective decision-making processes, those processes only begin following the identification of a problem or issue. Public policy and political ecology scholars have demonstrated the extraordinary power held by those who are able to define problems and set agendas (Bardach and Patashnik, 2019; Corson et al., 2014) and the extended effects of agendas that often carry forward beyond the initial decision-making process (Hughes and Vadrot, 2019; MacDonald and Corson, 2012). The resulting problem definitions, agendas and venues influence which actors and issues engage and are privileged in the process. Attending to the ways in which different power hierarchies and inequalities inform the phase before decision-making on a given problem begins (the "decision-impetus phase") is critical for advancing procedural justice (Marion Suiseeya, 2020).

One example of the importance of the decision-impetus phase is the problem of biodiversity itself. The framing of the biodiversity problem was initially driven largely by conservation biologists (Haila, 2017; Takacs, 1996). The idea of biodiversity subsequently gained wider acceptance but still carries certain connotations that affect power relations and may not resonate with certain groups, e.g. seeing biodiversity loss as the depletion of a resource rather than as the disruption of a harmonious relationship between humans and nonhumans (see Chapter 9; Martin et al., 2013). This example highlights that while inclusion of affected actors in established decision-making processes is a critical element of transformative governance, just transformation requires earlier and broader attention to procedural justice and recognition.

8.5 How Should Resources Be Mobilized and Allocated?

While transformative governance is likely to yield net economic benefits over the longer term, it will require large-scale mobilization of financial resources and a shift away from financing activities that harm biodiversity (CBD, 2020; Chaffin et al., 2016; McCarthy et al., 2012; Chapter 6). However, given vast disparities in incomes worldwide, capacity to mobilize resources domestically varies widely. Justice requires that higher-capacity countries support those with more limited capacity (Armstrong, 2019).

The CBD obliges developed countries to "provide new and additional financial resources" to enable developing countries to meet their obligations under the Convention (UN, 1992: Article 20.2). Subsequently, Aichi Target 20 aimed for the mobilization of financial resources to "increase substantially from the current levels." A high-level panel of the CBD (2012) estimated the cost of meeting the Aichi targets globally at US$150–$440 billion per year, and it is likely that the cost of meeting more ambitious post-2020 targets will be at least within this range (CBD, 2020). Accordingly, resource mobilization has emerged as a key priority for the post-2020 framework.

In this section, we focus on two key questions that a just scale-up of resources for transformative biodiversity governance must address:

1. How should the global effort of mobilizing resources be shared among nation-states and nonstate actors?
2. How should resources be allocated across countries and communities?

Justice and Equity in Biodiversity Governance

Both questions raise complex issues of distributive justice but have been addressed far less in the literature on biodiversity finance than in literature on development assistance and climate finance. The discussion below draws on relevant findings from these other bodies of work.

8.5.1 Effort-Sharing

Recognizing the differentiated capabilities of its parties, the CBD notes "the importance of burden-sharing" among contributing parties in providing finance for developing countries (Article 20.2). This leaves open the question of which actors (whether states, international organizations, civil society or private actors) should contribute the most toward a scaled-up international financing effort: is it those who have contributed the most to biodiversity loss, those with the greatest capacity to mobilize resources or those who stand to gain the most (economically or otherwise) from conservation? These three principles – sometimes referred to as the contributor/polluter pays principle, the capacity to pay principle and the beneficiary pays principle – have been widely debated in the literature on climate justice (see e.g. Dellink et al., 2009; Page, 2011) but have so far received only modest attention in the literature on biodiversity finance (for notable contributions, see Armstrong, 2019; Balmford and Whitten, 2003).

While some argue that the extent to which actors will benefit from conservation should be the primary factor in distributing costs (Balmford and Whitten, 2003), others argue that a pluralist approach combining all three principles is necessary, not least because those who stand to benefit most – e.g. forest communities – may have little capacity to pay for additional conservation efforts, even though they are often the most active participants in existing conservation practices (Armstrong, 2019). Moreover, it would be unfair to expect potential beneficiaries to pay the most when others (e.g. consumers in other countries) may be driving biodiversity loss in those areas despite the availability of less destructive alternatives (Dowie, 2011).

To date, parties to the CBD have not been able to agree on how to translate principles of equity into transparent, quantified effort-sharing measures. Sharing the overall conservation financing effort typically operates more informally.[3] However, improved transparency about how much parties are providing could help to clarify which parties are fulfilling their obligations, and inclusive deliberation could help to build shared understandings about broad parameters for effort-sharing (Pickering et al., 2015).

8.5.2 Allocation

Evidence indicates that conservation spending is more effective in lower-income countries than higher-income ones (Waldron et al., 2017), suggesting potential synergies between just allocation and effective ecological outcomes. However, when it comes to the question of

---

[3] While the Global Environment Facility (GEF) has an established system of burden-sharing, this is not based on a strict formula derived from equity principles.

allocating finance *among* lower-income countries, justice and effectiveness could pull in different directions.

Allocation according to need is a prominent justice-based principle for determining distribution, but in practice it competes with other principles of allocation. Existing patterns of allocation for environmental aid reflect a mix of donors' interests (e.g. supporting neighboring countries or trade partners) and equity considerations such as recipients' needs (e.g. national income and extent of the environmental problem) (Hicks et al., 2008). Miller et al. (2013) find that a country's biodiversity need (measured using indicators such as the number of threatened species or species richness) and quality of governance are strong predictors of the level of biodiversity aid it receives; income is negatively but weakly correlated with levels of biodiversity aid.

Trade-offs may arise in allocation decisions because the countries with the greatest levels of need may not be the ones with the greatest capacity to manage funds effectively, for example where low-income status coincides with limited institutional capacity. Managing these trade-offs is further complicated by different interpretations of need (e.g. degree of risk of biodiversity loss or capacity for domestic resource mobilization: Miller et al., 2013).

A massive scale-up of biodiversity finance would place considerable stress on existing institutional capacity to manage resources, particularly in countries with more constrained capacity (Presbitero, 2016). While this needs to be taken into account in efforts to maximize effective use of biodiversity finance, there is a risk that low-income countries could be further marginalized if the lion's share of funding goes to middle-income countries with stronger institutional capacity (Arndt and Tarp, 2017). Demand-driven mechanisms for allocating biodiversity finance may help to manage (if not fully resolve) these trade-offs, as recipient countries' level of demand for finance may reflect a mix of need and institutional capacity. Enhancing recipient countries' control over subnational allocation of biodiversity finance could enhance the effectiveness of implementation as well as furthering principles of procedural justice (Duus-Otterström, 2015).

8.6 How Can Transformative Governance Be Implemented Equitably?

In this section we discuss concerns arising for two prominent strategies that aim to address the drivers of biodiversity loss: (1) scaling up area-based conservation, and (2) mainstreaming biodiversity considerations across all sectors of decision-making.

8.6.1 Equitably Scaling Up Area-Based Conservation Initiatives

There is considerable debate regarding the expansion of area-based conservation and visions to achieve this, including whether expansion should comprise protected areas or "other effective area-based conservation measures" (OECMs) (Büscher et al., 2017; Dudley et al., 2018; Chapters 11 and 12). Here we focus on two key questions of justice that arise in scaling up conserved areas: (1) redistributive effects and (2) questions of procedural justice and recognition in decision-making.

Efforts to expand protected areas commonly curtail existing patterns of resource use in those areas. Recent pledges by many world leaders involve expanding protected areas to cover 30 percent of the Earth's land and ocean surface by 2030. Proposals to expand this to 50 percent – e.g. the Half Earth Project (inspired by Wilson, 2016) and Nature Needs Half (Kopnina et al., 2018) – could impact as many as one billion people (Schleicher et al., 2019). Such efforts could meet considerable political resistance from rural populations, particularly if they ignore the legacy of colonial land reallocations, displacement of IPLC and "green grabs" (Büscher et al., 2017). Equally, resistance may emerge from powerful groups (e.g. resource extraction or infrastructure industries) that are exploiting areas slated for protection. Although the redistributive effects of protected area expansion are often understood in human terms, an ecological justice perspective – which extends compassion, caring and rights to the entire living community – draws attention to the ways in which protected area expansion redistributes the Earth's resources between humans and nonhumans (Kopnina et al., 2018). A perspective on justice that encompasses both human and nonhuman concerns could highlight possible areas of convergence between ecocentric conservationists and social justice activists. In the Amazon, for example, coalitions have formed between conservation biologists and social scientists, or between grassroots popular movements and environmental organizations, that have resulted in the creation of protected areas that combine zones for sustainable use (encompassing subsistence or commercial exploitation) and conservation (Inoue and Franchini, 2020). The more ambitious the protected area target, the more challenging it is likely to be to achieve such convergence.

Protected area expansion raises complex governance issues relating to rights, access and control, such that the question of *how* protected areas are managed is as important as *what* is to be protected (Büscher et al., 2017; Coolsaet et al., 2020). In implementing international commitments on protected areas (such as Aichi Target 11), governments have tended to focus on the "headline" numbers of how much area is protected, with less emphasis on qualitative factors such as Aichi Target 11's call for protected areas and OECMs to be "equitably managed" (CBD, 2010). This is partly due to practical and conceptual difficulties of measuring equity. Some impact assessment and evaluation tools (see e.g. Schreckenberg et al., 2016; Zafra-Calvo et al., 2017) and conceptual frameworks (Pascual et al., 2017) incorporating the three dimensions of justice have been developed and adopted by the CBD as voluntary guidance (CBD, 2018). However, barriers remain both to the adoption of these tools and to the achievement of equitable management, particularly where national legal frameworks do not recognize customary land rights.

International recognition of the global network of Indigenous and community conserved areas (ICCAs), along with evidence of their mutual benefits for human wellbeing and nature, offers an example of an emergent transformative change in biodiversity governance (Armitage et al., 2020; IPBES, 2019: chapter 6: 61; Tran et al., 2020). Establishment of an ICCA or "territory of life" requires the autonomy of local people to govern and manage their territories. In many instances, this necessitates an overhaul of land and other laws or policies to transfer power to local institutions, in addition to redressing discriminatory social and political norms. Thus, while a transformative scale-up of area-based conservation will pose significant challenges to existing power relations, it also offers an opportunity to redress a range of injustices (Tauli-Corpuz et al., 2020).

8.6.2 Justice and Equity in Mainstreaming Transformative Governance

Transformative governance beyond protected areas remains essential, as the main direct and indirect drivers of biodiversity loss emanate from outside these areas (Chapter 1; Büscher et al., 2017). Here we address risks of injustice when conservation interventions adversely affect marginalized groups. In implementing biodiversity governance, just transformation requires at a minimum (a) careful assessment to identify implementation options that avoid or minimize adverse effects on marginalized groups; and (b) where adverse effects cannot reasonably be avoided, incorporating additional measures to ensure that the wellbeing of these groups is protected.[4] As outlined in the Introduction, injustice may arise not only from practices that adversely impact biodiversity but also from measures to address biodiversity loss.

Taking the example of subsidies harmful to biodiversity (which are addressed in Aichi Target 3), some subsidies (e.g. for fossil fuel extraction) may benefit wealthy interests at the expense of disadvantaged groups, so dismantling them could yield a double dividend for biodiversity and social justice. However, other subsidies (e.g. for fuel or fertilizer) may be designed to benefit disadvantaged groups, so dismantling those subsidies may adversely affect those groups. More broadly, policies that seek to shift people's livelihoods away from practices that degrade biodiversity can exacerbate inequalities of gender, education, ethnicity or socioeconomic status (Bidaud et al., 2017; Blythe et al., 2018). While in some cases unequal impacts can be avoided by choosing an alternative option, in other cases there may be no reasonable alternatives, in which case supporting measures are required to mitigate those impacts.

We highlight four types of additional measures: monetary compensation, localized in-kind support for livelihoods, broader social protection mechanisms and a wider-ranging reconfiguration of social and political relations.

First, economic theories of reform often emphasize monetary transfers to alleviate or compensate for adverse impacts (or conversely monetary incentives to adopt sustainable practices). International biodiversity finance, as outlined in Section 8.5, may help to reduce the risk that conservation efforts will impede the ability of developing countries to address other pressing development priorities. Similarly, payment for ecosystem services (PES) initiatives may enable communities to participate in conservation without endangering their livelihoods (IPBES, 2019). However, there remains the risk that a compensatory perspective will fail to recognize the

---

[4] Note that these principles could also apply to area-based conservation measures.

incommensurability of different values attached to nature, the agency of affected groups and other options for enhancing their wellbeing (Lliso et al., 2020).

A second option is localized support for livelihoods, such as through integrated conservation and development projects (ICDPs). Most case studies report that local integrated approaches to conservation have yielded very little benefit to people, even in cases that led to more effective conservation (Lund and Saito-Jensen, 2013; Twinamatsiko et al., 2014). The emerging understanding from this experience is that conservation effectiveness should be conceived as linked to social justice, rather than to a narrow economic understanding of development. In other words, for biodiversity governance to be transformative it is necessary to shift from an "integrated conservation and development" model to one of "integrated conservation and justice" (Martin, 2017; Vucetich et al., 2018). This would include, for example, stronger recognition of local visions of nature in decision-making processes and support for local environmental stewardship instead of separating local livelihoods from ecosystems or resources of conservation value.

The need to scale up and mainstream biodiversity objectives beyond individual projects points to the importance of exploring a third kind of measure: broad-based social protection mechanisms. These could take the form of unemployment insurance, welfare payments or cash transfers for low-income families (e.g. the Bolsa Floresta program in Brazil), universal basic income or other instruments (de Haan, 2014). Unlike project-specific support, these measures would help to safeguard communities against a wider range of risks to their wellbeing. However, broader redistributive measures may be difficult to implement effectively – particularly in low-income countries – and may need to be supplemented by international support.

Finally, a fourth strategy is to combine conservation measures with broader systemic reform that advances all dimensions of justice, particularly for marginalized groups and environment and human rights defenders (Bille Larsen et al., 2021; Scheidel et al., 2020). This could occur through formal recognition of the rights of IPLC (e.g. through constitutional recognition, parliamentary representation or treaty processes), strengthening social safeguards in conservation policy (to address concerns of displacement and impacts on livelihoods), reform of land tenure legislation, or other measures (Tran et al., 2020). This fourth strategy underscores the importance of thinking well beyond the conventional policy toolkit of financial transfers if just transformation is to be achieved.

8.7 Policy Implications

Our review confirms that action is required at multiple levels to reinterpret and uphold justice in transformative biodiversity governance across diverse geographic, temporal and spatial scales. Key areas for policy innovation emerging from the preceding sections that could enhance justice in transformative governance – especially through the implementation of the Global Biodiversity Framework – are outlined in Box 8.1.

> **Box 8.1: Policy options for advancing justice in transformative biodiversity governance**
>
> - **Norm development and fulfillment:** Further development of international norms of equity and justice in global sustainability governance could take the form of new norms (e.g. just biodiversity governance) or further diffusion or expansion of existing norms (e.g. the application of human rights to biodiversity governance, or entrenchment of the principle of equity across all three objectives of the CBD). However, norm development by itself is insufficient: indeed, it could be argued that the CBD already has a range of well-developed norms to work with, and that the key issue is *compliance* with or *fulfillment* of those norms – an issue that we address in the subsequent points in this list.
> - **Policy integration:** There is a need for stronger integration of justice concerns in biodiversity policy-making, policy implementation and policy review at all levels of governance. One option for doing so would be to build on the Sustainable Development Goals (SDGs) framework – which includes goals on biodiversity and on social and economic equity, along with other socioeconomic objectives – and associated tools for mapping and managing synergies and trade-offs across goals (e.g. ICSU, 2017).
> - **Decision-making:** Greater attention to how existing approaches to decision-making can exacerbate injustices could be coupled with further entrenchment of procedural rights (including through the Aarhus Convention and related international agreements), practices and measures (e.g. FPIC) to ensure that marginalized groups can shape and influence collective decision-making.
> - **Resource mobilization:** This could take the form of credible, time-bound, multilateral, national and nonstate commitments to scale up resource mobilization to support biodiversity policy in developing countries – including meaningful progress on the long-discussed idea of a multilateral benefit-sharing mechanism (Nagoya Protocol, Article 10; see Chapters 6, 10 and 15) – along with efforts to build shared understandings about equitable effort-sharing and allocation of resources.
> - **Implementation:** Alongside more conventional measures to alleviate the impacts of conservation initiatives on marginalized groups (including social impact assessment and financial transfers), just transformation is likely to require strengthening broad-based social safety nets, international recognition of ICCAs and other measures to remedy unjust asymmetries of power in political systems (e.g. land reform and recognition of Indigenous rights).
> - **Monitoring, evaluation and accountability:** Meaningful mechanisms for monitoring and evaluating equity in conservation, sustainable use and benefit-sharing need to be developed, incorporated into decision-making, and used in reporting on national and collective performance under the post-2020 framework. Existing voluntary guidance for assessing equity in protected area management could be implemented as standard, used to hold decision-makers accountable and extended to other areas of biodiversity governance. Stronger measures are required to ensure that policy-makers and other actors are held accountable for their commitments to transformative change, and that legal sanctions are strengthened for those who persecute environmental defenders or wantonly destroy biodiversity on a large scale.

8.8 Conclusion

This chapter has demonstrated that in both conceptualizing and implementing transformative biodiversity governance, issues of justice need urgent attention. Justice is at the core of how to envision and achieve transformative change, and how to maintain a desired future state. Failure to take account of preexisting unjust conditions – or the potential for misguided governance strategies to create further injustice – may not only result in morally reprehensible decisions but may also provoke resistance that ultimately blocks transformative change and results in a failure to address the underlying causes of biodiversity loss. Transformative governance requires not only *inclusive* governance but a broader *integrative* vision of justice and sustainability, exemplified by the idea of just transformation.

The literature reviewed in this chapter emphasizes the need for a multidimensional view of justice – comprising not only distributive justice but also procedural justice and recognition – as well as attention to global, intergenerational and interspecies aspects, while also remaining cognizant of diverse social values and local circumstances. The depth, scale and urgency of transformative change underscore the importance of a multidimensional perspective. Achieving a simultaneous transformation toward justice and sustainability remains a daunting challenge replete with complex trade-offs. Nevertheless, it remains vital to strive for a just transformation in which everyone – especially those most often excluded in society – is able to participate in, influence and benefit from more just and sustainable biodiversity governance.

References

Adam, R. (2014). *Elephant treaties: The colonial legacy of the biodiversity crisis*. Hanover; London: University Press of New England.
Agyeman, J., Bullard, R. D., and Evans, B. (Eds.). (2003). *Just sustainabilities: Development in an unequal world*. Cambridge, MA: MIT Press.
Alvarez, L., and Coolsaet, B. (2020). Decolonising environmental justice studies: A Latin-American perspective. *Capitalism Nature Socialism* 31, 50–69.
Armitage, D., Mbatha, P., Muhl, E. K., Rice, W., and Sowman, M. (2020). Governance principles for community-centered conservation in the post-2020 global biodiversity framework. *Conservation Science and Practice* 2, e160.
Armstrong, C. (2019). Sharing conservation burdens fairly. *Conservation Biology* 33, 554–560.
Arndt, C., and Tarp, F. (2017). Aid, environment and climate change. *Review of Development Economics* 21, 285–303.
Balmford, A., and Whitten, T. (2003). Who should pay for tropical conservation, and how could the costs be met? *Oryx* 37, 238–250.
Bardach, E., and Patashnik, E.M. (2019). *A practical guide for policy analysis: The eightfold path to more effective problem solving*: Thousand Oaks, CA:CQ Press.
Bell, D., and Carrick, J. (2017). Procedural environmental justice. In: *The Routledge Handbook of Environmental Justice*. R. Holifield, J. Chakraborty and G. Walker (Eds.), pp. 101–112. Abingdon: Routledge.
Bennett, N. J., Blythe, J., Cisneros-Montemayor, A. M., Singh, G. G., and Sumaila, U. R. (2019). Just transformations to sustainability. *Sustainability* 11(14), 3881.
Bennett, N. J., and Satterfield, T. (2018). Environmental governance: A practical framework to guide design, evaluation, and analysis. *Conservation Letters* 11, e12600.
Bidaud, C., Schreckenberg, K., Rabeharison, M., et al. (2017). The sweet and the bitter: Intertwined positive and negative social impacts of a biodiversity offset. *Conservation and Society* 15, 1–13.

Bille Larsen, P., Le Billon, P., Menton, M., et al. (2021). Understanding and responding to the environmental human rights defenders crisis: The case for conservation action. *Conservation Letters* 14, e12777.

Blythe, J., Silver, J., Evans, L., et al. (2018). The dark side of transformation: Latent risks in contemporary sustainability discourse. *Antipode* 50, 1206–1223.

Brechin, S. R., Fortwangler, C. L., Wilshusen, P. R., and West, P. C. (2003). *Contested nature: Promoting international biodiversity with social justice in the twenty-first century*. Albany: SUNY Press.

Broggiato, A., Dedeurwaerdere, T., Batur, F., and Coolsaet, B. (2015). Access benefit-sharing and the Nagoya Protocol: The confluence of abiding legal doctrines. In: *Implementing the Nagoya Protocol. Comparing Access and Benefit-sharing Regimes in Europe*. B. Coolsaet, F. Batur, A. Broggiato, J. Pitseyes and T. Dedeurwaerdere (Eds.), pp. 1–29. Leiden: Brill/Martinus Nijhoff

Büscher, B., Fletcher, R., Brockington, D., et al. (2017). Half-Earth or Whole Earth? Radical ideas for conservation, and their implications. *Oryx* 51, 407–410.

CBD. (2010). The strategic plan for biodiversity 2011–2020 and the Aichi Biodiversity Targets. Decision X/2. UNEP/CBD/COP/DEC/X/2 (October 29, 2010).

(2018). Protected areas and other effective area-based conservation measures. Decision adopted by the Conference of the Parties to the Convention on Biological Diversity. CBD/COP/DEC/14/8 (November 30, 2018).

(2020). Estimation of resources needed for implementing the post-2020 global biodiversity framework: Preliminary second report of the panel of experts on resource mobilization. CBD/SBI/3/5/Add.2 (June 18, 2020).

Celermajer, D., Schlosberg, D., Rickards, L., et al. (2021). Multispecies justice: Theories, challenges, and a research agenda for environmental politics. *Environmental Politics* 30, 119–140.

Chaffin, B. C., Garmestani, A. S., Gunderson, L. H., et al. (2016). Transformative environmental governance. *Annual Review of Environment and Resources* 41, 399–423.

Ciplet, D., and Harrison, J. L. (2020). Transition tensions: Mapping conflicts in movements for a just and sustainable transition. *Environmental Politics* 29, 435–456. DOI: 10.1080/09644016.2019.1595883

Colchester, M., and Ferrari, M. F. (2007). *Making FPIC–free, prior and informed consent–work: Challenges and prospects for indigenous peoples*. Moreton-in-Marsh: Forest Peoples Programme. Available from https://bit.ly/3HiV2rt.

Cooke, B., and Kothari, U. (2001). *Participation: The new tyranny?* London; New York: Zed Books.

Coolsaet, B., Dawson, N., Rabitz, F., and Lovera, S. (2020). Access and allocation in global biodiversity governance: A review. *International Environmental Agreements: Politics, Law and Economics* 20, 359–375.

Coolsaet, B., and Pitseys, J. (2015). Fair and equitable negotiations? African influence and the international access and benefit-sharing regime. *Global Environmental Politics* 15, 38–56.

Cordonier Segger, M. C., and Phillips, F. K. (2015). Indigenous traditional knowledge for sustainable development: The biodiversity convention and plant treaty regimes. *Journal of Forest Research* 20, 430–437.

Corson, C., Gruby, R., Witter, R., et al. (2014). Everyone's solution? Defining and redefining protected areas at the convention on biological diversity. *Conservation and Society* 12, 190–202.

Davoudi, S., and Brooks, E. (2014). When does unequal become unfair? Judging claims of environmental injustice. *Environment and Planning A* 46, 2686–2702.

Dawson, N., Coolsaet, B., and Martin, A. (2018a). Justice and equity: Emerging research and policy approaches to address ecosystem service trade-offs. In: *Ecosystem services and poverty alleviation. trade-offs and governance*. K. Schreckenberg, G. Mace and M. Poudyal (Eds.), pp. 22–38. London: Routledge.

Dawson, N. M., Martin, A., and Danielsen, F. (2018b). Assessing equity in protected area governance: Approaches to promote just and effective conservation. *Conservation Letters* 11, e12388. DOI: 10.1111/conl.12388

de Haan, A. (2014). The rise of social protection in development: Progress, pitfalls and politics. *European Journal of Development Research* 26, 311–321.
Dellink, R., Den Elzen, M., Aiking, H., et al. (2009). Sharing the burden of financing adaptation to climate change. *Global Environmental Change* 19, 411–421.
Dempsey, J. (2016). *Enterprising nature: Economics, markets, and finance in global biodiversity politics.* Chichester, UK: Wiley.
Dowie, M. (2011). *Conservation refugees: The hundred-year conflict between global conservation and native peoples.* Cambridge, MA: MIT Press.
Dryzek, J. S., and Pickering, J. (2019). *The politics of the Anthropocene.* Oxford: Oxford University Press.
Dudley, N., Jonas, H., Nelson, F., et al. (2018). The essential role of other effective area-based conservation measures in achieving big bold conservation targets. *Global Ecology and Conservation* 15, e00424.
Duus-Otterström, G. (2015). Allocating climate adaptation finance: Examining three ethical arguments for recipient control. *International Environmental Agreements: Politics, Law and Economics* 16, 655–670.
Elmhirst, R. (2011). Introducing new feminist political ecologies. *Geoforum* 42, 129–132.
FAO. (2001). *International Treaty on Plant Genetic Resources for Food and Agriculture.* Rome: Food and Agriculture Organization of the United Nations. https://www.fao.org/3/i0510e/i0510e.pdf
Finnemore, M. (1996). *National interests in international society.* Ithaca, NY: Cornell University Press.
Fraser, N. (1995). From redistribution to recognition? Dilemmas of justice in a post-socialist age. *New Left Review* 212, 68–93.
Gavin, M. C., McCarter, J., Mead, A., et al. (2015). Defining biocultural approaches to conservation. *Trends in Ecology & Evolution* 30, 140–145.
Gellers, J. C. (2017). *The global emergence of constitutional environmental rights.* London: Routledge.
Godden, L., and O'Connell, E. (2015). Biodiversity justice in a climate change world: Offsetting the future. In: *The Search for Environmental Justice.* P. Martin, S. Z. Bigdeli, T. Daya-Winterbottom, W. du Plessis and A. Kennedy (Eds.), pp. 62–82. Cheltenham: Edward Elgar.
Gonzalez, C. G. (2012). Environmental justice and international environmental law. In: *The Routledge handbook of international environmental law.* E. J. Techera (Ed.), pp. 107–128. London: Routledge.
Gunnarsson-Östling, U., and Svenfelt, A. (2018). Sustainability discourses and justice: Towards social-ecological justice. In: *The Routledge handbook of environmental justice.* R.Holifield, J. Chakraborty and G. Walker (Eds.), pp. 160–171. Abingdon: Routledge.
Haila, Y. (2017). Biodiversity: Increasing the political clout of nature conservation. In: *Conceptual innovation in environmental policy.* J. Meadowcroft and D. J. Fiorino (Eds.), pp. 207–232. Cambridge, MA: MIT Press.
Hicks, R. L., Parks, B. C., Timmons Roberts, J., and Tierney, M. J. (2008). *Greening aid: Understanding the environmental impact of development assistance.* Oxford: Oxford University Press.
Holland, B. (2017). Procedural justice in local climate adaptation: Political capabilities and transformational change. *Environmental Politics* 26, 391–412. DOI: 10.1080/09644016.2017.1287625
Holland, T. G., Peterson, G. D., and Gonzalez, A. (2009). A cross-national analysis of how economic inequality predicts biodiversity loss. *Conservation Biology* 23, 1304–1313.
Hotes, S., and Opgenoorth, L. (2014). Trust and control at the science–policy interface in IPBES. *BioScience* 64, 277–278.
Howe, C., Suich, H., Vira, B., and Mace, G. M. (2014). Creating win-wins from trade-offs? Ecosystem services for human well-being: A meta-analysis of ecosystem service trade-offs and synergies in the real world. *Global Environmental Change* 28, 263–275.
Hughes, H., and Vadrot, A. B. M. (2019). Weighting the world: IPBES and the struggle over biocultural diversity. *Global Environmental Politics* 19, 14–37.
Human Rights Council. (2017). Report of the Special Rapporteur on the issue of human rights obligations relating to the enjoyment of a safe, clean, healthy and sustainable environment. A/HRC/34/49 (19 January 2017).

ICSU. (2017). *A guide to SDG interactions: From science to implementation* (D. J. Griggs, M. Nilsson, A. Stevance, and D. McCollum [Eds.]). Paris: International Council for Science.

Inoue, C. Y. A., and Franchini, M. (2020). Socio-environmentalism. In: *International relations from the global south: Worlds of difference*. A. B. Tickner and K. Smith (Eds.), pp. 159–184. Abingdon; New York: Routledge.

Intergovernmental Science-Policy Platform on Biodiversity and Ecosystem Services (IPBES). (2019). *Global assessment report of the Intergovernmental Science-Policy Platform on Biodiversity and Ecosystem Services*. E. S. Brondízio, J. Settele, S. Díaz and H. T. Ngo (Eds.). Bonn: IPBES secretariat.

Jones-Walters, L., and Çil, A. (2011). Biodiversity and stakeholder participation. *Journal for Nature Conservation* 19, 327–329.

Kashwan, P., Biermann, F., Gupta, A., and Okereke, C. (2020). Planetary justice: Prioritizing the poor in earth system governance. *Earth System Governance* 6, 100075.

Kaswan, A. (2020). Distributive environmental justice. In: *Environmental justice: Key issues*. B. Coolsaet (Ed.), pp. 21–36. Abingdon; New York: Routledge.

Kopnina, H. (2016). Half the Earth for people (or more)? Addressing ethical questions in conservation. *Biological Conservation* 203, 176–185.

Kopnina, H., Washington, H., Gray, J., and Taylor, B. (2018). The "future of conservation" debate: Defending ecocentrism and the Nature Needs Half movement. *Biological Conservation* 217, 140–148.

Leach, M., Reyers, B., Bai, X., et al. (2018). Equity and sustainability in the Anthropocene: A social–ecological systems perspective on their intertwined futures. *Global Sustainability* 1 Article e13, 1–13. https://doi.org/DOI: 10.1017/sus.2018.12

Liu, J., Hull, V., Batistella, M., et al. (2013). Framing sustainability in a telecoupled world. *Ecology and Society* 18 Article 26, 1–19. https://doi.org/10.5751/es-05873-180226

Lliso, B., Pascual, U., Engel, S., and Mariel, P. (2020). Payments for ecosystem services or collective stewardship of Mother Earth? Applying deliberative valuation in an indigenous community in Colombia. *Ecological Economics* 169, 106499.

Lund, J. F. and Saito-Jensen, M. (2013). Revisiting the issue of elite capture of participatory initiatives. *World Development* 46, 104–112.

MacDonald, K. I., and Corson, C. (2012). "TEEB begins now": A virtual moment in the production of natural capital. *Development and Change* 43, 159–184.

Malin, S. A., and Ryder, S. S. (2018). Developing deeply intersectional environmental justice scholarship. *Environmental Sociology* 4, 1–7.

Marion Suiseeya, K. R. (2016). Transforming Justice in REDD+ through a politics of difference approach. *Forests* 7, article 300. DOI: 10.3390/f7120300

(2017). Contesting justice in global forest governance: The promises and pitfalls of REDD+. *Conservation and Society* 15, 189–200.

(2020). Procedural justice matters: Power, representation, and participation in environmental governance. In: *Environmental Justice: Key Issues*. B. Coolsaet (Ed.), pp. 38–51. Abingdon; New York: Routledge.

Marion Suiseeya, K. R, and Zanotti, L. (2019). Making influence visible: Innovating ethnography at the Paris Climate Summit. *Global Environmental Politics* 19, 38–60.

Martin, A. (2017). *Just conservation: Biodiversity, wellbeing and sustainability*. Abingdon: Routledge.

Martin, A., Armijos, M. T., Coolsaet, B., et al. (2020). Environmental justice and transformations to sustainability. *Environment: Science and Policy for Sustainable Development* 62, 19–30.

Martin, A., Coolsaet, B., Corbera, E., et al. (2016). Justice and conservation: The need to incorporate recognition. *Biological Conservation* 197, 254–261.

Martin, A., McGuire, S., and Sullivan, S. (2013). Global environmental justice and biodiversity conservation. *Geographical Journal* 179, 122–131.

Martinez-Alier, J. (2002). *The environmentalism of the poor: A study of ecological conflicts and valuation*. Cheltenham: Edward Elgar.

McCarthy, D. P., Donald, P. F., Scharlemann, J. P., et al. (2012). Financial costs of meeting global biodiversity conservation targets: Current spending and unmet needs. *Science* 338, 946–949.

McShane, T. O., Hirsch, P. D., Trung, T. C., et al. (2011). Hard choices: Making trade-offs between biodiversity conservation and human well-being. *Biological Conservation* 144, 966–972.

Miller, D. C., Agrawal, A., and Roberts, J. T. (2013). Biodiversity, governance, and the allocation of international aid for conservation. *Conservation Letters* 6, 12–20.

Morrison, T. H., Adger, W. N., Brown, K., et al. (2017). Mitigation and adaptation in polycentric systems: Sources of power in the pursuit of collective goals. *Wiley Interdisciplinary Reviews: Climate Change* 8. https://onlinelibrary.wiley.com/doi/full/10.1002/wcc.479

Newell, P., and Mulvaney, D. (2013). The political economy of the "just transition." *Geographical Journal* 179, 132–140.

Niemeyer, S. (2014). A defence of (deliberative) democracy in the Anthropocene. *Ethical Perspectives* 21, 15–45.

Okereke, C. (2006). Global environmental sustainability: Intragenerational equity and conceptions of justice in multilateral environmental regimes. *Geoforum* 37, 725–738.

(2008). Equity norms in global environmental governance. *Global Environmental Politics* 8, 25–50.

Paavola, J. (2004). Protected areas governance and justice: Theory and the European Union's Habitats Directive. *Environmental Sciences* 1, 59–77.

Page, E. (2011). Climatic justice and the fair distribution of atmospheric burdens: A conjunctive account. *The Monist* 94, 412–432.

Pascual, U., Balvanera, P., Díaz, S., et al. (2017). Valuing nature's contributions to people: The IPBES approach. *Current Opinion in Environmental Sustainability* 26, 7–16.

Patterson, J., Schulz, K., Vervoort, J., et al. (2017). Exploring the governance and politics of transformations towards sustainability. *Environmental Innovation and Societal Transitions* 24, 1–16.

Pellow, D. N. (2017). *What is critical environmental justice?* New York: John Wiley & Sons.

Pickering, J., Jotzo, F., and Wood, P. J. (2015). Sharing the global climate finance effort fairly with limited coordination. *Global Environmental Politics* 15, 39–62.

Presbitero, A. F. (2016). Too much and too fast? Public investment scaling-up and absorptive capacity. *Journal of Development Economics* 120, 17–31.

Rawls, J. (1999). *A theory of justice.* revised ed. Cambridge, MA: Belknap Press of Harvard University Press.

Reimerson, E. (2013). Between nature and culture: Exploring space for indigenous agency in the Convention on Biological Diversity. *Environmental Politics* 22, 992–1009. DOI: 10.1080/09644016.2012.737255

Roe, D. (2008). The origins and evolution of the conservation-poverty debate: A review of key literature, events and policy processes. *Oryx* 42, 491–503.

Scheidel, A., Del Bene, D., Liu, J., et al. (2020). Environmental conflicts and defenders: A global overview. *Global Environmental Change* 63, 102104.

Schleicher, J., Zaehringer, J. G., Fastré, C., et al. (2019). Protecting half of the planet could directly affect over one billion people. *Nature Sustainability* 2, 1094–1096.

Schlosberg, D. (2007). *Defining environmental justice: Theories, movements, and nature.* Oxford; New York: Oxford University Press.

Schlosberg, D., and Carruthers, D. (2010). Indigenous struggles, environmental justice, and community capabilities. *Global Environmental Politics* 10, 12–35.

Schlosberg, D., Collins, L. B., and Niemeyer, S. (2017). Adaptation policy and community discourse: Risk, vulnerability, and just transformation. *Environmental Politics* 26, 413–437. DOI: 10.1080/09644016.2017.1287628

Schreckenberg, K., Franks, P., Martin, A., and Lang, B. (2016). Unpacking equity for protected area conservation. *Parks* 22, 11–26.

Schroeder, H. (2010). Agency in international climate negotiations: The case of indigenous peoples and avoided deforestation. *International Environmental Agreements: Politics, Law and Economics* 10, 317–332.

Sen, A. (2009). *The idea of justice.* London: Allen Lane.

Shelton, D. (2007). Equity. In: *Oxford handbook of international environmental law.* D. Bodansky, J. Brunnée and E. Hey (Eds.), pp. 640–662. Oxford: Oxford University Press.

Sikor, T. (2013). *The justices and injustices of ecosystem services.* London: Routledge.

Sikor, T., and Newell, P. (2014). Globalizing environmental justice? *Geoforum* 54, 151–157.
Snyder, H. (2019). Literature review as a research methodology: An overview and guidelines. *Journal of Business Research* 104, 333–339.
Stevis, D. (2020). Labour unions and environmental justice: The trajectory and politics of just transitions. In: *Environmental Justice: Key Issues.* B. Coolsaet (Ed.), pp. 249–265. Abingdon; New York: Routledge.
Stone, C. D. (1996). La convention de Rio de 1992 sur la diversité biologique. In: *Le droit international face à l'éthique et à la politique de l'environnement.* I. Rens (Ed.), pp. 119–131. Geneva: Georg.
Swanson, T. (1999). Why is there a biodiversity convention? The international interest in centralized development planning. *International Affairs* 75, 307–331.
Swilling, M. (2019). *The age of sustainability: Just transitions in a complex world.* London: Routledge.
Takacs, D. (1996). *The idea of biodiversity: Philosophies of paradise.* Baltimore, MD: Johns Hopkins University Press.
Tauli-Corpuz, V. (2016). Report of the Special Rapporteur of the Human Rights Council on the rights of indigenous peoples. United Nations General Assembly. A/71/150 (29 July 2016). Available from https://bit.ly/34a58N0.
Tauli-Corpuz, V., Alcorn, J., Molnar, A., Healy, C., and Barrow, E. (2020). Cornered by PAs: Adopting rights-based approaches to enable cost-effective conservation and climate action. *World Development* 130, 104923.
Tengö, M., Hill, R., Malmer, P., et al. (2017). Weaving knowledge systems in IPBES, CBD and beyond – Lessons learned for sustainability. *Current Opinion in Environmental Sustainability* 26, 17–25. https://doi.org/https://doi.org/10.1016/j.cosust.2016.12.005
Tran, T. C., Ban, N. C., and Bhattacharyya, J. (2020). A review of successes, challenges, and lessons from Indigenous protected and conserved areas. *Biological Conservation* 241, 108271.
Twinamatsiko, M., Baker, J., Harrison, M., et al. (2014). *Linking conservation, equity and poverty alleviation: Understanding profiles and motivations of resource users and local perceptions of governance at Bwindi Impenetrable National Park, Uganda.* London: International Institute for Environment and Development. Available from https://bit.ly/34jMli7.
UN. (1992). Convention on Biological Diversity. United Nations Treaty Series 1760, 79-307. Available from https://bit.ly/3s86Eam.
Vucetich, J. A., Burnham, D., Macdonald, E. A., et al. (2018). Just conservation: What is it and should we pursue it? *Biological Conservation* 221, 23–33.
Waldron, A., Miller, D. C., Redding, D., et al. (2017). Reductions in global biodiversity loss predicted from conservation spending. *Nature* 551, 364–367.
Walker, G. (2012). *Environmental justice: Concepts, evidence and politics.* Abingdon; New York: Routledge.
Whyte, K. P. (2011). The recognition dimensions of environmental justice in Indian country. *Environmental Justice* 4, 199–205.
  (2018). The recognition paradigm of environmental injustice. In: *The Routledge handbook of environmental justice.* R.Holifield, J. Chakraborty and G. Walker (Eds.), pp. 113–123. Abingdon: Routledge.
Williams, M. (1993). Re-articulating the third world coalition: The role of the environmental agenda. *Third World Quarterly* 14, 7–29.
Wilson, E. O. (2016). *Half-Earth. Our planet's fight for life.* New York: Norton.
Witter, R., Marion Suiseeya, K. R., Gruby, R. L., et al. (2015). Moments of influence in global environmental governance. *Environmental Politics* 24, 894–912. DOI: 10.1080/09644016.2015.1060036
Zafra-Calvo, N., Pascual, U., Brockington, D., et al. (2017). Towards an indicator system to assess equitable management in protected areas. *Biological Conservation* 211, 134–141.

9

Mainstreaming the Animal in Biodiversity Governance: Broadening the Moral and Legal Community to Nonhumans

ANDREA SCHAPPER, INGRID J. VISSEREN-HAMAKERS, DAVID HUMPHREYS AND CEBUAN BLISS

9.1 Introduction

The individual animal has often been neglected in biodiversity governance debates, with animals mainly considered in terms of species, biodiversity, wildlife or natural resources. Indeed, and somewhat counterintuitively, biodiversity governance is not always animal-friendly. Think, for example, of the issues of wildlife management, ("sport") hunting, captive breeding, reintroduction and relocation of endangered species, and the use of animal testing in conservation research (De Mori, 2019). For some issues, the relationship is more complex, for example the "management" of Invasive Alien Species (IAS), which is detrimental to the individuals of the species considered "invasive" but beneficial to native species and habitats (Barkham, 2020). Elsewhere, economic development and incentives impact both biodiversity and animal concerns, such as the negative effects of animal agriculture (see Visseren-Hamakers, 2018a; 2020 for more detailed overviews of these relationships). How can we transform biodiversity governance in order to incorporate individual animal interests (Bernstein, 2015)? That is the central question of this chapter.

To answer this question, we apply an integrative governance perspective to link animal and biodiversity governance systems. Integrative governance can be defined as the theories and practices focused on the relationships between governance instruments (policies and rules) and systems (the entirety of instruments on a specific issue at a certain level of governance, from the global to the local) (Visseren-Hamakers, 2015; 2018a; 2018b). Our main argument focuses on integrating the interests of the individual animal in order to enable a shift from dominant anthropocentric ontologies to a more ecocentric approach, thereby improving human–nonhuman relationships and preventing further biodiversity loss without compromising our ethical obligations. The chapter argues that transformative biodiversity governance requires integrating animal rights and rights of nature approaches to enable a shift from dominant anthropocentric ontologies to a more ecocentric approach.

We review relevant literature and policy developments through an integrative governance perspective (Visseren-Hamakers, 2015) that brings together debates which, to date, have remained rather disconnected, including those in philosophy, political science, law and veterinary sciences. We also discuss attempts to integrate these debates. We have organized the review into academic and policy debates around: animal rights; animal welfare; rights of nature and integrative approaches, including One Health, One Welfare and compassionate conservation. Our literature review outlining academic debates is based primarily on secondary sources, but also includes gray literature and documents including legislative texts, policy papers, and reports by international and civil society organizations. The chapter does not provide a comprehensive overview of animal and biodiversity governance around the world, but rather aims to show how different concepts are operationalized in various contexts. Below, we first review the different debates and practices. The discussion section integrates the debates and reflects on their transformative potential, and the conclusion reflects on their implications for transformative biodiversity governance.

9.2 Animal Rights

9.2.1 The Academic Debate

The idea that animals are rights-holders has origins in political theory, philosophy and law. Until today, the discourse and practice on animal rights, including the animal rights movement, has been inspired by normative thinking on interspecies justice, in other words justice for and between human and nonhuman animals (Donaldson and Kymlicka, 2011; Nussbaum, 2006; Regan, 1983; Singer, 1975). Cavalieri, for example, proposes deleting the word "human" from human rights (Cavalieri, 2001), thus expanding our understanding of rights to other species.

Two influential monographs on animal ethics were published in the 1970s and early 1980s: *Animal Liberation* by Peter Singer (1975); and *The Case for Animal Rights* by Tom Regan (1983). Singer proposes a more sophisticated account of equality, extending it to all beings irrespective of gender, ethnicity or, indeed, species. He builds on the concept of speciesism (Ryder, 1971), which, analogous with racism, discriminates against species other than one's own. Following the eighteenth/nineteenth century philosopher Jeremy Bentham, who suggests that we should not ask whether animals can reason or talk but whether animals can suffer, Singer proposes we consider their sentience. He argues that the capacity to suffer gives one the right to equal consideration with others. To avoid vast suffering of nonhuman animals, humans need to make radical changes not only to their diet, farming methods, scientific experiments, practices of hunting, trapping and wearing fur, but also to entertainment, including circuses, zoos and rodeos (Singer, 1975). Singer is not against using animals but argues that their interests should be considered on an equal basis to those of humans.

Regan (1983) agrees with Singer that speciesism is unjust and wrong. However, what he conceives as wrong is to view animals as human resources, that is, to eat them, to exploit them for entertainment, sport, or any commercial activity, or to surgically manipulate them for medical research. Regan denies that animal husbandry methods should become "more humane"; he supports the complete abolition of commercial animal agriculture (Regan, 1983: 337). He also criticizes the utilitarian perspective of Singer: the value of animals cannot be reduced to their usefulness for the greater good of others (Regan, 1983: 343). It is our duty to recognize their rights and, as such, Regan views the animal rights movement as part of the human rights movement. Thus, in animal ethics one can differentiate between interest theories of rights for eliminating animal suffering, such as Singer's, and anti-use theories supported by Regan (Regan, 1983; see also Ahlhaus and Niesen, 2015: 16).

More recently, in *Zoopolis: A Political Theory of Animal Rights* (2011), Sue Donaldson and Will Kymlicka argue for a more comprehensive approach to animal rights that varies according to the relational nature of specific groups of animals to humans. Such an approach integrates universal negative rights, such as the absence of suffering, with differentiated positive rights, such as healthcare for domesticated animals, depending on the character of the human–animal relationship (Donaldson and Kymlicka, 2011: 11; see also Ahlhaus and Niesen, 2015: 18). Donaldson and Kymlicka argue that liberalism today combines universal human rights with more relational, bounded and group-differentiated rights. Upon this base, they claim, citizenship theory can be fruitfully used to "combine traditional animal rights theory with a positive and relational account of obligations" (Donaldson and Kymlicka, 2011: 14).

When referring to human–animal relationships, Donaldson and Kymlicka differentiate between: (a) animals living in the wild forming sovereign communities in their own territories, (b) animals that, similar to migrants or denizens, move into areas of human habitation and (c) domesticated animals that have been bred over generations to coexist with human beings. Domesticated animals, Donaldson and Kymlicka argue, should enjoy citizenship rights (Donaldson and Kymlicka, 2011: 14). Acknowledging domestic animals as citizens with rights is a moral obligation that arises from their integration into human societies, which removes their independence and ability to survive in the wild. Wild animals, in contrast, should be conceived as citizens of their own sovereign communities whose autonomy and territory should be respected. Non-domesticated "liminal" animals living among humans are compared to denizens. They need to be respected as coresidents of urban spaces but are not included in the citizenship scheme of humans and domesticated animals (Donaldson and Kymlicka, 2011: 15).

By employing political concepts, such as citizenship, denizenship, sovereignty, territory, migration and membership, and exploring their use or adaptation in the context of animals, Donaldson and Kymlicka make a clear attempt to promote animal rights beyond mere justifications for rights and justice for animals. While this has been criticized by some scholars because it challenges the distinctive meanings of concepts like citizenship or denizenship (Ladwig, 2015; Seubert, 2015; Stein, 2015), it has also given fresh impetus to the debate on animal rights. If animals are citizens, they are perceived as actors that can directly participate in political communities and be represented through institutions (Donaldson and Kymlicka, 2011). Especially in democratic political systems, Peter Niesen (2019) argues, there is consensus that those affected by laws should be able to influence the process of making these laws. If institutions neglect certain perspectives and interests, they are undemocratic. We therefore need to rethink our relationship with (and domination over) animals (Niesen, 2019: 381). This is reiterated from a post-humanist perspective, which deconstructs species supremacy and anthropocentrism to acknowledge animals' own agency (Braidotti, 2013). This perspective leads us to question whether humans have the "right" to grant animals rights at all.

9.2.2 Political Practice

The modern animal rights movement has been heavily influenced by the work of the philosophers Singer and Regan (Wise, 2016). Additionally, lawyers, scientists, academics, veterinarians, theologians and psychologists have influenced the movement. Consequently, since the beginning of the twenty-first century, a number of lawsuits have been brought forward to protect the interests and rights of animals. Legal scholars have attempted to advance basic animal rights in political practice, often accompanied by scientific evidence that provides a better understanding of the capacities and behavior of animals (Wise, 2016).

An increasing number of animal rights groups have raised awareness of the abusive conditions in which animals are kept, including on factory farms and in medical research laboratories. Rights groups are active at various levels, from local animal shelters to international groups such as PETA (People for the Ethical Treatment of Animals).

At the national level, the animal rights movement has succeeded in achieving stronger legal protection of animals by lobbying for the inclusion of animal rights in national constitutions. Two prominent examples are Switzerland and Germany. Animal protection has long been an issue of debate in the Swiss parliament. The "dignity of creatures" ("die Würde der Kreatur") was first mentioned in the constitution of the canton Aargau in 1980. It initiated a wider debate about the need to include animal welfare and dignity in the federal constitution (Goetschel, 2000: 12). The discourse on animal protection in Switzerland has been strongly linked to debates about the legal boundaries of genetic engineering. On the basis of a successful animal rights campaign, a constitutional amendment was passed in 1992 that stated that researchers need to respect the "dignity of creatures" (Jaber, 2000). In the course of creating a new constitution in 1998, animal activists tried to strengthen this amendment but were unsuccessful. However, in 2000, the wording of the 1992 amendment was included in the revised constitution (Evans, 2010: 239).

In Germany, a decade-long battle between campaigners and conservative politicians ended with paragraph 20a of the German constitution stating that animals have to be respected and protected by the state (Connolly, 2002). The campaign was started because the basic law protected freedom of research and freedom of profession. As a consequence, courts usually ruled in favor of researchers, even if they conducted experiments that caused animal suffering (Evans, 2010: 235). A political opportunity arose when a Social Democrat/ Green government coalition was in power from 1998 until 2002, after animal activists' efforts to include animal rights in the constitution were blocked by the Christian-Democrat majority in parliament during the 1990s. In 2002, activists increased public awareness after the Supreme Court granted permission to practice a traditional religious slaughter ritual that – according to many campaigners – involved unnecessary cruelty (Judd, 2003: 122). Public opinion against this decision and the support of the Green Party led to a successful constitutional amendment that year. Article 20a of German Basic Law now reads:

"(t)he state protects, in the interest of future generations, the natural basis of life, and the animals, within the framework of constitutional laws and through the making of laws and in accordance with ordinances and through judicial decision." (German Basic Law, Art. 20a).

Here, we can see the strong link between animal rights, rights of nature and intergenerational justice. Even though the German Animal Protection League was hoping that this constitutional amendment would lead to a number of relevant changes protecting animals in Germany (Connolly, 2002), there are still many problems, mostly relating to animals kept in factory farms and live animal transport. However, legislative changes at the federal and state level following the constitutional amendment of 2002 have almost completely eliminated inhumane research practices involving animals, and keeping animals for fur farming.

> **Box 9.1: Oostvaardersplassen: Animal Welfare and Rights Versus Conservation**
>
> In the Netherlands, the Oostvaardersplassen rewilding project has been subject to controversy after large herbivores (Konik horses, Heck cattle and red deer) introduced by humans starved when they exceeded the carrying capacity of the fenced-in nature reserve. There was a political debate among the Dutch public and animal protection NGOs, who felt responsibility for the welfare of these animals and the duty to prevent unnecessary suffering, and managers stressing the importance of noninterference and allowing natural processes to occur (Kopnina et al., 2019; Lorimer and Driessen, 2014; Ohl and van der Staay, 2012). These animals straddle the divide between wild and domesticated and raise questions regarding our level of responsibility for their welfare, and indeed what their rights are.

9.3 Animal Welfare

9.3.1 The Academic Debate

Over the last three decades, animal welfare has accelerated as a field of scientific study. There is no universally accepted definition of animal welfare and the various conceptions in use lead to different ways of assessing the welfare of animals (Weary and Robbins, 2019). Most definitions, however, differentiate between physical elements contributing to, or impeding, the welfare of animals, including malnutrition, exposure, disease and injury, on the one hand, and affective elements like thirst, hunger, discomfort, pain, fear and distress, on the other hand (Mellor, 2016: 8). Challenges to animal welfare can originate in natural and unnatural environments, and to assess the welfare of an individual animal or collective species one needs to consider not only fitness and health, biological needs and wants, but also animals' sensory or emotional experiences, feelings or affective states (Mellor, 2016: 14).

Important ideas on animal welfare originate in the 1965 *Report of the Technical Committee to Enquire into the Welfare of Animals Kept under Intensive Livestock Husbandry Systems*, also known as the Brambell report, published in the UK. The report highlighted that farm animals should be guaranteed five freedoms: to "stand up, lie down, turn around, groom themselves and stretch their limbs" (FAWC, 2009). In reaction to the Brambell report, the UK Farm Animal Welfare Advisory Committee (FAWAC) was established, and subsequently the Farm Animal Welfare Council (FAWC). John Webster, a former Professor of Animal Husbandry at the University of Bristol, helped develop the five freedoms. In his book *Animal Welfare: A Cool Eye Towards Eden*, he explains the usefulness of this framework in order to assess animal welfare:

Preserving the concept of the "Five Freedoms", I attempted to produce a logical, comprehensive method for first analysis of *all* the factors likely to influence the welfare of farm animals, whether on the farm itself, in transit or at the point of slaughter. *(Webster, 1994: 11).*

Minimum standards based on the five freedoms have been modified by the FAWC, which in 2019 was renamed the Animal Welfare Committee (AWC), and were supplemented by five provisions detailing how to implement them. The AWC today classifies animals' quality of life as a good life, a life worth living and a life not worth living (FAWC, 2009: iii). Furthermore, in 2018, the UK Government acknowledged animal sentience, which it defines as "the capability to experience pain, distress and harm" (FAWC, 2018), reiterating its commitment to Article 13 of the Lisbon Treaty of the European Union (EU), which recognizes animal sentience. Such recognition paves the way for the acknowledgment of the individual animal in biodiversity governance.

Considerations on animal welfare, including relevant welfare and assessment schemes in the UK and beyond, are still guided by the five freedoms and respective provisions (Mellor, 2016: 2). The 2009 FAWC report includes:

- *Freedom from hunger and thirst*, by ready access to water and a diet to maintain health and vigour;
- *Freedom from discomfort*, by providing an appropriate environment;
- *Freedom from pain, injury and disease*, by prevention or rapid diagnosis and treatment;
- *Freedom to express normal behavior*, by providing sufficient space, proper facilities and appropriate company of the animal's own kind;
- *Freedom from fear and distress*, by ensuring conditions and treatment, which avoid mental suffering
(FAWC, 2009: 2).

Even though the language of "freedom" is akin to the human rights language employed in international agreements (e.g. in the Universal Declaration of Human Rights), there is a clear distinction between animal rights and animal welfare approaches. Whereas animal rights proponents emphasize that it is morally wrong for humans to use or exploit animals, animal welfarists are concerned with reducing suffering. Welfarists' acceptance of the instrumental use of animals by humans is in accordance with anthropocentric thinking, and in line with arguments brought forward by Singer. This utilitarian perspective is contrary to the philosophical ideas of Regan and animal rights proponents, who argue against using animals as a resource to be exploited by humans at all. Still, the five freedoms and pertinent animal welfare schemes are criticized by others for being normative and too idealistic to serve as a code of recommendation for welfare assessment (McCulloch, 2013).

Furthermore, the five freedoms have been criticized for being tailored to contexts of animal exploitation (Haynes, 2011), and focused on "negative freedoms" in which "freedom from" (e.g. hunger, disease and fear) is emphasized. The exception is "freedom to express normal behavior" (FACW, 2009). Scholars have suggested that this focus on negative experiences may not be sufficient because animal welfare should also comprise positive elements, such as being housed in species-relevant environments and encouraging animal-to-animal interaction (Mellor, 2016: 2). A more subjective measure of welfare, qualitative behavior assessment (QBA), goes some way to countering the criticisms on the five freedoms. QBA proposes an integrative measurement to assess the behavior of an animal and its interaction with its environment (Wemelsfelder and Lawrence, 2001).

9.3.2 Political Practice

Conceptualizations of animal welfare, and in particular the five freedoms, have had considerable impact on policy development from the global to the national level.

The World Organisation for Animal Health (OIE), established in 1924, is an intergovernmental organization with 182 member states. It focuses mainly on the health of domesticated animals kept for food. It has developed animal welfare standards, included in the regularly updated *Terrestrial Animal Health Code*, and an animal welfare strategy in 2017 that covers standards related to transport, slaughter and the use of animals in research (OIE, 2020a). The organization supports member countries in the implementation of the standards (OIE, 2020b; Visseren-Hamakers, 2018a). Pertinent to this chapter, the OIE's revised mandate to improve animal health and welfare worldwide extends its scope to wild animals (OIE, 2002). Nevertheless, the focus of its dedicated wildlife working group, created in 1994, is almost exclusively on wildlife diseases, rather than welfare (OIE, 2020c). In sum, the OIE remains predominately anthropocentric in its aims.

Additionally, the creation of a *United Nations Convention on Animal Health and Protection* (UNCAHP) is currently under consideration. The draft convention is an initiative of the Global Animal Law Project (2018). The 2018 draft affirms that animals are sentient beings and acknowledges the five freedoms in its preamble. It proposes general measures in relation to non-cruelty and good treatment, and recommends the creation of a United Nations (UN) institution on animal health, welfare and protection. Another development at the global level concerning animal welfare advocacy was the launch in 2021 of the World Federation for Animals (WFA), a collaboration of animal protection organizations (WFA, 2021).

At the regional level, in Europe, the five freedoms are reflected in the welfare assessment criteria of the European Welfare Quality® scheme. The criteria established are used as assessment standards to determine levels of animal welfare and inform EU citizens on meat products (McCulloch, 2013). The EU Strategy for Protection and Welfare of Animals (2012–2015) was evaluated between 2019 and 2020 to assess whether its objectives were delivered. The final report states that the uneven level of protection for different animal species is at odds with the recognition by the EU of animal sentience and that EU citizens' concerns for animal welfare have increased since 2012 (EU, 2020). The African Union established its Animal Welfare Strategy in Africa (AWSA) in 2017, which refers to One Health and One Welfare approaches and includes all animals, including kept animals and animals in the wild (AU-IBAR, 2017). Meanwhile, the Association of Southeast Asian Nations (ASEAN) has established Good Animal Husbandry Practices (GAHP), currently focused on livestock important to the region, namely chickens and pigs (ASEAN, 2020).

An overview of animal welfare policies of different countries, as developed by the animal welfare NGO World Animal Protection (2020), shows a tremendous difference in the manner in which animal welfare is recognized and operationalized around the world. In its ratings of welfare policies, not one country receives an A, the highest possible score, with a handful of European countries (Austria, Denmark, the Netherlands, Sweden, Switzerland and the UK) receiving a B. In the UK, for example, the 2006 UK Animal Welfare Act includes duties of animal owners that are based on the five freedoms, including protection from pain, suffering, injury and disease, as well as the duty to provide a suitable environment, an appropriate diet and adequate housing, and to enable normal behavior patterns (UK Animal Welfare Act, 2006). The five freedoms are also an integral part of a number of welfare codes and schemes in the UK. Examples are various Department of Environment, Food and Rural Affairs (Defra) codes of recommendations for the welfare of livestock, for instance for meat chickens and breeding chickens (2002), pigs (2003) and cattle (2003).

Box 9.2: Combining Animal Rights and Welfare Approaches in India

An interesting example on how a combination of animal rights and welfare can be realized is the country case of India. Its constitution stipulates that "...compassion for living creatures" is considered a duty of every citizen (The Constitution of India 1950, amended 2019). Supreme Court decisions, like the 2014 ruling banning the use of bulls for Jallikattu events, directly refer to the dignity of animals, animal rights and animal welfare – and the court considered itself as the guardian of the rights of animals. Court rulings even recognize a transition from anthropocentric perspectives to ecocentric approaches in animal welfare legislation (Animal Welfare Board of India, 2014). Respect for animals' dignity and intrinsic value is the basis for a number of specific practices, such as prohibition of hunting, reduced meat production and consumption, and encouraging ethically tenable global conservation practices that do not inflict unnecessary harm (Wallach et al., 2018).

9.4 Earth Jurisprudence and Rights of Nature

The idea that nature has rights is recognized in many indigenous cultures in the Americas (Gill, 1987; Weaver, 1996; see also Chapter 2), resonating particularly strongly in the Andes mountains. *Pachamama*, or Mother Earth, is an Andean goddess who, as the giver of life, has rights irrespective of human desires. A concept related to *Pachamama* (sometimes written as *Pacha Mama*) is *buen vivir*. The term is usually translated into English as "living well" or "good living." *Buen vivir* articulates a notion of community and citizenship that embraces all life, with collective rights, including those of nature, prevailing over individual rights (Villalba, 2013).

9.4.1 The Academic Debate

The idea of "rights of nature" has gained tentative acceptance in the United States through Christopher Stone's landmark paper "Should trees have standing?" (Stone, 1972). Stone extended the concept of standing (*locus standi*) to insist that it is unfair for trees to be denied legal protection because they cannot speak and concludes that guardians who wish to defend the rights of trees should be permitted to bring legal action against those whose actions would harm them (Stone, 1972). Stone's paper led to a dissenting opinion in the US Supreme Court. In *Sierra Club v. Morton*, the Sierra Club opposed a development in the Sequoia National Forest on ecological grounds. The court ruled that the Sierra Club had no standing in the case as neither the club nor its members would be harmed by the development (Baude, 1973). However, Justice William Douglas dissented, citing Stone's paper to argue that natural objects should have legal standing, thereby giving guardians the ability to sue for their preservation (Hogan, 2007).

Roderick Nash (1989) saw the extension of rights to other species and natural objects as a broadening of liberal political theory. He argued that freedom of human action should be limited to prevent people from interfering with the rights of other species. Thomas Berry argued that healthy communities cannot be defined solely in terms of the health of people; the health of the natural environment within which a community of people lives also needs to be considered. To Berry, any part of the Earth community has "the right to be, the right to habitat, and the right to fulfil its role in the ever-renewing processes of the Earth community" (Berry, 2011, 229). So a river has the right to flow, a tree has the right to grow, a wild animal has the right to roam free in nature and ecosystems have the right to evolve and adapt.

Proponents of Earth jurisprudence argue that nature should be treated as a subject that requires transformative change to secure legally guaranteed rights, rather than an object owned through property rights to satisfy the instrumental needs of humans. There are diverse conceptions of "environmental personhood" (Gordon, 2019). Legal scholar Cormac Cullinan builds on the work of Berry to argue that modifying contemporary legal systems will not protect nature. Instead, a thorough transformation of the law, in which humans are recognized as just one species in the Earth community, is needed (Cullinan, 2011). Humans should limit their actions in order to uphold nature's rights both for moral reasons (it is right to do so) and for instrumental reasons (human rights ultimately depend on the conservation of nature). Under Earth jurisprudence, therefore, obligations are owed not only to humans but to other species and natural features (Burdon, 2015).

An important academic debate on the relevance of Earth jurisprudence for biodiversity conservation concerns property rights. The liberal notion of private property is essentially individualistic, often emphasizes rights rather than duties and privileges the legal property owner while excluding other stakeholders. Proponents of Earth jurisprudence argue that contemporary property rights are inconsistent with biodiversity conservation. Peter Burdon distinguishes between two approaches to private property. In one view, private property is "inconsistent with ecocentric ethics and ought to be discarded as a social institution" (Burdon, 2015: 101). In this view, private property establishes a hierarchy, with humans having ownership and dominion over nature. The second, reformist, approach sees private property as an "evolving social institution" that needs to be reconceptualized to take into account the impacts of property use on other people and nature (Burdon, 2015). In the case of biodiversity governance, nature's limits should be respected in order to avoid the devastation that humans can cause when property rights are unconditional and unrestricted.

Much contemporary biodiversity policy is based on private property rights and recognizes, implicitly or explicitly, that property owners are entitled to use nature without restrictions, including degrading it. The policy of payments for ecosystems services (PES), for example, rests on the notion that if landowners voluntarily give up a measure of free use in order to provide ecosystem services for the community then payment should be made by that community. PES makes sense in a neoliberal policy context, where owners are free to "sell" on markets the ecosystem services they "provide" to those who benefit from them (see also Chapters 4 and 6).

Earth jurisprudence disputes this logic, arguing that private property is an evolving social construct that needs redefining to take into account our responsibilities to other people and to the community of life. While this runs counter to the liberal notion of property, it is central to the intimate relationship with the land of many indigenous communities, who recognize custodianship as well as ownership. Earth jurisprudence, therefore, articulates a very different notion of property, one in which ethical responsibility to other species is integral and that regulates not just relations between people, but between people and the Earth community.

9.4.2 Political Practice

In 1982 the United Nations General Assembly (UNGA) adopted the World Charter for Nature (Wood, 1984). The charter contains twenty-four principles, some of which are now invoked in Earth jurisprudence, including the statements that "Nature shall be respected and its essential processes shall not be impaired" (United Nations, 1982: article 1) and "The genetic viability on the earth shall not be compromised; the population levels of all life forms, wild and domesticated, must be at least sufficient for their survival, and to this end necessary habitats shall be safeguarded" (United Nations, 1982: principle 2). The charter contains the first political recognition by the UN of "harmony with nature," a phrase that has been repeated in subsequent international environmental declarations, including the 1992 Rio Declaration on Environment and Development (United Nations, 1992).

In 2008, Ecuador became the first country to include rights of nature in its constitution, article 71 of which declares:

Nature, or Pacha Mama, where life is reproduced and occurs, has the right to integral respect for its existence and for the maintenance and regeneration of its life cycles, structures, functions and evolutionary processes. All persons, communities, peoples and nations can call upon public authorities to enforce the rights of nature *(Constitution of the Republic of Ecuador, 2008: Article 71).*

The Ecuadorian constitution allows any individual or group to take legal action to uphold nature's rights, a provision that is consistent with Stone's idea of guardians. Indigenous peoples were represented in the drafting process by the Confederation of Indigenous Nationalities of Ecuador (CONAIE), which paved the way for the inclusion of rights of nature in the constitution. In 2011, the first court case to uphold the rights of nature was brought, namely *Wheeler v. Director de la Procuraduria General Del Estado de Loja*. The court ruled that the dumping of road debris into the Vilcabamba River violated nature's rights and ordered the removal of the debris in order to restore the right of the river to flow (CELDF, 2015; Daly, 2012).

In 2009, Bolivia adopted a new constitution stipulating that Bolivians have a duty to "protect and defend an adequate environment for the development of living beings" (Constitution of the Plurinational State of Bolivia, 2009: Article 108.16). The following year, the Bolivian legislature passed the Law of the Rights of Mother Earth, which recognizes seven rights of nature:

- the right to life and to exist;
- not to be genetically altered or structurally modified in an artificial way;
- to pure water;
- to clear air;
- to balance;
- to restoration; and
- not to be polluted.

While the federal government in the United States does not recognize rights of nature, there has been some recognition at the subfederal level. In Tamaqua Borough, Pennsylvania, in 2006 an ordinance was issued that recognized natural ecosystems within the borough as "legal persons" for the purpose of preventing sewage sludge dumping on wild land (Tamaqua Borough Sewage Sludge Ordinance, 2006). The ordinance, which represents the first instance a public body in the United States granted personhood to nature, stipulated that corporations causing environmental degradation will lose the rights of personhood. Also, in November 2010, the city of Pittsburgh issued an ordinance banning natural gas drilling and fracking, elevating community rights and the rights of nature over and above those of corporate personhood (Pittsburgh Pennsylvania Code of Ordinances, 2013).

The examples of Ecuador and Bolivia (at the national level) and the United States (at the subnational level) have inspired rights of nature movements in other countries, with rivers being granted legal rights in three other countries, namely Colombia, India (including the Ganges and Yamuna) and New Zealand (the Whanganui River) (Pecharroman, 2018).

In 2010, the World People's Conference on Climate Change and the Rights of Mother Earth met in Cochabamba, Bolivia and agreed a Universal Declaration of the Rights of Mother Earth. The declaration is the most important set of Earth jurisprudence principles produced by civil society groups, although as yet it has no legal status. It aspires to a fundamentally different form of human society in which the rights of nature prevail over other rights: "The rights of each being are limited by the rights of other beings and any conflicts between their rights must be resolved in a way that maintains the integrity, balance and health of Mother Earth" (UDRME, 2010: article 1). Article 1 also states that "Mother Earth is a living being" and "The inherent rights of Mother Earth are inalienable in that they arise from the same source as existence" (UDRME, 2010: article 1). This has similarities to

Gaia theory, which conceives of the Earth as a self-regulating and holistic system of living organisms (Lovelock, 1990). A further civil society initiative is the International Rights of Nature Tribunal, established in 2014. This tribunal hears cases brought by aggrieved parties and those who seek to defend nature's rights in line with the principles of Earth jurisprudence. Prosecutors and judges are appointed by the Global Alliance for the Rights of Nature (Boyd, 2017).

In 2011, the UN established an annual interactive dialogue on "harmony with nature" (UN, 2020), and in 2012 the expression "rights of nature" appeared for the first time in a UNGA resolution. Resolution 66/288, endorsing the "The future we want," the main outcome from the United Nations Conference on Sustainable Development (Rio +20), notes:

> We recognize that planet Earth and its ecosystems are our home and "Mother Earth" is a common expression in a number of countries and regions, and we note that some countries recognize the rights of nature in the context of the promotion of sustainable development.
>
> *(United Nations, 2012: para.39)*

Also in 2012, the eleventh Conference of the Parties (COP 11) to the Convention on Biological Diversity (CBD) passed a decision noting that "biodiversity and development processes can be achieved taking into account non-market-based approaches and respect for 'Mother Earth' and the concept of the rights of nature, and that the valuation of biodiversity and ecosystem services is one, among other, tools available" (CBD, 2012). This decision represents a broadening of the range of approaches that the CBD is prepared to endorse and a recognition that market valuation and PES policy approaches are not always the most effective. Additionally, the International Union for the Conservation of Nature (IUCN) Congress adopted a resolution recognizing the rights of nature "as a fundamental and absolute key element for planning, action and assessment at all levels and in all areas of intervention" (IUCN, 2012). The Summary for Policymakers of the Global Assessment of the Intergovernmental Science-Policy Platform on Biodiversity and Ecosystem Services (IPBES) also mentions rights-based approaches and animal welfare (Razzaque et al., 2019).

Relatedly, the crime of ecocide for violating rights of nature is gaining traction. Ecocide is defined as "extensive damage to, destruction or loss of ecosystem(s) of a given territory, whether by human agency or by other causes to such an extent that peaceful enjoyment by the inhabitants of that territory has been or will be severely diminished" (Higgins et al., 2013: 257). Legge and Brooman (2020) propose that international animal law should recognize "animal ecocide" through an amendment to the Rome Statute, which they argue would significantly advance wild animal welfare. They see animal ecocide as the "unnecessary killing or slaughter of a wild or wild-caught animal, by any human agency, or allowing such killing or slaughter to be so caused by any governmental organisation, to such an extent that an animal, or group of animals, lose their sentient capacity to live a natural life according to their species" (Legge and Brooman, 2020: 212). Speciesism is seen as the root cause of ecocide. Recognizing the value and rights of other species would help to prevent such destruction (Jer, 2019; Sollund, 2020).

9.5 Integrative Approaches to Animal and Biodiversity Governance

9.5.1 One Health and One Welfare

Whereas ideas on animal welfare and animal rights focus on the relationship between human beings and nonhuman animals, while rights of nature focuses on the relationship between humans and nature, conceptions of *One Health* (discussed in detail in Chapter 5) emphasize how all three, namely human, animal and environmental health, are interlinked (Galaz et al., 2015; Zinsstag et al., 2006; 2011). While the idea has been discussed for decades (Cook et al., 2004), the outbreak of avian influenza in the early 2000s considerably strengthened discussions relating to the *One Health* concept. The required cooperation between different international organizations, including the World Health Organization (WHO), the Food and Agriculture Organization (FAO) and the World Organisation for Animal Health (OIE), with oversight from the UN System Influenza Coordination Office, emphasized the need for an integrated, intersectoral, interinstitutional and interdisciplinary response (Galaz et al., 2015: 3). In 2020, the COVID-19 pandemic, almost certainly caused by a novel coronavirus that was transmitted to humans from animals, led to renewed calls to recognize the interrelationship between environmental, animal and human health.

However, the idea of equally integrating human, animal and environmental health has proven difficult to implement in practice. In addition to hierarchies between professional disciplines, institutional preferences for single-sector approaches, the paucity of funding, capacity-building, education and training that hamper One Health implementation, there is also the critique that the One Health agenda is geographically Northern-dominated, is top-down and lacks consideration of local experiences and knowledge (Galaz et al., 2015). Thus, the concept is still weak in its practical application, lacking institutional capacities and interdisciplinary collaboration between the natural and social sciences as well as a fruitful exchange between research and policymakers (Valeix, 2014).

The nascent concept of One Welfare, which is not currently applied at the international level, extends the approach of One Health and highlights the interconnections between animal welfare, human well-being and the environment (Garcia Pinillos et al. 2016; One Welfare, 2020).

9.5.2 Compassionate Conservation

Proponents of another approach connecting animal and biodiversity concerns, namely compassionate conservation, argue that conservation objectives need to go beyond protecting species and ecological processes to include animal ethics and a concern for animal welfare (Wallach et al., 2018). This implies not only considering species as a collective but also the interests of individual animals as sentient beings. Hence, compassionate conservationists suggest combining compassion for individuals with conservation of collectives. This can be relevant, for instance, in wildlife management programs or in other areas of conservation practice that opt for killing individual animals from one species to preserve individuals from another species, killing predators to save endangered prey animals, killing introduced or "invasive" species to save native megafauna, killing individual animals for conservation research, or breeding animals in zoos for conservation and education (Wallach et al., 2018: 1261). According to compassionate conservation, these practices will have to be fundamentally reviewed and reformed in order not to compromise individual animals' well-being for the sake of their own, or another, species (Bekoff, 2013).

Compassionate conservationists propose transforming human–animal interaction in an ethically appropriate and sustainable way based on four principles: do no harm, individuals matter, inclusivity and peaceful coexistence (Wallach et al., 2018). Acknowledging the intrinsic value of individual animals requires moving away from instrumentalist thinking, in which animals have material value for human beings, toward valuing them in their own right, irrespective of benefits to humans. This means decentering humans, giving equal consideration to animals and biodiversity as integral parts of an ecosystem and overcoming the human–nature dichotomy. Empirical evidence suggests that nonanthropocentric perspectives, and a stronger focus on the well-being of animals, are increasingly supported within society. Thus, there has been a profound shift toward acknowledging the intrinsic, as opposed to instrumental, value of animals (Bruskotter et al., 2017).

9.6 Discussion: Integrating the Different Debates

This chapter has reviewed different literatures and policy developments to make the argument for integrating animal rights and rights of nature approaches in biodiversity governance. With many human practices neither sustainable nor ethically sound, it is clear that all of the approaches discussed above, namely animal rights, animal welfare, Earth jurisprudence, One Health and One Welfare, and compassionate conservation, in different ways require a significant rethinking of the relationship between humans, nonhuman animals and nature. At the heart of these approaches is the idea that nature and animals should not merely be treated as objects managed by humans but have equal moral and legal standing with them. The perspectives we have examined vary in terms of how radical their proposals are: Whereas some advocate fundamentally restructuring the relationship between humans and animals, such as animal rights approaches, others suggest the need to diminish inequalities in this relationship, such as animal welfare perspectives.

While academic discussions on animal rights have been ongoing for decades, and their transformative potential is significant, their impact on policy practices has to date been relatively limited due to the prevailing dominance of anthropocentric policy-making. In contrast to animal rights approaches, policies and practices on animal welfare are established and implemented in many countries but often merely reproduce the status quo whereby humans manage and govern animals, albeit with some limited improvements in their living conditions. The transformative potential of animal welfare approaches is therefore limited compared to those on protecting and promoting animal rights. In recent years, rights of nature have been increasingly adopted and implemented internationally and domestically. Ecosystem rights have significant transformative potential, especially if they can be protected by guardians and implemented in court decisions. Integrative perspectives like One Health, One Welfare and compassionate conservation encourage holistic policy development recognizing the link in human, animal and ecosystem health and well-being, and also hold transformative potential, but have not yet had large-scale effects on the ground.

Based on the review, we argue that in a world that is severely threatened by sustainability challenges such as biodiversity loss, we need to refocus our understanding of governance to acknowledge rights as the basis for conflict resolution, peace and just sustainable development. What we can observe in discourse and practice is a changing understanding of human rights: from individual civil and political rights (with its origins in the 1215 Magna Carta and internationally in the 1948 Universal Declaration on Human Rights), to collective intergenerational rights (such as the 2007 UN Declaration on the Rights of Indigenous Peoples or the International Human Right to a Healthy Environment recognized by the UN Human Rights Council in 2021). The emerging discourses and practices on promoting ecosystem integrity, animal rights and interspecies justice, as discussed in this chapter, can be seen as further steps in this ongoing process of increasingly recognizing rights.

Table 9.1 summarizes the main developments discussed in the chapter. It illustrates that many of these debates have been ongoing for decades, with the integrative approaches developing later. Our integrative perspective highlights that different discourses are actually part of the same process of expanding the moral and legal community to include species, individual animals and nature. With this, the chapter has contributed to inclusive governance debates by making the case for the emancipation of those whose interests are not yet being met (see Chapter 1).

Table 9.1 *Overview of important developments*

Year	Event
1924	World Organisation for Animal Health (OIE) founded
1965	Publication of the UK Brambell report on animal welfare and introduction of the "Five Freedoms"
Early 1970s	Richard Ryder coins the term speciesism
1972	Christopher Stone publishes "Should trees have standing?"
1975	Peter Singer publishes *Animal Liberation*
1980	Founding of People for the Ethical Treatment of Animals (PETA)
1982	World Charter for Nature adopted at United Nations (UN) General Assembly
1983	Tom Regan publishes *The Case for Animal Rights*
2000	Swiss constitution includes respect of "dignity of creatures"
2002	German constitution includes the protection of animals
2004	Wildlife Conservation Society conference launches One World, One Health
2005	First OIE global animal welfare standards
2006	Tamaqua Borough in Pennsylvania, USA, recognizes natural ecosystems as legal persons
2008	Ecuador includes rights of nature in its constitution

Table 9.1 (*cont.*)

Year	Event
2009	UN General Assembly declares April 22 International Mother Earth Day
2009	European Union recognizes animal sentience in Article 13 of Lisbon Treaty
2010	Bolivia adopts Law of the Rights of Mother Earth
2010	Draft Universal Declaration on the Rights of Mother Earth
2010	Pittsburgh, USA, passes an ordinance recognizing rights of natural communities and ecosystems
2011	Court case on rights of Vilcabamba River, Ecuador
2011	Draft Universal Declaration on Animal Welfare
2011	First UN interactive dialogue on harmony with nature
2012	Rights of nature acknowledged in UN General Assembly resolution
2013	Marc Bekoff introduces concept of compassionate conservation
2016	Idea of One Welfare published
2017	OIE Global Animal Welfare Strategy
2017	African Union Animal Welfare Strategy
2017	Legal rights for rivers in Colombia, India and New Zealand
2018	UK Government acknowledges animal sentience
2021	World Federation for Animals launched

9.7 Conclusion: Toward Ecocentric Animal and Biodiversity Governance

In this chapter we have analyzed the transformative potential of mainstreaming animal rights and rights of nature in biodiversity governance. We have done so based on an integrative analysis of ongoing academic and policy debates on animal rights, animal welfare, rights of nature and approaches that integrate these debates.

One of the most important insights derived from our review is the recognition of the differences between the discourses on animal rights and rights of nature. The animal rights discourse focuses on animals, arguing that all individual animals have rights, but is silent on the rights of flora and inanimate natural objects such as mountains, which feature prominently in rights of nature discourses that focus on collective rights but are silent on the rights of individual animals. We therefore argue that integrating animal rights and rights of nature approaches is necessary to fully enable ecocentric approaches in biodiversity governance.

Our analysis has several implications for transformative biodiversity governance, in the context of the Post-2020 Global Biodiversity Framework. Rights of nature played a prominent role in the negotiations of the framework. We argue that an integrative approach to rights of nature and animal rights should be included in the (implementation of the) framework.

Mainstreaming the individual animal entails designing conservation practices that are more ethically sound and acknowledging human obligations to nature (Burdon, 2020). Trade-offs between the lives of individual animals and species are not inevitable, but where there is conflict, for example with species deemed invasive, conservation actions can be implemented in ways that respect individuals. This would for example entail choosing management methods that minimize suffering (Barnhill-Dilling and Delborne, 2021). A further implication of mainstreaming the individual animal would mean taking wild animals into account as individuals in their own right, rather than just thinking of them as resources or disease vectors. In terms of integrative governance, as exemplified by the OIE's tripartite+ collaboration (WHO, FAO, OIE, UNEP), which is particularly focused on One Health, this entails shifting the current anthropocentric focus and not automatically prioritizing the interests of humans. There is already evidence that respect for the lives of individual animals will become increasingly important in the future, such as with the launch of the World Federation for Animals to influence international policy-making.

Transformative change, defined as fundamental change including in terms of paradigms, goals and values (Díaz et al., 2019), in our view requires fundamentally rethinking the relationship between human beings, individual animals and nature, thereby reorienting biodiversity governance from an anthropocentric to an ecocentric perspective. Expanding the moral and legal community to include not only humans, but also nonhuman animals and nature, is an explicit and essential part of the transformative change required to halt biodiversity loss. Such an ecocentric perspective also requires a foundational rethinking of the concept of sustainable development to incorporate proper acknowledgment of the individual animal (see Visseren-Hamakers, 2020), species and entire ecosystems.

Only a fundamental shift to ecocentric approaches, considering ecosystems holistically and recognizing the rights of individual animals and nature, will allow for the establishment of alternative institutions, structures and processes as part of a broader transformative governance for biodiversity and sustainable development (Chaffin et al., 2016; Visseren-Hamakers, 2018a). The shift also requires rethinking core elements of democracy, such as representation, considering theoretical and practical implications of ecological democracy (Kopnina et al., 2021). This will, ultimately, benefit the lives of humans and nonhumans alike, and this approach is embraced in new debates on ecosystem justice, interspecies justice (Nussbaum, 2006) and multispecies justice (Celermajer et al., 2021).

References

Ahlhaus, S., and Niesen, P. (2015). What is animal politics? An outline of a new research agenda. *Historical Social Research* 40, 7–31.
Animal Welfare Board of India. (2014). Supreme Court's Judgement on Jallikattu. Available from http://awbi.in/awbi-pdf/sc_judgement_jallikattu_7-5-14.pdf.
ASEAN. (2020). Establishment of ASEAN Good Animal Husbandry Practices (ASEAN GAHP). Available from https://bit.ly/3Kclnts.
AU-IBAR. (2017). *The Animal Welfare Strategy in Africa*. Nairobi, Kenya: African Union – Inter-African Bureau for Animal Resources (AU-IBAR).
Barkham, P. (2020). Should we cull one species to save another? *The Guardian*. Available from: https://bit.ly/3Gm33eS.
Barnhill-Dilling, S. K., and Delborne, J. A. (2021). Whose intentions? what consequences? interrogating "intended consequences" for conservation with environmental biotechnology. *Conservation Science and Practice* 3, 1–11.

Baude, P. L. (1973). Sierra Club v. Morton: Standing trees in a thicket of justiciability. *Indiana Law Journal* 48, 197–215.

Bekoff, M. (2013). Compassionate conservation and the ethics of species research and preservation: Hamsters, black-footed ferrets, and a response to Rob Irvine. *Bioethical Inquiry* 10, 527–529.

Bernstein, M. H. (2015). *The moral equality of humans and animals.* Basingstoke: Palgrave Macmillan.

Berry, T. (2011). Rights of the earth: We need a new legal framework which recognises the rights of all living beings. In *Exploring wild law: The philosophy of Earth jurisprudence.* P. Burdon (Ed.), pp. 227–229. Wakefield, Kent Town, Australia: Wakefield.

Boyd, D. R. (2017). *The rights of nature: A legal revolution that could save the world.* Toronto: ECW Press.

Braidotti, R. (2013). *The posthuman.* Cambridge: Polity Press.

Bruskotter, J., Vucetich, J., and Nelson, M. (2017). Animal rights and wildlife conservation: Conflicting or compatible? *The Wildlife Professional* 7–8, 40–43.

Burdon, P. D. (2015). *Earth jurisprudence: Private property and the environment.* Abingdon: Routledge.

(2020). Obligations in the Anthropocene. *Law and Critique* 31, 309–328.

Cavalieri, P. (2001). *The animal question: Why non-human animals deserve human rights.* Oxford: Oxford University Press.

Celermajer, D., Schlosberg, D., Rickards, L., et al. (2021). Multispecies justice: Theories, challenges, and a research agenda for environmental politics. *Environmental Politics* 30, 119–140.

Chaffin, B. C., Garmestani, A. S., Gunderson, L. H., et al. (2016). Transformative environmental governance. *Annual Review of Environment and Resources* 41, 399–423.

Community Environmental Legal Defense Fund (CELDF). (2015). Rights of Nature Going Global. Available from https://celdf.org/rights/rights-of-nature/.

Connolly, K. (2002). *German animals given legal rights.* Available from www.theguardian.com/world/2002/jun/22/germany.animalwelfare.

Constitution of the Plurinational State of Bolivia. (2009). Translated by Luis Francisco Valle V. Printed in San Bernadino, California. Spanish version. Available from https://bit.ly/34PBOeX.

Constitution of the Republic of Ecuador. (2008). English translation from Political Database of the Americas, Georgetown University. Available from https://bit.ly/311D8JP. Spanish version available from https://bit.ly/3noJv1T.

Convention on Biological Diversity (CBD). (2012). UNEP/CBD/COP/DEC/XI/22, Decision adopted by the Conference of the Parties to the Convention on Biological Diversity at its Eleventh Meeting, XI/22 Biodiversity for poverty eradication and development, 5 December. Available from www.cbd.int/doc/decisions/cop-11/cop-11-dec-22-en.pdf.

Cook, R. A., Karesh, W. B., and Osofsky, S. A. (2004). *The Manhattan principles on "One World, One Health": Building interdisciplinary bridges to health in a globalized world.* New York: Wildlife Conservation Society.

Cullinan, C. (2011). *Wild law: A manifesto for Earth justice.* 2nd ed. Chelsea, VT: Chelsea Green Publishing.

Daly, E. (2012). The Ecuadorian exemplar: The first ever vindications of constitutional right of nature. *Review of European Community and International Environmental Law (RECIEL)* 21, 63–66.

De Mori, B. (2019). Animal testing: The ethical principle of the 3Rs from laboratories to "field" research with wild animals. *Etica & Politica / Ethics & Politics* 21, 553–570.

Díaz, S., Settele, J., Brondízio, E., et al. (2019). Pervasive human-driven decline of life on Earth points to the need for transformative change. *Science* 366, 6471. DOI: 10.1126/science.aax3100.

Donaldson, S., and Kymlicka, W. (2011). *Zoopolis: A political theory of animal rights.* Oxford: Oxford University Press.

EU. (2020). Study to support the evaluation of the European Union strategy for the protection and welfare of animals 2012–2015. https://bit.ly/3qn6ix0.

Evans, E. (2010). Constitutional inclusion of animal rights in Germany and Switzerland: How did animal protection become an issue of national importance? *Society and Animals* 18, 231–250.

FAWC. (2009). Farm Animal Welfare in Great Britain: Past, Present and Future. Available from https://bit.ly/3ocpZpS.
  (2018). FAWC advice on animal sentience. Available from https://bit.ly/3qpAGqP.
Galaz, V., Leach, M., Scoones, I., and Stein, C. (2015). *The political economy of One Health research and policy.* STEPS Working Paper 81. Brighton: STEPS Centre.
Garcia Pinillos, R., Appleby, M., Manteca, X., et al. (2016). One welfare – A platform for improving human and animal welfare. *Veterinary Record* 179, 412–413.
Gill, S. D. (1987). *Mother Earth: An American story.* Chicago, IL:University of Chicago Press.
Global Animal Law Project. (2018). UN Convention on Animal Health and Protection. www.globalanimallaw.org/database/universal.html.
Goetschel, A. F. (2000). Animal cloning and animal welfare legislation in Switzerland. Report by the Foundation for the Animal in the Law, Presented at a symposium at the University of Lüneburg, Germany.
Gordon, G. J. (2019). Environmental personhood. *Columbia Journal of Environmental Law* 43, 49–91.
Haynes, R. P. (2011). Competing conceptions of animal welfare and their ethical implications for the treatment of non-human animals. *Acta Biotheoretica* 59, 105–120.
Higgins, P., Short, D., and South, N. (2013). Protecting the planet: A proposal for a law of ecocide. *Crime, Law and Social Change: An Interdisciplinary Journal* 59, 251–266.
Hogan, M. (2007). Standing for nonhuman animals: Developing a guardianship model from the dissents in *Sierra Club v. Morton*. *California Law Review* 95, 513–534.
IUCN. (2012). Resolution WCC-2012-Res-100-EN Incorporation of the Rights of Nature as the organizational focal point in IUCN's decision making. Available from https://bit.ly/338SvBD.
Jaber, D. (2000). Human dignity and the dignity of creatures. *Journal of Agricultural and Environmental Ethics* 13, 29–42.
Jer, S. B. (2019). Ecocide or environmental self-destruction? *Environmental Ethics* 41, 237–247.
Judd, R. (2003). The politics of beef: Animal advocacy and the kosher butchering debates in Germany. *Jewish Social Studies* 10, 117–150.
Kopnina, H., Leadbeater, S., and Cryer, P. (2019). Learning to rewild: Examining the failed case of the Dutch "new wilderness" Oostvaardersplassen. *International Journal of Wilderness* 25, 72–89.
Kopnina, H., Spannring, R., Hawke, S., et al. (2021). Ecodemocracy in practice: Exploration of debates on limits and possibilities of addressing environmental challenges within democratic systems. *Visions for Sustainability* 15, 9–23.
Ladwig, B. (2015). Animal rights – Politicised, but not humanised. An interest-based critique of citizenship for domesticated animals. *Historical Social Research* 40, 32–46.
Legge, D., and Brooman, S. (2020). Reflecting on 25 years of teaching animal law: Is it time for an international crime of animal ecocide? *Liverpool Law Review* 41, 201–218.
Lorimer, J., and Driessen, C. (2014). Wild experiments at the Oostvaardersplassen: Rethinking environmentalism in the Anthropocene. *Transactions of the Institute of British Geographers* 39, 169–181.
Lovelock, J. E. (1990). Hands up for the Gaia hypothesis. *Nature* 344, 100–102.
McCulloch, S. P. (2013). A critique of FAWC's five freedoms as a framework for the analysis of animal welfare. *Journal of Agricultural and Environmental Ethics* 26, 959–975.
Mellor, D. (2016). Updating animal welfare thinking: Moving beyond the "five freedoms" towards "a life worth living." *Animals* 6, 1–20.
Nash, R. F. (1989). *The rights of nature: A history of environmental ethics.* Madison: University of Wisconsin.
Niesen, P. (2019). Menschen und Tiere – ein politisches Verhältnis. In *Haben Tiere Rechte? Bundeszentrale für politische Bildung.* E. Diehl and J. Tuider (Eds.), pp. 379–383. Bonn: Germany.
Nussbaum, M. (2006). *Frontiers of justice: Disability, nationality, species membership.* Cambridge, MA: Harvard University Press.
Ohl, F., and van der Staay, F. J. (2012). Animal welfare: At the interface between science and society. *The Veterinary Journal* 192, 13–19.

OIE. (2002). Resolution No. XIV: Animal Welfare Mandate of the OIE. www.oie.int/app/uploads/2021/03/a-reso-2002.pdf.

(2020a). OIE animal welfare standards. www.oie.int/en/what-we-do/standards/.

(2020b). World Organization for Animal Health. www.oie.int.

(2020c). Working Group on Wildlife. https://bit.ly/33w5ahR.

One Welfare. (2020). www.onewelfareworld.org/ (accessed August 12, 2020).

Pecharroman, L. C. (2018). Rights of nature: Rivers that can stand in court. *Resources* 7, 1–14.

Pittsburgh Pennsylvania Code of Ordinances. (2013). Code of Ordinances City of Pittsburgh Pennsylvania. Codified through Ordinance No. 12–2013, April 19, 2013. Available from https://bit.ly/3FqPeuz.

Razzaque, J., Visseren-Hamakers, I., Prasad Gautam, A., et al. (2019). Options for decision makers. In *Global assessment report of the Intergovernmental Science-Policy Platform on Biodiversity and Ecosystem Services*. E. Sonnewend Brondízio, J. Settele, S. Díaz and H. T. Ngo (eds.), pp. 875–1028. Bonn: IPBES Secretariat. https://doi.org/10.5281/zenodo.3832107

Regan, T. (1983). *The case for animal rights*. Berkeley: University of California Press.

Ryder, R. (1971). Experiments on animals. In *Animals, men and morals: An enquiry into the maltreatment of non-humans*. S. Godlovitch, R. Godlovitch and J. Harris (eds.). New York: Taplinger Publishing Company.

Seubert, S. (2015). Politics of inclusion. Which conception of citizenship for animals? *Historical Social Research* 40, 63–69.

Singer, P. (1975). *Animal liberation*. New York: HarperCollins.

Sollund, R. (2020). Wildlife management, species injustice and ecocide in the Anthropocene. *Critical Criminology* 28, 351–369.

Stein, T. (2015). Human rights and animal rights: Differences matter. *Historical Social Research* 40, 55–62.

Stone, C. D. (1972). Should trees have standing? Towards legal rights for natural objects. *Southern California Law Review* 45, 450–501.

Tamaqua Borough Sewage Sludge Ordinance. (2006). Tamaqua Borough, Schuylkill County, Pennsylvania Ordinance No. 612, 2006. Available from http://files.harmonywithnatureun.org/uploads/upload666.pdf.

UK Animal Welfare Act. (2006). Available from www.legislation.gov.uk/ukpga/2006/45/contents.

United Nations (UN). (1982). A/RES/37/7. World Charter for Nature. Available from https://bit.ly/3rfqKzb.

(1992). Rio Declaration on Environment and Development. Available from https://bit.ly/3Hodjnx.

(2012). The future we want: Outcome document of the United Nations Conference on Sustainable Development. Available from https://bit.ly/3od4bdB.

(2020). Harmony with Nature. www.harmonywithnatureun.org/.

Universal Declaration of the Rights of Mother Earth (UDRME). (2010). World People's Conference on Climate Change and the Rights of Mother Earth. Cochabamba, Bolivia. Available from https://bit.ly/3ghsJy4.

Valeix, S. (2014). *Toward One Health? Evolution of international collaboration networks on Nipah virus research from 1999–2011*. STEPS Working Paper 74. Brighton: STEPS Centre.

Villalba, U. (2013). Buen Vivir vs development: A paradigm shift in the Andes? *Third World Quarterly* 34, 1427–1442.

Visseren-Hamakers, I. J. (2015). Integrative environmental governance: Enhancing governance in the era of synergies. *Current Opinion in Environmental Sustainability* 14, 1–16.

(2018a). A framework for analyzing and practicing integrative governance: The case of global animal and conservation governance. *Environment and Planning C: Politics and Space* 36(8), 1391–1414.

(2018b). Integrative governance: The relationships between governance instruments taking center stage. *Environment and Planning C: Politics and Space* 36, 1341–1354.

(2020). The 18th sustainable development goal. *Earth System Governance* 3, 100047.

Wallach, A., Bekoff, M., Batavia, C., Nelson, M., and Ramp, D. (2018). Summoning compassion to address the challenges of conservation. *Conservation Biology* 32, 1255–1265.

Weary, D. M., and Robbins, J. A. (2019). Understanding the multiple conceptions of animal welfare. *Animal Welfare* 28, 33–40.

Weaver, J. (Ed.). (1996). *Defending Mother Earth: Native American perspectives on environmental justice*. Maryknoll NY: Orbis Books.

Webster, J. (1994). *Animal welfare: A cool eye towards Eden*. Oxford: Blackwell Publishing.

Wemelsfelder, F., and Lawrence, A. B. (2001). Qualitative assessment of animal behaviour as an on-farm welfare-monitoring tool. *Acta Agriculturae Scandinavica*, 51, 21–25.

Wise, S. (2016). *Animal Rights*. Available from www.britannica.com/topic/animal-rights.

Wood, H. W. (1984). The United Nations World Charter for Nature: The developing nations' initiative to establish protections for the environment. *Ecology Law Quarterly* 12, 977–996.

World Animal Protection. (2020). https://api.worldanimalprotection.org.

World Federation for Animals (WFA). (2021). About us. https://wfa.org/about-us/.

Zinsstag, J., Schelling, E., Waltner-Toews, D., and Tanner, M. (2011). From "one medicine" to "one health" and systemic approaches to health and well-being. *Preventive Veterinary Medicine* 101, 148–156. DOI: 10.1016/j.prevetmed.2010.07.003.

Zinsstag, J., Schelling, E., Wyss, K., and Mahamat, M. B. (2006). Potential of cooperation between human and animal health to strengthen health systems. *The Lancet* 9503, 2142–2145.

10

Industry Responses to Evolving Regulation of Marine Bioprospecting in Polar Regions

KRISTIN ROSENDAL AND JON BIRGER SKJÆRSETH

10.1 Introduction

A central question in biodiversity governance is how the international community will regulate the conservation and equitable sharing of benefits from the utilization of marine genetic resources in areas beyond national jurisdiction (ABNJ). The equity question concerns how to secure benefits from global commons resources for all, not only for financially and technologically strong actors. The access and benefit-sharing (ABS) principles set out in the Convention on Biological Diversity (CBD) (CBD, 1993) and elaborated in its 2014 Nagoya Protocol are decisive rules on these equity concerns. ABS is central to the CBD Post-2020 Global Biodiversity Framework deliberations, which have been delayed due to COVID-19, but which are expected to be adopted in 2022. Along with the pandemic, ABS also put health and biodiversity relationships more prominently on the political agenda, as nature can be both a resource (genetic resources as sources of medicines) and pose threats through zoonoses, depending on how biodiversity is governed (UNEP, ILRI, 2020). Legal regulation of the utilization of genetic material from ABNJ in the polar regions is currently subject to negotiation within the framework of the United Nations Convention on the Law of the Sea (UNCLOS), based on UN General Assembly decision 72/249, 2017. The ABNJ remain among the few unregulated areas of the world in which bioprospecting is taking place, as the ABS principles of the CBD do not apply directly outside national jurisdiction. The growing focus on the value of marine genetic resources, not least for medicinal development, is likely to be affected by the evolving legal conditions for access and rights to use this material. Addressing a central theme of this volume, we examine the potential effects of the options on the negotiating table in terms of transformative biodiversity governance (TBG).

Here we investigate various aspects of the equity questions, taking stock of evolving regulatory regimes for dealing with the technological aspects of marine bioprospecting, with emphasis on the bioprospectors themselves. First, as we examine legal processes in the making, this study addresses the anticipatory dimension of transformative governance, where the options are still open and malleable. Second, as bioprospectors are central in the utilization of genetic resources, a better understanding of their role and positions is an

---

Research project funded by the Research Council of Norway (project number 257631). We are grateful to all reviewer comments and also to Susan Høivik for language editing.

important element in governing the equity issues of biodiversity conservation and use. By focusing on actors we can examine stakeholder participation, which is central to the TBG debate discussed in this volume. Third, studying the responses and behavior of corporate bioprospecting actors allows an often-neglected focus on technological development as a driver and underlying cause of biodiversity loss. Questions of how to deal with digital sequence information (DSI) and synthetic biology are at the core of international governance of genetic resources (see Box 10.1). Thus, this chapter speaks to Chapter 7 on DSI in this volume, from a more empirically oriented angle.

In multilateral environmental cooperation, issues of North–South divides and equity usually focus on technology transfer and capacity building whenever technology is addressed. Technological developments may also have direct economic and distributional ramifications for poorer countries. We examine how vested interests in biotechnology could challenge transformative change by undermining the principle of equitable sharing of benefits arising from utilization of genetic resources, the ABS regime, as this is central to transformative biodiversity governance (see Chaffin et al., 2016).

"Bioprospecting" refers to the systematic search for biochemical and genetic information in nature, in order to develop commercially valuable products for pharmaceutical, agricultural, cosmetic and other applications (Svenson, 2013). Marine organisms may be more likely than terrestrial species to contain useful natural compounds, partly because they have evolved in response to extreme environments (see e.g. Bodnar, 2016). However, less than 1 percent of marine organisms have been explored scientifically, and little is known about their rarity or vulnerability. Recent technological advances are making the marine genetic resources of the Arctic and Antarctic Oceans increasingly available and of commercial interest. Collecting biological material from these regions is still very costly and conducted predominantly by a small number of state-funded, oceangoing vessels (see Leary, 2018; Müller and Schøyen, 2021). In view of the high levels of public funding that go into infrastructure, collections in biobanks, and delivery of ready bioactive compounds – all of which is necessary to develop commercial products – this has raised questions of *cost-sharing* as well as benefit-sharing in bioprospecting (Rosendal et al., 2016). A handful of multinational corporations are behind more than 80 percent of the patent applications on this material, with BASF alone filing almost half of the patent applications on marine genetic resources since 1988 (Blasiak et al., 2018). As bioprospecting is largely conducted by private (often multinational) corporations, we must ask whether and how these bioprospectors respond to emerging measures in the ABS legislation.

There are very few studies of bioprospectors, except for some cases of terrestrial medicinal plants (Wynberg et al., 2009). Also, ABS issues regarding utilization of genetic resources in ABNJ are less explored in social scientific terms than are those lying within national territories. For ABNJ, most of the literature available is from the legal field (Arico, 2010; Drankier et al., 2012; Greiber, 2011; Jørem and Tvedt, 2014; Tvedt, 2020). ABS-related studies within the aquaculture and agriculture breeding sector have shown that both commercial and noncommercial breeders alike would prefer aquatic and plant genetic resources to be freely (affordably) accessible, although commercial breeders also need to ensure revenues (royalties) from their own innovations and breeding results through some form of intellectual property rights (Greer and Harvey, 2004; Olesen et al., 2007; Rosendal et al., 2006; 2013). Similar dilemmas are likely to emerge among marine bioprospectors, as many will need to seek access to genetic material through biobank collections, and many will seek to patent the material. With this chapter, we aim to fill the knowledge gaps concerning the ABS strategies of marine bioprospectors in order to inform the debate on TBG.

The ABS debate has a history of conflict regarding accusations of biopiracy; therefore, ABS strategies may be sensitive data for the corporations. Moreover, corporations rarely provide position papers in international negotiations. In order to disclose information and compensate for the lack of position papers to the UNCLOS negotiations, we have examined bioprospector responses to public hearings on two draft proposals for Norwegian ABS legislation. Further, in-depth, semi-structured interviews have been conducted with key actors in two corporations, to complement the analysis (see Yin, 2003). Data have also been collected from public records, secondary literature and interviews with seven key actors from ministries, R&D institutes and international scientific organizations (see footnotes for details). Most of the interview materials were collected between 2012 and 2018, but data collection on the international negotiation process has continued to spring 2021.

We have chosen Norway as a case for three reasons, in addition to easy access. First, resources from marine and polar areas are traditional core Norwegian interests. Second, Norway is investing heavily in marine research and innovation: marine bioprospecting, samples collections to marine biobanks and high-cost oceangoing vessels (about €150 million in public funding to the most recent vessel, *Kronprins Haakon*, alone) (see Müller and Schøyen, 2021). Third, Norway has a long history of advocating the access and equitable benefit-sharing regime of the CBD, further specified in its Nagoya Protocol, but ABS regulations at home are still stalling, with long and controversial debates and hearings. All this makes Norway a relevant case for examining the political scope between norms expressed internationally and concern for domestic interests (Rosendal et al., 2016).[1]

We begin by presenting an analytical framework for assessing and explaining corporate strategies and responses to evolving regulations, outlining the main conflicts of the ABS debate. Next, after explaining what marine bioprospecting entails, we turn to the international legal debate on such activities and the current governance of genetic resources in the polar regions. In Sections 10.4 and 10.5 we present findings from our embedded case study of Norwegian actors engaged in polar marine bioprospecting, based on the analytical models for assessing corporate strategies. In the concluding section we offer inputs to the debate on transforming biodiversity governance based on our analysis.

10.2 Analytical Framework on Corporate Strategies

Our examination concerns responses to the new ABS regulation as regards bioprospecting companies and industry associations – here broadly defined as actors with commercial interests. As such regulation is still evolving, we focus mainly on *political responses* that

---

[1] https://bit.ly/3nrd03g.

may, or may not, lead to actual *market adaptation* (Kolk and Pinkse, 2004). "Political responses" refer here to strategic company support of (proactive) or opposition to (reactive) emerging regulation. These strategies are ideal-typical opposite poles: Real-life companies engaged in a wide range of activities cannot be expected to fit perfectly with such opposing extremes. Our aim is to assess the degree of fit between expectations and observations in the content and direction of corporate strategies in relation to ABS regulations.

We focus on three "ideal type" models for explaining company responses (Skjærseth and Eikeland, 2013; 2019). The first model sees companies as reactive and "reluctant adapters" to strengthened regulations. This "reactive" model is grounded in the traditional economic view of the firm as a unitary, rational, profit-maximizing agent that develops strategies based on full information on the relative costs of various alternatives (Ambec et al., 2011; Gravelle and Rees, 1981). As new ABS regulations (in ABNJ) would charge companies for previously free access to genetic material and impose administrative and compliance costs that could erode profits, regulation is held to divert capital away from other investments, thus threatening a firm's competitiveness. We expect political responses that seek to minimize new regulatory costs by opposition to the ABS regime: saying "no" to all kinds of monetary benefit-sharing and resisting expanding its legal scope. Opposition expressed in interviews and lobby papers will be in line with this expectation.

The second model views companies as "proactive innovators." This model is based on bounded rationality and the search for new market opportunities. The "proactive" response model assumes that firms are "boundedly rational" (Simon, 1976). Profit maximization is seen as central, with strategic managerial choices influenced by the design of regulations, organizational practices and operating procedures, perceptions of risks and opportunities, and information constraints, habits or routines (Cyert and March, 1963; Delmas and Toffel 2008; Sanchez 1997).

Anchored in these assumptions, environmental regulation does not necessarily represent a threat to profits and competitiveness; on the contrary, it may contribute to innovation, improved performance and competitive advantages (Esty and Winston, 2006). According to Porter and van der Linde (1995a; 2005b), "appropriately" designed regulation may spur learning about resource inefficiencies and technological improvements, reduce uncertainty about future investment and stimulate innovations that can offset the costs of compliance. Adjusting to appropriately designed regulations, a company may support regulation and view compliance as a rational way to improve profits and attract new customers. Promoting a profile of green equity can also help companies avoid accusations of "biopiracy," and hence secure access to resources and collaboration with partners that can promote and increase such access.

"Appropriate" in this case can be assumed to imply adhering to the basic principles of ABS, while not condoning any kind of expansion in its legal scope. "Proactive response" to the ABS regime means accepting *monetary* benefit-sharing with "provider" countries, but excludes *derivatives* (see Box 10.1), excludes monitoring through *disclosure* of origin of genetic material through patent application systems and limits the *time scope* to the entry into force of the CBD's Nagoya Protocol in 2014 rather than to the CBD itself (1993) (ENB, 2018; Oberthür and Rosendal, 2014). The idea behind *disclosure* is that intellectual property

> **Box 10.1: The derivative debate**
>
> The CBD, Article 2, defines "genetic resources" as genetic material of actual or potential value, and "genetic material" as any material of plant, animal, microbial or other origin containing functional units of heredity. The second definition has given rise to dispute, as genetic sequences and enzymes applied in synthetic biology do not necessarily contain functional units of heredity. The real value of genetic resources lies, however, in their information. Hence developing countries argue that such derivatives of genetic resources must remain part of the ABS regime even when this material does not contain functional units of heredity. Technological developments in synthetic biology have produced large quantities of biological data, which are stored online in databanks. This digital sequence information on genetic resources is increasingly replacing the need to access biological samples of genetic resources in nature and this has major implications for the CBD architecture on ABS (see Chapter 7). If access to derivatives, necessary to foster scientific research, is not accompanied by benefit-sharing modalities, the CBD's third objective on equitable sharing may become increasingly undermined. Similarly, it may be argued that all new drugs that enter the market still originate from the natural world, and that excluding derivatives would also exclude incentives for biodiversity conservation. Industry actors, coordinating their views on DSI through the International Chamber of Commerce (ICC), would strongly oppose the expansion of the CBD and the Nagoya Protocol to cover DSI (ICC, 2017).

rights (IPR) systems are most useful for monitoring ABS, hence the proposal to include the origin of genetic resources in patent applications (Jørem and Tvedt, 2014; Morgera et al., 2013; Prip et al., 2014). Some of these elements have been included in the 2014 EU ABS regulation, which accepts the basic principles of monetary benefit-sharing and derivatives, but not disclosure and extended time scope. We will use acceptance or support as expressed in interviews and government consultations to check whether these elements are in line with expectations.

The third "social responsibility" model assumes that company managers can have mixed motivations that may include social norms of responsibility, in addition to profit maximization. This perspective builds on the tentative assumptions that managers evaluate options broadly in terms of social, economic and political aspects, and that their response to regulation is affected by social norms of responsibility. Regulation can affect such norms of responsibility for companies operating in a complex political and social environment where consumers and civil society organizations play an important role.

Norm-guided behavior has increasingly been incorporated into economic studies of responses to governmental regulation (Esty and Winston, 2006) and is discussed in the vast literature on corporate social responsibility (CSR). Companies can contribute to providing public goods, for instance through voluntary CSR principles and measures. However, since voluntary contributions are rarely deemed sufficient to provide important public goods, like conserving biodiversity, additional state regulation is normally viewed as necessary (Barth and Wolff, 2009). In the context of ABS and bioprospecting, expected responses here are full acceptance of ABS: accepting *monetary* benefit-sharing, accepting the inclusion of *derivatives*, linking monitoring to *disclosure* through IPR/patent systems and setting the *time scope* to the entry into force of the CBD (1993). This position accepts an ABS design broadly in line with what developing countries have generally fronted in ABS negotiations. However, empirical assessment of this perspective may prove challenging, as corporate norm-guided behavior is difficult to distinguish from other motivations.

10.3 Governing Bioprospecting in the Polar Regions

10.3.1 Marine Bioprospecting

There is increasing economic interest in genetic material from marine bacteria, sponges, krill, corals and seaweeds. Marine biotechnology research includes aquaculture, novel products such as Omega 3, fatty acids from fish oil, carotenoids, pigments, flavorings and nutritional supplements (Blunt et al., 2011). The total value is difficult to assess and may be overrated (Leary, 2018), but Blasiak et al. (2018) estimate the value of global marine bioprospecting in 2025 at $6.4 billion. However, although the number of patent applications based on marine genetic resources is increasing, only 1–2 percent of preclinical candidates become commercial products (Leary, 2018). Patent applications merely indicate a demand for patent rights, not actual control, and there is yet little information regarding patents granted.

Bioprospecting the high seas is a cost-intensive, high-risk activity. Apart from possible legal constraints on bioprospecting, collectors also face economic and biological challenges. Economic: Only a few research vessels are equipped to access and collect samples in the polar regions. Some of the collected material is already known and has been analyzed, isolated and characterized, as most species studied have a large geographical distribution (Svenson, 2013). Biological: The high-cost, high-risk nature of collecting makes resampling difficult; hence the motivation to stock up as much as the vessel's freezing capacity allows, which gives rise to issues of sustainability in harvesting (Svenson, 2013).

Reflecting the high costs, marine bioprospecting and patent applications come predominantly from a few developed nations and their industries (Arnaud-Haond et al., 2011; Müller and Schøyen, 2021; Oldham and Kindness, 2020). Beside Australia, Germany, Norway, Russia, the USA and UK, China is currently preparing to join this exclusive club,[2] having established a large marine science center in Qindao, Shandong province, aiming to study the extreme marine environments of the Polar regions, and building ocean-going vessels specifically rigged for collecting marine samples.[3]

Bioprospecting takes place by directly collecting organic material from nature, and through genetic sequencing of such material that has already entered biobanks. At the time of harvesting (collecting from the wild), the material typically includes all kinds of living specimens or samples from organisms. On return to shore, the marine material is usually stored in biobanks in various forms, from living organisms through dried material to prepared laboratory samples. From these, new expressions can be made, including

---

[2] China was accepted as an observer in Arctic Council in 2013; Beijing sees the Arctic as part of its Belt and Road project, with interests in transport, oil and gas, and marine natural resources (Rottem and Soltvedt, 2020).
[3] Personal communication, Erlend Ek, Norwegian embassy, Beijing, October 9, 2018. www.xinhuanet.com/english/2018-02/08/c_136959522.htm. See also http://www.qnlm.ac/en/index.

taxonomic information, ready-made assays, biochemical compositions, DNA sequencing, DSI, screened genomes and synthesized enzymes (biological molecules) copying those found (Tvedt, 2020). Enzymes are a central part of polar marine bioprospecting for their function in catalyzing chemical reactions in living organisms (respiration, digestion, etc.).[4] Bacteria for antibiotics and anticancer agents form another major group (Oldham and Kindness, 2020). As shown in Box 10.1, synthetic biology and the use of DSI are increasingly affecting the ABS debate, and also invoking and intensifying the derivative debate in UNCLOS (Lai et al., 2019; Wolman, 2016).

10.3.2 Governing Bioprospecting

Three levels of law are relevant for polar bioprospecting. At the *regional level*, Antarctica is governed by the Antarctic Treaty System (ATS). The legal overlap between ATS and UNCLOS regarding Antarctica is subject to some controversy; however, neither of these regimes has regulations relating directly to marine bioprospecting. The *international level* includes the rules in UNCLOS, treaties on patent law harmonization and the general rules concerning ABS in the CBD. A general ruling in UNCLOS, Article 118, states that the parties shall cooperate in the conservation and management of living resources in the areas of the high seas. The ABS regime of the CBD is more specific about ABS conduct in bioprospecting, and might take precedence over UNCLOS through "lex specialis" (as more specific legal acts tend to take precedence over less specific ones). The ABS regime is also more inclusive, with 196 ratifying member states, as against 168 UNCLOS ratifications. The USA has not ratified either. The CBDs ABS regime demands prior informed consent (PIC) and mutually agreed terms (MAT) about where genetic material is found and under what conditions the material has been appropriated. Unlike UNCLOS, the ABS regime is not directly applicable outside of national jurisdiction, however.

In the ongoing UNCLOS negotiations (based on UN General Assembly decision 72/249, 2017), developing countries have advocated an ABS regime, whereas the developed countries' main concern has been with open access to the high seas (Blasiak et al., 2016). Developing countries favor an ABS regime along the lines of the CBD, which may involve mandatory, monetary benefit-sharing upon commercialization, the inclusion of derivatives, linking monitoring of ABS to patent systems (with mandatory disclosure of origin of genetic material in patent applications) and that the time scope for collected material is reckoned from the entry into force of the CBD (1993). Most of the developed countries, and increasingly China, are opposed to a fully fledged ABS regime along the lines of the CBD.[5] Unlike the case in the CBD, FAO, WIPO and WTO, corporations and industry associations are hardly represented in the UNCLOS preparatory committee meetings where ABS and marine bioprospecting are discussed. They are, however, active in lobbying state actors on UNCLOS agenda issues.[6] The ICC coordinates industry views on use of genetic resources within national borders, but this strategy does not address ABNJ directly (ICC, 2018). The

---

[4] https://www.britannica.com/science/enzyme [5] https://enb.iisd.org/vol25/enb25129e.html.
[6] Personal communication at Antarctic Conference in Tromsø, May 7, 2018, with Professor Steven Chown, Director of SCAR (Scientific Commission on Antarctic Research), of the Antarctic Treaty System.

biotechnology industry is concerned with reducing legal uncertainty, which, many argue, is hampering innovation and the development of products from marine habitats (ICC, 2018).[7]

The UNCLOS negotiating parties are split between the principle of the freedom of the high seas versus principles on ABS from use of marine genetic resources. This is predominantly a North–South conflict, exacerbated by the diverging norms embedded in internationally harmonized patent regimes (IPR[8]), and the ABS regime of the CBD with its Nagoya Protocol, respectively (Oberthür and Rosendal, 2014). In the polar regions the IPR–ABS discussion assumes new aspects as it relates to regulations of resources beyond national jurisdiction. The debate here can be seen as about striving to fill a legislative gap in the governance of genetic resources, as genetic resources in ABNJ are not directly covered by the CBD ABS regime. This concerns the access to and equitable sharing of benefits arising from the use of what may be regarded as a global commons resource, traditionally conceived of as a Common Heritage of Mankind (the CHM principle) (De Lucia, 2019). The central argument linked to the global commons nature of these resources concerns the need to maintain affordable access to the resources also for those without the financial means to conduct bioprospecting on the high seas. Such access might, for instance, be achieved through common pool collections of marine genetic samples (Jørem and Tvedt, 2014; Tvedt, 2020). The transparency necessary to realize benefit-sharing could also be achieved by notification through a clearinghouse mechanism: Prip (2021) argues that such a notification system should cover not only marine genetic resources (MGRs) collected in the sea, but also those held *ex situ*, as in gene banks, as well as DSI on MGRs. Another advantage of a common pool collection and a clearinghouse mechanism concerns sustainability in harvesting: Duplicates would be accessible to all, instead of each collector needing to collect their own sample, which might reduce the pressure on potentially rare marine specimens.

The third legal level refers to private rights subject to domestic legislation, contracts and patents. Updated information on national ABS legislation can be found at the CBD Clearing House site.[9] The majority of developing countries have enacted, or are in the process of enacting, ABS legislation, whereas this is less widespread in developed, typical "user" countries. In 2014, the EU issued ABS legislation that is in support of mandatory monetary benefit-sharing and includes acceptance of derivatives and disclosure. The EU legislation timeframe is limited to 2014 (with the entry into force of the Nagoya Protocol) and does not cover utilization of genetic resources back to the establishment of the ABS regime in 1993. Iceland regulates bioprospecting in relation to microbes isolated from their geothermal areas (Leary, 2008), and Queensland, Australia has ABS for commercial bioprospecting at home (Prip et al., 2014). Sweden and Denmark have determined that for the time being they do not intend to regulate ABS of genetic resources within their own national borders; similarly, Russia has no ABS regulation. The USA, which is not party to the CBD, has ABS-like regulations for bioprospecting within its national parks. Norway and Denmark have

---

[7] Observation by Morten Walløe Tvedt at Brest meeting of biobank collections, May 14–15, 2018.
[8] Mainly the World Trade Organization's Agreement on Trade-Related Aspects of Intellectual Property Rights (TRIPS) and the World Intellectual Property Organization.
[9] https://absch.cbd.int/en/countries.

advanced ABS regimes for regulating Norwegian and Danish bioprospecting abroad. Both countries have modified their patent acts, requiring disclosure of where genetic material is found and under what conditions the material has been appropriated (PIC and MAT obligations). Norway and Finland are in the process of developing ABS regimes for regulating bioprospecting also at home. In Norway, an administrative order on how to regulate ABS and foreign bioprospectors at home has been subjected to two separate hearings (2012 and 2017). In our case study of Norway, we pay specific attention to this decision-making process, which was still pending at the time of writing (spring 2021).

10.4 The Case of Norway: Bioprospecting Policies and Positions

The polar regions (Antarctic and Arctic) are part of Norway's identity as a polar nation. Norway is one of the seven claimant Parties to Antarctica. The Norwegian government's marine bioprospecting strategy (White Paper, 2009: 7) aims to "strengthen bioprospecting activities in the High North by giving priority to the collection of marine organisms from the northern ocean region." Of the Arctic states, Norway has the most highly developed marine biotechnology sector and has territorial waters and an exclusive economic zone (EEZ) ranging from the North Sea and Skagerrak to the polar areas surrounding Svalbard, Jan Mayen and the Barents Sea.

Norwegian (and foreign) bioprospectors receive considerable public funding through the research programs under the Research Council of Norway and public funding of oceangoing research vessels collecting biological samples, as well as access to the marine samples deposited in the public marine biobank – Marbank in Tromsø (Svendsen, 2013). Most marine bioprospecting activities involve collaboration between academia and business, of which MabCent has been the largest in Norway (Greco and Cinquegrani, 2016; MabCent Report, 2015). A recurrent complaint associated with these public–private partnerships concerns the patent processes, which are necessary for commercial actors but tend to delay the publication of research results, on which the academic actors depend (MabCent Report, 2015, Prip et al., 2014; Rosendal et al., 2016). In 2015, MabCent was replaced by the Arctic Biodiscovery Centre at the Arctic University, which is not contractually linked to any specific commercial partner.[10]

About one third of the materials and samples in Marbank have their origin in ABNJ. This makes it pertinent to examine bioprospector positions in the UNCLOS debate as well as on relevant Norwegian policies. The lack of position papers and plenary statements in UNCLOS caused us to look elsewhere to identify the specific industry interests in marine bioprospecting. We gained some indications of bioprospector positions and strategies by evaluating the hearing responses from the two consultation processes (2012 and 2017) on the Norwegian draft ABS administrative order. These hearings appear highly relevant for our purposes, for two reasons: First, it is difficult to distinguish between marine material collected from areas *within* national jurisdiction and in ABNJ. Researchers on board the vessels collecting materials will know where the samples have been collected, but the

---

[10] https://bit.ly/34snBnM.

sampled organisms may well occur in many locations both *within and beyond* national jurisdiction. Second, the hearings cover the same issues, relevant at international and domestic levels, concerning the regulation of accessing samples from marine biobanks and regulating access and use of genetic digital sequence information and derivatives such as enzymes. Third, the corporate actors involved have an interest in marine resources from locations both within national jurisdiction and in ABNJ.

In examining the hearing responses, we distinguish among positions according to the three corporate models described above. Both drafts aim to comply with the CBD/Nagoya Protocol objectives for ABS. The first draft of the Norwegian ABS administrative order (2012) included monetary benefit-sharing, and defined derivatives (enzymes, digital sequences) as part of genetic resources. The hearing revealed strong support for this ABS model among public actors and NGO respondents, whereas industry actors were critical of what they feared could become a cumbersome, expensive access model (Rosendal et al., 2016).[11] Seven of the eight commercially oriented actors opposed the draft ABS legislation, albeit with conditional support, as they pointed to pending ABS legislation in the EU. One of the eight commercial respondents supported the full text of the ABS draft, citing the need to secure equitable sharing of benefits from the use of genetic resources.

In 2017, a revised draft administrative order on ABS was circulated. To accommodate industry responses to the first round of hearings, the revised draft did not mention monetary compensation except as a voluntary fee for access to public biobank collections. Also, the draft excluded enzymes – in other words, derivatives.

In response to these changes in the ABS design, the commercial respondents welcomed the 2017 draft administrative order. In general, their responses moved from what we expected in the first to the second model, apparently due mainly to the announced adjustment on excluding enzymes. In 2012, nearly all commercial respondents had referred to forthcoming EU legislation on ABS as a reason for stalling, but EU legislation was not mentioned by them in the 2017 hearings. This is hardly surprising, as the EU's ABS legislation in 2014 came out in support of mandatory monetary benefit-sharing and acceptance of derivatives and disclosure (setting the timeframe to the Nagoya Protocol [2014], and not the CBD [1993], though).[12] The 2017 hearing received twenty-nine responses, including nine from actors with commercial interests in marine bioprospecting, such as ArcticZymes (part of the MabCent consortium).

As monetary benefit-sharing was dropped to accommodate industry interests, the issue of access-fees (or "cost-sharing" [Rosendal et al., 2016]) in biobanks became relevant. The 2017 draft proposed that public (but not private) biobank collections should allow free access, and here the university museums and Marbank were critical. Marbank argued that the revised draft order might dissuade private collectors from sharing and depositing their material with Marbank, while having free access to Marbank's material and being free to patent innovations based on this material.[13] In Marbank's view, the public collectors that provide marine genetic

---

[11] This is based on the authors' reading of all the consultation responses to the 2012 draft administrative order.
[12] EU No. 511/2014. See also SEPA, 2018: 10: "The Regulation also applies to derivatives which were acquired at the same time as the genetic resource. Derivatives are defined as naturally occurring biochemical compositions that result from the genetic expression or metabolism of biological or genetic resources, although they do not contain functional units of genetic material. Examples of these are enzymes, proteins and essential oils."
[13] Interview / personal communication with Kjersti Lie Gabrielsen, Director of Marbank, May 8, 2018.

material would be left with no rights, whereas the commercial users of the material would have no obligations. In practice, access to Marbank has usually taken place through academia–industry consortia, but corporate actors are currently not allowed access, in anticipation of the new legislation.[14] Similar criticism came from noncommercial actors, who argued that the revised draft was no longer in compliance with the CBD's ABS obligations, and warned that monopolization might follow from the lack of restraints on patenting.[15] Critics pointed at the flaw in the 2017 draft: Unlike ABS regulations in the EU, it does not include enzymes and derivatives and may hence undermine the CBD's ABS regime and be poorly equipped to deal with synthetic biology activities.[16] As noted, enzymes constitute about one third of marine bioprospecting. This may partly explain why ArcticZymes reacted positively to the revised draft order.

According to the Norwegian Ministry of Trade, Industry, and Fisheries, which is responsible for the ABS administrative order, however, enzymes will not be excluded from the (still pending) ABS legislation.[17] The reason is that excluding enzymes would not be compatible with EU legislation, let alone the CBD (see note 12, on enzymes [as derivatives] being part of the EU definition of genetic resources).

10.5 Variation in Corporate Strategies

We have examined the ABS positions of two bioprospecting corporations in further detail (see Table 10.1). Novozymes is a multinational corporation, headquartered in Denmark, and among the world's largest producers of industrial enzymes. ArcticZymes, based in Tromsø, Norway, is a smaller company that is part of a multinational pharmaceutical corporation, Biotec Pharmacon (thus also part of MabCent); now known as ArcticZymes Technologies.

10.5.1 Novozymes

Novozymes (part of NovoNordic until 2000) is actively engaged in the ABS issue, with the explicit policy of adhering to the ABS principles of the CBD. Going further than the EU's ABS legislation, Novozymes holds that ABS starts with the entry into force of the CBD in 1993. Further, the corporation is set on avoiding accusations of biopiracy. According to its explicit policy:

Novozymes endorses the globally recognized principles in the CBD and ABS. As a part of our obligation towards the CBD, we only take samples in agreement with all relevant laws and regulations in the countries we operate in. In addition, we have stringent internal procedures including a database system for traceability of genetic resources to ensure that we live up to our commitments.[18]

---

[14] Interview / personal communication with Kjersti Lie Gabrielsen, Director of Marbank, May 8, 2018.
[15] This view was central in the hearing letters from the Research Council of Norway, the National Ethical Research Committees, the Norwegian Institute of Marine Research, the Ministry of Agriculture and the Norwegian Coastal Administration.
[16] This is based on the authors' reading of all the consultation responses to the 2017 draft administrative order.
[17] Personal communication with NN, of the Ministry of Trade and Fisheries and responsible for the second draft administrative order, August 29, 2018, at the FNI Genetic Resources Seminar, Lysaker, Norway. At the time of writing, the administrative order is still pending.
[18] www.novozymes.com/en/about-us/positions-policies.

The strong ABS policy is often linked to Novozymes' and NovoNordic's first CEO, Steen Riisgaard – due partly to his background from the NGO sector, which included Friends of the Earth and WWF, Denmark, and due partly to his many official statements on how enzymes technology can contribute to a more environmentally friendly world.[19]

Still, the ABS regime seems to have had more of a hampering effect on Novozymes' bioprospecting than boosting it. As a direct result of the CBD principles on ABS, most of its bioprospecting collaboration with university partners in developing countries stopped in the mid 1990s. According to our interviewees at Novozymes, this is because ABS legislation is sometimes inappropriately designed, leaving too much legal uncertainty regarding documentation of PIC and MAT about where genetic material is found and under what conditions the material has been appropriated.[20] And indeed, when Brazil changed its ABS legislation (2015–2017) to a simpler system, Novozymes reengaged in cooperation.[21] The Brazilian example indicates that it is possible to design "appropriate" ABS legislation that provides bioprospectors with enough legal certainty to trust in collaboration.

At present, Novozymes may be largely self-sufficient in genetic resources through its own collections, but the company acknowledges that marine genetic resources from the deep sea may be interesting and necessary in the future. Novozymes already has roughly 50,000 bacteria and fungi in its collection, which dates back over sixty years; however, as put by Peter Falholt, head of R&D at Novozymes, "I'm a little bit skeptical of synthetic biology [as being able to provide sufficient genetic material for bioprospecting], because you cannot beat four billion years of evolution" (quoted in Peplow, 2015). Their patent filings on polar, marine material date back to 1986 and 1992 for *candida Antarctica* (an enzyme, lipase), with applications ranging from food and fuels to detergents and medicine (Oldham and Kindness, 2020). Novozymes ranks as number one by a considerable margin for first patent filings involving Antarctic organisms (Oldham and Kindness, 2020): Novozymes tops the list with 300 filings, with BASF coming second with 113 filings. However, Oldham points out that in patent filing registrations, Novozymes is more likely to appear prominently because it is far more likely (than any of its competitors) to state the origin of the material in its patent applications: This is in line with the corporation's formal policy and guidelines to abide by the CBD's ABS regime. Further, BASF may appear less prominently because it is less explicit in stating the origin of its material, so the sources will not be registered in the patent filings.[22] This shows how ABS compliance could also expose a corporation to criticism, as Novozymes "appears" to have more patent filings involving marine organisms.

Novozymes does not participate directly in UNCLOS but coordinates its positions with the ICC. Although Novozymes might seem to fit into our third model given its strong language on ABS, there is agreement within the ICC group to lobby against applying ABS to genetic digital sequence information (ICC, 2017; 2019). The ICC is explicit in strongly opposing any expansion of the scope of the ABS regime to apply to digital sequence

---

[19] https://en.wikipedia.org/wiki/Steen_Riisgaard.
[20] Interview with representative from Novozymes, Copenhagen, October 2018.
[21] www.cbd.int/abs/ABNJ-views/2019/Brazil-DSI.pdf (on the new ABS legislation in Brazil).
[22] Personal communication with Dr. Paul Oldham of Lancaster University, UK. May 11, 2020.

information and genetic resources in ABNJ (ICC, 2017: 1). That places Novozymes closer to our second model on appropriate design, as they argue against closing the legal gap, hence against increasing the scope of ABS.

10.5.2 ArcticZymes / Biotec Pharmacon

ArcticZymes describe itself as follows: "we use access to the marine Arctic to identify novel cold-adapted enzymes for use in molecular research, in vitro diagnostics, and manufacturing."[23] It has a history of collaboration with Norwegian universities through the MabCent project, as part of holding company Biotec Pharmacon. The academic collaboration has allowed free and open access to Marbank's collections and to Marbio's ready-made assays. Access to Marbank is now a thing of the past, due to legal uncertainty linked to the fate of the Marbank material in the draft Norwegian ABS order. However, losing access to Marbank is not seen as a problem for ArcticZymes as it can find what it needs in international biobank collections and databases, where digital genetic sequences may be purchased online.[24]

Biotec Pharmacon ranks as the largest holder of patent filings on Arctic marine materials[25] and, according to MedNous (2019), "Biotec Pharmacon's inventive step was to scour the marine environment for solutions that were not already on the market and patent them." Several patented products based on cold-water enzymes are presented on ArcticZymes' online website.[26] These include proteinase, which is an unspecific endopeptidase (an enzyme) originating from an Arctic marine microbial source,[27] and glycosylase, which belongs to a family of enzymes involved in DNA repair and stemming from Atlantic cod.[28] Compared to Novozymes, ArcticZymes is a small firm that attracts scant public attention and might hence be less worried about possible accusations of biopiracy. When enzymes were excluded from the Norwegian ABS draft legislation, the company came out in favor of the ABS proposal. If enzymes were to be redefined as subject to the ABS legislation, ArcticZymes could be expected to oppose it. Hence, it is hard to judge whether it fits into our first or second model (Table 10.1).

10.6 Discussion

The hearing responses to the two draft ABS administrative orders reveal how the Norwegian authorities have found it hard to adjust domestic legislation to the global ABS regime of the CBD, for which they were strong advocates at the time. The government's response has been to change the wording of the ABS administrative order from including enzymes (2012), to excluding enzymes (2017), and then possibly to include enzymes in the regulatory scope once more. The Norwegian ABS regulation is still pending, nearly

---

[23] https://arcticzymes.com/company/about-us/.
[24] Interview with representative from ArcticZymes, Tromsø, September 2018. Corroborated in Hearing from ArcticZymes, Tromsø, October 3, 2017.
[25] Oldham and Kindness (2020).
[26] https://arcticzymes.com/products/enzymes/. See also ArcticZymers Technologies. 2020. Q4 Report 2020. Tromsø, Norway.
[27] https://bit.ly/3HojXKv.   [28] https://bit.ly/3KdVedG.

Table 10.1 *Positions on ABS: ArcticZymes and Novozymes*

COMPANIES/ ORGANIZATIONS	SUPPORT OR OPPOSITION TO ABS	ARGUMENTS BASED ON MODEL 1	ARGUMENTS BASED ON MODEL 2	ARGUMENTS BASED ON MODEL 3
ARCTICZYMES	Conditional support	Oppose benefit-sharing on material from open publications and databases.	Accept monetary benefit-sharing, but not if derivatives (incl. enzymes) are defined as genetic resources.	
NOVOZYMES	Support		Accept monetary benefit-sharing. As part of ICC: do not wish to close the legal geographical and technological gaps.	Bases corporate strategy on strong language on ABS. In favor of EU ABS legislation on derivatives and goes further than EU by accepting CBD (1993) as timeline for ABS.

a decade after its conception and despite the intensity with which Norway advocated the ABS regime of the CBD. This would seem to be a classic example of how policies may change when internationally agreed policy obligations are to be translated into domestic policies, if these policies prove to entail explicit costs to specific subnational target groups. Similarly, Norway is no longer a strong advocate of ABS principles in the current process on the Post-2020 Global Biodiversity Framework.

Corporate actors were far less skeptical of the second draft administrative order, which excluded enzymes from the definition of genetic resources. The increased acceptance was due mainly to this (potentially short-lived) adjustment on excluding enzymes. The change in responses suggests an effect of adjusting the regulatory design, which is in line with our Model 2 expectations.

Model 3 responses would be closer to transformative biodiversity governance but are difficult to assess. On the rhetorical level, Novozymes' history and links to the NGO sector have made a deep impact on its policy to support ABS. Moreover, its history, not least its visibility as a large corporation, has made Novozymes cautious, as well as vulnerable to being associated with accusations of biopiracy. In effect, Novozymes has backed away from bioprospecting collaboration with countries with (arguably) unclear ABS legislation that is claimed to engender legal uncertainty. When Brazil simplified its ABS legislation, Novozymes resumed collaboration – indicating that deeds followed words, but also showing how the company's responses may be more in line with Model 2. Further, a possible problem for bioprospectors in complying with the ABS principle of disclosure is that this may expose the corporation to criticism: This is exemplified by Novozymes' reporting of origin of material in patent applications, possibly to a much larger extent than BASF reporting. Moreover, as part of the ICC collaboration in UNCLOS, Novozymes is apprehensive about any expansion of the ABS scope ("don't close the legal gap").

The takeaway message is that policymakers have legal and political room to maneuver in adjusting ABS to get bioprospecting corporations on board. The political feasibility room here might not fully correspond to the aims of transformative biodiversity governance, however.

Finally: How do corporate actors plan and strategize regarding their own access to marine biological material from ABNJ? Access may be affected if only a handful of multinational corporations come to monopolize the bulk of collected material through patent applications. Granted patents have been few, but if the majority of patent applications succeed, that would clearly undermine ABS efforts while also severely restricting access for other bioprospectors. This aspect indicates an interesting potential for common ground between ABS principles and corporate interests, which might increase the political feasibility room.

10.7 Conclusions

Future transformative biodiversity governance should heed two regulatory "gaps" in current legislation: one primarily of a *technological* nature, the second *geographical*. Developments in new biotechnologies may widen the technology gap in ABS regulations: This gap is likely to remain open to corporate actors unless states decide to define derivatives (DSI, synthetic enzymes, etc.) as part of genetic resources. As noted, ABS legislation in the EU has included derivatives (enzymes and DSI) in its definition of genetic resources. Will industry become more inclined to accept the EU design for ABS as appropriate? And will the EU's approach to ABS have a bearing on how the UNCLOS debate deals with derivatives? (Bio)technological developments make it difficult to monitor bioprospecting, as bioprospectors can now access, sample and develop a large range of digital genetic sequences from online databanks.

With the evolving technological potential, technology has direct and significant implications for the global governance of biodiversity and genetic resources, as the ABS regime is more readily undermined by genetic resources expressed as digital sequence information. This challenge to the equity principles of the ABS regime indicates how vested interests in biotechnology might obstruct central elements in TBG.

Second, while the UNCLOS debate continues, uncertainty remains as to whether the regulatory geographical gap is likely to remain open to bioprospectors. One way of closing this gap would be to subject marine genetic resources from ABNJ to ABS regulation as global commons resources by making it mandatory to share duplicates of collected samples in common pool biobanks. This proposal features centrally on the UNCLOS agenda, along with the proposed clearinghouse mechanism. Sharing duplicates openly would, in addition, reduce pressure on rare marine species – an important point for transforming biodiversity governance.

Our study has revealed important elements and differences between formal and informal stakeholder participation and inclusion, the latter being central to the TBG debate dealt with in this volume. There are indications that multinational corporations may exert strong influence on legislative processes and policymaking, albeit being formally absent from decision-making forums, internationally and nationally. Turning to the domestic level in Norway, despite the small number of industry actors engaged in the ABS hearing processes there, they had a deep impact on the output of the first hearing and may have influenced the stalling of the regulation. On the international arena, the ABS regime clearly represents a normative victory for developing countries, who continue to advocate ABS principles in international forums also outside the CBD. However, it remains to be seen whether the ABS norms will succeed in steering UNCLOS' governance of common, marine resources in a more equitable direction. The most inclusive suggestions currently on the UNCLOS negotiation table would seem to involve a combination of establishing common pool collections for marine genetic resources and a clearinghouse mechanism.

References

Ambec, S., Cohen, M. A., Elgie, S., and Lanoie, P. (2011). *The Porter hypothesis at 20: Can environmental regulation enhance innovation and competitiveness?* Washington, DC: Resources for the Future.

Arico, S. (2010). Marine genetic resources in areas beyond national jurisdiction and intellectual property rights. In *Law, technology and science for oceans in globalisation: IUU fishing, oil pollution, bioprospecting, outer continental shelf.* D. Vidas (Ed.), pp. 383–396. Leiden: Martinus Nijhoff.

Arnaud-Haond, S., Arrieta, J. M., and Duarte, C. M. (2011). Marine biodiversity and gene patents. *Science* 331, 1521–1522. DOI: 10.1126/science.1200783

Barth, R., and Wolff, F. (2009). Corporate social responsibility and sustainability impact: Opening up the arena. In *Corporate social responsibility and sustainability impact: Rhetoric and responsibilities*. R. Barth and F. Wolff (Eds.), pp. 3–25. Cheltenham: Edward Elgar.

Blasiak, R., Jouffray, J. B., Wabnitz, C. C., Sundström, E., and Österblom, H. (2018). Corporate control and global governance of marine genetic resources. *Science Advances* 4, 1–7. DOI: 10.1126/sciadv.aar5237

Blasiak, R., Pittman, J., Yagi, N., and Sugino, H. (2016). Negotiating the use of biodiversity in marine areas beyond national jurisdiction. *Frontiers in Marine Science* 3, 1–10.

Blunt, J. W, Copp, B. R, Munro, M. H. G, Northcote, P. T, and Prinsep, M. R. (2011). Marine natural products. *Natural Products Reports* 28, 196–268.

Bodnar, A. (2016). Marine animals provide models for biomedical research. *Environment* 58, 16–25.

Chaffin, B. C., Garmestani, A. S., Gunderson, L. H., et al. (2016). Transformative environmental governance. *Annual Review of Environment and Resources* 41, 399–423.

Cyert, R. M., and March, J. G. (1963). *A behavioral theory of the firm.* Englewood Cliffs, NJ: Prentice Hall.

Delmas, M. A, and Toffel, M. W. (2008). Organizational responses to environmental demands: Opening the black box. *Strategic Management Journal* 29, 1027–1055.

De Lucia, V. (2019). Ocean commons, law of the sea and rights *for* the sea. *Canadian Journal of Law and Jurisprudence* 32, 45–57.

Drankier, P., Oude Elferink, A. G., Visser, B., and Takács, T. (2012). Marine genetic resources in areas beyond national jurisdiction: Access and benefit-sharing. *The International Journal of Marine and Coastal Law* 27, 375–433.

ENB (Earth Negotiation Bulletin) (2018). Summary of UN Biodiversity Conference, November 14–29, 2018, ENB 9 (725), 25. Available from https://enb.iisd.org/events/2018-un-biodiversity-conference.

Esty, D., and Winston, A. S. (2006). *Green to gold: How smart companies use environmental strategy to innovate, create value and build competitive advantage.* New Haven, CT: Yale University Press.

Gravelle, H., and Rees, R. (1981). *Microeconomics.* London: Longman.

Greco, G., and Cinquegrani, M. (2016). Firms plunge into the sea. Marine biotechnology industry, a first investigation. *Frontiers in Marine Science* 2. https://doi.org/10.3389/fmars.2015.00124

Greer, D., and Harvey, B. (2004). *Blue genes: Sharing and conserving the world's aquatic genetic resources.* London: Earthscan.

Greiber, T. (2011). *Access and benefit sharing in relation to marine genetic resources from areas beyond national jurisdiction: A possible way forward.* Bonn: IUCN.

ICC. (2017). Digital sequence information and the Nagoya Protocol. Prepared by the ICC task force on ABS. DYE/abs 14.09.2017. 450/1111.

(2018). Encouraging ABS agreements and sustainable use of genetic resources. Available from https://bit.ly/33AwriS.

(2019). Digital sequence information and benefit sharing. Prepared by the ICC task force on ABS. 02.05.2019. 450/1121.

Jørem, A., and Tvedt, M. W. (2014). Bioprospecting in the high seas: Existing rights and obligations in view of a new legal regime for marine areas beyond national jurisdiction. *The International Journal of Marine and Coastal Law* 29, 321–343.

Kolk, A., and Pinkse, J. (2004). Market strategies for climate change. *European Management Journal* 22, 304–314.

Lai, H. E., Canavan, C., Cameron, L., et al. (2019). Synthetic biology and the United Nations. *Trends in Biotechnology. Science and Society* 37, 1146–1151. https://doi.org/10.1016/j.tibtech.2019.05.011

Leary, D. (2008). Bi-polar disorder? Is bioprospecting an emerging issue for the Arctic as well as for Antarctica? *Review of European Community and International Environmental Law* 17, 41–55.

(2018). Marine genetic resources in areas beyond national jurisdiction: Do we need to regulate them in a new agreement? *MarSafeLawJournal* 5, 39.
MabCent Report. (2015). Arctic marine bioprospecting in Norway. *Phytochemistry Reviews* 12, 567–578.
MedNous. (2019). *Exploiting the Arctic ecosystem. Biotec Pharmacon enables modern medicines*. London: Evernow Publishing Ltd.
Morgera, E., Buck, M., and Tsioumani, E. (Eds.). (2013). *The 2010 Nagoya Protocol on Access and Benefit-Sharing in perspective: Implications for international law and implementation challenges*. Leiden: Martinus Nijhoff.
Müller, F., and Schøyen, H. (2021). Polar research and supply vessel capabilities – An explorative study. *Ocean Engineering* 224, 108671. https://doi.org/10.1016/j.oceaneng.2021.108671
Oberthür, S., and Rosendal, G. K. (Eds.). (2014). Global governance of genetic resources: Access and benefit sharing after the Nagoya Protocol. London: Routledge.
Oldham, P., and Kindness, J. (2020). Biodiversity research and innovation in Antarctica and the Southern Ocean. *bioRxiv*. https://doi.org/10.1101/2020.05.03.074849
Olesen, I., Rosendal, G. K., Bentsen, H. B., Tvedt, M. W., and Bryde, M. (2007). Access to and protection of aquaculture genetic resources: Strategies and regulations. *Aquaculture* 272, S47–S61.
Peplow, M. (2015). The enzyme hunters. Available from https://bit.ly/3rinPWt.
Porter, M., and van der Linde, C. (1995a). Toward a new conception of the environment–competitiveness relationship. *Journal of Economic Perspectives* 9, 97–118.
Porter, M., and van der Linde, C. (1995b). Green and competitive: Ending the stalemate. *Harvard Business Review* 73, 120–134.
Prip, C. (2021). Virtual progress towards a new global treaty on marine biodiversity in areas beyond national jurisdiction. Available from https://bit.ly/3FylZWp.
Prip, C., Rosendal, G. K., Andresen, S., and Tvedt, M. W. (2014). *The Australian ABS framework: A model case for bioprospecting?* FNI Report 1/2014. Lysaker: Fridtjof Nansen Institute.
Rosendal, G. K., Myhr, A. I., and Tvedt, M. W. (2016). Access and benefit sharing legislation for marine bioprospecting: Lessons from Australia for Norway. *Journal of World Intellectual Property Rights* 19, 86–98. DOI: 10.1111/jwip.12058
Rosendal, G. K., Olesen, I., Bentsen, H. B., Tvedt, M. W., and Bryde, M. (2006). Access to and legal protection of aquaculture genetic resources: Norwegian perspectives. *Journal of World Intellectual Property* 9, 392–412.
Rosendal, G. K., Olesen I., Tvedt, M. W. 2013. Evolving legal regimes, market structures and biology affecting access to and protection of aquaculture genetic resources. *Aquaculture* 402–403, 97–105. DOI: 10.1016/j.aquaculture.2013.03.026
Rottem, S., and Soltvedt, I. F. (2020). *Arctic governance: Norway, Russia and Asia*. Vol. 3. London: I. B. Tauris.
Sanchez, C. M. (1997). Environmental regulation and firm-level innovation: The moderating effects of organizational- and individual-level variables. *Business and Society* 36, 140–168.
Simon, H. A. (1976). *Administrative behavior: A study of decision-making processes in administrative organization*, 3rd ed. London: The Free Press/Collier Macmillan.
Skjærseth, J. B., and Eikeland, P. O. (Eds.). (2013). *Corporate responses to EU emissions trading: Resistance, innovation or responsibility?* London: Routledge.
Skjærseth, J. B., and Eikeland, P. O. (2019). *The politics of low-carbon innovation: The EU strategic energy technology plan*. Cham: Palgrave Macmillan.
Svenson, J. (2013). MabCent: Arctic marine bioprospecting in Norway. *Phytochemistry Reviews* 12, 567–578.
Swedish Environmental Protection Agency (SEPA). (2018). *The Nagoya Protocol and the EU ABS regulation*. Stockholm: Swedish Environmental Protection Agency.
Tvedt, M. W. (2020). Marine genetic resources. A practical legal approach to stimulate research, conservation, and benefit sharing. In *The law of the seabed: Access, uses, and protection of seabed resources*. C. Banet (Ed.), pp. 238–254. Leiden: Brill Nijhoff.

UNEP, ILRI. (2020). *Preventing the next pandemic: Zoonotic diseases and how to break the chain of transmission*. Nairobi, United Nations Environment Programme and International Livestock Research Institute.

Wolman, D. (2016). Humanity's health may depend on what sits on the Arctic seabed. Available from https://bbc.in/3tsU8oy.

Wynberg R., Schroeder, D., and Chennells, R. (Eds.). (2009). *Indigenous peoples, consent and benefit sharing*. Dordrecht: Springer.

Yin, R. K. (2003). *Case study research: Design and methods*. 3rd ed. London: SAGE.

Part IV

Transforming Biodiversity Governance in Different Contexts

11

Transformative Biodiversity Governance for Protected and Conserved Areas

JANICE WEATHERLEY-SINGH, MADHU RAO, ELIZABETH MATTHEWS, LILIAN PAINTER, LOVY RASOLOFOMANANA, KYAW T. LATT, ME`IRA MIZRAHI AND JAMES E. M. WATSON

11.1 Introduction

This chapter analyzes the potential for transformative change for biodiversity conservation in the governance of protected areas and other conserved areas (which incorporates other effective area-based conservation measures or OECMs). This is achieved by analyzing efforts to achieve Aichi Target 11 under the UN Convention on Biological Diversity (CBD) strategic plan to 2020, and discussing the need for a new outcome-based approach under the CBD's Post-2020 Global Biodiversity Framework (GBF), which is under discussion at the time of writing but expected to be adopted during 2022. Under Aichi Target 11,[1] governments collectively agreed to designate 17 percent of terrestrial and inland waters and 10 percent of coastal and marine areas as protected areas and OECMs that are effectively and equitably managed, ecologically representative, well connected and integrated into the wider landscape and seascape. It is widely considered to be the Aichi Target that governments have made most progress on delivering (UNEP-WCMC et al., 2018).

The issue is discussed here through the conceptual lens of transformative governance, which is defined in Chapter 1 of this book and is understood to address the underlying causes of biodiversity loss through governance mechanisms that are *integrative*, *inclusive*, *transdisciplinary*, *anticipatory* and *adaptive* (Chaffin et al., 2016; Guston, 2014). How to follow up Aichi Target 11 with a new area-based target has formed a key part of the CBD discussions in advance of the adoption of a new Post-2020 GBF. A range of perspectives emerged during these discussions. These can be summarized as: full implementation of Aichi Target 11; more ambitious area-based targets (such as a 30 percent area-based target or "half-earth" approach); "new conservation," which intends to integrate conservation with neoliberal economic approaches; and a "whole earth" approach, which aims to find a balance between human and nonhuman needs (Bhola et al., 2020, see also Chapter 12). This chapter does not fit neatly within any of these categories but contributes to the discussion by recognizing the valuable role played by protected and conserved areas, and the need for their continued prioritization at the policy level, and provides recommendations for their implementation under the Post-2020 GBF.

---

[1] The full text of CBD Aichi Target 11 is: "By 2020, at least 17 per cent of terrestrial and inland water, and 10 per cent of coastal and marine areas, especially areas of particular importance for biodiversity and ecosystem services, are conserved through effectively and equitably managed, ecologically representative and well connected systems of protected areas and other effective area-based conservation measures, and integrated into the wider landscapes and seascapes."

We understand transformative change in the context of protected and conserved area governance as referring to their contribution to the effective conservation of existing biodiversity, as well as its restoration, where possible. This chapter thus begins with an introduction to protected and conserved area governance, before examining the extent to which Aichi Target 11 has stimulated action to achieve effective biodiversity conservation outcomes. Outcomes are understood as being the consequences of project interventions and provide reliable indicators of long-term conservation impacts, either success or failure (Howe and Milner-Gulland, 2012; Kapos et al., 2009). Biodiversity outcomes as used in this chapter refer to the status of biodiversity elements such as species and ecosystems. Equity outcomes refer to the fair sharing of power, responsibility and benefits in natural resource management, as well as strengthening governance arrangements including legal entitlements and making decisions more transparent, inclusive and equitable.

We therefore first review the academic literature on protected and conserved areas through the lens of transformative governance (in Section 11.2), including recent literature analyzing the strengths and weaknesses of policy efforts to reach Aichi Target 11 (in Section 11.3). In Section 11.4, we use three case studies through which to explore the transformative change needed, before drawing conclusions related to the potential for a new outcome-based approach to protected and conserved area governance. The three case studies were selected to encompass different continents (Africa, Asia and Latin America), different ecosystems (terrestrial and marine) and different governance approaches.

11.2 Governance of Protected and Conserved Areas

Protected areas have been viewed as a mainstay of actions to conserve biodiversity and have long been at the fore of conservation and research efforts (Andam et al., 2008; Rands et al., 2010). It has been shown that well-managed protected areas are effective in conserving biodiversity and can reduce habitat loss and maintain species populations (Bruner et al., 2001; Leverington et al., 2010; Watson et al., 2014), as well as provide a range of societal benefits (Stolton and Dudley, 2010). The widely accepted International Union for the Conservation of Nature (IUCN) definition of protected areas is "a clearly defined geographical space, recognized, dedicated and managed, through legal or other effective means, to achieve the long-term conservation of nature with associated ecosystem services and cultural values" (Dudley, 2008). This places nature conservation objectives firmly at the center. They have a prominent position within global environmental governance fora, such as the CBD, which has a dedicated program of work on protected areas, and have long been promoted as an important conservation tool by IUCN and its member organizations through the World Commission on Protected Areas (WCPA). Other international environmental conventions have also placed a high importance on designating and managing protected areas, such as the Ramsar Convention on Wetlands of International Importance and the World Heritage Convention concerning the Protection of the World's Cultural and Natural Heritage (agreed in 1971 and 1972, respectively).

The concept of other effective area-based conservation measures (or OECMs) was introduced for the first time in the international policy arena in 2010, as an additional way by which national governments could meet Aichi Target 11. At the time, there was no accepted definition, and it was not until 2018 that the CBD adopted a decision that defined OECMs as "a geographically defined area other than a Protected Area, which is governed and managed in ways that achieve positive and sustained long-term outcomes for the in situ conservation of biodiversity, with associated ecosystem functions and services and where applicable, cultural, spiritual, socio–economic, and other locally relevant values" (CBD, 2018). OECMs are expected to involve a wider array of stakeholders in governance arrangements, particularly IPLCs, and spiritual and religious groups (Laffoley et al., 2017), and provide an opportunity to engage rights-holders and promote equitable and diverse partnerships in conservation efforts (IUCN-WCPA, 2019). Their governance arrangements are therefore expected to be more complex than those of traditional protected areas and are likely to require strengthening or the gaining of official recognition of informal arrangements (Dudley et al., 2018). OECMs are now commonly referred to as "conserved areas," which is the term we adopt throughout the rest of this chapter.

Protected area governance is well-documented in the academic literature and has been influenced by integrative governance (IG) concepts (Visseren-Hamakers, 2015; 2018). This includes polycentric governance, under which there are several centers of decision-making (Carlisle and Gruby, 2019), and multilevel governance, in which decision-making takes place at different scales in support of common goals (Bennett and Satterfield, 2018). Terrestrial protected areas, for example, have been heavily influenced by multilevel collaborative governance with growing interest in scaling-up to the landscape level with an increased focus on transboundary and connectivity issues (Lockwood, 2010). Forests, in particular, have been valued for assets other than biodiversity, notably for timber, and more recently for carbon sequestration and, as a consequence, governance arrangements and stakeholder engagement for forest areas are often complex (Nagendra and Ostrom, 2012; Reinecke et al., 2014). Forest governance also reflects IG approaches and has been heavily influenced by the concept of networked governance, which recognizes diverse configurations of stakeholders interacting at multiple levels, with a diffusion of authority (Jedd and Bixler, 2015). Marine protected areas (MPAs) have similarly been highly influenced by shared governance approaches between governments and local communities (Bown et al., 2013). Locally managed marine areas (LMMAs), in which nearshore waters are actively managed by communities, have been widely adopted across the tropical western Pacific, for example, as a way of achieving biodiversity conservation and fisheries management objectives simultaneously (Jupiter et al., 2014).

Protected area governance has also been influenced by particular concepts under the transformative governance agenda. In addition to adopting increasingly *integrative* approaches, protected and conserved area governance has also become more *inclusive* and *adaptive*. Recent efforts to scale-up both terrestrial and marine protected areas to the landscape/seascape scale, for example, including through the newer emphasis now being given to conserved areas, have moved toward more *inclusive* forms of governance, including by the increased and more formal involvement of IPLCs in governance mechanisms

(Premauer and Berkes, 2015). Studies indicate that *adaptive* management approaches that have integrated local communities in co-governance arrangements have been the most successful for terrestrial protected areas (Dawson et al., 2018; Premauer and Berkes, 2015). Similarly, the most successful approaches to MPA governance have found a balance between top-down and bottom-up approaches with a diversity of institutions involved (McCay and Jones, 2011) that take an *adaptive* approach with room for experimentation in management strategies (Bown et al., 2013). A global analysis of both marine and terrestrial protected areas by Oldekop et al. (2015) suggested that conservation benefits for biodiversity were highest when protected areas also delivered positive socioeconomic outcomes for local people, and that a co-management approach between local communities and conservation organizations delivered the greatest benefits to both local people and biodiversity.

Although the study by Oldekop et al. (2015) supports the joint achievement of biodiversity and socioeconomic outcomes, it is important to recognize that in practice protected area governance is impacted by debates concerning the ownership of, access to and governance of natural resources (Ostrom, 1990). Tensions often exist over competing objectives to be achieved (Anthony and Szabo, 2011) and have increased in complexity due to expectations that protected areas will achieve a wide range of objectives (Watson et al., 2014). In high-income countries this tends to reflect an increasing move toward achieving multiuse areas for a wide range of social and economic goals and ensuring the continuation of a range of ecosystem services (Hammer et al., 2012). In low-income countries, such tensions are often more concerned with how to enable local, and often marginalized, communities to achieve social and economic justice and livelihood goals, alongside nature conservation goals (McShane et al., 2011; Shahabuddin and Rao, 2010). This situation is likely to increase in complexity with the more recent inclusion of conserved areas, as by their very definition biodiversity conservation is not necessarily the main objective but is rather one of a number of objectives or an outcome resulting from management that is primarily for another purpose other than conservation. The management of trade-offs is, therefore, potentially even more complex in the case of conserved areas and there are still no commonly agreed minimum criteria of accepted biodiversity outcomes for potential OECMs that can be managed for purposes other than biodiversity conservation. Discussion is ongoing, for example, regarding the appropriate balance between biodiversity conservation and fisheries management objectives for conserved areas (Diz et al., 2018).

Resolving such issues and achieving conservation and socioeconomic outcomes is related to both the quality of protected and conserved area governance as well as management effectiveness, and there is often blurring and confusion between these two issues (Lockwood, 2010). While protected area management is concerned with the means and actions to achieve given objectives, protected area governance is concerned with decisions on what the objectives are, how decisions are taken and who has power, authority and responsibility and should be held accountable (Borrini-Feyerabend and Hill, 2015). In the last few decades there has been a shift from mostly state-driven, top-down governance approaches to a range of approaches to protected area governance, summarized in a typology adopted by the IUCN (Lockwood, 2010). It encompasses four main types: state governance; shared governance, which is more collaborative in nature between state and nonstate actors; private governance (i.e. governance by nonstate actors) and governance by IPLCs (Borrini-Feyerabend et al., 2013). This IUCN categorization by governance-type is distinct from but complementary to the more widely cited IUCN protected area categories system, which classifies protected areas according to their management objectives (Dudley, 2008).

Increasing attention has been given to assessing the management effectiveness of protected areas (for example, Bruner et al., 2001 and Leverington et al., 2010), but this has not been matched by efforts to examine whether protected area management and governance are leading to positive outcomes for biodiversity. This is despite the recent boom in satellite remote sensing tools that can provide relatively cheap and rapid assessments of terrestrial biodiversity (Luque et al., 2018), including for tropical forest ecosystems (Mulatu et al., 2017). In particular, little attention has been paid to the role of protected area governance in achieving effective conservation (and, where relevant, restoration) of biodiversity, and the role of conserved areas in this regard needs further examination. It is particularly important to examine biodiversity outcomes in the context of increasing pressures from a range of underlying drivers linked to unsustainable, global patterns of consumption and trade (Folke et al., 2019). There has been limited examination, for example, of how effective the approach set by the overarching international biodiversity governance agenda has been in stimulating action toward achieving biodiversity outcomes and addressing underlying drivers of loss. In the next section, we thus turn to considering the overall strengths and weaknesses of protected area governance approaches agreed at the international policy level, by examining the implementation of the Aichi Targets (2011–2020) through a transformative governance lens and further analyzing the new and growing role of conserved areas.

11.3 Strengths and Weaknesses of International Policy Approaches to Protected and Conserved Area Governance

11.3.1 CBD Aichi Target 11

Although the prominence of protected areas within the CBD and other international environmental conventions has ensured that high-level goals exist to stimulate government action, there has been a clear gap between such aspirational targets and the realization of actual outcomes on the ground. Lack of national level implementation is commonly cited as a problem and has been referred to as "perhaps the most significant factor in the failure of international biodiversity law" (Jóhannsdóttir et al., 2010:146), including the implementation of provisions under the CBD that are soft and open-ended in nature. There has been a lack of implementation of national biodiversity strategies and action plans (NBSAPs) in many countries, with limited progress made toward achieving many of the Aichi Biodiversity Targets set in 2010 (Buchanan et al., 2020; Maxwell et al., 2020; also see Chapter 3). CBD Aichi Target 11, however, provided a quantified target for the percentage of terrestrial (17 percent) and marine (10 percent) areas to be conserved and has been more successful than the other Aichi targets in stimulating government action. The coverage of terrestrial protected areas, for example, increased from 10.9 percent in 2011[2] to 14.9 percent in 2018 (UNEP-WCMC et al., 2018). The growth of MPAs has been particularly dramatic in recent years, with a fifteen-fold increase since 1993 when the CBD came into force, to a total of 16.8 percent of national waters having been designated by 2018 (UNEP-WCMC et al., 2018). According to the World Database on Protected Areas, the global coverage of terrestrial and marine protected areas as of January 2021 is 16 percent and 8 percent respectively.[3] Although Aichi Target 11 brought the designation of protected areas to the fore, much less attention has been paid to implementing the second half of the target, which is concerned with ensuring that such protected areas are *effectively and equitably managed*, and *ecologically representative* (Maxwell et al., 2020; Watson et al., 2016a), as we next discuss.

Effective Management

Efforts have been made to assess the management effectiveness of protected areas in the academic literature (for example analyses by Bruner et al., 2001 and Leverington et al., 2010), by IUCN and through publicly funded partnership initiatives. One such example is the European Union (EU) funded BIOPAMA program, managed by IUCN, which aims to support data management and analysis in Africa, Caribbean and Pacific (ACP) countries to facilitate better decision-making for protected areas.[4] A range of new tools have been developed, such as the Protected Areas database on Protected Area Management Effectiveness (PAME),[5] the Integrated Management Effectiveness Tool (IMET), the Management Effectiveness Tracking Tool (METT) (Hockings et al., 2018), and the IUCN green list of protected and conserved areas (IUCN and WCPA, 2017), a certification program and standard for the effective management and fair governance of protected and conserved areas. One weakness of some of these approaches has been the focus on management structures and procedures, with less attention given to assessing whether protected areas are effectively conserving and restoring biodiversity, which cannot necessarily be inferred from PAME assessments (Coad et al., 2015; Maxwell et al., 2020).

Furthermore, efforts made to assess management effectiveness have not necessarily been matched by efforts by national governments to support work on the ground to protect biodiversity within sites (Geldmann et al., 2021). Studies that intersect these Earth observation data with networks of protected areas (so as to assess their effectiveness) show that many protected areas slow, but fail to halt, human pressures and biodiversity loss within their borders (Verma et al., 2019). Protected area management is unable to address many of the underlying drivers of biodiversity loss linked to unsustainable production and consumption patterns and global trade, which is placing increasing pressure on biodiversity around the world (Folke et al., 2019). At least a third of protected areas are reported to be facing intense human pressure (Jones et al., 2018b),

---

[2] www.cbd.int/gbo/gbo4/gbo4-draft2-tech-doc-chapter-11-en.pdf
[3] UN Environment – WCMC World Database of Protected Areas. www.protectedplanet.net/en.   [4] www.biopama.org.
[5] https://pame.protectedplanet.net.

and around 40 percent of protected areas globally are estimated to face major deficiencies in management (Leverington et al., 2010), making it difficult to resist and adapt to such pressures. The study by Jones and colleagues (2018b) also found that human pressure had increased in 55 percent of protected areas between 1993 and 2009. Deficiencies in m*anagement effectiveness* are partly due to a lack of allocation of the necessary finance and resources, including of well-trained staff responsible for site management (Geldmann et al., 2018). The general lack of finance for protected areas is well-documented (for example, McCarthy et al., 2012; Waldron et al., 2013), with the Global South facing the greatest shortfalls in budgets and staffing (Coad et al., 2019).

Equitable Management

Governments have also given insufficient attention to achieving the *equitable management* component of Aichi Target 11 (Hagerman and Pelai, 2016). This is despite an increase in co-governance arrangements with an increasing trend of participation in biodiversity governance from nonstate actors and the emergence of multistakeholder partnerships, such as the Congo Basin Forest Partnership (CBFP) (Pattberg et al., 2017), which could represent a shift toward more *inclusive* forms of governance. The EU, for example, has been supporting public–private partnerships (PPPs) as a key tool for protected area governance, through its development aid programs (European Commission, 2014). In the case of MPAs, a key challenge is how to ensure that local communities remain meaningfully involved in governance with a greater focus on increasing the coverage of areas under designation and on scaling-up (Gruby and Basurto, 2014; McCay and Jones, 2011). Progress has also been limited toward achieving CBD Aichi Target 18, which was dedicated to the full and effective participation of IPLCs, but which did not provide a measurable target (Fajardo et al., 2021).

The importance of equity considerations is not to be underestimated, particularly as the scale and global importance of involvement by Indigenous Peoples in protected area governance is now becoming apparent. A recent study showed such communities impact the governance of approximately 40 percent of sites worldwide (Garnett et al., 2018), and the Intergovernmental Science-Policy Platform on Biodiversity and Ecosystem Services (IPBES) Global Assessment recognized their involvement as critical to achieving transformative governance (IPBES, 2019). The valuable contribution made by Indigenous People to conservation is particularly significant in the case of forests (Fa et al., 2020), which harbor around 75 percent of global terrestrial biodiversity (FAO, 2016). The need to halt tropical deforestation is recognized as being one of the most pressing and urgent global environmental challenges (Franklin and Pindyck, 2018), with CBD Aichi Target 5 aiming to halve, and where feasible bring to zero, the loss of natural habitats including forests. Primary tropical forest is of disproportionate value for biodiversity and ecosystem services (Morales-Hidalgo et al., 2015), but forest loss continues unabated in low-income tropical countries (Keenan et al., 2015). Although the conservation outcomes achieved by community conserved areas (CCAs) vary widely depending on the context (Rao et al., 2016), there is increasing evidence that securing land rights for Indigenous Peoples over forest land is an effective and important conservation management strategy (Fa et al., 2020; Oliveira et al., 2007; Watson et al., 2018).

Furthermore, governance by Indigenous Peoples is having a positive conservation impact outside of protected areas, with studies showing that areas under their governance harbor as much biodiversity as protected areas (Schuster et al., 2019; Sheil et al., 2015). The high focus on site designation contrasts strongly with the very limited attention given to delivering Aichi Targets concerned with conserving biodiversity outside of protected areas, including lands governed by Indigenous People (Hagerman and Pelai, 2016). Although many of the Aichi Targets are concerned with areas outside of protected areas, very little attention has been given to the 83 percent and 90 percent of undesignated terrestrial and marine areas, respectively. This is despite the significant biodiversity and ecosystem services they harbor and provide, which if lost would be an unmitigated disaster for both nature and people (Jones et al., 2018a; Maron et al., 2018).

Ecological Representation

Limited attention has also been given to ensuring the *ecological representativeness* component of Aichi Target 11. There is evidence that protected area designation has not necessarily targeted areas with high levels of threatened species but has instead been established in areas that minimize conflicts with agriculturally suitable land (Venter et al., 2017). The same is true of MPAs, which have failed to include all ecoregions, with area selection being influenced by socioeconomic factors (Jantke et al., 2018). Existing tools such as the global standard for identifying key biodiversity areas (KBAs) (IUCN, 2016), for example, could help governments identify the most valuable areas for biodiversity, but no such standard has been formally adopted by governments globally to help guide designation of protected and conserved areas (Visconti et al., 2019). Although the CBD Aichi Targets are global, they are often interpreted at a national level, with an assumption that all governments will try to achieve the 17 percent and 10 percent targets within their countries. Within the EU, an approach was adopted under the Birds and Habitats Directives that enabled a network of protected areas (known as Natura 2000) to be selected based on ecological representativeness at the continental level (Maiorano et al., 2015). No equivalent framework exists at the global level by which to incentivize countries with disproportionately high levels of biodiversity, such as megadiverse countries (Yang et al., 2020), to designate a larger percentage area of their territories. This lack of incentive is compounded by the overlap of key areas for biodiversity with areas facing high levels of poverty (Fisher and Christopher, 2007), with low-income countries facing the greatest relative shortfalls in site designation that would ensure ecological representativeness (Butchart et al., 2015).

11.3.2 Influence of the International Climate and Forest Governance Agendas

Protected area governance has been strongly influenced during the past decade by the climate regime, not least due to the introduction of the REDD+ initiative under the UN Framework Convention on Climate Change (UNFCCC), which aims to reduce carbon emissions from deforestation and forest degradation and encourage the conservation and enhancement of forest carbon stocks (UNFCCC, 2007). Although REDD+ was intended to target areas where deforestation is highest, in many cases this has included protected areas with large expanses of forest habitat (Scharlemann et al., 2010). This new focus on carbon as the main value to be conserved represented a departure from previous forest governance approaches, which had tended to focus on finding a balance between biodiversity conservation and economic use, notably timber extraction (McDermott, 2014). The involvement of new stakeholders from a climate perspective has brought an extra level of complexity and stimulated networked forest governance arrangements (Reinecke et al., 2014; Visseren-Hamakers et al., 2012).

In its early days, REDD+ was mainly implemented through projects that targeted specific forest areas, and in many cases such projects were spearheaded by conservation organizations (see, for example, Ferguson, 2009). As REDD+ has evolved, however, there has been a shift toward implementation through *integrative* and *transdisciplinary* governance approaches at the landscape level that engage stakeholders from different land-use sectors. Agribusiness companies seeking to reduce their impacts on deforestation, for example, have engaged in initiatives such as deforestation-free supply chains, sustainable commodity roundtables and certification schemes (Boucher and Elias, 2013). This is often supported by provincial governments working to deliver emission reductions as part of nested efforts to deliver nationally appropriate mitigation actions (NAMAs), for example, through low emission rural development (LED-R) activities (Nepstad et al., 2013). Such initiatives, which have developed subsequent to the adoption of the Aichi Biodiversity Targets, may overlap with efforts to scale-up terrestrial protected areas to the landscape scale and to designate conserved areas. This is expected to lead to more complexity and confusion in governance mandates due to the diversity of stakeholders involved who represent different interests, and bring different perspectives of "landscape" as either ecosystems and habitats, commodity production areas, administrative areas or territories with land rights (Weatherley-Singh and Gupta, 2017).

In sum, under the framework of the Aichi Targets, the considerable progress made by national governments in designating sites has not been matched by efforts to ensure effective and equitable management, nor ecological representativeness, combined with limited consideration as to how to achieve conservation outcomes for biodiversity. Efforts made to achieve Target 11 have been undermined by the lack of progress in achieving other Aichi Targets, which are complementary and necessary to fully address the drivers of biodiversity loss but more difficult to measure and achieve. This includes targets that are concerned with conserving biodiversity outside of protected areas, for example Target 5, which aims to "at least halve, and where feasible bring close to zero, the rate of loss of all natural habitats"; Target 10, which is concerned with decreasing pressures on coral reefs; Target 7, on the sustainable management of areas under agriculture, forestry or aquaculture; Target 15, on enhancing ecosystem resilience; and Target 18, on the full and effective participation of IPLCs (Fajardo et al., 2021; Hagerman and Pelai, 2016; Watson et al., 2016b). There has also been little effort by high-income and importing countries to achieve

Aichi Targets that focus on underlying drivers of biodiversity loss, such as Target 3 on the phasing out of harmful subsidies, and Target 4 on sustainable consumption and production.

There is, therefore, a need to ensure that protected and conserved area governance approaches (including international targets) achieve a better balance between site designation, and equitable and effective management and ecological representativeness. Furthermore, the impact on governance of newer landscape approaches, both due to the relatively recent inclusion of conserved areas as well as the engagement of new stakeholders from the climate sector, is still unclear. To inform this discussion, in the next section, we analyze the links between governance and biodiversity outcomes at the field level, from which lessons can be learned to inform recommendations for a transformative GBF.

11.4 A Transformative Policy Agenda: An Outcome-Based Approach to Protected and Conserved Area Governance

The last decade under the policy framework of the Aichi Targets has not provided a transformative governance agenda with clear outcomes for biodiversity conservation. We therefore discuss some of the policy and governance changes needed to redress this issue through an approach based on achieving biodiversity outcomes, in the light of the Post-2020 GBF and the growing importance of conserved areas. We draw on three case study examples that highlight how incorporating different transformative governance approaches can work in practice.

11.4.1 Case Studies

In this section, we present three case studies of protected areas and/or conserved areas with different forms of governance, in which the Wildlife Conservation Society (WCS) has been working, in order to draw some common lessons to inform a transformative policy agenda. The three selected case studies provide examples from different continental regions, varying ecosystem types and a range of governance scales and approaches. The first is Makira National Park and REDD+ project in Madagascar, the second is Kyeintali marine fisheries OECM in Myanmar, and the third is the Madidi-Tombopata Landscape in Bolivia and Peru (an area which encompasses both protected areas and OECMs).

Makira National Park and REDD+ Project

Makira National Park in the MaMaBay landscape in northeast Madagascar makes up the largest remaining intact humid rainforest in Madagascar, a country known for its unique endemic biodiversity. Containing half of Madagascar's remaining coastal forest, a quarter of its lowland forest, 50 percent of all its flowering plant species, as well as coral reefs, mangroves and wetlands, the MaMaBay forest landscape receives some of the highest rainfall rates in the country. Despite its size and importance, the forests of Makira remain under threat from deforestation and unsustainable natural resource extraction. As the human population grows, traditional hillside rice cultivation (known as "*tavy*") has become a major driver of forest loss.

The Makira National Park is managed collaboratively by the WCS as a "delegated manager" with "local community managers" of natural renewable resources,[6] thereby providing an example of *inclusive* governance. This institutional arrangement is based on Madagascar's 1996 Secured Local Managed Forests (GELOSE) Law and the 2001 Contracted Management of Forests Decree, which delegate the management of some natural resources to Community Based Groups (COBAs). These rules and regulations underline local communities as the main actors in forest management and restore the legitimacy of local management of common resources (Sarrasin, 2009). The Makira project is one of the world's first forest carbon mitigation projects (thereby demonstrating an *integrative* approach), and at times REDD+ has provided a financial mechanism through which to fund the activities of COBAs and provide benefits to communities, although it has been necessary to secure supplementary income from official development assistance (ODA).

Combined efforts by WCS and the COBAs has strengthened the overall management structure of the Makira National Park. Park staff now work with communities to promote the sustainable use of natural resources through awareness-raising of COBA rules and regulations and environmental education activities. This is resulting in a decrease of anthropic pressure on the forests, demonstrated by the reduction of slash and burn, illegal settlements and clearing. For example, there was a decrease in the areas cleared from 834 to 605 hectares between 2016 and 2018. Improving the situation in future will require the reinforcement of joint patrols and enhancing access to justice for local communities, including by tackling some of the underlying drivers of biodiversity loss, such as corruption.

Kyeintali, Marine Fisheries Co-management Area

Kyeintali is located in the southern Rakhine state of Myanmar, one of the country's poorest regions. Seventy-eight percent of the population live in poverty and over 80 percent are largely dependent on small-scale fishing for their livelihoods and subsistence. Traditional fishers (primarily men) and fish-workers who process the fish (primarily women) are rarely involved in decision-making or planning processes (Matthews et al., 2020). These coastal households are highly susceptible to impacts from the evident fisheries depletion. Recent interviews suggest that catches have more than halved in the past few years and provide evidence of bycatch of threatened species, though information is guarded and poorly documented (WCS, 2018).

The Kyeintali Inshore Fisheries Co-Management Area is now governed by the Kyeintali Inshore Fisheries Co-Management Association (KIFCA), which includes local community members (one man and one woman from each nearby village). Advisory and working committees composed of representatives from the government, police and Rakhine Fisheries Partnership support KIFCA. These groups were formed after a lengthy participatory process facilitated by WCS, which included the collection of scientific data on fishing

---

[6] Annual report of Makira National Park and Annual Operational Plan.

activities, biodiversity and socioeconomic needs; detailed consultations with the fishing dependent communities and management planning in which the communities proposed their own no-take zones, seasonally closed areas, gear-restricted zones and protected turtle nesting beaches (Exeter et al., 2021; WCS, 2018). Following this process, the area was officially accepted by the government in 2018 as Myanmar's first marine fisheries co-management area (Latt, 2019).

Factors critical to success include *inclusive* and *adaptive* governance approaches. The engagement and recognition of the needs of all local stakeholders, combined with coordination of activities among the fisheries department, local coastal conservation association and local communities, for example, has been very important. Support from the Rakhine Fisheries Partnership helped secure strong relationships with Kyeintali fishers. This engagement was further supported by fair and open elections to select members of the management association, with efforts made to deliberately include women from the communities. Management decisions by the participants and stakeholders are also being supported by scientific evidence (primarily through GPS-based tracking of fishing activity, and household and market surveys), with the potential for *adaptive* management based on the outcomes of scientific surveys.

The process for developing the co-management area was slower than anticipated because this decentralized form of management is very new in this national context. However, other communities are now interested in developing similar management schemes. As the zoned areas were proposed by the communities themselves, it is expected that levels of compliance will be high. Such compliance will be key, as one of the greatest limitations to achieving sustainable fishing in coastal Myanmar is a lack of enforcement of marine-related regulations. In areas where enforcement is low, compliance must be won through local support, therefore a co-management area in which communities have a strong voice can be an appropriate strategy to recover local fish stocks, while also achieving biodiversity outcomes as a complementary goal.

Greater Madidi-Tombopata Landscape

The Madidi-Tambopata landscape is found in northwestern Bolivia and neighboring Peru and stretches from the High Andes to the tropical lowlands. It covers 14 million hectares, five national protected areas, three subnational protected areas and eight indigenous territories, as well as communities of ten indigenous groups, providing an example of *inclusive* governance. Connectivity and overlap between protected areas and indigenous lands across the Amazon is critical to maintaining intact forests, which are necessary for wide ranging species, such as the jaguar, as well as for maintaining globally important ecosystem services such as climate mitigation and freshwater provision (Painter et al., 2017). The WCS has been working in the Greater Madidi-Tambopata landscape in Bolivia for two decades to support efforts by Indigenous People to secure legal recognition of their ancestral territories and increase their capacity to manage their lands and waters.

This is partly being achieved by the development of Indigenous Life Plans (or territorial management plans) for 1.8 million hectares of titled and claimed Indigenous territory. These plans enable Indigenous People to protect their lands, as well as using and managing natural resources in line with environmental, social and economic sustainability criteria, reflecting an *inclusive* and *integrative* approach. Such plans also contribute to the preservation of indigenous cultural identity and the revalorization of ancestral knowledge. They identify areas for achieving integrated conservation and development objectives, as well as connectivity corridors that link protected areas and Indigenous territories, to enhance the conservation of intact forest and healthy wildlife populations.

Management capacity-building processes have resulted in increased awareness among IPLCs of the environmental, economic and sociocultural value of their territories and have helped to secure local land rights. Local Indigenous People have worked together in the ordering and titling of their territories and benefit from increased security in access to and use of natural resources and the development of productive enterprises. The lives of Amazonian Indigenous Peoples depend on maintaining a harmonious relationship with nature for their spiritual, social, cultural and economic development. The Indigenous territorial management model has been developed by Indigenous People from their perspective and cultural identity, and also strengthens their commitment to conservation.

These three case studies demonstrate the critical importance of incorporating elements of transformative governance (particularly *inclusive*, *integrative* and *adaptive* approaches) to long-term success at the local level. In the next section, we draw some recommendations and discuss how these can benefit the development and implementation of policies at the international level under the Post-2020 GBF.

11.4.2 *Moving toward a Transformative, Outcome-Based Approach to Conservation*

As described in Section 11.3.1, implementation of Aichi Target 11 over the past decade has mainly focused on site designation, with limited attention given to achieving the second half of the target, relating to effective and equitable management and ecological representativeness. Instead of a quantified target that reflects the size of the area designated combined with measures of effective management, a transformative governance agenda for protected areas under the Post-2020 GBF needs to be based on achieving measurable outcomes for biodiversity. Discussions on the Post-2020 GBF, which are still ongoing at the time of writing, have still tended to focus on extending the coverage of protected and conserved areas (Bhola et al., 2020; Woodley et al., 2019), but parallel discussions on measurable biodiversity targets (Díaz et al., 2020; Geldmann et al., 2021), combined with increasing recognition of the need to integrate IPLCs, means there is scope for the implementation of an outcome-based goal.

As shown by the three preceding case studies, achieving equitable management goals by involving IPLCs in the governance and management of protected and conserved areas and landscapes is a slow and time-consuming process but absolutely vital to ensuring conservation goals are achieved in the long term. The elements of transformative governance that have been incorporated within the case studies have not only been *inclusive* of IPLCs, but have worked toward ensuring the recognition and enforcement of their rights. The case studies demonstrate that a range of actions are needed, depending on the context, to ensure ownership and buy-in of IPLCs as well as achieving conservation outcomes. These actions may involve environmental education, strengthening of community access to and rights over land, enforcement measures, spatial planning and tools that enable scientific findings to be combined with local knowledge. As highlighted in the example from Myanmar, an *inclusive* approach is associated with increased compliance with regulations in areas with limited capacity for enforcement, thereby tackling a driver of biodiversity loss.

Ensuring new governance arrangements are equitable and *inclusive* is, therefore, of paramount importance. Although we advocate here a new governance approach based on the achievement of biodiversity outcomes, this must be accompanied by the achievement of equitable outcomes for IPLCs. The inclusion of conserved areas can assist in this regard, as such areas do not have to have biodiversity conservation as their primary purpose and can instead be managed for socioeconomic, cultural or other purposes. As mentioned, the role played by IPLCs has been undervalued and under-recognized until recently (see, for example, IPBES, 2019). IPLC-led governance will be crucial in enabling conserved areas to contribute to biodiversity conservation as demonstrated in the three case studies described here. A greater role for IPLCs in decision-making and policy-setting at regional, national and international levels is only likely to facilitate the achievement of biodiversity outcomes and the management of potential economic and social trade-offs. The importance of properly including IPLCs has gained traction within discussions on the Post-2020 GBF, but much work remains to ensure this is fully embedded and implemented (Fajardo et al., 2021).

There is also a need for greater equity at the global scale, particularly in terms of the distribution of financial resources for biodiversity conservation. Although the Global South harbors most of the world's important biodiversity, and the Global Environment Facility (GEF) was in part established to facilitate finance to these areas, the amount of financing under this mechanism is still inadequate. Discussions on CBD resource mobilization have moved toward increasing consideration of private sector sources of finance, to complement development aid and public sector support (OECD, 2020). New governance arrangements for conserved areas and landscape-level approaches under the climate regime may facilitate the resource mobilization agenda by involving the private sector and ensuring that climate finance simultaneously achieves biodiversity outcomes, although, as shown by the Makira National Park case, it often needs to be complemented by other types of finance, including ODA. An approach based around biodiversity and equity outcomes could be accompanied by a financing framework, under which the areas or countries with the highest biodiversity values are identified and prioritized due to the increased focus on achieving ecological representation and biodiversity outcomes. This could facilitate transfers of finance from high-income, importing countries to low-income countries, and the establishment of new, innovative forms of financing mechanisms, which could even include performance-related payments, based on the achievement of biodiversity outcomes.

At a global level, considerable work is underway to better assess management effectiveness (and to a limited extent, governance effectiveness) of protected areas (Geldmann et al., 2021). Conservation areas or OECMs are, by their nature, considered to be "effective" conservation measures and, according to the accepted definition, they must result in biodiversity outcomes, regardless of the primary objective for which there are managed. Rather than attempting to assess the effectiveness of protected area management with a relatively small amount of monitoring effort dedicated to monitoring outcomes, a monitoring approach can be adopted for both protected and conserved areas (i.e. OECMs) that focuses more strongly on biodiversity outcomes. This will enable more responsive to *adaptive* and *anticipatory* governance responses. Post-2020, implementing an outcome-based approach will be more practical and cost-effective than it has been in the preceding decade, at least for terrestrial sites, due to the rapid advances being made in the area of remote sensing tools and the availability of proxy data on which to base an estimate of biodiversity outcomes (Watson and Venter, 2019).

Rapid advances in remote sensing that can monitor biodiversity outcomes can also assist in the future designation of terrestrial protected and conserved areas by ensuring that areas are selected for their biodiversity value. More attention needs to be given, however, to the development of equivalent tools for marine areas. This would ensure that an outcome-based approach is taken to site designation as well as to management and governance. In the case of conserved area designation, criteria need to be developed that provide a common global understanding of what constitutes an accepted biodiversity outcome. This would prevent a situation in which, for example, a marine area managed primarily for fisheries is designated because it results in positive outcomes for one or two specific species, despite little discernible benefit for a wider assemblage of species, and even potential harm due to bycatch. Recent discussions to designate much larger areas for biodiversity, such as the "half-earth" approach to set aside half of the earth for nature, as proposed by Wilson (2016), would indeed help to achieve greater ecological representation in some areas. The concept has received some criticism, however, from those who view this as a land-grab by the conservation community (Dudley et al., 2018), and for some ecoregions not enough natural habitat remains to meet this goal without substantial restoration efforts (Dinerstein et al., 2017; Mappin et al., 2019). A transformative agenda needs to go beyond target-setting and aim for ecological representation and achievement of biodiversity outcomes at the global level, accompanied by equity outcomes. As noted previously, this should be complemented with a mechanism for increasing financial resources to low-income countries that are high in biodiversity. It would also need to provide scope for restoration of ecoregions that have been most depleted.

As shown by the case study from Bolivia, new outcome-based approaches to transformative governance are necessary where protected areas with the primary goal of biodiversity conservation are located in large landscape and seascape areas where biodiversity conservation and equity are being achieved even in the areas that are not managed for biodiversity per se. It also gives space for *adaptive, integrative* and *anticipatory* governance responses by decreasing some of the pressures on biodiversity. Given the increasing pressure on land from many sectors, conserved areas with the primary purpose of food production and carbon storage will necessarily play an increasingly important role. Such an approach will, however, require the development of an accepted global standard, with associated criteria to determine what constitutes positive biodiversity outcomes. This will be facilitated by new cost-effective monitoring based on satellite remote sensing.

The move toward landscape approaches under REDD+ and other policy initiatives by provincial governments under the climate regime, such as LED-R initiatives, present a potential opportunity for integration with the emerging landscape-based approach to protected and conserved area governance. New points of intersection between the climate, forest, agriculture and biodiversity policy agendas could facilitate innovative, *integrative* forms of governance and financing for large areas. There may be potential for conserved areas to be designated and managed primarily for their carbon values, with biodiversity benefits as a major outcome. The involvement of new stakeholders from different interest groups in governance arrangements, particularly from the agricultural and climate sectors, and the potential complications that can arise from competing objectives, however, should not be underestimated. The establishment of an outcome-based approach with agreed minimum standards and criteria for achieving biodiversity outcomes could assist in managing and agreeing such trade-offs by ensuring that a certain threshold of biodiversity value must be met.

To be truly transformative, the implementation of this approach also needs to be combined with efforts by high-income or consuming countries to address some of the underlying drivers of biodiversity loss linked to unsustainable consumption and global trade that negatively impacts biodiversity in low-income countries (Lenzen et al., 2012). This type of *integrative* governance approach is now gaining traction within forest policy discussions, as governments in consuming countries consider how they ensure that their imports of agricultural commodities are deforestation-free (Weatherley-Singh and Gupta, 2018). Such policies need to be advanced to ensure that supply chains do not cause biodiversity loss in producer countries. This would also help the Post-2020 GBF to address the pressures on biodiversity, and not just the responses to those pressures (OECD, 2019).

A review and revision of the IUCN management and governance categories would also be necessary to enable a new approach based around biodiversity and equity outcomes. Notwithstanding the crucial need to retain a number of protected areas that are managed primarily for biodiversity conservation, these could be incorporated within a set of IUCN protected and conserved area categories that is based around the achievement of biodiversity and equity outcomes. The current IUCN categorization of governance types could also be reviewed and expanded to encapsulate and include the full range of conserved areas and their outcomes for biodiversity, to better reflect their governance and management structures.

11.5 Conclusion

Although the international policy framework, and particularly CBD Aichi Target 11, has stimulated further progress in protected and conserved area governance, especially in site designation, this falls short of meeting the criteria for transformative biodiversity governance. Efforts have been made in recent years to ensure such areas become more *inclusive* (through, for example, co-governance arrangements that engage IPLCs). The valuable role played by IPLCs is starting to gain recognition, including in new discussions around conserved areas, but governance needs to go beyond including them as beneficiaries, to recognizing and strengthening their rights as active stakeholders. This will require a considerable investment of time and resources at local and landscape levels to conduct *inclusive* consultations, build capacity where needed, especially in terms of access to technology, and to find solutions that meet the needs of IPLCs and that reflect their own visions for their territories, which will ultimately be more sustainable.

There has also been a focus on achieving *adaptive* management that allows for some experimentation in management approaches. This principle can be used to quickly integrate new scientific findings, which are now providing more timely, up to date information on species and habitats and the human pressures they are facing. This increase in information is occurring from the local to global level, enabling decision-making to be better informed and potentially also *anticipatory*, for example, by modeling the impacts of human pressures and facilitating future site designations.

Greater information on biodiversity outcomes will also enable finance to be directed to the areas of greatest biodiversity value, thereby helping to achieve greater ecological representation. The potential scaling-up of protected areas to become part of decision-making governance structures at landscape and seascape scales, including conserved areas, is expected to open the door to greater *integrative* approaches and new forms of financing, although such arrangements bring a new level of complexity.

In conclusion, a new approach based around the delivery of biodiversity outcomes could help drive forward a transformative governance agenda. Its success will also depend on the long-term engagement of IPLCs and the achievement of equity outcomes. Reviewing the IUCN governance and management categories would be an additional small first step toward building a more supportive policy framework at the international level that facilitates transformative change. If such efforts are combined with actions taken by high-income, importing countries to increase the sustainability of their consumption and trade patterns, and thereby tackle some of the underlying drivers of biodiversity loss, this would be even more transformative.

References

Andam, K. S., Ferraro, P. J., Pfaff, A., Sanchez-Azofeifa, G. A., and Robalino, J. A. (2008). Measuring the effectiveness of protected area networks in reducing deforestation. *Proceedings of the National Academy of Sciences* 105, 16089–16094.

Anthony, B. P., and Szabo, A. (2011). Protected areas: Conservation cornerstones or paradoxes? Insights from human-wildlife conflicts in Africa and Southeastern Europe. In *The importance of biological interactions in the study of biodiversity*. J. Lopez-Pujol (Ed.), pp. 255–282. Rijeka, Croatia: InTech.

Bennett, N. J., and Satterfield, T. (2018). Environmental governance: A practical framework to guide design, evaluation, and analysis. *Conservation Letters* e12600. https://doi.org/10.1111/conl.12600

Bhola, N., Klimmek, H., Kingston, N., et al. (2020). Perspectives on area-based conservation and what it means for the post-2020 biodiversity policy agenda. *Conservation Biology* 35, 168–178. https://doi.org/10.1111/cobi.13509

Borrini-feyerabend, G., Dudley, N., Jaeger, T., et al. (2013). *Governance of protected areas: From understanding to action, best practice protected area guidelines*. Gland: IUCN.

Borrini-Feyerabend, G., and Hill, R. (2015). Governance for the conservation of nature. In *Protected area governance and management*. G. L. Worboys, M. Lockwood, A. Kothari, S. Feart and I. Pulsford (Eds.), pp. 169–206. Canberra: ANU.

Boucher, D., and Elias, P. (2013). From REDD to deforestation-free supply chains: The persistent problem of leakage and scale. *Carbon Management* 4, 473–475. https://doi.org/10.4155/cmt.13.47

Bown, N. K., Gray, T. S., and Stead, S. M. (2013). Co-management and adaptive co-management: Two modes of governance in a Honduran marine protected area. *Marine Policy* 39, 128–134. https://doi.org/10.1016/j.marpol.2012.09.005

Bruner, A. G., Gullison, R. E., Rice, R. E., and Fonseca, G. A. B. (2001). Effectiveness of parks in protecting tropical biodiversity. *Science* 291, 125–128.

Buchanan, G. M., Butchart, S. H. M., Chandler, G., and Gregory, R. D. (2020). Assessment of national-level progress towards elements of the Aichi Biodiversity Targets. *Ecological Indicators* 116, 106497.

Butchart, S. H. M., Clarke, M., Smith, R. J., et al. (2015). Shortfalls and solutions for meeting national and global conservation area targets. *Conservation Letters* 8, 329–337. https://doi.org/10.1111/conl.12158

Carlisle, K., and Gruby, R. L. (2019). Polycentric systems of governance: A theoretical model for the commons. *Policy Studies Journal* 47, 927–952. https://doi.org/10.1111/psj.12212

CBD. (2018). Decision adopted by the COP to the CBD 14/8. Protected areas and other effective area-based conservation measures (No. CBD/COP/DEC/14/8).

Chaffin, B., Garmestani, A. S., Gunderson, L., et al. (2016). Transformative environmental governance. *Annual Review of Environment and Resources* 41, 399–423. https://doi.org/10.1146/annurev-environ-110615-085817

Coad, L., Leverington, F., Knights, K., et al. (2015). Measuring impact of protected area management interventions: Current and future use of the Global Database of Protected Area Management Effectiveness. *Philosophical Transactions of the Royal Society B: Biological Sciences* 370, 1–10.

Coad, L., Watson, J. E. M., Geldmann, J., et al. (2019). Widespread shortfalls in protected area resourcing undermine efforts to conserve biodiversity. *Frontiers in Ecology and the Environment* 17, 259–264. https://doi.org/10.1002/fee.2042

Dawson, N., Martin, A., and Danielsen, F. (2018). Assessing equity in protected area governance: Approaches to promote just and effective conservation. *Conservation Letters* 11, 1–8. https://doi.org/10.1111/conl.12388

Díaz, S., Zafra-Calvo, N., Purvis, A., et al. (2020). Set ambitious goals for biodiversity and sustainability. *Science* 370, 411–413. https://doi.org/10.1126/science.abe1530

Dinerstein, E., Olson, D., Joshi, A., et al. (2017). An ecoregion-based approach to protecting half the terrestrial realm. *Bioscience* 67, 535–545. https://doi.org/10.1093/biosci/bix014

Diz, D., Johnson, D., Riddell, M., et al. (2018). Mainstreaming marine biodiversity into the SDGs: The role of other effective area-based conservation measures (SDG 14. 5). *Marine Policy* 93, 251–261. https://doi.org/10.1016/j.marpol.2017.08.019

Dudley, N. (2008). *Guidelines for applying protected area management categories*. Gland: IUCN.

Dudley, N., Jonas, H., Nelson, F., et al. (2018). The essential role of other effective area-based conservation measures in achieving big bold conservation targets. *Global Ecology and Conservation* 15, 1–7. https://doi.org/10.1016/j.gecco.2018.e00424

European Commission. (2014). A stronger role of the private sector in achieving inclusive and sustainable growth in developing countries. COM(2014) 263 final. Available from https://bit.ly/3gjERyg.

Exeter, O. M., Htut, T., Kerry, C. R., et al. (2021). Shining light on data-poor coastal fisheries. *Frontiers in Marine Science* 7, 625766. DOI: 10.3389/fmars.2020.625766

Fa, J. E., Watson, J. E. M., Leiper, I., et al. (2020). Importance of Indigenous Peoples' lands for the conservation of intact forest landscapes. *Frontiers in Ecology and the Environment* 18, 135–140. https://doi.org/10.1002/fee.2148

Fajardo, P., Beauchesne, D., Carbajal-lópez, A., et al. (2021). Aichi Target 18 beyond 2020: Mainstreaming traditional biodiversity knowledge in the conservation and sustainable use of marine and coastal ecosystems. *PeerJ* 9, e9616. https://doi.org/10.7717/peerj.9616

FAO. (2016). State of the world's forests. Forests and agriculture: Land-use challenges and opportunities. Rome. Available from www.fao.org/3/i5588e/i5588e.pdf.

Ferguson, H. B. (2009). REDD in Madagascar: An Overview of Progress. Independent Report, 5th November 2009.

Fisher, B., and Christopher, T. (2007). Poverty and biodiversity: Measuring the overlap of human poverty and the biodiversity hotspots. *Ecological Economics* 62, 93–101. https://doi.org/10.1016/j.ecolecon.2006.05.020

Folke, C., Österblom, H., Jouffray, J., et al. (2019). Transnational corporations and the challenge of biosphere stewardship. *Nature Ecology & Evolution* 3, 1396–1403. https://doi.org/10.1038/s41559-019-0978-z

Franklin, S. L., and Pindyck, R. S. (2018). Tropical forests, tipping points, and the social cost of deforestation. *Ecological Economics* 153, 161–171. https://doi.org/10.1016/j.ecolecon.2018.06.003

Garnett, S. T., Burgess, N. D., Fa, J. E., et al. (2018). A spatial overview of the global importance of Indigenous lands for conservation. *Nature Sustainability* 1, 369–374. https://doi.org/10.1038/s41893-018-0100-6

Geldmann, J., Coad, L., Barnes, M. D., et al. (2018). A global analysis of management capacity and ecological outcomes in terrestrial protected areas. *Conservation Letters* 11, e12434. https://doi.org/10.1111/conl.12434

Geldmann, J., Deguignet, M., Balmford, A., et al. (2021). Essential indicators for measuring site-based conservation effectiveness in the post-2020 global biodiversity framework. *Conservation Letters* 14, e12792. https://doi.org/10.1111/conl.12792

Gruby, R. L., and Basurto, X. (2014). Multi-level governance for large marine commons: Politics and polycentricity in Palau's protected area network. *Environmental Science & Policy* 36, 48–60. https://doi.org/10.1016/j.envsci.2013.08.001

Guston, D. H. (2014). Understanding "anticipatory governance." *Social Studies of Science* 44, 218–242. https://doi.org/10.1177/0306312713508669

Hagerman, S. M., and Pelai, R. (2016). "As far as possible and as appropriate": Implementing the aichi biodiversity targets. *Conservation Letters* 9, 469–478. https://doi.org/10.1111/conl.12290

Hammer, T., Mose, I., Scheurer, T., Siegrist, D., and Weixlbaumer, N. (2012). Societal research perspectives on protected areas in Europe. Eco.mont 4, 5–12.

Hockings, M., Stolton, S., Dudley, N., and Deguignet, M. (2018). Protected Area Management Effectiveness (PAME): Report on a training course for protected area staff in Myanmar. Available from https://bit.ly/3ul3JOj.

Howe, C., and Milner-Gulland, E. J. (2012). Evaluating indices of conservation success: A comparative analysis of outcome- and output-based indices. *Animal Conservation* 15, 217–226. https://doi.org/10.1111/j.1469-1795.2011.00516.x

IPBES. (2019). Summary for policymakers of the global assessment report on biodiversity and ecosystem services of the Intergovernmental Science-Policy Platform on Biodiversity and Ecosystem Services (IPBES). S. Díaz, J. Settele, E. S. Brondízio, et al. (Eds.). Bonn: IPBES Secretariat. https://doi.org/10.1111/padr.12283. Available from https://bit.ly/3gk5kMa.

IUCN. (2016). A global standard for the identification of key biodiversity areas. Gland: IUCN. Available from https://bit.ly/3J2W0Jb.

IUCN and WCPA. (2017). IUCN green list of protected and conserved areas: Standard, Version 1.1. Gland: IUCN. Available from https://bit.ly/35CgDNH.

IUCN-WCPA. (2019). Recognising and reporting other effective area-based conservation measures. Gland, Switzerland: IUCN. https://doi.org/10.2305/iucn.ch.2019.patrs.3.en

Jantke, K., Jones, K. R., Allan, J. R., et al. (2018). Poor ecological representation by an expensive reserve system: Evaluating 35 years of marine protected area expansion. *Conservation Letters* 11: e12584. https://doi.org/10.1111/conl.12584

Jedd, T., and Bixler, R. P. (2015). Accountability in networked governance: Learning from a case of landscape-scale forest conservation. *Environmental Policy and Governance* 25, 172–187. https://doi.org/10.1002/eet.1670

Jóhannsdóttir, A., Cresswell, I., and Bridgewater, P. (2010). The current framework for international governance of biodiversity: Is it doing more harm than good? *Review of European Community & International Environmental Law* 19, 139–149.

Jones, K. R., Klein, C., Halpern, B. S., et al. (2018a). The location and protection status of Earth's diminishing marine wilderness. *Current Biology* 28, 2506–2512.

Jones, K. R., Venter, O., Fuller, R. A., et al. (2018b). One-third of global protected land is under intense human pressure. *Science* 360, 788–791.

Jupiter, S. D., Cohen, P. J., Weeks, R., Tawake, A., and Goven, H. (2014). Locally-managed marine areas: Multiple objectives and diverse strategies. *Pacific Conservation Biology* 20, 165–179.

Kapos, V., Balmford, A., Aveling, R., et al. (2009). Outcomes, not implementation, predict conservation success. *Oryx* 43, 336–342. https://doi.org/10.1017/S0030605309990275

Keenan, R. J., Reams, G. A., Achard, F., et al. (2015). Forest ecology and management dynamics of global forest area: Results from the FAO global forest resources assessment 2015. *Forest Ecology and Management* 352, 9–20. https://doi.org/10.1016/j.foreco.2015.06.014

Laffoley, D., Dudley, N., Jonas, H., et al. (2017). An introduction to "other effective area – based conservation measures" under Aichi Target 11 of the Convention on Biological Diversity: Origin, interpretation and emerging ocean issues. *Aquatic Conservation: Marine and Freshwater Ecosystems* 27, 130–137. https://doi.org/10.1002/aqc.2783

Latt, K. T. (2019). Ensuring a Blue Future for Myanmar's Coastal Communities. National Geographic Society Newsroom. Available from https://bit.ly/3ugoVFr.

Lenzen, M., Moran, D., Kanemoto, K., et al. (2012). International trade drives biodiversity threats in developing nations. *Nature* 486, 109–112. https://doi.org/10.1038/nature11145

Leverington, F., Lemos, K., Pavese, H., Lisle, A., and Hockings, M. (2010). A global analysis of protected area management effectiveness. *Environmental Management* 46, 685–698. https://doi.org/10.1007/s00267-010-9564-5

Lockwood, M. (2010). Good governance for terrestrial protected areas: A framework, principles and performance outcomes. *Journal of Environmental Management* 91, 754–766. https://doi.org/10.1016/j.jenvman.2009.10.005

Luque, S., Pettorelli, N., Vihervaara, P., and Wegmann, M. (2018). Improving biodiversity monitoring using satellite remote sensing to provide solutions towards the 2020 conservation Targets. *Methods in Ecology and Evolution* 9, 1784–1786. https://doi.org/10.1111/2041-210X.13057

Maiorano, L., Amori, G., Montemaggiori, A., et al. (2015). On how much biodiversity is covered in Europe by national protected areas and by the Natura 2000 network: Insights from terrestrial vertebrates. *Conservation Biology* 29, 986–995. https://doi.org/10.1111/cobi.12535

Mappin, B., Chauvenet, A. L. M., Adams, V. M., et al. (2019). Restoration priorities to achieve the global protected area Target. *Conservation Letters* 12, 1–9. https://doi.org/10.1111/conl.12646

Maron, M., Simmonds, J. S., and Watson, J. E. M. (2018). Bold nature retention Targets are essential for the global environment agenda. *Nature Ecology & Evolution* 2, 1194–1195. https://doi.org/10.1038/s41559-018-0595-2

Matthews, E., Mizrahi, M., Boyd, C., et al. (2020). Tailoring a business skills training programme for self-employed women in coastal fishing communities in Myanmar. *Women in Fisheries Information Bulletin* 32, 19–23.

Maxwell, S. L., Cazalis, V., Dudley, N., et al. (2020). Area-based conservation in the 21st century. Preprints. https://doi.org/10.20944/preprints202001.0104.v1

Mccarthy, D. P., Donald, P. F., Scharlemann, J. P. W., et al. (2012). Financial costs of meeting global current spending and unmet needs. *Science* 338, 946–950.

McCay, B. J., and Jones, P. J. S. (2011). Marine protected areas and the governance of marine ecosystems and fisheries. *Conservation Biology* 25, 1130–1133. https://doi.org/10.1111/j.1523-1739.2011.01771.x

McDermott, C. L. (2014). REDDuced: From sustainability to legality to units of carbon – The search for common interests in international forest governance. *Environmental Science & Policy* 35, 12–19. https://doi.org/10.1016/j.envsci.2012.08.012

McShane, T. O., Hirsch, P. D., Trung, T. C., et al. (2011). Hard choices: Making trade-offs between biodiversity conservation and human well-being. *Biological Conservation* 144, 966–972. https://doi.org/10.1016/j.biocon.2010.04.038

Morales-Hidalgo, D., Oswalt, S. N., and Somanathan, E. (2015). Forest ecology and management status and trends in global primary forest, protected areas, and areas designated for conservation of biodiversity from the Global Forest Resources Assessment 2015. *Forest Ecology and Management* 352, 68–77. https://doi.org/10.1016/j.foreco.2015.06.011

Mulatu, K. A., Mora, B., Kooistra, L., and Herold, M. (2017). Biodiversity monitoring in changing tropical forests: A review of approaches and new opportunities. *Remote Sensing* 9, 1–22. https://doi.org/10.3390/rs9101059

Nagendra, H., and Ostrom, E. (2012). Polycentric governance of multifunctional forested landscapes. *International Journal of the Commons* 6, 104–133.

Nepstad, D., Irawan, S., Bezerra, T., et al. (2013). More food, more forests, fewer emissions, better livelihoods: Linking REDD+, sustainable supply chains and domestic policy in Brazil, Indonesia and Colombia. *Carbon Management* 4, 639–658. https://doi.org/10.4155/cmt.13.65

OECD. (2019). The Post-2020 Biodiversity Framework: Targets, indicators and measurability implications at the global and national level. November version. Available from https://bit.ly/3HnvQ3g.

(2020). A comprehensive overview of global biodiversity finance: Initial results. Interim report made available for the thematic workshop on resource mobilisation for the post-2020 global biodiversity framework, January 14–16, 2020.

Oldekop, J. A., Holmes, G., Harris, W. E., and Evans, K. L. (2015). A global assessment of the social and conservation outcomes of protected areas. *Conservation Biology* 30, 133–141. https://doi.org/10.1111/cobi.12568

Oliveira, P. J. C., Asner, G. P., Knapp, D. E., et al. (2007). Land-use allocation protects the Peruvian Amazon. *Science* 317, 1233–1237.

Ostrom, E. (1990). *Governing the commons: The evolution of institutions for collective action.* Cambridge:Cambridge University Press.

Painter, L., Montoya, M., and Varese, M. (2017). Territorial management, as a mechanism for mitigation and adaptation to climate change. In *Secretariat of the Convention on Biological Diversity (2017) The Lima Declaration on Biodiversity and Climate Change: Contributions from science to policy for sustainable development.* Technical Series No.89. L. Rodríguez and I. Anderson (Eds.), pp. 109–115. Montreal: Secretariat of the Convention on Biological Diversity. www.cbd.int/doc/publications/cbd-ts-89-en.pdf

Pattberg, P., Kristensen, K., and Widerberg, O. (2017). *Beyond the CBD: Exploring the institutional landscape of governing for biodiversity.* Amsterdam: Institute for Environmental Studies/IVM.

Premauer, J. M., and Berkes, F. (2015). A pluralistic approach to protected area governance: Indigenous Peoples and Makuira National Park, Colombia. *Ethnobiology and Conservation* 4, 1–16. https://doi.org/10.15451/ec2015-5-4.4-1-16

Rands, M. R. W., Adams, W. M., Bennun, L., et al. (2010). Biodiversity conservation: Challenges beyond 2010. *Science* 329, 1298–1303.

Rao, M., Nagendra, H., Shahabuddin, G., and Carrasco, L. R. (2016). Integrating community-managed areas into protected area systems: The promise of synergies and the reality of trade-offs. In Protected areas: Are they safeguarding biodiversity? L. N. Joppa, J. E. M. Baillie and J. G. Robinson (Eds.), pp. 169–189. Chichester: John Wiley & Sons.

Reinecke, S., Pistorius, T., and Pregernig, M. (2014). UNFCCC and the REDD+ Partnership from a networked governance perspective. *Environmental Science & Policy* 35, 30–39. https://doi.org/10.1016/j.envsci.2012.09.015

Sarrasin, B. (2009). La Gestion LOcale SÉcurisée (GELOSE): L'expérience malgache de gestion décentralisée des ressources naturelles. Etudes Caribeenne 1–13. https://bit.ly/33Fh2Oi.

Scharlemann, J. P. W., Kapos, V., Campbell, A., et al. (2010). Securing tropical forest carbon: The contribution of protected areas to REDD. *Oryx* 44, 352–357. https://doi.org/10.1017/S0030605310000542

Schuster, R., Germain, R. R., Bennett, J. R., Reo, N. J., and Arcese, P. (2019). Vertebrate biodiversity on indigenous-managed lands in Australia, Brazil, and Canada equals that in protected areas. *Environmental Science & Policy* 101, 1–6. https://doi.org/10.1016/j.envsci.2019.07.002

Shahabuddin, G., and Rao, M. (2010). Do community-conserved areas effectively conserve biological diversity? Global insights and the Indian context. *Biological Conservation* 143, 2926–2036. https://doi.org/10.1016/j.biocon.2010.04.040

Sheil, D., Boissiere, M., and Beaudoin, G. (2015). Unseen sentinels: Local monitoring and control in conservation's blind spots. *Ecology and Society* 20, 39. http://dx.doi.org/10.5751/ES-07625-200239

Stolton, S., and Dudley, N. (2010). *Arguments for protected areas: Multiple benefits for conservation and use*. London; Washington, DC: Earthscan.

UNEP-WCMC, IUCN, NGS (2018). *Protected planet report 2018*. Cambridge; Gland; Washington, DC: UNEP-WCMC, IUCN and NGS. Available from https://bit.ly/3HnA7DQ.

UNFCCC (2007). Report of the Conference of the Parties on its thirteenth session, held in Bali from 3 to 15 December 2007. Available from https://unfccc.int/documents/5078.

Venter, O., Magrach, A., Outram, N., et al. (2017). Bias in protected-area location and its effects on long-term aspirations of biodiversity conventions. *Conservation Biology* 32, 127–134. https://doi.org/10.1111/cobi.12970

Verma, M., Jones, K. R., Rheindt, F. E., et al. (2019). Severe human pressures in the Sundaland biodiversity hotspot. *Conservation Science and Practice* 2, e169. https://doi.org/10.1111/csp2.169

Visconti, B. P., Stuart, H. M., Brooks, T. M., et al. (2019). Protected area targets post-2020. *Science* 364, 239–242.

Visseren-Hamakers, I. J. (2015). Integrative environmental governance: Enhancing governance in the era of synergies. *Current Opinion in Environmental Sustainability* 14, 136–143. https://doi.org/10.1016/j.cosust.2015.05.008

(2018). Integrative governance: The relationship between governance instruments taking center stage. *Environment and Planning C: Politics and Space* 36, 1341–1354.

Visseren-Hamakers, I. J., Mcdermott, C., Vijge, M. J., and Cashore, B. (2012). Trade-offs, co-benefits and safeguards: Current debates on the breadth of REDD+. *Current Opinion in Environmental Sustainability* 4, 646–653.

Waldron, A., Mooers, A. O., Miller, D. C., et al. (2013). Targeting global conservation funding to limit immediate biodiversity declines. *Proceedings of the National Academy of Sciences* 110, 12144–12148. https://doi.org/10.5061/dryad.p69t1

Watson, J. E. M., Darling, E. S., Venter, O., et al. (2016a). Bolder science needed now for protected areas. *Conservation Biology* 30, 243–248. https://doi.org/10.1111/cobi.12645

Watson, J. E. M., Dudley, N., Segan, D. B., and Hockings, M. (2014). The performance and potential of protected areas. *Nature* 515, 67–73. https://doi.org/10.1038/nature13947

Watson, J. E. M., Evans, T., Venter, O., et al. (2018). The exceptional value of intact forest ecosystems. *Nature Ecology & Evolution* 2, 599–610. https://doi.org/10.1038/s41559-018-0490-x

Watson, J. E. M., Jones, K. R., Fuller, R. A., et al. (2016b). Persistent disparities between recent rates of habitat conversion and protection and implications for future global conservation targets. *Conservation Letters* 9, 413-421. https://doi.org/10.1111/conl.12295

Watson, J. E. M., and Venter, O. (2019). Mapping the continuum of humanity's footprint on land. *One Earth* 1, 175–180. https://doi.org/10.1016/j.oneear.2019.09.004

WCS. (2018). *Characterization of fisheries and marine wildlife occurrence in southern Rakhine State and western Ayayarwady Region, Myanmar*. Yangon, Myanmar: Wildlife Conservation Society.

Weatherley-Singh, J., and Gupta, A. (2017). An ecological landscape approach to REDD + in Madagascar: Promise and limitations? *Forest Policy and Economics* 85, 1–9. https://doi.org/10.1016/j.forpol.2017.08.008

Weatherley-Singh, J., and Gupta, A. (2018). "Embodied deforestation" as a New EU policy debate to tackle tropical forest loss: Assessing implications for REDD+ Performance. *Forests* 9, 1–21. https://doi.org/10.3390/f9120751

Wilson, E. O. (2016). *Half Earth: Our planet's fight for life*. New York: Liveright Publishing Corporation.

Woodley, S., Locke, H., Laffoley, D., et al. (2019). A review of evidence for area-based conservation targets for the post-2020 global biodiversity framework. *Parks* 25, 31–46. https://doi.org/10.2305/IUCN.CH.2019.PARKS-25-2SW2.en

Yang, R., Cao, Y., Hou, S., et al. (2020). Cost-effective priorities for the expansion of global terrestrial protected areas: Setting post-2020 global and national Targets. *Science Advances* 6, 1–9. https://doi.org/10.1126/sciadv.abc3436

12

The Convivial Conservation Imperative: Exploring "Biodiversity Impact Chains" to Support Structural Transformation

BRAM BÜSCHER, KATE MASSARELLA, ROBERT COATES, SIERRA DEUTSCH, WOLFRAM DRESSLER, ROBERT FLETCHER, MARCO IMMOVILLI AND STASJA KOOT

12.1 Introduction

News on the state of the environment does not seem to be improving. Despite some holding on to "conservation optimism,"[1] the general conclusion in the academic and policy literature is that global biodiversity, the global climate and the state of other environmental indicators are bad, and getting worse (CBD, 2020; European Environment Agency, 2019; IPBES, 2019; Lenton et al., 2019; Newbold et al., 2016; Tucker et al., 2018; Watson et al., 2016; WWF, 2018). This has resulted in growing calls for transformative change in the way we govern biodiversity, and the environment more broadly (Bennett et al., 2019; Scoones et al., 2020). Making incremental, adaptive changes to the current system and structures is no longer considered sufficient to move us to a sustainable future; rather, deeper, more *fundamental* transformation is needed (Blythe et al., 2018). In relation to biodiversity conservation, an important example of this new emphasis is the 2019 Intergovernmental Science-Policy Platform on Biodiversity and Ecosystem Services (IPBES) report, which argues that "nature can be conserved, restored and used sustainably while simultaneously meeting other global societal goals through urgent and concerted efforts fostering transformative change" (IPBES, 2019: 7). The report realizes this is not easy, but insists:

> Since current structures often inhibit sustainable development and actually represent the indirect drivers of biodiversity loss, such fundamental, structural change is called for. By its very nature, transformative change can expect opposition from those with interests vested in the status quo, but such opposition can be overcome for the broader public good. *(IPBES, 2019: 9)*

Clearly, "transformative change" is an extremely complex proposition, and precisely what it means is widely debated and contested (Brown et al., 2013; Scoones et al., 2020). The IPBES (2019: 9) report, however, provides many suggestions, including a particularly important one: "A key constituent of sustainable pathways is the evolution of global financial and economic systems to build a global sustainable economy, steering away from the current limited paradigm of economic growth." The European Environment Agency, likewise, states that economic growth should no longer be pursued at the expense

---

[1] www.conservationoptimism.org.

of the environment and urges governments to "deliver transformative change in the coming decade" (European Environment Agency, 2019: 10).

Coming from major international reports, these are not just "regular" transformative suggestions; they are *radically* transformative suggestions that go to the roots ("radix") of the problem of contemporary unsustainability.[2] Demand for such change is echoed by a growing number of civil society groups, networks and social movements battling the myriad environmental and social conflicts caused by unfettered economic growth and consumption.[3] And while the global COVID-19 pandemic had many governments and institutions scrambling to get back to "normal," it also amplified the demands for transformative change. The key questions, then, become: How do we act on these demands and suggestions? What do they imply for environmental governance and biodiversity conservation?

In this chapter, we support and advance arguments for a fundamental *structural* transformation that envisions radically different institutional and societal structures. This view is in line with the current volume and increasingly shared by many calling for transformative change (e.g. Chaffin et al., 2016; Martin et al., 2020; Massarella et al., 2021). At the same time, many actors still believe that transformative change can happen without directly and explicitly challenging the capitalist underpinning of contemporary institutional and societal structures (Feola, 2020). We argue that many of the solutions put forward for transforming biodiversity conservation follow this belief. More specifically, we argue that even seemingly "radical" new approaches, such as "neoprotectionism" (focused on creating space for protected areas) and "new conservation" or natural capital approaches (championing the use of market-based mechanisms to integrate people and nature) are not actually transformative. Although they call for radical shifts – both in symbolism and how we govern biodiversity on a global scale – they do not sufficiently address or challenge the main driver of biodiversity loss: the neoliberal capitalist model that dominates our global political economy. In fact, by not responding holistically and critically to the global challenges we are facing, including currently disturbing authoritarian trends in global governance systems and an increasing concentration of corporate governance, these proposals for transformative change may even set us back.

We therefore argue that the only way to properly conceptualize transformative change is to combine radical reformism in the short term with an intermediate to long-term vision for fundamental *structural* transformation that directly challenges our contemporary capitalist political economic model and its newfound turn to authoritarianism. In doing so, we emphasize, following Scoones et al. (2020), that our structural approach can and should be seen in conjunction with – not necessarily against – what they call systemic and enabling approaches that focus more on complex system change and values and actions of different actors. The latter, however, can only gain (appropriate) direction through a critique of the dominant political economy and hence why we emphasize *structural* transformation. In what follows, we contribute to the current volume by presenting a vision for structural

---

[2] O'Brien et al. (2013) define this as moving from processes of circular change (repeatedly adjusting the existing system) to axial change (moving to a new way of thinking and being).
[3] See https://ejatlas.org/about.

transformation under the banner of "convivial conservation." Convivial conservation is a vision, a politics and a set of transformative governing principles that moves biodiversity governance beyond market-based mechanisms and a central focus on protected areas (PAs). We outline and analyze these three elements and propose the idea of "biodiversity impact chains" (BICs) to operationalize some of the transformative governance aspects of convivial conservation in practice.

BICs, in essence, aim to *politicize* transformative environmental governance by drawing more concrete connections between differentiated actors, and their variegated impacts on biodiversity, in a highly uneven conservation field. This allows us not only to understand that those with the largest footprints must change their lives the most in order to redress biodiversity loss, but also that spatial proximity to conservation areas should be of less concern to conservation action than is often the case (see also Chapter 14 of this volume). BICs, therefore, help us to gain a clearer view of the structural pressures on biodiversity, and how these need to be mediated or challenged in order to achieve structural transformation. In the penultimate section of the paper, we develop this perspective in more detail in order to explain, in the conclusion, how a convivial transformation may be our most realistic chance to respond positively to the global biodiversity crisis. First, however, we summarize the arguments for fundamental structural transformation and what we believe this should entail.

12.2 Authoritarian Currents and the State of Biodiversity (Conservation)

As is apparent from the preceding discussion, the necessity for fundamental transformation is becoming increasingly obvious in global environmental governance circles and, indeed, within global governance more generally. As shown by many other chapters in this volume, most environmental indicators around climate, oceans, biodiversity, forests and more are so alarming that even most mainstream commentators now call for forms of change beyond mere nudges within the general parameters of the current system. Much evidence from the current transformations literature could be presented here, but for an overview we refer to Chapters 1 and 4 in this volume and Massarella et al. (2021). What we want to add is a more sociological analytic, namely that the mainstream system in which global environmental governance approaches have been operating is increasingly leading to forms of authoritarian populism and right-wing extremism. Prominent examples include the recent Trump regime in the United States, the Bolsonaro regime in Brazil and the Modi regime in India, among others (Kiely, 2021; Saad-Filho and Boffo, 2021). All of these regimes articulate narrow versions of both nation and nature, to the extent that Indigenous and other minority groups are frequently cast as the enemies of national economic progress, often violently so. Indeed, one key constant across these regimes is that they have come to power with the support of major extractive industries and have, in turn, unapologetically exercised their power in support of these industries to directly attack and dismantle forms and institutions of environmental protection that stand in their way (Kiely, 2021; McCarthy, 2019; Saad-Filho and Boffo, 2021).[4]

---

[4] Another indicator for this on the global scale is the rise, over the past twenty years, of the killing of environmental defenders; see: https://bit.ly/3Id8EEG.

We argue that these worrying trends need to be acknowledged and challenged directly for transformative conditions to arise (Mason, 2019). After all, as Polanyi (1957) argued as far back as the 1950s, the rise of authoritarianism is the ultimate response to the threat that social and environmental protection poses to the continued advancement of neoliberal capitalism. As crises of capitalism are increasingly accompanied by crises of legitimation (of the continuation of "business-as-usual"), authoritarianism offers a solution to both. This "authoritarian fix" allows for capital accumulation to continue by removing barriers to the exploitation of natural resources and labor, while simultaneously removing the need for legitimation (Bruff, 2014: 125; Poulantzas, 1978). Thus, it is hardly coincidental that many of the new authoritarians have sought to undermine or withdraw from global institutions focused on climate change mitigation, at the precise moment of a growing political tension between environmental protection and economic business-as-usual. The dissolution of restrictions on agriculture and mining have gone hand-in-hand with a denial of the scientific truth of environmental degradation, and widespread attacks on agencies producing spatial data on deforestation and defaunation (Neimark et al., 2019). By undermining protections at all levels, new authoritarian regimes thus act to sustain a capitalist economy that demands continuous growth in order to remain stable (Büscher and Fletcher, 2020). From this perspective, the fight against environmental catastrophe is also a fight against authoritarianism, given how the latter is directly implicated in the defense of the current capitalist political economy (Kiely, 2021; McCarthy, 2019; Saad-Filho and Boffo, 2021; Scoones et al., 2018).[5]

Given this context and these threats, it is little surprise that many in the conservation community feel great anxiety and pressure. And while they do often agree that transformation is needed, it seems very difficult to break out of the neoliberal consensus-mold many organizations embraced in the 1980s and 1990s. As documented in the literature (Adams, 2017; Fletcher, 2014; MacDonald, 2010; MacDonald and Corson, 2012), since the 1980s conservation organizations have increasingly conformed to the general, consensus-oriented "sustainable development" models that have thoroughly neoliberalized biodiversity conservation (Fletcher et al., 2019). Indeed, Büscher (2013) identifies consensus and anti-politics as two of three foundational elements of a general neoliberal conservation politics that pervaded the 1990s and early 2000s (with "marketing" being the third). Since the late 2000s, and especially triggered by the 2007/2008 financial crisis, the international political context has changed rapidly, leading – inter alia – to the abovementioned authoritarian developments. One would expect that, from the imperative to oppose these forces, a more political and less consensus-oriented approach to environmental governance would ensue. Yet, this has only marginally proven to be the case thus far.

For example, the WWF flagship Living Planet report, released two days after Bolsonaro was elected in November 2018, calls for a "new global deal for nature and people" and urges "decision-makers at every level" to "make the right political, financial and consumer choices to achieve the vision that humanity and nature thrive in harmony on our only

---

[5] Some may argue that eco-authoritarianism is the only way out of the failure of liberal-democratic societies to prevent environmental catastrophe, but it should be clear from our line of argumentation that we are adamantly against such an approach.

planet." To operationalize this "ambitious pathway," WWF, together with other organizations, will launch a new research initiative based around "systems modelling" to help "us determine the best integrated and collective solutions and to help understand the 'trade-offs' we may need to accept to find the best path ahead" (WWF, 2018: 8). Similarly, the European Environment Outlook 2020 paints a grim picture of prospects for European biodiversity and argues that its "message of urgency cannot be overstated." At the same time, it states that "transformative change will require that all areas and levels of government work together and harness the ambition, creativity and power of citizens, businesses and communities" (European Environment Agency, 2019: 7; 17). On a superficial level, this may be correct, but it leaves out which businesses, types of activities and communities (such as the oil, coal, infrastructure, large-scale agriculture and other communities) will inevitably have to "lose" (that is, to *degrow, and rapidly so*) in order to reach a more sustainable overall state.

To a degree, we can understand that conservation and government organizations want to be careful politically. But a big problem with this conciliatory, mainstream approach is that it is easy to ignore for alt-right and authoritarian (-leaning) politicians and movements, and their corporate backers. Another problem is that it often does not lead to more political or politicized action to demand structural change, and may – unintentionally – lead other actors to take politicized action into dubious terrains of increasingly militarized, even ecofascist, forms of environmental protection that often further marginalize local communities (Duffy et al., 2019). As a result, we have seen more direct-action movements such as Extinction Rebellion, Fridays for Future and others rapidly take center stage in environmental politics, while, from within the conservation community, we have also seen more radical alternatives emerge to challenge mainstream approaches.

Two of the more prominent conservation communities espousing discontent at the status quo are "neoprotectionists" and "new conservationists." New conservationists have been quite radical in a sense, as they have started criticizing the key elements on which the global conservation movement has been built since the nineteenth century: protected areas and the ideas of "pristine" nature and wilderness. Instead, they suggest a full integration of conservation into dominant, capitalist political economic systems for conservation to stand a chance in the future and maintain or retain legitimacy (Kareiva et al., 2012). In this way, they build on a growing trend within mainstream dominant approaches to conservation, represented, among others, by the Capitals Coalition, which aims to turn nature and natural resources into a form of "capital" that can be traded on markets and used to offset more regular forms of development (Fletcher, 2014; Fletcher et al., 2019).

Yet another community of conservationists – "neoprotectionists" – strongly contest the new conservationists. Neoprotectionists believe that the new conservation strategy would not only be the death of conservation, but of the entire planet (Wuerthner et al., 2014; 2015; Wilson, 2016). Basing their conservation objectives and strategies on conservation biology science, neoprotectionists believe that to ensure long-term viability of an ecosystem, nature must be set aside from the influence of people (Locke, 2015). Such ideas have important lineages to colonial conservation strategies, in which fences, fines and ideas about "pristine wilderness" were crucial tools to evict people from protected areas and to keep them out. According to neoprotectionists, we need to go back to protected areas and wilderness protection, but on a scale hitherto unseen. Some even argue that only if at least half the planet becomes a system of nature reserves can the ecological processes critical to human and planetary survival persist (Wilson, 2016).

Since earlier versions of these movements were suggested, they have also morphed, nuanced and developed. Neoprotectionist approaches, for instance, have reduced their emphasis on "protected areas only" somewhat to focus also on other conservation measures. They have also given more attention to social goals related to conservation, seemingly embracing a "social turn" that aims to bridge nonhuman nature and people (Ellis, 2019; Locke et al., 2019). While inclusion of Indigenous knowledge in conservation is now discussed by neoprotectionists (Locke, 2018), it still remains quite vague, with sparse and somewhat superficial references to land rights and integration of Indigenous knowledge in policymaking, while separating humans from nature via protected areas is maintained (Locke, 2018). It is unclear how this emergent "social turn" will manifest and be integrated into neoprotectionist visions on protected areas, a concern further highlighted by recent research finding that protecting half of the Earth might negatively affect over one billion people and result in widespread social and environmental injustices (Schleicher et al., 2019). Of particular importance, climate mitigation and adaptation are now widely discussed and tentatively integrated into protected area targets in order to accommodate broader regimes, such as around the sustainable development goals (SDGs). In this regard, Dinerstein et al. (2019) proposed a "Global Deal for Nature" including half Earth approaches that they believe should be paired with the 2015 Paris Climate Agreement.

Clearly, the debate on biodiversity governance is dynamic, diverse and rapidly changing in response to ongoing socioecological dynamics. Within these diverse dynamics, however, two core issues remain central: how to relate people to the rest of nature and how to situate conservation vis-à-vis the political economy of neoliberal capitalism. And despite more recent iterations that nuance earlier and more radical proposals to mix people and nonhuman nature through "natural capital" valuation, or separate people and nature on an unprecedented global scale, it is doubtful whether the dominant options currently on the table can provide a productive way forward. As argued in Büscher and Fletcher (2020), none of the current approaches will provide the fundamental structural transformations needed, as they do not directly confront the drive for continual accumulation of capital at the heart of the neoliberal capitalist economy. Neither do they sufficiently engage with the social injustices that have historically plagued both protectionist and market-based approaches to environmental governance (Martin et al., 2013). Nor do they take into account the vast differences in ways of knowing nature, environmental values and perspectives on what makes "good governance" (Sikor et al., 2013). We therefore need a different approach to transformation that can bring about the "substantial, profound and fundamental change" required (Massarella et al., 2021). We outline one pathway to transformative change through the alternative approach of convivial conservation.

12.3 Convivial Conservation: Vision, Politics, Governance

Convivial conservation emphasizes the vision, politics and governance mechanisms needed for a realistic, *structural* transformation of biodiversity protection. This is because convivial conservation is founded on a political ecology approach that is critical of contemporary capitalism, the global and unsustainable political economy it has spawned over the last centuries and the recent increase in global authoritarianism (Büscher and Fletcher, 2020). This makes convivial conservation itself a political economic approach to environmental governance, characterized by questions such as: How can we understand political economy and international development from the perspective of integrated socioecological dynamics around biodiversity? Or, how can a concern for biodiversity become central to the ways we (need to) rethink the relationship between political economy and development generally? And how does this lead to the implementation of concrete policies and measures at all levels that are sustainable, equitable and just? In short, convivial conservation is a critical-constructive approach that, contrary to practice-oriented, consensus and neoliberal approaches, bases its strategy on a critique of the structural context within which actors and organizations maneuver.

While a fuller elaboration of the convivial conservation vision has been published elsewhere (Büscher and Fletcher, 2020), it can be summarized as a postcapitalist, political economic approach to conservation that aims to integrate and reconnect people and nature in landscapes across different scales, spaces and times. The convivial conservation vision functions within the broader transformative vision of degrowth: an overall quantitative downsizing of economic throughput to ecologically sustainable levels coupled with widespread wealth redistribution to make this reduction "socially sustainable" (D'Alisa et al., 2015; Hicks et al., 2016; Holland et al., 2009; Kallis, 2011; Raworth, 2017; Wilkinson and Pickett, 2010). Within these overarching contexts, convivial conservation defines specific parameters for a fundamentally different form of conservation that does not separate people and nature. This means that protected areas and urban centers, as the two quintessential "end-points" of traditional human–nature dichotomies, have to be connected more, with the ultimate aim of achieving a better balance between human and nonhuman lives and needs across urban and rural spaces.

Convivial conservation envisions five fundamental shifts for conservation: moving from protected to promoted areas; from a framing of saving nature to one of celebrating human *and* nonhuman nature; from touristic voyeurism to engaged visitation; from a focus on spectacle to a focus on everyday environmentalisms and from privatized expert technocracy to common democratic engagement (Büscher and Fletcher, 2020: 163–174). In line with the themes of the book, this chapter focuses on how this vision is also a politics and form of governance. Central to convivial conservation is the fact that it politicizes conservation – meaning that it explicates the interests of different actors and how they may or may not be compatible, and always function within broader frameworks of power. Convivial conservation, therefore, is not focused on achieving consensus and does not believe that all actors with widely differential interests can or want to come together to promote biodiversity conservation. Rather, it conceptualizes biodiversity conservation as a *political struggle* caught up in histories and contexts of power that provide structural and agentic challenges and barriers. In this struggle, commonalities need to be sought and created, but not at the expense of the overall *political* direction of the convivial conservation vision, which, as mentioned, entails (moving toward and encouraging) degrowth, wealth redistribution and, ultimately, postcapitalism. In this sense, convivial conservation also aligns with environmental justice movements that conceptualize political struggle as an imperative part of radical transformation (Pellow, 2017; Temper et al., 2018).

This brings us, finally, to governance, or the way that actors steer, direct and influence affairs in particular directions. Governance mechanisms include, among others, legal regimes, state and other forms of organization, (formal *and* informal) institution-building or breaking, and more, in both material and discursive forms. What constitutes biodiversity governance is very broad and encapsulates a wide range of actors, activities and approaches. However, the concept of "transformative governance" established in Chapter 1 of this book is more specific, and it is this concept that is at the center of convivial conservation. Transformative biodiversity governance is understood as a product of deliberate and political acts that directly challenge embedded power structures, dominant agendas and framings, and mainstream approaches to conservation (see Chapter 1). In order to disrupt embedded hierarchies and power structures and bring about this transformative governance, we must first critically interrogate the (historical and contemporary) framings, responsibilities and roles of different actors within biodiversity conservation.

Table 12.1 provides a heuristic basis for such an analysis, depicting our conceptualization of four broad categories of conservation actors and organizations. We regard rural lower classes (category 4) as those actors who often live in or with biodiversity and who (still) depend on the land for subsistence, especially in tropical countries. They are often (seen as) poor and the ones who have least contributed to global problems of biodiversity loss (historically and contemporarily). Yet they are most often targeted in conservation interventions and forced or "incentivized" to change their livelihoods to meet biodiversity targets. Category 3 actors comprise urban, semiurban or semirural middle and lower classes

Table 12.1 *Generic categorization of classes important for conservation*

1. Upper classes	- Political, economic and other elites, inherited wealth
	- At the helm of the global capitalist system
	- Multiple properties, including in wealthy urban neighborhoods and (biodiverse) estates or areas
2. Land-owning capitalist classes	- Commercial farmers, large plantation or otherwise productive landowners
	- Responsible for / implicated in much land-use change, soil depletion, biodiversity loss, etc.
3. Middle and lower classes	- Urban, peri-urban, peri-rural working classes
	- Non-subsistence: dependent on wage labor, market-based commodity consumption
4. Lower rural classes	- Rural/forest communities, residents, dwellers
	- Partially or wholly dependent on subsistence activities
	- At the bottom of global capitalist system

(*source:* Büscher and Fletcher, 2020: 182).

throughout the world, who are not directly land-dependent for subsistence and who participate and rely on local and global labor and consumer markets. Through their consumption and place in global markets, they do heavily influence biodiversity in many places, but are often not part of or specifically targeted by conservation interventions, except as potential donors.

Category 2 actors are land-owning capitalist classes such as major capitalist farmers and/or landholders for large agro-industry. They are often targeted by conservation, not as part of community-based interventions, but as partners in the conservation effort or as targets of (so-called) activist interventions or forms of resistance. In many places (Indonesia, Brazil, Central Africa and so forth), these classes are also part of violent frontiers of land conversion, and hence difficult to target and engage with (Campbell, 2015). Finally, category 1 actors comprise the global upper classes that are, politically, economically or otherwise, at the helm of the global capitalist system (often referred to as the "transnational capitalist class"; see Sklair, 2001). These elite actors are often both urban and rural – owning multiple properties, including in rich residential areas in cities to be close to elite political-economic circles, but also with second, third or even more properties in rural, semirural and biodiversity-rich spaces, including large estates and private reserves (Holmes, 2012). Upper-class elites are often recruited as funders or included on boards of conservation organizations, but rarely targeted as part of conservation initiatives aiming at behavioral or livelihood change, as they are often either seen as unreachable or as doing good for the environment through their philanthrocapitalism or other forms of conservation-related charity (including through the privatization of nature/parks, etc.). Hence the upper classes have a strange double role, as they are at the helm of the system that keeps the pressure on biodiversity intense and high, while also considered either untouchable or even to be championing conservation through their large donations to conservation causes, NGOs and more (Edwards, 2008; Ramutsindela et al., 2011).

While empirical reality is much more complex than this table can depict, its point is that currently dominant conservation paradigms focus mostly on category 4 actors in terms of whose lives need to change. Convivial conservation would change this and target actors according to their differential responsibilities and accountabilities in relation to both the direct and indirect impacts of their actions on biodiversity, as well as the relative power these actors possess within broader structures of capital accumulation. Paraphrasing Moore (2016), it is about identifying, targeting and "shutting down the relations" that produce biodiversity loss, not just about geographical proximity.

In this way, we might reverse the model of "polycentric" governance proposed by Ostrom and others (e.g. Ostrom and Cox, 2011). In this standard model, governance is seen to start with local people and then must consider their embeddedness within overarching structures of governance with which they must contend to assert their space for self-governance. In our vision, by contrast, effective conservation governance would start by addressing actors in these superordinate levels in order to first target their actions, then work down toward the local people in direct contact with the biodiversity in question. In this way, the pressures exerted on local conservation initiatives can be proactively addressed at their source rather than merely retrospectively in relation to their impacts.

We should clarify that this governance model pertains only to the ways that conservationists frame and confront threats to conservation, not to how decision-making regarding effective conservation should proceed. The latter must embody deeply democratic forms of engagement in which local actors, those generally affected heaviest by conservation measures, *are* placed at center stage (see also Chapter 8). A convivial conservation politics, therefore, must simultaneously center local people as key decision-makers in conservation planning and decenter them as the central targets of interventions aimed at fostering behavioral change. This analysis gives rise to a number of questions, and in a short chapter it is not possible to work out all the details of the convivial conservation vision and the politics it necessitates. Our analysis does, however, point to the need for transformative governance mechanisms that disrupt this conservation class structure, "trigger regime shifts" and ultimately alter the "structures and processes that define the system" (Chaffin et al., 2016: 400).

We have previously put forward some suggestions for transformative governance mechanisms, including a program of historical reparations directed at category four actors, developing "integrated conservation landscapes" that prioritize human and nonhuman coexistence (Büscher and Fletcher, 2020), and alternative finance mechanisms such as "conservation basic income" for those living close to areas of high biodiversity (Fletcher and Büscher, 2020). In the remainder of this chapter, we discuss the rationale for BICs as both a political methodology and a transformational governance mechanism. The basic idea behind BICs is simple: to better understand and politicize the relationships between different actors and the impacts that their livelihoods and consumption choices have on the conservation of particular forms of biodiversity. BICs challenge many of the embedded assumptions that we have previously outlined in this section by refocusing attention onto those with the largest footprints – likely to be in class 1 and 2 – while challenging the problematic focus on class 4 actors. In doing so we open up the potential for transformative change in biodiversity governance, as the focus of conservation discourses, actions and interventions shifts onto those with the biggest footprints.

12.4 Biodiversity Impact Chains

The idea of BICs is partly inspired by the value chain literature, which studies value supply chains to see how commodities are produced, distributed and consumed, and to study social, political and environmental issues along the way (Bair, 2009). The value chain literature has developed in numerous directions, including how value chains relate to forms of more sustainable production or the tracing of knowledge as a valuable commodity in its own right (Büscher, 2014; Guthman, 2008; Ponte, 2019). A classic example comes from Hartwick's (1998: 426) focus on gold, where she shows how production, processing and consumption dimensions are connected through "vertical" long-distance relationships but also consist of "horizontal" dimensions of local interrelationships along various points on the chain. She contends that the production of one commodity can imply multiple chains, while along points on a singular chain "halo-effects" can occur. In this way, wider social and environmental effects are brought about by particular activities along the chain.

A major critique in much of the literature on value chains is that they have quite a linear understanding of the chains they describe and a very simplistic or instrumental idea of the "value" they envision. According to Starosta (2010: 435):

[W]hat commodity chain studies do is simply to offer, through an essentially inductive-empirical methodology, a typological description of the immediate outer manifestations of the determinations at stake. This failure firmly to explain the nature of GCCs [*Global Commodity Chains*] is expressed, for instance, in the disjuncture between the portrayal of the particular dynamics internal to each industry and the general dynamics of the "system as a whole."

Like others (Ponte, 2019), Starosta (2010: 455) argues that we should pay more attention to irregular circulatory dynamics of value, rather than "captive governance structures" that work according to linear models of how value is produced. The same lessons apply for how we should study the idea of "*impact* chains." Like value or commodity chains, the last decades have seen a major literature develop around the idea of impact, including in relation to "cross-sectoral cumulative impacts" that we draw upon and are inspired by (Baird and Barney, 2017).

Building on these important considerations, we imagine BICs as a *political methodology* and a *governance mechanism* to (further) study, map and steer political economic activities in particular bioregions (both urban and rural, and everything in between) and how they relate to specific ecosystems and biodiversity that provide the (raw) materials for these activities. In many cases, this is impossible to establish given the complex considerations above. Hence, we consider starting with specific ecosystems wherein this dependency can be most directly established. These could include (fresh) water, as the distances between water and their use – although they can be large – are often local or regional. As the important case of the drought in Cape Town, South Africa, in 2018 shows – a recent example of a major global city facing an acute water crisis[6] – the conservation of water sources is critically important, and depends on complex political-ecological factors, some of which can be directly controlled and some not (such as climate change). But once the availability and sustainable supply of water are more-or-less known, needs and interests can be renegotiated accordingly, which is precisely what happened in Cape Town, where more pressure was put on major water users in particular to conserve.[7]

Other examples could relate to locally specific biodiversity and their needs vis-à-vis inhabited (urban or rural) landscapes. But all of these are still, in many ways, local or regional. Given the thoroughly global nature of today's value and impact chains, it is critical to also map and study global connections so as to more directly highlight the political implications and biodiversity impacts of richer lifestyles. There are two ways to do this, both of which are already being explored in practice: first, to start from a specific and important ecosystem or species and "work up" toward the main actors or economic sectors that impact it; or, second, to "work down" from particular actors and economic sectors to show their cumulative impacts on different biodiversity and ecosystems. In what follows,

---

[6] See www.capetowndrought.com for more information. Accessed 25 February 2018, two months before alleged "day zero" was projected, the day that water will no longer come from Cape Town taps.
[7] See https://bit.ly/3ttJRIC.

we provide some first tentative examples of both, after which we wrap up the section by suggesting how we can take this concept forward as part of a broader move to operationalize the transformative governance of convivial conservation.

12.4.1 Working down the Biodiversity Impact Chain

Conservation areas and biodiversity are often – and rather self-evidently – said to be impacted mostly by "local people" aiming to fulfill their livelihood needs by utilizing surrounding natural resources. This is, among other factors, the basis of much of the "community-based conservation" literature (Dressler et al., 2010), as well as an explicit assumption of many elite actors involved in conservation. One example concerns famous Virgin billionaire-entrepreneur Richard Branson. In a video supporting conservation in Africa, he asks the question, "what is Africa?" and answers bluntly that "Africa is its animals. That is the beauty of Africa, that's what makes it different from the rest of the world. And to lose those animals would be catastrophic." Branson blames "dwindling wildlife numbers" on "Africa's increasing (human) populations" and argues that Africa should "increase the amount of land for the animals and by increasing the amount of land for the animals, that will help human beings."[8]

Unfortunately, this neocolonial discourse is not uncommon when it comes to conservation in Africa (Mbaria and Ogada, 2017). Convivial conservation challenges colonizing discourses and practices by more clearly identifying the impacts of extra-local actors, and especially global elites who have the largest footprints. In the case of Branson, his environmental impacts are quite well-documented and provide a pertinent example. Branson, after all, owns several luxury game reserves around the world and has voiced some of the largest climate commitments of any elite actor. Together, these could constitute quite an environmental legacy were it not for the fact that scholars have thoroughly debunked these commitments. Naomi Klein (2015: 251–252), for example, argues that "Branson set out to harness the profit motive to solve the climate crisis – but the temptation to profit from practices worsening the crisis proved too great to resist. Again and again, the demands of building a successful empire trumped the climate imperative." Scott Prudham (2009), similarly argued that Branson's environmentalism did nothing to limit further capitalist expansion, including the resource extraction and use this entails. However, while these authors may show that Branson is far from an environmental hero, his precise impact on biodiversity is unclear and needs more research.

At the same time, this research also needs to be extended to aggregate sectors instead of (only) individuals. Our own research on the high-end tourism sector in South Africa provides a short example of how a BIC analysis could work by analyzing the impact of all four conservation classes (Table 12.1) on biodiversity. Adjacent to the world-famous Kruger National Park, philanthrocapitalists such as Richard Branson have their own residences on private protected lands ("upper class," category 1), while lodge operators and large tourism companies own enormous tracts of private lands ("land-owning capitalist class," category 2)

---

[8] www.youtube.com/watch?v=F0LhU4XFHAM.

for relatively wealthy tourists to enjoy ("upper class," category 1 and "land-owning capitalist class," category 2). Furthermore, some wealthy South Africans, Europeans and others own properties on so-called "wildlife estates," sometimes as a permanent residence but often also as "second homes" (again categories 1 and 2, but also 3) (Koot et al., 2019).

Meanwhile, the inequality between these classes and the "middle and lower classes" (category 3) and "lower rural classes" (category 4) remains enormous, and people from the latter two categories are often associated with causing most of the problems of conservation, including poaching (Duffy et al., 2019). However, these people also provide substantial "conservation labor" (needed for the first two class categories to enjoy nature) and, through the tourism industry, are increasing the value of private land, thereby reducing the chances of the middle and lower and lower rural classes to claim land for other purposes (Ramutsindela, 2015; Sodikoff, 2009), perpetuating and fortifying socioeconomic inequality. Despite a variety of such negative social and environmental consequences, the tourism industry often champions itself for its sustainable contribution to conservation (including much support for militarized anti-poaching conservation initiatives) and community development. However, initial research from several of this chapter's authors suggests that tourism's contributions are actually quite meager. More research is needed to accurately evaluate the impacts that all of the classes outlined here have on the national park and its biodiversity, and we posit that BICs as political methodology would enable such an analysis (see also Mugo et al., 2020).

12.4.2 Working up the Biodiversity Impact Chain

The other way to operationalize impact chains is to work "up" from specific biodiverse spaces, and document the direct and indirect pressures on these areas. Unlike the aforementioned top-down impact-chain mapping, this is an area where a lot of work is already being done. NGOs like Greenpeace, Friends of the Earth, the Rainforest Action Network and many others are well known not just for their (direct) actions but also for their research linking environmental impacts on specific areas to specific actors. The Rainforest Action Network, for example, published a report in 2017 tracking the impact chains on Southeast Asian rainforest, especially those in the Leuser Ecosystem in Sumatra, Indonesia (RAN, 2017). According to the report, it

> profiles key environmental, social and governance (ESG) performance issues of 8 companies operating in Southeast Asia's tropical forest-risk commodity sectors. The 8 companies profiled – Felda Global Ventures Holdings, Indofood Sukses Makmur, IOI Corporation, Wilmar International, Asia Pulp and Paper Group, Oji Holdings Corporation, Marubeni Corporation, and Itochu Corporation – were found to have had a range of serious ESG violations in their own operations or direct supply chains. These violations include: use of child and forced labour; conflicts with local communities over violations of their tenure rights; tropical deforestation and destruction of carbon-rich peatlands; threats to biodiversity; corruption; and illegality.
> *(RAN, 2017: 3)*

But the report doesn't just highlight the responsibility of the companies directly involved in the destruction of biodiversity and other misdemeanors; it goes all the way up to specific institutional investors, which they argue are equally responsible for the impacts on biodiversity:

> The forest-risk commodity sector operations of the 8 companies profiled in this report have been enabled by at least 6.38 billion USD in bond- and shareholdings at the most recent filing date in May 2017 by institutional investors (asset managers, insurance companies, pension funds) and have received more than 32.67 billion USD in loans and underwriting facilities between 2010 and 2016.
> (RAN, 2017: 3)

They then list the investors and bank and highlight that these "have both a moral and corporate responsibility, and a fiduciary duty to understand and address the harmful ESG impacts ... which they are connected to" (RAN, 2017: 3).

This type of work is critical and puts the spotlight where it belongs: on the wealthy, often extra-local actors that have disproportionate (negative) impact on biodiversity. A similar "working up" approach was also recently applied by Amazon Watch to destruction of the Amazon and Cerrado biomes in Brazil, in their report entitled "Complicity in Destruction" (Amazon Watch, 2019). Home to 10 percent of the world's biodiversity and 20 percent of its flowing freshwater, it is hard to imagine a convivial conservation transition without a concerted international effort to curb rapid deforestation and land conversion that has increased by more than 50 percent since 2016 (Amazon Watch, 2019). The report echoes research implicating soy and beef production for over 80 percent of forest land conversion in Brazilian Amazonia, and while noting the difficulty in following the exact trail to consumption destinations, it outlines clearly the global financial sources underwriting local and multinational companies implicated in the commodity chain. Among the largest creditors and equity investors in companies active in the Amazon and Cerrado, including those fined for illegal practices, were Barclays, Capital Group, BlackRock, Bank of America, Citigroup, JPMorgan Chase, BNP Paribas, Santander, HSBC, Credit Suisse, Vanguard, Morgan Stanley and Fidelity Investments (Amazon Watch, 2019: 19–24). Illegal timber supply chain links were also found with major importers in France, Belgium, the Netherlands, Denmark, the UK and the USA. Ultimately, Amazon Watch calls for a no-deforestation policy by global financiers, which are effectively underwriting the rapid decline of the world's most biodiverse region, and sees scope for targeting EU and North American governments, given their accounting for 18.3 percent and 11 percent of Brazilian agricultural exports, respectively.

The importance of viewing the soy and beef industries together in this conservation impact chain is not incidental. Research has shown that despite the primary driver of Amazon deforestation by far being cattle production, this has occurred partly as a result of displacement of medium and smaller cattle ranchers from land now occupied by soy (Barona et al., 2010). Perhaps even more salient has been Brazil's efforts to "flex" its soy crop for animal feed processing and biofuel production in order to maintain a degree of domestic control – and significant revenues – as China monopolized Brazilian whole bean exports after 2008 (Oliveira and Schneider, 2016). Maintaining a Brazilian soy-crushing and animal feed production capability effectively depends on constantly expanding domestic cattle production, or else losing out to global competition.

With China now crushing the bulk of Brazilian soy to make chicken, pig, salmon and cattle feed for markets worldwide, the "working up" of Amazonian biodiversity destruction simultaneously results in a "working down" to numerous examples of agro- and aqua-industrial pollution and ecosystem decline across worldwide cases from Norwegian salmon to Vietnamese shrimp, and beef industrial expansion across much of Asia. In Brazil itself, then, the conversion of some 200,000 square kilometers of highly biodiverse Cerrado forest and savanna for monocrop GM soy, with associated intensive pesticide use and seed consolidation by a tiny list of corporate players, has meant a wholesale collapse of pre-existing nature and agrarian livelihoods, while also enabling biodiversity destruction associated with agribusiness around the globe (Oliveira and Hecht, 2016). Arguments that we need to continually expand food production to feed a growing population are quickly countered by deeply uneven global access, distribution and profiteering from corporate-led food systems that themselves increasingly depend on ecological catastrophe and the undermining of local food production in favor of export markets (McMichael, 2014).

Finally, with regard to Amazonian biodiversity decline, "scaling up" also highlights the complicity of the global financial and market connections already identified in the rise of authoritarian government. The close association of extractivism with the new Latin American far right is well covered in the literature (Arsel et al., 2016; McCarthy, 2019; Saad-Filho and Boffo, 2021), yet often understated are the simultaneous attacks on protected areas in the Amazon and elsewhere – especially those managed by Indigenous Peoples – that the expansion of mining and the cattle–soy nexus necessitates. The dismantling of Brazilian government ministries for Indigenous Peoples and the environment is effectively now preventing *any* regulation of Amazonian conservation. The same list of global financiers noted above thus profit from the authoritarian *enforcement* of biodiversity decline, a fact further highlighting the urgent need for institutional control of global finance.

More examples can be mentioned, but what is clear is that the transformation to convivial conservation would rely on a dramatic extension and normalization of such research and exposure endeavors. In doing so, the precise details of the impact-mapping in the above examples should be as important as the sociocultural and political-economic process that accompanies it. Again: we see this methodology and governance mechanism as a *politicization tool* that connects different actors from Table 1 in relation to how biodiversity is conserved or not. This political process can then further map the needs and interests of stakeholders in the short term, and also how these needs might change as the overall economy shifts toward degrowth, sharing the wealth and convivial conservation. In addition, the planning process could start to create awareness of how people in bioregions can contribute to degrowth and sharing of wealth. This is how an active process of shifting needs and interests (and hence, ultimately, human nature itself), and challenging the vested interests associated with the creation of capitalist needs and interests, might start or be further encouraged. Moreover, "impact chains" can never do justice to all the different types of impacts generated through activities, especially the complicated climate-related impacts. The point is therefore not to get one-on-one impacts "measured" precisely but rather to complicate, and politicize, the capitalist governance of biodiversity by incorporating direct and indirect pressures and by targeting and challenging these from two sides (bottom-up and top-down).

12.5 Conclusion

Along with climate change, inequality and, more recently, a global pandemic, biodiversity loss is considered to be one of the world's most pressing challenges. As such, calls for transformative change in the ways biodiversity is governed and conserved are growing. However, major differences on how to approach transformative change exist, and some prominent responses to the biodiversity crisis that consider themselves transformative do not actually address underlying structural drivers of destruction. We therefore argue that these responses, including neoprotectionism and new conservation, should not be considered transformative in the way we have defined the term. Instead, and in line with a growing number of academics, social movements and civil society groups, we contend that fundamental structural transformation is needed to achieve the biodiversity and wider environmental governance capable of adequately addressing the growing biodiversity crisis. In this chapter we have built on the vision of convivial conservation, put forward as a necessary and realistic alternative – one that has fundamental structural transformation at its core.

We have also outlined a practice tool – biodiversity impact chains – as an example of a transformative governance mechanism that reframes perspectives on biodiversity conservation by politicizing the uneven relationships and impacts that different actors have with and on biodiversity. BICs can be seen as a tool for governance *of* transformations (Chapter 1) as they aim to steer the transformative change outlined in this chapter as part of the convivial conservation vision. Two characteristics of transformative governance highlighted in Chapter 1 are reiterated here as particularly important in relation to BICs. First, BICs are *inclusive* as they emphasize the interests of different actors and how such interests impact biodiversity. Second, BICs are *integrative* as they connect actions and solutions across scales. BICs also demonstrate the need for transformative governance to expand yet further and provide a mechanism through which the very framing of biodiversity and its conservation is politicized, challenged and disrupted. Local communities are still typically conceptualized as the recipients, or targets, of biodiversity governance interventions – even in cases where this governance is thought to be transformative. BICs support an alternative approach – one that could support policymakers in better targeting interventions in a more impactful and transformative way.

BICs are just one tool in the convivial conservation toolbox that we and other diverse actors are developing, and in line with other transformative movements such as degrowth. The convivial conservation vision, however, goes beyond the use of individual tools, and the focus, we argue, must be on broader "whole earth" transformation (Büscher et al., 2017). This requires what Wark (2015) calls "alternative realism," in contrast to "capitalist realism," asserting that there is no viable alternative to the existing order – and a questioning of many of the assumptions that underpin conservation as we know it.

This may seem impossible, but if, as Olsson et al. (2010: 280) argue, "transformational change is most likely to occur at times of crisis, when enough stakeholders agree that the current system is dysfunctional," then this moment could be the opportunity to make the fundamental, structural changes that are needed.

References

Adams, W. (2017). Sleeping with the enemy? Biodiversity conservation, corporations and the green economy. *Journal of Political Ecology* 24, 243–257.

Amazon Watch. (2019). *Complicity in destruction II: How Northern consumers and financiers enable Bolsonaro's assault on the Amazon*. Oakland, CA: Amazon Watch. Available from https://bit.ly/3FDyvnL.

Arsel, M., Hogenboom, B., and Pellegrini, L. (2016). The extractive imperative in Latin America. *The Extractive Industries and Societies* 3, 880–887.

Bair, J. (2009). Global commodity chains: Genealogy and review. In *Frontiers of Commodity Chain Research*. J. Bair (Ed.), pp. 1–34. Stanford, CA: Stanford University Press.

Baird, I. G., and Barney, K. (2017). The political ecology of cross-sectoral cumulative impacts: Modern landscapes, large hydropower dams and industrial tree plantations in Laos and Cambodia. *The Journal of Peasant Studies* 44, 769–795.

Barona, E., Ramankutty, N., Hyman, G., and Coomes, O. T. (2010). The role of pasture and soybean in deforestation of the Brazilian Amazon. *Environmental Research Letters* 5, 024002.

Bennett, N. J., Blythe, J., Cisneros-Montemayor, A. M., Singh, G. G., and Sumaila, U. R. (2019). Just transformations to sustainability. *Sustainability* 11, 3881.

Blythe, J., Silver, J., Evans, L., et al. (2018). The dark side of transformation: Latent risks in contemporary sustainability discourse. *Antipode* 50, 1206–1223.

Brown, K., O'Neill, S., and Fabricius, C. (2013). Social science understandings of transformation. In *World social science report 2013: Changing global environments*. OECD (Ed.), pp. 100–106. Paris: OECD Publishing.

Bruff, I. (2014). The rise of authoritarian neoliberalism. *Rethinking Marxism* 26, 113–129.

Büscher, B. (2013). *Transforming the frontier. Peace parks and the politics of neoliberal conservation in Southern Africa*. Durham, NC: Duke University Press.

  (2014). Selling success: Constructing value in conservation and development. *World Development* 57, 79–90.

Büscher, B., and Fletcher, R. (2020). *The conservation revolution. Radical ideas for saving nature beyond the Anthropocene*. London: Verso.

Büscher, B., Fletcher, R., Brockington, D., et al. (2017). Half-Earth or whole Earth? Radical ideas for conservation and their implications. *Oryx* 51, 407–410.

Campbell, J. (2015). *Conjuring property: Speculation and environmental futures in the Brazilian Amazon*. Seattle, WA: University of Washington Press.

CBD (Convention on Biological Diversity). (2020). *Global biodiversity outlook 5*. Montreal: CBD Secretariat.

Chaffin, B, Garmestani, A., Gunderson, L., et al. (2016). Transformative environmental governance. *Annual Review of Environment and Resources* 41, 399–423.

D'Alisa, G., Demaria, F., and Kallis, G. (Eds.). (2015). *Degrowth. A vocabulary for a new era*. Abington: Routledge.

Dinerstein, E., Vynne, C., Sala, E., et al. (2019). A global deal for nature: Guiding principles, milestones, and targets. *Science Advances* 5, eaaw2869.

Dressler, W., Büscher, B., Schoon, M., et al. (2010). From hope to crisis and back again? A critical history of the global CBNRM narrative. *Environmental Conservation* 37, 5–15.

Duffy, R., Massé, F., Smidt, E., et al. (2019). Why we must question the militarisation of conservation. *Biological Conservation* 232, 66–73.

Edwards, M. (2008). *Just another emperor? The myths and realities of philanthrocapitalism*. New York: Demos.

Ellis, E. C. (2019). To conserve nature in the Anthropocene, half earth is not nearly enough. *One Earth* 1, 163–167.
European Environment Agency. (2019). *The European environment – State and outlook 2020.* Copenhagen: EEA.
Feola, G. (2020). Capitalism in sustainability transitions research: Time for a critical turn? *Environmental Innovation and Societal Transitions* 35, 241–250.
Fletcher, R. (2014). Orchestrating consent: Post-politics and intensification of NatureTM Inc. at the 2012 World Conservation Congress. *Conservation and Society* 12, 329–342.
Fletcher, R., and Büscher, B. (2020). Conservation basic income: A non-market mechanism to support convivial conservation. *Biological Conservation* 244, 108520.
Fletcher, R., Dressler, W., Anderson, Z., and Büscher, B. (2019). Natural capital must be defended: Green growth as neoliberal biopolitics. *Journal of Peasant Studies* 46, 1068–1095.
Guthman, J. (2008). Unveiling the unveiling: Commodity chains, commodity fetishism, and the "value" of voluntary, ethical food labels. In *Frontiers of commodity chain research*. J. Bair (Ed.), pp. 190–106. Stanford, CA: Stanford University Press.
Hartwick, E. (1998). Geographies of consumption: A commodity-chain approach. *Environment and Planning D: Society and Space* 16, 423–437.
Hicks, C. C., Levine, A., Agrawal, A., et al. (2016). Engage key social concepts for sustainability. *Science* 352, 38–40.
Holland, T. G., Peterson, G. D., and Gonzalez, A. (2009). A cross-national analysis of how economic inequality predicts biodiversity loss. *Conservation Biology* 23, 1304–1313.
Holmes, G. (2012). Biodiversity for billionaires: Capitalism, conservation and the role of philanthropy in saving/selling nature. *Development and Change* 43, 185–203.
IPBES (Intergovernmental Science-Policy Platform on Biodiversity and Ecosystem Services). (2019). *Global assessment report of the Intergovernmental Science-Policy Platform on Biodiversity and Ecosystem Services*. E. S. Brondízio, J. Settele, S. Díaz and H. T. Ngo (Eds.). Bonn: IPBES secretariat.
Kallis, G. (2011). In defence of degrowth. *Ecological Economics* 70, 873–880.
Kareiva, P., Marvier, M., and Lalasz, R. (2012). Conservation in the Anthropocene: Beyond solitude and fragility. Available from https://bit.ly/3Golley.
Kiely, R. (2021). Conservatism, neoliberalism and resentment in Trumpland: The 'betrayal' and 'reconstruction' of the United States. *Geoforum* 124, 334–342.
Klein, N. (2015). *This changes everything. Capitalism vs the climate*. London: Allen Lane.
Koot, S., Hitchcock, R., and Gressier, C. (2019). Belonging, Indigeneity, land and nature in Southern Africa under neoliberal capitalism: An overview. *Journal of Southern African Studies* 42, 341–355.
Lenton, P., Rockström, J., Gaffney, O., et al. (2019). Climate tipping points – Too risky to bet against. *Nature* 575, 592–595.
Locke, H. (2015). Nature needs (at least) half. In *Protecting the wild. Parks and wilderness, the foundation for conservation*. G. Wuerthner, E. Crist and T. Butler (Eds.), pp. 3–15. London: Island Press.
  (2018). The International Movement to Protect Half the World: Origins, Scientific Foundations, and Policy Implications. *Reference Module in Earth Systems and Environmental Sciences*. https://doi.org/10.1016/B978-0-12-409548-9.10868-1
Locke, H., Ellis, E. C., Venter, O., et al. (2019). Three global conditions for biodiversity conservation and sustainable use: An implementation framework. *National Science Review* 6, 1080–1082.
MacDonald, K. I. (2010). The devil is in the (bio)diversity: Private sector "engagement" and the restructuring of biodiversity conservation. *Antipode* 42, 513–550.
MacDonald, K. I., and Corson, C. (2012). "TEEB begins now": A virtual moment in the production of natural capital. *Development and Change* 43, 159–184.
Martin, A., McGuire, S., and Sullivan, S. (2013). Global environmental justice and biodiversity conservation. *The Geographical Journal* 179, 122–131.
Martin, A., Teresa Armijos, M., Coolsaet, B., et al. (2020). Environmental justice and transformations to sustainability. *Environment: Science and Policy for Sustainable Development* 62, 19–30.
Mason, P. (2019). *Clear bright future. A radical defence of the human being*. London: Allen Lane.

Massarella, K., Nygren, A., Fletcher, R., et al. (2021). Transformation by conservation? How critical social science can contribute to transformative change in biodiversity conservation. *Current Opinion in Environmental Sustainability* 49, 79–87.

Mbaria, J., and Ogada, M. (2017). *The big conservation lie*. Auburn, WA: Lens&Pens.

McCarthy, J. (2019). Authoritarianism, populism, and the environment: Comparative experiences, insights, and perspectives. *Annals of the American Association of Geographers* 109, 301–313.

McMichael, P. (2014). *Food regimes and agrarian questions*. Halifax: Fernwood Publishing.

Moore, J. W. (2016). The rise of cheap nature. In *Anthropocene or Capitalocene? Nature, history, and the crisis of capitalism*. J. W. Moore (Ed.), pp. 78–115. Oakland, CA: PM Press.

Mugo, T. N., Visseren-Hamakers, I. J., and Van der Duim, V. R. (2020). Landscape governance through partnerships: Lessons from Amboseli, Kenya. *Journal of Sustainable Tourism*. DOI: 10.1080/09669582.2020.1834563

Neimark, B., Childs, J., Nightingale, A., et al. (2019). Speaking power to "post-truth": Critical political ecology and the new authoritarianism. *Annals of the American Association of Geographers* 109, 613–623.

Newbold, T., Hudson, L., Arnell, A., et al. (2016). Has land use pushed terrestrial biodiversity beyond the planetary boundary? A global assessment. *Science* 353, 288–291.

O'Brien, K., Reams, J., Caspari, A., et al. (2013). You say you want a revolution? Transforming education and capacity building in response to global change. *Environmental Science & Policy* 28, 48–59.

Oliveira, G., and Hecht, S. (2016). Sacred groves, sacrifice zones and soy production: Globalization, intensification and neo-nature in South America. *The Journal of Peasant Studies* 43, 251–285.

Oliveira, G. de L. T., and Schneider, M. (2016). The politics of flexing soybeans: China, Brazil and global agroindustrial restructuring. *The Journal of Peasant Studies* 43, 167–194.

Olsson, P., Bodin, Ö., and Folke, C. (2010). Building transformative capacity for ecosystem stewardship in social–ecological systems. In *Adaptive capacity and environmental governance*. D. Armitage and R. Plummer (Eds.), pp. 263–285. Berlin: Springer.

Ostrom, E., and Cox, M. (2011). Moving beyond panaceas: A multi-tiered diagnostic approach for social-ecological analysis. *Environmental Conservation* 37, 451–463.

Pellow, D. N. (2017). *What is critical environmental justice?* New York: John Wiley & Sons.

Polanyi, K. (1957). *The great transformation*. Boston, MA: Beacon Press.

Ponte, S. (2019). *Business, power and sustainability in a world of global value chains*. London: Zed.

Poulantzas, N. (1978). *State, power, socialism*. Trans. P. Camiller. London: New Left Books.

Prudham, S. (2009). Pimping climate change: Richard Branson, global warming, and the performance of green capitalism. *Environment and Planning A* 41, 1594–1613.

Rainforest Action Network (RAN). (2017). *Every investor has a responsibility. A Forests&Finance dossier*. San Francisco, CA: RAN.

Ramutsindela, M. (2015). Extractive philanthropy: Securing labour and land claim settlement in private nature reserves. *Third World Quarterly* 36, 2259–2272.

Ramutsindela, M., Spierenburg, M., and Wels, H. (2011). *Sponsoring nature: Environmental philanthropy for conservation*. New York: Routledge.

Raworth, K. (2017). *Doughnut economics. Seven ways to think like a 21st-century economist*. London: Penguin.

Saad-Filho, A., and Boffo, M. (2021). The corruption of democracy: Corruption scandals, class alliances, and political authoritarianism in Brazil. *Geoforum* 124, 300–309.

Schleicher, J., Zaehringer, J., Fastré, C., et al. (2019). Protecting half of the planet could directly affect over one billion people. *Nature Sustainability* 2, 1094–1096.

Scoones, I., Edelman, M., Borras, S., et al. (2018). Emancipatory rural politics: Confronting authoritarian populism. *Journal of Peasant Studies* 45, 1–20.

Scoones, I., Stirling, A., Abrol, D., et al. (2020). Transformations to sustainability: Combining structural, systemic and enabling approaches. *Current Opinion in Environmental Sustainability* 42, 65–75.

Sikor, T., Fischer, J., Few, R., Martin, A., and Zeitoun, M. (2013). The justices and injustices of ecosystem services. In *The justices and injustices of ecosystem services*. T. Sikor (Ed.), pp. 187–200. New York: Routledge.

Sklair, L. (2001). *The transnational capitalist class*. Oxford: Blackwell.
Sodikoff, G. (2009). The low-wage conservationist: Biodiversity and perversities of value in Madagascar. *American Anthropologist* 111, 443–455.
Starosta, G. (2010). Global commodity chains and the Marxian law of value. *Antipode* 42, 433–465.
Temper, L., Walter, M., Rodriguez, I., Kothari, A., and Turhan, E. (2018). A perspective on radical transformations to sustainability: Resistances, movements and alternatives. *Sustainability Science* 13, 747–764.
Tucker, M. A., Böhning-Gaese, K., Fagan, W. F., et al. (2018). Moving in the Anthropocene: Global reductions in terrestrial mammalian movements. *Science* 359, 466–469.
Wark, M. (2015). *Molecular red. Theory for the Anthropocene*. London: Verso.
Watson, J., Shanahan, D., Di Marco, M., et al. (2016). Catastrophic declines in wilderness areas undermine global environment targets. *Current Biology* 26, 2929–2934.
Wilkinson, R., and Pickett, K. (2010). *The spirit level: Why more equal societies almost always do better*. London: Penguin.
Wilson, E. O. (2016). *Half-Earth. Our planet's fight for life*. London: Lifereight Publishing.
Wuerthner, G., Crist, E., and Butler, T. (Eds.). (2014). *Keeping the wild: Against the domestication of the Earth*. New York: Island Press.
Wuerthner, G., Crist, E., and Butler, T. (2015). *Protecting the wild. Parks and wilderness, the foundation for conservation*. London: Island Press.
WWF. (2018). *Living planet report 2018*. Gland: WWF.

13

Transformative Biodiversity Governance in Agricultural Landscapes: Taking Stock of Biodiversity Policy Integration and Looking Forward

YVES ZINNGREBE, FIONA KINNIBURGH, MARJANNEKE J. VIJGE, SABINA J. KHAN AND HENS RUNHAAR

13.1 Introduction

Agricultural land systems, covering about 40 percent of the world's ice-free terrestrial surface, are the single largest contributor to biodiversity loss worldwide (Chapin et al., 2000; IPBES, 2018a; 2019). Agricultural practices have been linked to staggering losses in critical ecosystems such as tropical forests and ecologically functional species such as pollinators, raising concerns of losing biodiversity as both an intrinsic global value and as a central pillar of food security and ecosystem functions (IPBES, 2016; Laurance et al. 2014; Ramankutty et al., 2018). Conserving biodiversity in this sector is crucial beyond this intrinsic value (see Chapter 2), since biodiversity in agricultural landscapes supports ecosystem services that sustain human well-being through provisioning services such as food production, regulating services including flood and climate control or stabilization, and supporting services such as pollination and soil fertility (IPBES, 2016; 2018b; 2019; Scherr and McNeely, 2008; Tscharntke et al., 2012). There are a wide range of approaches proven to enhance synergies and reduce conflicts between biodiversity, food production and livelihood objectives, such as agroecology, permaculture, organic agriculture, agroforestry and "nature-inclusive" agriculture (Bouwma et al., 2019; Chapin et al., 2000; Chappell and LaValle, 2011; Runhaar, 2017; Scherr and McNeely, 2008). Climate change, the projected rise in global food demand and changing diets are projected to further increase pressures on food systems and land use (FAO, 2017a). The challenge for transformational policies is to disincentivize unsustainable practices while incentivizing biodiversity-friendly food production approaches. While healthy diets (Chapter 5) and animal welfare (Chapter 9) are also fundamental components of future food systems, this chapter focuses on governance of agricultural land use.

Conserving and enhancing biodiversity in agriculture is central to some of the most prominent international environmental agreements and conventions. The Convention on Biological Diversity (CBD) aims to ensure sustainable management and biodiversity conservation (Aichi Target 7 of the 2011–2020 Strategic Plan) and keep resource extraction within sustainable limits (Aichi Target 4). The impending Post-2020 CBD Global Biodiversity Framework (GBF), which is expected to be approved in 2022, is also expected to reflect the importance of sustainable agriculture. The importance of agricultural biodiversity has been reconfirmed by the 2015 United Nations Sustainable Development Goals (SDGs), particularly SDG15 (Life on Land), SDG2 (Zero Hunger) and SDG8 (Sustainable Production and Consumption). In 2017, the UN Framework Convention on Climate Change also initiated a work stream aiming to promote sustainable agricultural systems (UNFCCC, 2017).

Within these international conventions, as well as in national-level governance frameworks, an increasingly important way to promote biodiversity conservation in agricultural landscapes is through the *mainstreaming* of biodiversity[1] into public and private governance of the agricultural sector, a strategy that was specifically advocated in the CBD's 2011–2020 Strategic Plan. This chapter analyzes the progress in mainstreaming biodiversity into public and private sector agricultural policies worldwide by employing the concept of *biodiversity policy integration* (BPI). BPI analyzes the consideration of biodiversity in all sectors and levels of policymaking and implementation, providing a conceptual approach to identify leverage points for transformative change. In this chapter, we analyze BPI in agricultural landscapes, which adds to the toolbox of the transformative biodiversity governance framework. We review available literature on BPI in agricultural policies in developed countries (with a focus on the European Union [EU]) and developing countries (with a focus on tropical countries). Recognizing the important role of nonstate actors in biodiversity governance, we also include private sector governance in our analysis, defined here as rules and standards developed and monitored by firms or nongovernmental organizations (Grabs et al., 2020).

This chapter proceeds as follows. We first provide an overview of trends and threats to biodiversity, highlighting the necessity to integrate biodiversity in the governance and management of agricultural landscapes (Section 13.2). We then introduce our analytical approach (BPI) and how it relates to the broader literature on environmental policy integration and mainstreaming (Section 13.3), before analyzing to what extent and how biodiversity is integrated into agricultural governance in developed and developing countries (Section 13.4). Based on these analyses, we discuss four central leverage points for transformative biodiversity governance in agricultural landscapes and reflect them with the analytical dimensions of this book (Section 13.5), before concluding with key lessons (Section 13.6).

13.2 Current Trends and Key Threats to Biodiversity

This section focuses on two principal mechanisms through which agriculture impacts biodiversity: land use change for agricultural expansion and management choices on agricultural land – that is, intensification, specialization and enlargement of farms (Ramankutty et al., 2018). After introducing these issues within the broader contemporary debate, we discuss central arguments for segregated ("land-sparing") versus integrated ("land-sharing") approaches.

---

[1] Article 6b of the Convention on Biological Diversity (CBD) requires parties to "Integrate, as far as possible and as appropriate, the conservation and sustainable use of *biological diversity* into relevant sectoral or cross-sectoral plans, programmes and policies" (my emphasis).

13.2.1 Land Use Change

Land use change for the production of feed, fuel, biofuels and livestock is one of the major drivers of biodiversity loss (IPBES, 2019; MEA, 2005). Between 2000 and 2010, 80 percent of deforestation worldwide was directly attributable to the agricultural sector (Hosonuma et al., 2012). Agriculture currently occupies 38 percent of the world's terrestrial land surface, with about 12 percent devoted to crops and about 25 percent to livestock rearing and grazing (Foley et al., 2011). Of the area used for cereal production, 31 percent is devoted to animal feed (Mottet et al., 2017). Although land clearing has slowed since the 1950s relative to the previous century in temperate latitudes, it has shifted to tropical highly biodiverse forests in Latin America, Southeast Asia and Africa (IPBES, 2019; Ramankutty et al., 2018). In addition to loss of ecosystems and their intrinsic value, deforestation of biodiverse, tropical forests reduces carbon sinks, which are important for mitigating climate change (Bunker et al., 2005; IPCC, 2014).

The causes of agricultural expansion into intact ecosystems differ by region. In Africa, subsistence and small-scale farming drives the majority of expansion and deforestation (IPBES, 2019; Seymour and Harris, 2019). In contrast, deforestation in South America (particularly in the Amazon) and Southeast Asia is primarily driven by commercial agriculture supplying international markets, most notably since the 1990s (Hosonuma et al., 2012; IPBES, 2019; Seymour and Harris, 2019). Though the majority of agricultural commodities are consumed domestically, global trade of a select few agricultural commodities – notably soybeans (of which the majority is used for animal feed globally), beef and palm oil – is a major external driver of ecosystem loss (DeFries et al., 2013; Green et al., 2019; Henders et al., 2015; Meyfroidt et al., 2013). As a prominent example, oil palm plantations supplying global markets have been responsible for over 80 percent of agricultural land expansion in South Asia since the 1990s (Gibbs et al., 2010). Countries that consume these commodities are thus contributing to ecosystem and biodiversity loss, as recognized in recent attempts to reduce "imported deforestation" (Bager et al., 2021). The long-term effects of land use change are often underestimated as – particularly in biodiversity-rich regions – species continue to be lost even if the agricultural land has been abandoned (Gibson et al., 2011).

13.2.2 Management Choices

Agriculture has undergone significant structural changes since the Second World War. New farming practices falling under the paradigm of "industrial agriculture" were strongly subsidized by governments, particularly in developed countries and in some developing countries, as part of the "Green Revolution." This "agricultural modernization" relied heavily on mechanization, genetic alterations of crops (e.g. hybridization, genetically modified organisms) and the use of chemical inputs to increase productivity (Bosc and Belières, 2015; Duru et al., 2015). Three overarching and interrelated trends can be distinguished: intensification, specialization and scale enlargement (Aubert et al., 2019; Poux and Aubert, 2018).

Intensification refers to increasing productivity on a given parcel of land through the heavy use of inputs (such as pesticides and fertilizers). Though this may increase profits, and in some cases also food security, it generally drives biodiversity loss as it is currently practiced (Batáry et al., 2017; Hendershot et al., 2020; Rasmussen et al., 2018). Studies point to the detrimental impacts on biodiversity in general, and on soil biodiversity and insects in particular, especially through mechanization and pesticide use (see, for example, Orgiazzi et al., 2016; Sanchez-Bayo and Wyckhuys, 2019; Seibold et al., 2019; Tsiafouli et al., 2015). Globally, pesticide sales and use continue to increase, with hundreds of older generation pesticides that are highly toxic to vertebrates and invertebrates still being used in developing countries, although banned in many developed countries (Schreinemachers and Tipraqsa, 2012). Through run-off, pesticides and fertilizers also have biodiversity impacts reaching far beyond the farm (Beketov et al., 2013; Van Dijk et al., 2013; Yamamuro et al., 2019). Solutions related to increasing efficiency, such as precision agriculture, can contribute to sustainability and food security through the reduction of inputs (IPCC, 2019). However, recent work shows that implementation remains a problem (Lindblom et al., 2017). Moreover, such solutions do not address many of the underlying problems of conventional intensification, including the need for energy-intensive inputs (Kremen, 2015).

Secondly, *specialization* describes a shift away from diversified crop production to monocultures and a separation of crops and livestock systems. At the macro level, specialization is driven by the logic of economies of scale and the creation of regional or national comparative advantages in trade (Abson, 2019). As a prominent example, Brazil has developed a significant comparative advantage in soybean production by using soybeans as a "flex crop" with multiple processing pathways that differentiate the product into a food grain, livestock feed or fuel (Oliveira, 2016). However, these regional advantages come at a cost – extreme specialization of food and agriculture is a major driver of the decline in biodiversity at genetic, species and ecosystem levels (FAO, 2019; IPBES, 2019). While agronomic research and technical expertise have focused on the production of a few key staple crops (wheat, corn and rice initially, now followed by oilseeds, e.g. soybeans and rapeseed), technical knowledge on other crops remains low (FAO, 2019; Magrini et al., 2016). Furthermore, specialization conflicts with the idea of multifunctional production and its potential for contributing to food security (Bommarco et al., 2018; Misselhorn et al., 2012), climate-smart landscapes (Scherr et al., 2012) and viable farming income, despite potential trade-offs in efficiency (Lakner et al., 2018).

Lastly, *scale* enlargement entails a trend toward fewer but larger farms. Although there is still a wide variety of farm types and sizes around the world, a productivist ideology has led farms to increase in size overall in order to benefit from economies of scale, which enables cost reductions and helps farmers remain competitive (Duffy, 2009). This strategy is capital- and input-intensive, requiring high investments in machinery and chemical inputs that are only considered worthwhile if farm output is high, lowering costs per unit of production (McIntyre et al., 2009). Concentration across the agri-food industry, and the resulting control exerted by a small number of companies on farmers, has further encouraged a consolidation and enlargement trend (Folke et al., 2019; IPES-Food, 2017). Scale enlargement contributes to biodiversity loss principally through the destruction of seminatural landscape features, such as hedges, field margins and permanent prairies, which maintain heterogeneity and connectivity of habitats at the landscape level (Poux and Aubert, 2018; Tscharntke et al., 2012).

13.2.3 Land-Sharing and Land-Sparing in a Telecoupled World

For many decades, the dominant global discourse on food security has resulted in the notion that there is direct competition for land between biodiversity conservation and agricultural production and that the two are incompatible (Butler et al., 2007; Henle et al., 2008; Steffan-Dewenter et al., 2007; Tscharntke et al., 2012). This has led to a simplified framing in which "land-sparing" (segregating intensive agriculture from conservation lands) and "land-sharing" (more extensive agriculture that contributes to conservation) are viewed as a dichotomy, though neither of them singularly has the full potential to address the challenge of sustainable agriculture (Kremen, 2015). Instead, we argue that a combined approach of both large, protected regions *and* wildlife-friendly farming areas is critical to conserving biodiversity (Kremen, 2015; Kremen and Merenlender, 2018).

The land-sparing logic argues that effective biodiversity conservation on nonagricultural land (see Chapter 11) depends on the separation of agricultural land from protected areas, necessitating the intensification of production on agricultural land to "free up" land for conservation. However, since the effectiveness of protected areas correlates with the pressures from its surroundings (Kremen and Merenlender, 2018; Watson et al., 2014), conservation in these designated areas will still depend on the management of external or internal pressures. Therefore, the idea of completely separating the interactions between biodiversity conservation and agricultural production areas is conceptually flawed, as landscape structures are shaped by cultural dynamics and human–nature interactions, as well as geographical and climate conditions, making ecological and productive systems mutually interdependent (Fischer et al., 2011; 2014). In addition to localized detrimental impacts of intensive farming, the land-sparing approach can also have far-reaching impacts on biodiversity: Land-sparing in one area can have spill-over effects that drive relocation and expansion of production in other regions, rather than leading to an overall reduction of biodiversity threats (Meyfroidt, 2018; Meyfroidt et al., 2013; Rudel et al., 2009). Even in regions where the extension of agricultural land use remains relatively constant (such as within the EU), the "imported land" needed to satisfy consumer demand continues to grow (Asici and Acar, 2016, Teixidó-Figueras and Duro, 2014; Yu et al., 2013). This shows that consumption decisions and agricultural management in a globalizing world are "telecoupled" (Friies et al. 2016; Sun et al., 2017). Therefore, while protected areas remain crucial to maintaining biodiversity, the land-sparing approach requires policy integration.

In contrast, land-sharing recognizes agriculture as "both the greatest cause of biodiversity loss *and the greatest opportunity for conservation*" (Hendershot et al, 2020: 393, emphasis added). Land-sharing approaches recognize the need and potential for agricultural land to help protect biodiversity through a range of practices, as agricultural expansion and its (inadequate) management drive biodiversity loss. While this is a good idea in theory, the above-described trajectories show that land conversion and management choices continue to invade important ecosystems and fail to produce sound ecological structures. At the same time, the separation of sufficiently large areas seems necessary for the conservation of certain ecosystem values and habitats (Kremen and Merenlender, 2018; Watson et al., 2014).

Hence, while a conceptual separation of land-sparing and land-sharing can help to identify socio-ecological trade-offs, it has largely failed in identifying solutions for addressing them (Fischer et al., 2014). We argue that in transformative biodiversity governance, area-based (land-sparing) *and* integrated (land-sharing) approaches offer a complementary toolkit to address direct and indirect drivers of biodiversity loss in agricultural landscapes, and that biodiversity policy integration is crucial in both of these approaches.

13.3 Conceptual Framework for Biodiversity Policy Integration

Biodiversity policy integration (BPI) is an analytical tool derived from the broader literature of environmental policy integration (EPI) (Zinngrebe, 2018). EPI can be defined as "the incorporation of environmental objectives in non-environmental policy sectors such as agriculture, energy and transport" and can be considered transformative because of its "aim to target the underlying driving forces, rather than merely symptoms, of environmental degradation" (Persson et al., 2018: 113). Governance elements and processes that support EPI have been widely studied, particularly in European and OECD countries (see e.g. Jordan and Lenschow, 2010; OECD, 2018; Persson et al., 2018; Runhaar, 2016; Runhaar et al., 2014; 2018; 2020, Visseren-Hamakers, 2015). This literature shows that no single instrument can realize policy integration, but rather, EPI needs a suite of complementary instruments and mechanisms (Persson and Runhaar, 2018; Runhaar et al., 2020).

In this chapter, we use BPI as an analytical tool deriving from EPI literature, with a focus on biodiversity (Zinngrebe, 2018). To date, empirical analyses of policy integration between agriculture and biodiversity are scarce. A Web of Science search for the terms "agriculture" AND "policy integration" AND "biodiversity" resulted in six articles, all of which are included in the analysis in this chapter (Karlsson-Vinkhuyzen et al., 2017; 2018; Söderberg and Eckerberg, 2013; Somorin et al., 2016; Zinngrebe, 2018, Zinngrebe et al., 2017). Other combinations of search terms were also explored: "biodiversity" OR "mainstreaming biodiversity" AND "production landscapes," "agricultural policy," "coherence," "inclusion," "social capital" and "capacity." These also returned few hits of direct relevance that included concrete examples. Redford et al. (2015) note that publications by practitioners involved in public and private biodiversity mainstreaming programs and projects are severely deficient in the peer-reviewed literature, particularly those focused on developing countries. Therefore, to capture relevant gray literature, we also applied the following Google searches. "mainstreaming biodiversity" AND "production landscapes" (yields sixty-seven results) and "mainstreaming biodiversity" AND "agricultural policy" (yields ninety results). Titles and abstracts were screened to select relevant publications.

In order to analyze the extent to which biodiversity considerations have been incorporated in agricultural policies, we distinguish five dimensions of BPI (see Figure 13.1) (Zinngrebe et al., 2018; for similar approaches see Kivimaa and Mickwitz, 2006 and Uittenbroek et al., 2013):

1. *Inclusion*: the extent to which the objective of biodiversity conservation is included in political sectors. This is measured by the extent to which a sector has reframed a biodiversity objective into sector-specific targets and specific biodiversity indicators.
2. *Operationalization*: the extent to which a sector has adopted or adjusted policy instruments and monitoring and enforcement mechanisms to implement biodiversity objectives (see also Runhaar, 2016), and the uptake of biodiversity values in internal evaluation processes.
3. *Coherence*: the extent to which objectives and policy instruments within a sector complement rather than contradict each other. This is measured by the extent to which policies within a sector are internally consistent and direct sector activities toward biodiversity objectives.
4. *Capacity*: the level of institutional development, available resources and political mechanisms that ensure the implementation of instruments identified in the "operationalization" dimension, as well as the extent to which other actors are supported by their organization ("social capital") (Zinngrebe et al., 2020).
5. *Weighting*: the importance given to biodiversity objectives in relation to other political objectives. Weighting further analyzes whether biodiversity, as natural capital, is regarded as substitutable by other forms of capital and whether ecological limits are recognized.

In the next section, we use this analytical framework to analyze the current state of BPI in agricultural governance along the five dimensions. However, we note that while the BPI framework assesses the level of integration at a specific point in time, transformative governance is adaptive, requiring dynamic policy design and institutional reconfigurations to iteratively improve BPI performance. In Section 5, we draw on our BPI analysis to reflect on enabling factors and barriers and discuss them in relation to the transformative governance analytical framework of this book.

13.4 Taking Stock: Assessing the Level of Biodiversity Policy Integration in Agricultural Governance

13.4.1 Inclusion

In many developing countries with available studies, biodiversity is not an explicit target in agricultural policies (Zinngrebe, 2018; Zinngrebe et al., 2020). While most Parties to the CBD identify the need for both ex-situ and in-situ biodiversity conservation, only 3 percent have mainstreamed biodiversity in their agricultural policies, plans and programs (Lapena et al., 2016). Among the exceptions is Kenya, where the Ministry of Agriculture in Busia County has set a performance target for establishing a biodiversity policy (Hunter et al.,

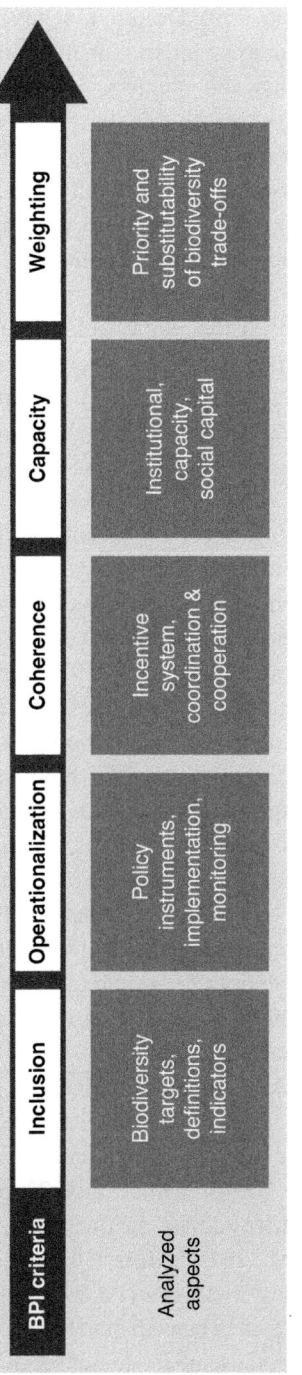

Figure 13.1 Five dimensions of biodiversity policy integration (reprinted from Zinngrebe, 2018).

2018). Similarly, Costa Rica has a biodiversity law setting general standards (although in rather generic terms) to also be considered in agricultural landscapes, which has been regarded as "one of the most comprehensive efforts to implement ... the Convention on Biological Diversity" (Miller, 2006: 359). Despite few government-led policy initiatives to advance BPI in developing countries, international organizations have been active in pushing for integrated instruments and planning procedures, which we include in the following sections.

In the EU, various policies have aimed to integrate biodiversity objectives into the agricultural sector to differing degrees. Most recently, the European Green Deal includes a "Farm to Fork" strategy that explicitly aims to reverse biodiversity loss by aiming for a "neutral or positive impact" within agri-food systems (EC, 2019; 2020a). As an additional element, the EU Biodiversity Strategy for 2030 includes area-based targets aimed at protecting 30 percent of its terrestrial area, with "at least 10 percent of utilized agricultural area under high diversity landscapes," and a life-cycle assessment assuming responsibility for outsourced environmental impacts as well as a reduction of the overall EU's global footprint (EC, 2020b, section 2.2.2). The key legal instruments underpinning the EU's conservation policies date back several decades: the Birds and Habitats Directives established the Natura 2000 network, which covers almost 18 percent of the EU's terrestrial surface area (Bouwma et al., 2019). Almost 90 percent of all Natura 2000 sites are subject to agriculture or forestry activities, making BPI highly relevant (Tsiafouli et al., 2013). The Habitats and Birds Directives do not, however, include targets or indicators related to land use systems or ecosystem services. Instead, they have the objective of maintaining healthy habitats for selected species (Bouwma et al., 2019). Similarly, the European Common Agricultural Policy (CAP) speaks more generally of "sustainable management of natural resources and climate action" in the 2013–2020 period and uses a farmland bird index and High Nature Value farmland index as proxies for biodiversity (EC, 2013). Since 2018, a proposal by the European Commission that includes a strategic objective on the protection of biodiversity, enhancement of ecosystem services and preservation of habitats and landscapes (Target F, EC, 2018) has been negotiated by EU institutions. While this proposal takes a comprehensive approach to envisioning sustainability in agriculture, the proposed indicators target farm management and land use in general and have been assessed as insufficient for monitoring biodiversity (Pe'er et al., 2020).

Overall, countries face challenges in translating international biodiversity targets into nationally determined targets (Chandra and Idrisova, 2011; Velázquez Gomar, 2014). In an analysis of 144 national biodiversity strategies and action plans (NBSAPs) developed by countries that signed the CBD, 72 percent of developing countries and 58 percent of developed countries acknowledge agriculture explicitly as a threat to biodiversity conservation (Whitehorn et al., 2019). Despite this, only 23 percent of the developing and 33 percent of the developed countries address the question of trade-offs between agriculture and conservation (Whitehorn et al., 2019). More tellingly, almost no national agricultural plan cross-references the countries' NBSAPs (Pe'er et al., 2019; Zinngrebe, 2018). This means that although these NBSAPs may be well developed by environmental ministries and include agriculture-related targets, these goals do not reach the actors they need to engage, such as agricultural ministries and the network of actors in the agricultural sector. In some agricultural policies, the need for considering "sustainability," the "environment" or certain land use practices are mentioned, but without linking it to specific ecological criteria or policy instruments (Zinngrebe, 2018).

13.4.2 Operationalization

The operationalization of biodiversity-related objectives into policies differs strongly between developing and developed countries. In many developing countries, operationalization of policy instruments is poorly executed (e.g. Carew-Reid, 2002; Huntley, 2014); regulatory frameworks are weak, poorly implemented or nonexistent (Huntley, 2014) and some countries have started to develop their environmental governance framework only in the past decade (e.g. Vijge, 2018). Nevertheless, some advancement in operationalization is visible, particularly in Latin America, including Costa Rica, Mexico, South Africa, Australia and Brazil (Harvey et al., 2008; Huntley, 2014; Somarriba et al., 2012).

Costa Rica made significant advancements in the institutionalization of payment for ecosystem services schemes, aimed at enhancing forest biodiversity on agricultural land (Sanchez-Azofeifa et al., 2007). However, these payment schemes are regarded as insufficiently funded in the long-term and to complement but not substitute regulatory interventions by governments (Schomers and Matzdorf, 2013; Wunder et al., 2008). In South Africa, the national Biodiversity Act sets bioregional plans, biodiversity assessments and biodiversity action plans as legal instruments for BPI operationalization at the regional spatial scale (Botts et al., 2020). Additionally, "conservation farming" is supported by stringent regulation, involvement of nongovernmental organizations and farmer communities, effective communication with farmers and scientific and technical support for farmers (Donaldson, 2012). In Brazil, operationalization focuses on specific tools such as national plans promoting agroecology and organic production (Biodiversity International 2016), an "agrobiodiversity index" assessing private sector performance (Tutwiler et al., 2017) and a national school food program mandating 30 percent of federal funds toward procurement from family farms using agroecological production approaches (Johns et al., 2013).

In the private sector, producers and companies have started responding to the demand for deforestation-free commodities. Initiatives such as the Consumer Goods Forum, Tropical Forest Alliance, the New York Declaration on Forests, the Amsterdam Declaration Partnership, various beef and soy moratoriums and voluntary commitments under the Business for Nature coalition are, however, nonbinding and coexist with nonsustainable policies (Stabile et al., 2020).

In Europe, the main biodiversity-related instruments of the 2014–2020 CAP are direct subsidies to farmers conditioned on fulfilling "greening" obligations (Ecological Focus Areas) and cross compliance, as well as voluntary agri-environmental and climate measures (AECMs). These specific "deep green measures" have been found to produce strong local impacts (Batáry et al., 2015; Pe'er et al., 2017). However, the weak performance of "greening" (Pe'er et al., 2016) and the low allocation of funding to AECMs are central arguments for identifying the CAP's toolbox as weak "green architecture" (Pe'er et al., 2019). The new Post-2020 CAP proposal will continue to link direct payments to weak, unspecific targets (similar to cross compliance), while allowing for EU member states to use voluntary "eco-schemes" to support specific landscape features (Pe'er et al., 2020). Simultaneously, area-based instruments linked to the EU Birds and Habitats Directives are being used. However, evaluations of Natura 2000 indicate that only about a third of the sites have developed specific management plans for biodiversity conservation and only 4 percent show an improvement of habitats (Bouwma et al., 2019; EEA, 2015). Literature suggests that effective implementation of Natura 2000 sites depends on a joint implementation with policies such as agri-environmental measures (Bouwma et al., 2019; Lakner et al., 2020).

13.4.3 Coherence

Even in cases where conservation is included as one of the targets in agricultural policies, and when policies have been appropriately reconfigured to achieve those targets, they may still run counter to specific biodiversity conservation policies in the environmental sector. Often, decisions about trade-offs between productivity and conservation are avoided or not explicitly addressed, and a patchwork of incoherent policies result in a lack of incentives for biodiversity-friendly farming.

One barrier to coherent agri-environmental policies is a lack of horizontal coherence, notably, a lack of coordination between ministries and agencies at the national level. Insights from Indonesia, Uganda, Peru and Honduras show that while different regulatory processes for agricultural landscapes exist for the governmental sphere and for sustainability markets in the private sector, they are incoherent and generally favor conventional practices, rather than biodiversity-sound management systems such as agroforestry (Zinngrebe et al., 2020). Even in Costa Rica, which has relatively strong environmental laws and regulations, incoherent policies have been reported (Brockett and Gottfried, 2002; Lansing, 2014). One general issue is that ministries of finance and planning – which generally hold decision-making power on large-scale investment allocations – are often not in regular consultation with the ministries responsible for biodiversity governance (Swiderska, 2002).

Besides a lack of horizontal coherence (i.e. between sectoral policies at one level of governance) there is also often a lack of vertical coherence (i.e. between national and subnational biodiversity strategies). Vertical coherence is especially pertinent in developing countries, since many are in the process of decentralizing their governance systems (Carew-Reid, 2002; Hunter et al., 2016; Swiderska, 2002). The few existing studies indicate that vertical integration across political levels for the implementation, enforcement and monitoring of biodiversity conservation in agricultural landscapes is generally low (e.g. Zinngrebe, 2018). Nevertheless, the example of local stakeholder networks in Ethiopia illustrated that despite low coherence at the national level, local collaboration can lead to coherent management approaches (Jiren et al., 2018). In Rwanda, the successes of watershed management plans in enabling dialogue and policy coordination across ministries of agriculture, fisheries and rural and social development at both local and national levels are another promising exception (FAO, 2017b). Based on selected case studies from countries within Africa and Latin America, the FAO (2017b) highlights that management models that take an ecosystem-based approach can serve as a lever for coordination, integration and synergies, though this has not been sufficiently applied to improve coherence. In South Africa for instance, bioregional plans enhance both coherence in local land use planning and across core sectoral strategies at the national level (Botts et al., 2020). Deliberations in trade-off options between conservation and other goals is part of the planning process for this purpose (Redford et al., 2015). The international Biodiversity for Food and Nutrition Project, funded by the Global Environment Facility, shows how, in Brazil, Kenya, Turkey and Sri Lanka, a sound evidence-base on how biodiversity supports nutritional outcomes, and the establishment of multistakeholder and multisectoral steering committees, improves coherence across agriculture and food policies (Beltrame et al., 2016; 2019).

The EU is a strong advocate of policy coherence across sectors, as acknowledged in a large number of official EU documents. However, while most EU policies are coherent at the level of objectives, they provide incoherent incentives at the implementation stage, and therefore have not managed to effectively or efficiently reverse declining biodiversity trends (Pe'er et al., 2017). For example, while the EU Birds and Habitats Directives aim to conserve biodiversity, the CAP's fundamental targets, defined by the Treaty of Rome in 1957, direct agricultural policy toward increased productivity, low food prices and supporting farmers' incomes. Another example of incoherence in the CAP is the aforementioned Ecological Focus Areas, which obligates each farm of more than fifteen hectares to dedicate 5 percent of its land to conservation activities. In reality, this instrument primarily results in measures with a low contribution to biodiversity, such as catch crops and nitrogen-fixing crops (Cole et al., 2020; Pe'er et al., 2017). Watering down ecological standards in federal implementation processes, as well as misconceptions about farmers' motivations to engage in biodiversity conservation, reduce the CAP's potential to contribute to conservation (Brown et al., 2020). In the EU proposal for a post-2020 CAP (EC, 2018), direct payments will continue to dominate and low ecological targets continue to persist (Pe'er et al., 2020). Overall, studies show that despite the EU's rhetoric for policy coherence, large inconsistencies in the instruments and implementation of EU policies remain (De Schutter et al., 2020; Nilsson et al., 2012).

Within the EU, there are also strong calls for enhancing coherence of EU policies with non-aid policies that impact developing countries. These calls have grown since the 1990s, when Europe's need for agricultural biodiversity and production land substantially increased and was therefore transferred to other parts of the world. This policy blind-spot results in the EU's contribution to tropical deforestation and biodiversity loss in developing countries (Fuchs et al., 2020). However, while the EU and member states such as Denmark, the Netherlands, Sweden and the UK (which was an EU member at the time of analysis) have tested approaches for policy coherence for development, implementation performance has been weak (Carbone, 2008; see also Pendrill et al., 2019). Civil society actors have created a proposal to streamline EU policies into a "Common Food Policy" for Europe (De Schutter et al., 2020; IPES-Food, 2019). Blueprints describe an integrated food policy framework that promotes healthy diets and sustainable food systems through coherence across policy areas and governance levels, including by aiming to relocalize food production and to reduce dependence on global food imports (De Schutter et al., 2020; IPES-Food, 2019). It remains to be seen to what extent the integrated approach of the European Green Deal, and its "Farm to Fork" strategy, can translate such suggestions into practice.

13.4.4 Capacity

While there is generally higher institutional capacity in developed countries relative to developing countries, the aforementioned division between the institutional processes of the environmental and agricultural sectors undermines social capital for BPI in most countries.

In developing countries, the capacities to develop biodiversity (and other environmental) policies are limited to environmental ministries or departments. In Indonesia, Uganda, Honduras and Peru, social capital and capacities for training, financial support and regulation exist, but are not targeted at ecologically sound forms of production (Zinngrebe et al., 2020). The availability of institutional capacities is further undermined by unclear mandates between government agencies, high turnover among government officials resulting in discontinuous policy formulation and execution, and a lack of experienced biodiversity research institutions or centers of excellence (Zinngrebe, 2018; Zinngrebe et al. 2020). In the public policy arena, there is a lack of knowledge on and awareness of the linkages between biodiversity and agriculture or food security (Beltrame et al., 2016; Chandra and Idrisova, 2011). This is largely due to lack of training, funding, incentives for experts to work in the environmental field (Chandra and Idrisova, 2011), biodiversity-focused science–policy interfaces, and institutionalized mechanisms for the participation of Indigenous Peoples and local communities (which hold critical local ecological knowledge) in monitoring, reporting and verification initiatives (Vanhove et al., 2017). Mexico tackles these issues via multistakeholder roundtables, consisting of agricultural, rural development and research agencies, Secretaries of States, academia, NGOs and private actors, which coordinate sector activities, financing and science-policy mechanisms at the national and state level (Tutwiler et al., 2017). In Uganda, the agricultural ministry, under the direction of the Ministry of Finance, Planning and Economic Development, has to allocate a portion of their budget to conservation activities (IIED, UNEP-WCMC, 2015). Their staff receive training and a dedicated conservation expert from the environmental ministry to help prepare plans, while policy actors use learning lessons from the ground to inform the national macroeconomic framework (IIED, UNEP-WCMC, 2015). In South Africa, implementation of the Biodiversity Act is supported by pilot projects, regular monitoring and a national science-policy institute and multiagency committees, which align partnerships and cofinancing (Botts et al., 2020).

Within the EU, implementation of agricultural and biodiversity policies is supported by institutions at the European, national and subnational levels. However, lack and variance of capacity among different members states has also been identified as a barrier to implementation of agricultural policy proposals that contribute to environmental protection (Erjavec et al., 2018). Political decision-making and implementation processes of theoretically synergistic policies are designed and implemented by separated policy regimes (Pe'er et al., 2020), undermining social capital and potential synergies. Capacity problems are further enhanced by budgetary imbalances between agricultural and environmental instruments. Although the CAP is the EU policy with the highest budget (€58.4 billion in 2020), the majority of this is dedicated to direct income support. As a result, most of the budget in the 2015–2020 CAP (approximately €40 billion in 2017) was spent on direct payments that support land-intensive and biodiversity-threatening forms of farming, such as intensive animal breeding and monocultures (Pe'er et al., 2019). Furthermore, though Natura 2000 has demonstrated improvements in biodiversity within agricultural areas, funding per hectare is considerably lower than for greening or agri-environment climate measures (Pe'er et al. 2017), hardly compensating farmers for resulting costs from forgone incomes due to management restrictions and lower rents, and thus not providing sufficient incentive for adoption by farmers (Bouwma et al., 2019). Additionally, contradictory technical advice by agricultural extension services and administrative hurdles have hampered effective implementation of biodiversity measures (Zinngrebe et al., 2017).

13.4.5 Weighting

Even where biodiversity policy objectives are present and have been operationalized through concrete instruments with allocated capacity, political discourses are dominated by productivist narratives. The political framing in which food production must increase above all else provides little incentive to phase out agricultural subsidies that support the dominant model but are harmful to biodiversity (Bouwma et al., 2019; Fouilleux et al., 2017; Roche and Argent, 2015). In 2015, OECD countries provided $100 billion in direct and indirect subsidies that stimulated intensive agricultural production (OECD, 2019: 73). Although certification and other schemes are partly driving growth in organic and sustainable practices, the overwhelming policy bias and dominance of conventional agricultural methods gives these practices limited scope for truly scaling-up (Aubert et al., 2018).

In developing countries, both policies and politics also prioritize agricultural intensification and expansion (Wilson and Rigg, 2003; Zinngrebe et al., 2020). Biodiversity narratives in Peru show that even conservationists do not dare to talk about limits to production carrying-capacity. Adverse impacts on ecological functionality and related pollution and water-management issues remain untargeted key drivers for biodiversity loss (Zinngrebe, 2016a; 2016b). Another example is China, where, though the Law of Agriculture provides for wetlands conservation, the priority is placed on the draining and cultivation of wetlands for food security, resulting in lower priority and trade-offs for biodiversity (Ongley et al., 2010). Despite successful instruments for supporting agrobiodiversity and integrated natural resource management, agricultural expansion and intensification dominates decision-making considerations (Laurance et al., 2014).

Similarly, in the EU, the political discourse and resulting policies are oriented toward increasing productivity for human nutrition (Erjavec et al., 2009; Freibauer et al., 2011; IPES-Food, 2019). Despite the emergence of new discourse elements targeting multifunctionality and liberal markets, central policy elements support productivity (Alons and Zwaan, 2016; Erjavec and Erjavec, 2015). Following this policy design, even the implementation of conservation mechanisms, such as Ecological Focus Areas, is biased toward measures supporting increased productivity of agricultural lands (e.g. cash crops and nitrogen-fixing crops) (Pe'er et al., 2016). This is one of the stated reasons for why the CAP has not managed to reverse biodiversity loss (Pe'er et al., 2017). Some argue that the CAP is also not likely to do so in the near future, considering the content of current proposals for a post-2020 CAP (Pe'er et al., 2019). This strongly conflicts with the European Green Deal, which explicitly aims to halt biodiversity loss due to agriculture (EC, 2019).

13.5 Looking Forward: Toward Transformative Biodiversity Governance in Agricultural Landscapes

The previous section highlighted the overall very modest advances of BPI in agricultural landscapes. Given that the majority of global and national biodiversity targets are vague and the agricultural sector is not held accountable for its biodiversity performance, there is little guidance for investments in operationalization and capacity-building. Likewise, biodiversity policies are mostly "added on" to regulations of agricultural landscapes, receiving a low share of support compared to that for conventional farming systems focused on productivity. Given the significant agri-food system lock-ins and incumbent power dynamics, more effective BPI will not be implemented spontaneously – rather, the required shifts will need leadership at various levels (Oliver et al., 2018; Runhaar et al., 2020). We argue that *political will* is required as a key driving force to overcome lock-ins and improve BPI performance (see Figure 13.2). In the following paragraphs, we present four central leverage points specifying the dimensions for the transformation of biodiversity governance for agricultural landscapes.

A first transformative factor is the creation of a coherent *sustainability vision based on inclusive biodiversity governance*, which will guide implementation and induce accountability among implementing agents. As we showed in the previous section, the BPI dimensions of *inclusion* and *coherence* suffer from a lack of clear orientation, and the *weighting* is geared toward specific production-oriented interests. Decisions on agricultural policy are often dominated by small but well-organized interest groups that marginalize values of biodiversity conservation and downplay societal mandates such as the biodiversity targets under the CBD (Brown et al., 2020, Pe'er et al., 2019). Stakeholder groups differ in the way they envision appropriate use of land and nature, leading to different, often disconnected, discourses that are not equally reflected in policy design and implementation processes (Velázquez Gomar, 2014; Zinngrebe, 2016a). Questions of accountability and legitimacy of planning will depend on the extent to which potentially conflicting values are acknowledged and diverse value systems and perceptions are reflected in democratic planning and participatory implementation processes

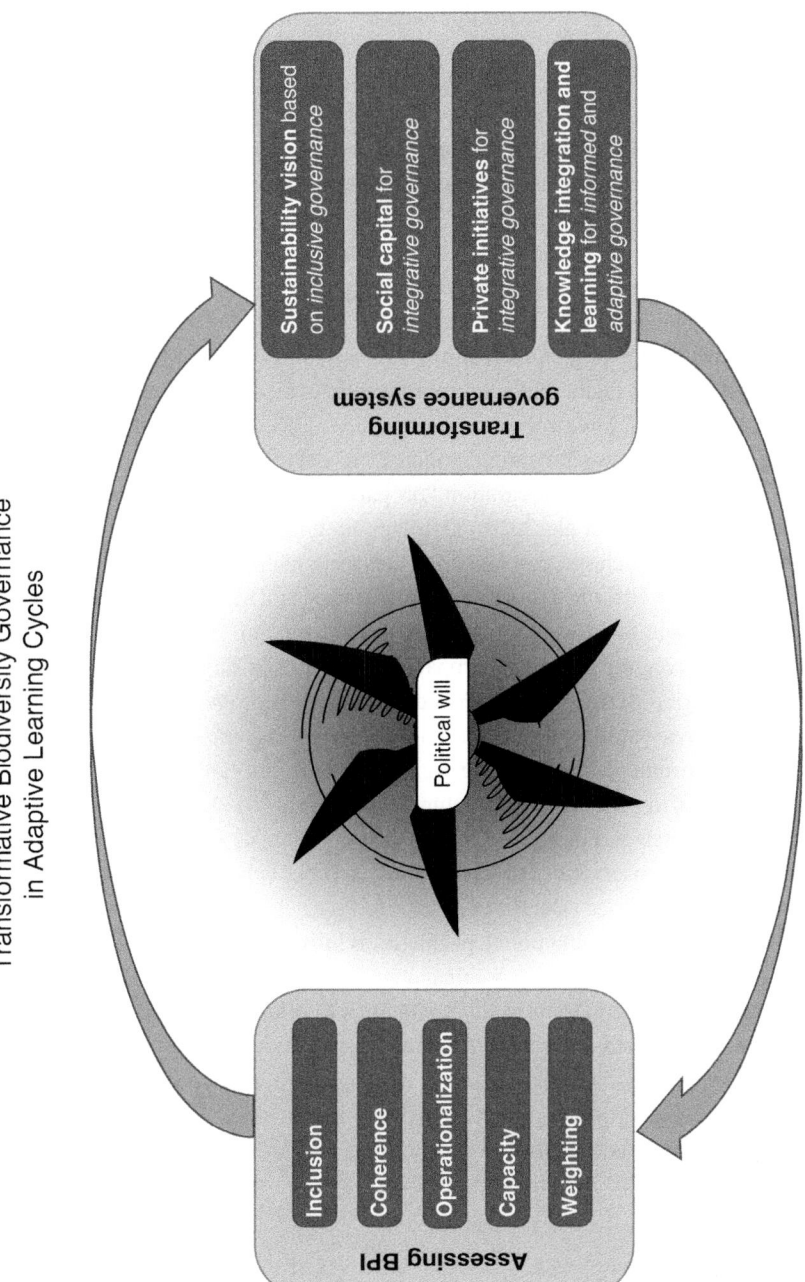

Figure 13.2 Improving the BPI level through transformative governance in adaptive learning circles.

(Díaz et al., 2018; Runhaar et al., 2020; Termeer et al., 2013; Zinngrebe, 2016b). Likewise, a positive perspective of what "sustainable agricultural landscapes" entail in a given context helps to orient the decisions and activities of political and nonpolitical actors. There are various alternatives to the dominant productivist model, including agroecology, sustainable intensification, agroforestry, and "nature-inclusive" agriculture (Brouder et al., 2015; IPCC, 2019; Loos et al., 2014; Perfecto and Vandermeer, 2010; Plieninger et al., 2020; Tscharntke et al., 2012; van Noordwijk, 2019; Zinngrebe et al., 2020). Agroforestry, as a specific example of an agroecological approach, has the potential to support ecosystem functions and biodiversity in both developed (Torralba et al., 2016) and developing countries (van Noordwijk, 2019). More concretely, objectives can be formulated around agroecological infrastructure such as hedges, trees and other seminatural habitats that protect multiple taxonomic groups and ecosystem services (Barrios et al., 2018; Fagerholm et al., 2016; Gonthier et al., 2014; Plieninger et al., 2019; 2020; Poux and Aubert, 2018; Torralba et al., 2018). Scenarios form an effective method for a participatory visioning process involving policymakers and other actors to deliberate options for land use and assess their implications for food security within a land-constrained world facing climate change (e.g. Aubert et al., 2019).

A second transformative factor that gives more weight to biodiversity in decision-making on trade-offs is *social capital for integrative governance*. Especially in developing countries, institutional *capacities* for implementing policies are severely lacking and often result in institutional gaps between policy integration "on paper" and the implementation of concrete policy instruments (Runhaar et al., 2020). Overlapping and unclear competences also create "responsibility gaps" in which no actor actually takes leadership in regulation or wider governance (Sarkki et al., 2016). Efforts to improve mainstreaming and fill these gaps have not resulted in institutional reconfigurations favoring effective implementation (Herkenrath, 2002; Prip and Pisupati, 2018). However, environmental impact assessments of large agricultural projects, or approval and monitoring of agroforestry concessions, can improve the operationalization of conservation objectives (Slootweg and Kolhoff, 2003; Zinngrebe, 2018). In Europe, both agricultural and environmental policies are well developed, but not institutionally connected in decision-making and implementation structures (Pe'er et al., 2019). Involving farmers in local implementation processes and partnerships with conservationists is an important strategy for improving biodiversity conservation leadership and outcomes in both developing (Harvey et al., 2008) and developed countries (Buizer et al., 2016; Pe'er et al., 2019; Persson et al., 2016). A collaborative process of aligning policy packages of information, regulation and finance can help overcome fragmentation between political actors and produce coherent incentive systems for conservation practices (Zinngrebe et al., 2020). Such a collaborative process should not only advance top-down implementation of (inter)national regulatory frameworks, but also cover a diverse range of locally based agricultural management practices. The IPBES Global Assessment (2019), for example, highlights a wide number of studies documenting the importance of small agricultural landholdings[2] in contributing to biodiversity conservation in different ecosystems (Batáry et al., 2017; Belfrage et al., 2015; Fischer et al., 2008).

---

[2] In this case, defined as under two hectares.

A third point of leverage is harnessing *private initiatives for integrative governance*. Private sector and market-based mechanisms can help with *operationalization*, provide new sources for institutional *capacity*, and increase *coherence* with farming interests (see Chapter 5). Engaging private actors is critical, particularly due to the rise and extent of private governance in the agricultural sector globally. Private actors can help incentivize biodiversity-friendly agriculture through various market opportunities, finance mechanisms, and public–private partnerships and other cooperative mechanisms. For example, numerous cases of the landscape approach have shown cooperation between governmental and private actors, such as co-funding from corporate actors in the maintenance of ecosystem services (Van Oosten, 2013). Private agricultural standards (including voluntary programs, such as various organic certifications) have become an integral part of agri-food chain governance (Henson and Reardon, 2005; Verbruggen and Havinga, 2017). Sustainability certifications (potentially) open new markets (FAO, 2017b) and provide opportunities for the scaling-up of environmental sustainability criteria, including for biodiversity (Runhaar et al., 2017). Particularly in countries that import large quantities of agricultural goods with high biodiversity impacts, government procurement of certified agricultural products can support and incentivize private sector actors in achieving biodiversity goals (Fransen, 2018). The use of economic instruments by firms, such as payment for ecosystem services, can also help provide financial incentives for other actors to engage in biodiversity-friendly farming and production processes (Donaldson, 2012; Harvey et al., 2008; Sanchez-Azofeifa et al., 2007).

However, to improve biodiversity outcomes, private initiatives need to be accompanied by political regulation and cooperation between private and public actors (Folke et al., 2019, Lambin et al., 2018; Runhaar et al., 2017; 2020). So far, land use change and management choices exercised by powerful transnational corporations have had a range of detrimental consequences for biodiversity (Folke et al., 2019). In the agri-food sector, consolidation is extremely high among corporations controlling fertilizers, agrochemicals and seeds, as well in the production of specific commodities such as coffee, bananas, soy, palm oil and cocoa (Folke et al., 2019). Private initiatives and certification schemes connecting consumer support for sustainable production systems have not yet proven effective in reversing detrimental environmental impacts (Dietz et al., 2019; Lambin et al., 2018; Pendrill et al., 2019). Experiences with green certification show that private standards need to be complemented with adequate regulatory frameworks to avoid deforestation and other detrimental effects to biodiversity, while simultaneously providing sufficient economic incentives for farmers (Dietz et al., 2019; Lambin et al., 2018).

Knowledge integration and learning for informed and adaptive governance is necessary to develop context-specific policy solutions for complex societal challenges. This can help in identifying suitable strategies for *operationalization* and (targeted) *capacity*-building. Experiences in participatory land use planning have shown how different knowledge systems can be integrated at the community level to build adaptive capacity and adopt more sustainable land use practices (Rodríguez et al., 2018). While the EU has a wide range of instruments for conservation in agricultural landscapes, it does not yet use all available knowledge to inform the improvement of these instruments from one funding period to the next (Pe'er et al.,

2020). Social capital can facilitate the input and reflection of available knowledge (Zinngrebe et al., 2020). Policy learning based on available experiences has the potential for overcoming complete policy failure and fragmentation (Feindt, 2010; Zinngrebe, 2018). Feindt (2010) argues that stronger institutionalized support for policy integration, balanced representation and wider societal engagement is needed to hold back powerful actors from dominating the policy arena to defend the status quo. Certain levels of flexibility and a complementary structure of CAP support and Natura 2000 instruments have shown synergistic effects in increasing the willingness of farmers to adopt conservation measures (Lakner et al., 2020). In addition, the integration of local knowledge has been shown to improve both farmers' engagement in reflexive learning processes and policy performance, in the EU context on the CAP's agri-environmental measures (Goldman et al., 2007; Prager et al., 2012) and in developing countries, for example in the context of conservation farming in South Africa (Donaldson, 2012) or in Mesoamerican landscapes (Harvey et al., 2008).

13.6 Conclusion

Low levels of biodiversity policy integration in agricultural policy in both developing and developed countries is a determining factor in the continued biodiversity loss within agricultural landscapes and beyond. While land-sparing approaches have proven to be indispensable for the conservation of certain components of biodiversity (Le Saout et al., 2013; Watson et al., 2014), a more integrated land-sharing approach is necessary to enable a transformation of current trajectories toward sustainable farming, in order to bend the curve of biodiversity loss while also ensuring food security, climate resilience, enhanced animal welfare and improved rural livelihoods.

With the exception of EU policies, in most countries, specific biodiversity-related objectives are missing in agricultural policies. Worldwide, the underlying drivers of biodiversity loss from agriculture are not sufficiently addressed. In particular, the objective of phasing out policies supporting threats to biodiversity and a strongly productivist-oriented agricultural sector overpowers the idea of sustainable agriculture. Instead of coherent targets and complementary institutional structures, conservation has generally been treated as an add-on to business-as-usual agricultural policy. Trade-offs considering biodiversity and ecological limits are seldom explicitly recognized in agricultural policies, and no country expresses a long-term vision for the development of sustainable agricultural landscapes. Political discourses remain centered on prioritizing intensive food production, thereby marginalizing the potential functions of agricultural landscapes for biodiversity conservation. Based on our BPI analysis, we extract the following recommendations for transformative biodiversity governance:

1. Inclusive governance needs to genuinely incorporate multiple stakeholder views and perceptions, and negotiate and develop clear, coherent visions and definitions of sustainable agriculture to legitimate policies and decision-making.

2. Integrative governance can be improved by building social capital as a means to creating favorable actor constellations and institutional structures incentivizing and prioritizing biodiversity-sound practices.
3. Integrative governance can benefit from complementing public and private initiatives in coherent governance structures.
4. Informed and adaptive governance requires a continuous and participatory reflection of governance systems to guide institutional learning processes toward sustainable agricultural landscapes.

We argue that the Post-2020 Global Biodiversity Framework should focus on the transformation of agricultural governance systems by concretely addressing key leverage points and providing specific guidance for member states to address country-specific drivers and potentials for sustainable innovation through biodiversity policy integration. Eventually, however, the dynamic of this transformative process will be conditioned by political will and active leadership at all levels.

References

Abson, D. J. (2019). The economic drivers and consequences of agricultural specialization. In *Agroecosystem diversity: Reconciling contemporary agriculture and environmental quality*. G. Lemaire, S. Kronberg, S. Recous and P. C. de Faccio Carvalho (Eds.), pp. 301–315. London: Academic Press.

Alons, G., and Zwaan, P. (2016). New wine in different bottles: Negotiating and selling the CAP post-2013 Reform. *Sociologia Ruralis* 56, 349–370.

Asici, A. A., and Acar, S. (2016). Does income growth relocate ecological footprint? *Ecological Indicators* 61, 707–714.

Aubert, P. M., Hege, E., Kinniburgh, F., et al. (2018). *Identification of solutions and first set of draft scenarios to strengthen the sustainability of European primary producers*. Brussels: Institute for Sustainable Development and International Relations (IDDRI) – Sustainable Finance for Sustainable Agriculture and Fisheries Project.

Aubert, P. M., Schwoob, M. H., and Poux, X. (2019). *Agroecology and carbon neutrality in Europe by 2050: What are the issues? Findings from the TYFA modelling exercise*. Institute for Sustainable Development and International Relations (IDDRI), Study N°02/2019. Available from https://bit.ly/3rDbU5R.

Bager, S. L., Persson, U. M., and dos Reis, T. N. (2021). Eighty-six EU policy options for reducing imported deforestation. *One Earth* 4, 289–306.

Barrios, E., Valencia, V., Jonsson, M., et al. (2018). Contribution of trees to the conservation of biodiversity and ecosystem services in agricultural landscapes. *International Journal of Biodiversity Science, Ecosystem Services & Management* 14, 1–16.

Batáry, P., Dicks, L. V., Kleijn, D., and Sutherland, W. J. (2015). The role of agri-environment schemes in conservation and environmental management. *Conservation Biology* 29, 1006–1016. https://doi.org/10.1111/cobi.12536

Batáry, P., Gallé, R., Riesch, F., et al. (2017). The former Iron Curtain still drives biodiversity–profit trade-offs in German agriculture. *Nature Ecology and Evolution* 1, 1279–1284.

Beketov, M. A., Kefford, B. J., Schäfer, R. B., and Liess, M. (2013). Pesticides reduce regional biodiversity of stream invertebrates. *Proceedings of the National Academy of Sciences* 110, 11039–11043.

Belfrage, K., Björklund, J., and Salomonsson, L. (2015). Effects of farm size and on-farm landscape heterogeneity on biodiversity – Case study of twelve farms in a Swedish landscape. *Agroecology and Sustainable Food Systems* 39, 170–188.

Beltrame, D., Eliot, G. E. E., Güner, B., et al. (2019). Mainstreaming biodiversity for food and nutrition into policies and practices: Methodologies and lessons learned from four countries. *ANADOLU Ege Tarımsal Araştırma Enstitüsü Dergisi* 29, 25–38.

Beltrame, D. M. O., Oliveira, C. N. S., Borelli, T., et al. (2016). Diversifying institutional food procurement – Opportunities and barriers for integrating biodiversity for food and nutrition in Brazil. *Raizes* 36, 55–69.

Biodiversity International. (2016). *Biodiversity for Food and Nutrition Initiative: Country Case-Study – Brazil*. Available from www.b4fn.org/countries/brazil.

Bommarco, R., Vico, G., and Hallin, S. (2018). Exploiting ecosystem services in agriculture for increased food security. *Global Food Security* 17, 57–63.

Bosc, P.-M., and Bélières, J. F. (2015). Transformations agricoles: un point de vue renouvelé par une mise en perspective d'approches macro et microéconomiques. *Cahiers Agricultures* 24, 206–214.

Botts, E., Skowno, A., Driver, A., et al. (2020). More than just a (red) list: Over a decade of using South Africa's threatened ecosystems in policy and practice. *Biological Conservation* 246, 108559.

Bouwma, I., Zinngrebe, Y., and Runhaar, H. (2019). Nature conservation and agriculture: Two EU policy domains that finally meet? In *EU bioeconomy economics and policies: Volume II*. L. Dries, W. Heijman, R. Jongeneel, K. Purnhagen and J. Wesseler (Eds.), pp. 153–175. Cham: Palgrave Macmillan.

Brockett, C. D., and Gottfried, R. R. (2002). State policies and the preservation of forest cover: Lessons from contrasting public-policy regimes in Costa Rica. *Latin American Research Review* 37, 7–40.

Brouder, P., Karlsson, S., and Lundmark, L. (2015). Hyper-production: A new metric of multifunctionality. *European Countryside* 7, 134–143.

Brown, C., Kovács, E., Herzon, I., et al. (2020). Simplistic understandings of farmer motivations could undermine the environmental potential of the Common Agricultural Policy. *Land Use Policy* 101, 105–136.

Bunker, D. E., DeClerck, F., Bradford, J. C., et al. (2005). Species loss and aboveground carbon storage in a tropical forest. *Science* 310, 1029–1031.

Buizer, M., Arts, B., and Westerink, J., 2016. Landscape governance as policy integration "from below": A case of displaced and contained political conflict in the Netherlands. *Environment and Planning C: Government and Policy* 34, 448–462.

Butler, S. J., Vickery, J. A., and Norris, K. (2007). Farmland biodiversity and the footprint of agriculture. *Science* 315, 381–384.

Carbone, M. (2008). Mission impossible: The European Union and policy coherence for development. *European Integration* 30, 323–342.

Carew-Reid, J. (2002). *Biodiversity planning in Asia: A review of national biodiversity strategies and action plans (NBSAPs)*. Gland; Cambridge: IUCN.

Chandra, A., and Idrisova, A. (2011). Convention on biological diversity: A review of national challenges and opportunities for implementation. *Biodiversity and Conservation* 20, 3295–3316.

Chapin, F. S., Zavaleta, E. S., Eviner, V. T., et al. (2000). Consequences of changing biodiversity. *Nature* 405, 234–242.

Chappell, M. J., and LaValle, L. A. (2011). Food security and biodiversity: Can we have both? An agroecological analysis. *Agriculture and Human Values* 28, 3–26.

Cole, L .J., Kleijn, D., Dicks, L. V., et al. (2020). A critical analysis of the potential for EU Common Agricultural Policy measures to support wild pollinators on farmland. *Journal of Applied Ecology* 57, 681–694.

De Schutter, O., Jacobs, N., and Clément, C. (2020). A "common food policy" for Europe: how governance reforms can spark a shift to healthy diets and sustainable food systems. *Food Policy* 96, 101849.

DeFries, R., Herold, M., Verchot, L., Macedo, M. N., and Shimabukuro, Y. (2013). Export-oriented deforestation in Mato Grosso: Harbinger or exception for other tropical forests? *Philosophical Transactions of the Royal Society B: Biological Sciences* 368, 20120173.

Díaz, S., Pascual, U., Stenseke, M., et al. (2018). Assessing nature's contributions to people. *Science* 359, 270–272. https://doi.org/10.1126/science.aap8826

Dietz, T., Estrella Chong, A., Grabs, J., and Kilian, B. (2019). How effective is multiple certification in improving the economic conditions of smallholder farmers? Evidence from an impact evaluation in Colombia's coffee belt. *The Journal of Development Studies* 5, 1141–1160.

Donaldson, J. S. (2012). Biodiversity and conservation farming in the agricultural sector. In *Mainstreaming biodiversity in development: Case studies from South Africa*. S. M. Pierce, R. M. Cowlings, T. Sandwith and K. MacKinnon (Eds.), pp. 43–55. Washington, DC: The World Bank.

Duffy, M. (2009). Economies of size in production agriculture. *Journal of Hunger and Environmental Nutrition* 4, 375–392.

Duru, M., Therond, O., and Fares, M. (2015). Designing agroecological transitions: A review. *Agronomy for Sustainable Development* 35, 1237–1257.

Erjavec, E., Lovec, M., Juvančič, L., Šumrada, T., and Rac, I. (2018). *Research for AGRI Committee – The CAP strategic plans beyond 2020: Assessing the architecture and governance issues in order to achieve the EU-wide objectives*. Brussels: European Parliament, Policy Department for Structural and Cohesion Policies.

Erjavec, K., and Erjavec, E. (2015). "Greening the CAP" – Just a fashionable justification? A discourse analysis of the 2014–2020 CAP reform documents. *Food Policy* 51, 53–62.

Erjavec, K., Erjavec, E., and Juvančič, L. (2009). New wine in old bottles: Critical discourse analysis of the current common EU agricultural policy reform agenda. *Sociologia Ruralis* 49, 41–55.

European Commission (EC). (2013). *Regulation (EU) No 1306/2013 of the European Parliament and of the Council of 17 December 2013 on the financing, management and monitoring of the common agricultural policy and repealing Council Regulations (EEC) No 352/78, (EC) No 165/94, (EC) No 2799/98, (EC) No 814/2000, (EC) No 1290/2005 and (EC) No 485/2008*. Brussels: European Commission.

(2018). *Proposal for a Regulation of the European Parliament and of the Council establishing rules on support for strategic plans to be drawn up by Member States under the Common agricultural policy (CAP Strategic Plans) and financed by the European Agricultural Guarantee Fund (EAGF) and by the European Agricultural Fund for Rural Development (EAFRD)*. COM (2018) 392. Brussels: European Commission. Available from https://bit.ly/34Y4baI.

(2019). *Communication from The Commission to The European Parliament, The European Council, The Council, The European Economic and Social Committee and The Committee of The Regions – The European Green Deal*. Brussels: European Commission.

(2020a). *Communication from The Commission to The European Parliament, The European Council, The Council, The European Economic and Social Committee and The Committee of The Regions – the Farm to Fork Strategy. For a fair, healthy and environmentally-friendly food system*. Brussels: European Commission.

(2020b). *Communication from The Commission to The European Parliament, The European Council, The Council, The European Economic and Social Committee and The Committee of The Regions – EU Biodiversity Strategy for 2030 Bringing nature back into our lives*. Brussels: European Commission.

European Environmental Agency (EEA). (2015). *State of nature in the EU. Results from reporting under the nature directives 2007–2012*. Luxembourg: Publications Office of the European Union. Available from www.eea.europa.eu/publications/state-of-nature-in-the-eu.

Fagerholm, N., Torralba, M., Burgess, P. J., and Plieninger, T. (2016). A systematic map of ecosystem services assessments around European agroforestry. *Ecological Indicators* 62, 47–65.

Feindt, P. H. (2010). Policy-learning and environmental policy integration in the Common Agricultural Policy, 1973–2003. *Public Administration* 88, 296–314.

Fischer, J., Abson, D. J., Butsic, V., et al. (2014). Land sparing versus land sharing: Moving forward. *Conservation Letters* 7, 149–157.

Fischer, J., Batáry, P., Bawa, K. S., et al. (2011). Conservation: Limits of land sparing. *Science* 334, 593.

Fischer, J., Brosi, B., Daily, G. C., et al. (2008). Should agricultural policies encourage land sparing or wildlife-friendly farming? *Frontiers in Ecology and the Environment* 6, 380–385.

Foley, J. A., Ramankutty, N., Brauman, K. A., et al. (2011). Solutions for a cultivated planet. *Nature* 478, 337–342.

Folke, C., Österblom, H., Jouffray, J. B., et al. (2019). Transnational corporations and the challenge of biosphere stewardship. *Nature Ecology and Evolution* 3, 1396–1403.

Food and Agriculture Organization of the United Nations (FAO). (2017a). *The future of food and agriculture trends and challenges. Annual report.* Rome: FAO. Available from www.fao.org/3/i6583e/i6583e.pdf.

(2017b). *Landscapes for life: Approaches to landscape management for sustainable food and agriculture.* Rome: FAO. Available from: www.fao.org/3/i8324en/i8324en.pdf.

(2019). The state of the world's biodiversity for food and agriculture. In *FAO Commission on Genetic Resources for Food and Agriculture Assessments.* J. Bélanger and D. Pilling (Eds.). Rome: FAO. Available from: www.fao.org/3/CA3129EN/CA3129EN.pdf.

Fouilleux, E., Bricas, N., and Alpha, A. (2017). "Feeding 9 billion people": Global food security debates and the productionist trap. *Journal of European Public Policy* 24, 1658–1677.

Fransen, L. (2018). Beyond regulatory governance? On the evolutionary trajectory of transnational private sustainability governance. *Ecological Economics* 146, 772–777.

Freibauer, A., Mathijs, E., Brunori, G., et al. (2011). *Sustainable food consumption and production in a resource-constrained world, the 3rd SCAR Foresight Exercise.* Brussels:European Commission.

Friis, C., Nielsen, J. Ø., Otero, I., et al. (2016). From teleconnection to telecoupling: Taking stock of an emerging framework in land system science. *Journal of Land Use Science* 11, 131–153.

Fuchs, R., Brown, C., and Rounsevell, M. (2020). Europe's Green Deal offshores environmental damage to other nations. *Nature* 586, 671–673.

Gibbs, H. K., Ruesch, A. S., Achard, F., et al. (2010). Tropical forests were the primary sources of new agricultural land in the 1980s and 1990s. *Proceedings of the National Academy of Sciences* 107, 1673216737.

Gibson, L., Lee, T. M., Koh, L. P., et al. (2011). Primary forests are irreplaceable for sustaining tropical biodiversity. *Nature* 478, 378–381.

Goldman, R. L., Thompson, B. H., and Daily, G. C. (2007). Institutional incentives for managing the landscape: Inducing cooperation for the production of ecosystem services. *Ecological Economics* 64, 333–343.

Gonthier, D. J., Ennis, K. K., Farinas, S., et al. (2014). Biodiversity conservation in agriculture requires a multi-scale approach. *Proceedings of the Royal Society B: Biological Sciences* 281, 20141358.

Grabs, J., Auld, G., and Cashore, B. (2020). Private regulation, public policy, and the perils of adverse ontological selection. *Regulation & Governance* 15, 1183–1208. doi:10.1111/rego.12354

Green, J. M., Croft, S. A., Durán, A. P., et al. (2019). Linking global drivers of agricultural trade to on-the-ground impacts on biodiversity. *Proceedings of the National Academy of Sciences* 116, 23202–23208.

Harvey, C. A., Komar, O., Chazdon, R., et al. (2008). Integrating agricultural landscapes with biodiversity conservation in the Mesoamerican hotspot. *Conservation Biology* 22, 8–15.

Henders, S., Persson, U. M., and Kastner, T. (2015). Trading forests: Land-use change and carbon emissions embodied in production and exports of forest-risk commodities. *Environmental Research Letters* 10, 125012.

Hendershot, J. N., Smith, J. R., Anderson, C. B., et al. (2020). Intensive farming drives long-term shifts in avian community composition. *Nature* 579, 393–396.

Henle, K., Alard, D., Clitherow, J., et al. (2008). Identifying and managing the conflicts between agriculture and biodiversity conservation in Europe: A review. *Agriculture, Ecosystems & Environment* 124, 60–71.

Henson, S., and Reardon, T. (2005). Private agri-food standards: Implications for food policy and the agri-food system. *Food Policy* 30, 241–253.

Herkenrath, P. (2002). The implementation of the convention on biological diversity: A non-government perspective ten years on. *Review of European, Comparative and International Environmental Law* 11, 29–37.

Hosonuma, N., Herold, M., de Sy, V., et al. (2012). An assessment of deforestation and forest degradation drivers in developing countries. *Environmental Research Letters* 7, 044009.

Hunter, D., Borelli, T., Olsen Lauridsen, N., Gee, E., and Nodar, G. R. (2018). *Biodiversity mainstreaming for healthy & sustainable food systems. A toolkit to support incorporating biodiversity into policies and programmes.* Rome: Biodiversity International. Available from https://hdl.handle.net/10568/98353.

Hunter, D., Özkan, I., Moura de Oliveira Beltrame, D., et al. (2016). Enabled or disabled: Is the environment right for using biodiversity to improve nutrition? *Frontiers in Nutrition* 3, 1–6.

Huntley, B. J. (2014). Good news from the South: Biodiversity mainstreaming – A paradigm shift in conservation? *South African Journal of Science* 110, 1–4.

IIED, UNEP-WCMC. (2015). *Mainstreaming biodiversity and development. Tips and tasks from African experience.* London: IIED. Available from http://pubs.iied.org/14650IIED.

IPBES (2016). *Assessment report on pollinators, pollination and food production.* S. G. Potts, V. L. Imperatriz-Fonseca and H. T. Ngo (Eds.). Bonn: Secretariat of the Intergovernmental Science-Policy Platform on Biodiversity and Ecosystem Services. https://doi.org/10.5281/zenodo.3402856

(2018a). *The assessment report on land degradation and restoration.* L. Montanarella, R. Scholes and A. Brainich (Eds.). Bonn: Secretariat of the Intergovernmental Science-Policy Platform on Biodiversity and Ecosystem Services. Available from https://ipbes.net/assessment-reports/ldr.

(2018b). *The regional assessment report on biodiversity and ecosystem services for Europe and Central Asia.* M. Rounsevell, M. Fischer, A. Torre-Marin Rando and A. Mader (Eds.). Bonn: Secretariat of the Intergovernmental Science-Policy Platform on Biodiversity and Ecosystem Services. Available from www.ipbes.net/assessment-reports/eca.

(2019). *The global assessment report on biodiversity and ecosystem services.* S. Díaz, J. Settele, E. Brondízio and H. T. Ngo (Eds.). Bonn: Secretariat of the Intergovernmental Science-Policy Platform on Biodiversity and Ecosystem Services. Available from https://ipbes.net/global-assessment.

IPCC (2014). *Climate Change 2014: Synthesis Report.* Contribution of Working Groups I, II and III to the Fifth Assessment Report of the Intergovernmental Panel on Climate Change (Core Writing Team, R. K. Pachauri and L. A. Meyer [Eds.]). Geneva: IPCC.

(2019). *Climate change and land: An IPCC special report on climate change, desertification, land degradation, sustainable land management, food security, and greenhouse gas fluxes in terrestrial ecosystems.* P. R. Shukla, J. Skea, E. Calvo Buendia, et al. (Eds.). IPCC. Available from https://bit.ly/3J4xG9y.

IPES-Food. (2017). *Too big to feed: Exploring the impacts of mega-mergers, concentration, concentration of power in the agri-food sector.* Brussels: International Panel of Experts on Sustainable Food Systems. Available from https://bit.ly/3FNwHsu.

(2019). *Towards a common food policy for the EU – The policy reform and realignment that is required to build sustainable food systems in Europe.* Brussels: International Panel of Experts on Sustainable Food Systems. Available from www.ipes-food.org/_img/upload/files/CFP_FullReport.pdf.

Jiren, T. S., Bergsten, A., Dorresteijn, I., et al. (2018). Integrating food security and biodiversity governance: A multi-level social network analysis in Ethiopia. *Land Use Policy* 78, 420–429.

Johns, T., Powell, B., Maundu, P., and Eyzaguirre, P. B. (2013). Agricultural biodiversity as a link between traditional food systems and contemporary development, social integrity and ecological health. *Science of Food and Agriculture* 93, 3433–3442.

Jordan, A., and Lenschow, A. (2010). Environmental policy integration: A state of the art review. *Environmental Policy and Governance* 20, 147–158.

Karlsson-Vinkhuyzen, S. I. S. E., Boelee, E., Cools, J., et al. (2018). Identifying barriers and levers of biodiversity mainstreaming in four cases of transnational governance of land and water. *Environmental Science and Policy* 85, 132–140.

Karlsson-Vinkhuyzen, S., Kok, M. T. J., Visseren-Hamakers, I. J., and Termeer, C. J. A. M. (2017). Mainstreaming biodiversity in economic sectors: An analytical framework. *Biological Conservation* 210, 145–156.

Kivimaa, P., and Mickwitz, P. (2006). The challenge of greening technologies: Environmental policy integration in Finnish technology policies. *Research Policy* 35, 729–744.

Kremen, C. (2015). Reframing the land-sparing/land-sharing debate for biodiversity conservation. *Annals of the New York Academy of Sciences* 1355, 52–76.

Kremen, C., and Merenlender, A. M. (2018). Landscapes that work for biodiversity and people. *Science* 362, eaau6020.

Lakner, S., Kirchweger, S., Hoop, D., Brümmer, B., and Kantelhardt, J. (2018). The effects of diversification activities on the technical efficiency of organic farms in Switzerland, Austria, and Southern Germany. *Sustainability* 10, 1304. https://doi.org/10.3390/su10041304

Lakner, S., Zinngrebe, Y., and Koemle, D. (2020). Combining management plans and payment schemes for targeted grassland conservation within the Habitats Directive in Saxony, Eastern Germany. *Land Use Policy* 97, 104642.

Lambin, E. F., Gibbs, H. K., Heilmayr, R., et al. (2018). The role of supply-chain initiatives in reducing deforestation. *Nature Climate Change* 8, 109–116.

Lansing, D. M. (2014). Unequal access to payments for ecosystem services: The case of Costa Rica. *Development and Change* 45, 1310–1331.

Lapena, I., Halewood, M., and Hunter, D. (2016). *Mainstreaming agricultural biological diversity across sectors through NBSAPs: Missing links to climate change adaptation, dietary diversity and the plant treaty.* CCAFS Info Note. CGIAR Research Program on Climate Change, Agriculture and Food Security, Copenhagen, Denmark. Available from https://cgspace.cgiar.org/handle/10568/78323.

Laurance, W. F., Sayer, J., and Cassman, K. G. (2014). Agricultural expansion and its impacts on tropical nature. *Trends in Ecology and Evolution* 29, 107–116.

Le Saout, S., Hoffmann, M., Shi, Y., et al. (2013). Protected areas and effective biodiversity conservation. *Science* 342, 803–805.

Lindblom, J., Lundström, C., Ljung, M., and Jonsson, A. (2017). Promoting sustainable intensification in precision agriculture: Review of decision support systems development and strategies. *Precision Agriculture* 18, 309–331.

Loos, J., Abson, D. J., Chappell, M. J., et al. (2014). Putting meaning back into "sustainable intensification." *Frontiers in Ecology and the Environment* 12, 356–361.

Magrini, M.-B., Anton, M., Cholez, C., et al. (2016). Why are grain-legumes rarely present in cropping systems despite their environmental and nutritional benefits? Analyzing lock-in in the French agrifood system. *Ecological Economics* 126, 152–162.

McIntyre, B., Herren, H. R., Wakhungu, J., and Watson, R. T. (2009). *International assessment of agricultural knowledge, science and technology for development (IAASTD): North America and Europe (NAE) report.* Washington, DC: IAASTD & Island Press.

Meyfroidt, P. (2018). Trade-offs between environment and livelihoods: Bridging the global land use and food security discussions. *Global Food Security* 16, 9–16.

Meyfroidt, P., Lambin, E. F., Erb, K. H., and Hertel, T. W. (2013). Globalization of land use: Distant drivers of land change and geographic displacement of land use. *Current Opinion in Environmental Sustainability* 5, 438–444.

Millennium Ecosystem Assessment (MEA). (2005). *Ecosystems and human well-being: Synthesis.* Washington, DC: Island Press.

Miller, M. J. (2006). Biodiversity policy making in Costa Rica: Pursuing indigenous and peasant rights. *The Journal of Environment & Development* 15, 359–381.

Misselhorn, A., Aggarwal, P., Ericksen, P., et al. (2012). A vision for attaining food security. *Current Opinion in Environmental Sustainability* 4, 7–17.

Mottet, A., de Haan, C., Falcucci, A., et al. (2017). Livestock: On our plates or eating at our table? A new analysis of the feed/food debate. *Global Food Security* 14, 1–8.

Nilsson, M., Zamparutti, T., Petersen, J. E., et al. (2012). Understanding policy coherence: Analytical framework and examples of sector–environment policy interactions in the EU. *Environmental Policy and Governance* 22, 395–423.

OECD. (2018). *Policy coherence for sustainable development 2018: Towards sustainable and resilient societies.* Paris: OECD Publishing. https://doi.org/10.1787/9789264301061-en

(2019). *Biodiversity: Finance and the economic and business case for action, report prepared for the G7 Environment Ministers' Meeting, 5–6 May 2019.* Available from https://bit.ly/3qHEVxY.

Oliveira, G. (2016). The geopolitics of Brazilian soybeans. *The Journal of Peasant Studies* 43, 348–372.

Oliver, T. H., Boyd, E., Balcombe, K., et al. (2018). Overcoming undesirable resilience in the global food system. *Global Sustainability* 1, e9.

Ongley, E., Rong, W., and Haohan, W. (2010). Semi-quantitative method for assessing "mainstreaming" of the regulatory framework in wetlands biodiversity conservation. *Water International* 35, 365–380.

Orgiazzi, A., Panagos, P., Yigini, Y., et al. (2016). A knowledge-based approach to estimating the magnitude and spatial patterns of potential threats to soil biodiversity. *Science of the Total Environment* 545, 11–20.

Pe'er, G., Bonn, A., Bruelheide, H., et al. (2020). Action needed for the EU Common Agricultural Policy to address sustainability challenges. *People and Nature* 2, 305–316.

Pe'er, G., Lakner, S., Müller, R., et al. (2017). *Is the CAP fit for purpose? An evidence-based fitness check assessment.* Leipzig: German Centre for Integrative Biodiversity Research (iDiv), Halle Jena-Leipzig. http://dx.doi.org/10.13140/RG.2.2.11705.26725

Pe'er, G., Zinngrebe, Y., Hauck, J., et al. (2016). Adding some green to the greening: Improving the EU's Ecological Focus Areas for biodiversity and farmers. *Conservation Letters* 10, 517–530.

Pe'er, G., Zinngrebe, Y., Moreira, F., et al. (2019). A greener path for the EU Common Agricultural Policy. *Science* 365, 449–451.

Pendrill, F., Persson, U. M., Godar, J., et al. (2019). Agricultural and forestry trade drives large share of tropical deforestation emissions. *Global Environmental Change* 56, 1–10.

Perfecto, I., and Vandermeer, J. (2010). The agroecological matrix as alternative to the land-sparing/agriculture intensification model. *Proceedings of the National Academy of Sciences* 107, 5786–5791. https://doi.org/10.1073/pnas.0905455107

Persson, Å., Eckerberg, K., and Nilsson, M. (2016). Institutionalization or wither away? Twenty-five years of environmental policy integration under shifting governance models in Sweden. *Environment and Planning C: Government and Policy* 34, 478–495.

Persson, Å., and Runhaar, H. (2018). Conclusion: Drawing lessons for environmental policy integration and prospects for future research. *Environmental Science and Policy* 85, 141–145.

Persson, Å., Runhaar, H., Karlsson-Vinkhuyzen, S., et al. (2018). Editorial: Environmental policy integration: Taking stock of policy practice in different contexts. *Environmental Science and Policy* 85, 113–115.

Plieninger, T., Muñoz-Rojas, J., Buck, L. E., and Scherr, S. J. (2020). Agroforestry for sustainable landscape management. *Sustainability Science* 15, 1255–1266. https://link.springer.com/article/10.1007/s11625-020-00836-4

Plieninger, T., Torralba, M., Hartel, T., and Fagerholm, N. (2019). Perceived ecosystem services synergies, trade-offs, and bundles in European high nature value farming landscapes. *Landscape Ecology* 34, 1565–1581.

Poux, X., and Aubert, P.-M. (2018). *Une Europe agroécologique en 2050 : une agriculture multifonctionnelle pour une alimentation saine. Enseignements d'une modélisation du système alimentaire européen.* Paris:IDDRI-AScA, Study N°09/18.

Prager, K., Reed, M., and Scott., A. (2012). Encouraging collaboration for the provision of ecosystem services at a landscape scale—Rethinking agri-environmental payments. *Land Use Policy* 29, 244–249.

Prip, C., and Pisupati, B. (2018). *Assessment of post-2010 National Biodiversity Strategies and Action Plans.* Narobi: UNEP. Available from https://bit.ly/3sjQsTN.

Ramankutty, N., Mehrabi, Z., Waha, K., et al. (2018). Trends in global agricultural land use: Implications for environmental health and food security. *Annual Review of Plant Biology* 69, 789–815.

Rasmussen, L. V., Coolsaet, B., Martin, A., et al. (2018). Social-ecological outcomes of agricultural intensification. *Nature Sustainability* 1, 275–282.

Redford, K. H., Huntley, B. J., Roe, D., et al. (2015). Mainstreaming biodiversity: Conservation for the twenty-first century. *Frontiers in Ecology and Evolution* 3, 137.

Roche, M., and Argent, N. (2015). The fall and rise of agricultural productivism? An antipodean viewpoint. *Progress in Human Geography* 39, 621–635.

Rodríguez, L. O., Cisneros, E., Pequeño, T., Fuentes, M. T., and Zinngrebe, Y. (2018). Building adaptive capacity in changing social-ecological systems: Integrating knowledge in communal land-use planning in the Peruvian Amazon. *Sustainability* 10, 511.

Rudel, T. K., Schneider, L., Uriarte, M., et al. (2009). Agricultural intensification and changes in cultivated areas, 1970–2005. *Proceedings of the National Academy of Sciences* 106, 20675–20680.

Runhaar, H. (2016). Tools for integrating environmental objectives into policy and practice: What works where? *Environmental Impact Assessment Review* 59, 1–9.

(2017). Governing the transformation towards "nature-inclusive" agriculture: Insights from the Netherlands. *International Journal of Agricultural Sustainability* 15, 340–349.

Runhaar, H., Driessen, P., and Uittenbroek, C. (2014). Towards a systematic framework for the analysis of environmental policy integration. *Environmental Policy and Governance* 24, 233–246.

Runhaar, H. A. C., Melman, T. C. P., Boonstra, F. G., et al. (2017). Promoting nature conservation by Dutch farmers: A governance perspective. *International Journal of Agricultural Sustainability* 15, 264–281.

Runhaar, H., Wilk, B., Driessen, P., et al. (2020). Policy Integration. In *Architectures of Earth system governance: Institutional complexity and structural transformation*. F. Biermann and R. Kim (Eds.), pp. 146–164. Cambridge: Cambridge University Press.

Runhaar, H., Wilk, B., Persson, A., Uittenbroek, C., and Wamsler, C. (2018). Mainstreaming climate adaptation: Taking stock about "what works" from empirical research worldwide. *Regional Environmental Change* 18, 1201–1210.

Sanchez-Azofeifa, G. A., Pfaff, A., Robalino, J. A., and Boomhower, J. P. (2007). Costa Rica's payment for environmental services program: Intention, implementation, and impact. *Conservation Biology* 21, 1165–1173.

Sánchez-Bayo, F., and Wyckhuys, K. A. (2019). Worldwide decline of the entomofauna: A review of its drivers. *Biological Conservation* 232, 8–27.

Sarkki, S., Niemelä, J., Tinch, R., et al. (2016). Are national biodiversity strategies and action plans appropriate for building responsibilities for mainstreaming biodiversity across policy sectors? The case of Finland. *Journal of Environmental Planning and Management* 59, 1377–1396.

Scherr, S. J., and McNeely, J. A. (2008). Biodiversity conservation and agricultural sustainability: Towards a new paradigm of "ecoagriculture" landscapes. *Philosophical Transactions of the Royal Society B: Biological Sciences* 363, 477–494.

Scherr, S. J., Shames, S., and Friedman, R. (2012). From climate-smart agriculture to climate-smart landscapes. *Agriculture and Food Security* 1, 1–15.

Schomers, S., and Matzdorf, B. (2013). Payments for ecosystem services: A review and comparison of developing and industrialized countries. *Ecosystem Services* 6, 16–30. https://doi.org/10.1016/j.ecoser.2013.01.002

Schreinemachers, P., and Tipraqsa, P. (2012). Agricultural pesticides and land use intensification in high, middle and low income countries. *Food Policy* 37, 616–626.

Seibold, S., Gossner, M. M., Simons, N. K., et al. (2019). Arthropod decline in grasslands and forests is associated with landscape-level drivers. *Nature* 574, 671–674.

Seymour, F., and Harris, N. L. (2019). Reducing tropical deforestation. *Science* 365, 756–757.

Slootweg, R., and Kolhoff, A. (2003). A generic approach to integrate biodiversity considerations in screening and scoping for EIA. *Environmental Impact Assessment Review* 23, 657–681.

Söderberg, C., and Eckerberg, K. (2013). Rising policy conflicts in Europe over bioenergy and forestry. *Forest Policy and Economics* 33, 112–119.

Somarriba, E., Beer, J., Alegre-Orihuela, J., et al. (2012). Mainstreaming agroforestry in Latin America. In: *Agroforestry – The future of global land ese*. P. K. R. Nair and D. Garrity (Eds.), pp. 429–454. Dordrecht:Springer.

Somorin, O. A., Visseren-Hamakers, I. J., Arts, B., Tiani, A. M., and Sonwa, D. J. (2016). Integration through interaction? Synergy between adaptation and mitigation (REDD+) in Cameroon. *Environment and Planning C: Government and Policy* 34, 415–432.

Stabile, M. C., Guimarães, A. L., Silva, D. S., et al. (2020). Solving Brazil's land use puzzle: Increasing production and slowing Amazon deforestation. *Land Use Policy* 91, 104362.

Steffan-Dewenter, I., Kessler, M., Barkmann, J., et al. (2007). Tradeoffs between income, biodiversity, and ecosystem functioning during tropical rainforest conversion and agroforestry intensification. *Proceedings of the National Academy of Sciences of the United States of America* 104, 4973–4978.

Sun, J., Tong, Y. X., and Liu, J. (2017). Telecoupled land-use changes in distant countries. *Journal of Integrative Agriculture* 16, 368–376.

Swiderska, K. (2002). *Mainstreaming biodiversity in development policy and planning: A review of country experience*. Biodiversity and Livelihoods Group International Institute for Environment and Development. Available from https://bit.ly/3HutbF5.

Teixidó-Figueras, J., and Duro, J. A. (2014). Spatial polarization of the ecological footprint distribution. *Ecological Economics* 104, 93–106.

Termeer, C., Stuiver, M., Gerritsen, A., and Huntjens, P. (2013). Integrating self-governance in heavily regulated policy fields: Insights from a Dutch farmers' cooperative. *Journal of Environmental Policy & Planning* 15, 285–302.

Torralba, M., Fagerholm, N., Burgess, P. J., Moreno, G., and Plieninger, T. (2016). Do European agroforestry systems enhance biodiversity and ecosystem services? A meta-analysis. *Agriculture, Ecosystems & Environment* 230, 150–161. https://doi.org/10.1016/j.agee.2016.06.002

Torralba, M., Fagerholm, N., Hartel, T., Moreno, G., and Plieninger, T. (2018). A social-ecological analysis of ecosystem services supply and trade-offs in European wood-pastures. *Science Advances* 4, eaar2176.

Tscharntke, T., Clough, Y., Wanger, T. C., et al. (2012). Global food security, biodiversity conservation and the future of agricultural intensification. *Biological Conservation* 151, 53–59.

Tsiafouli, M. A., Apostolopoulou, E., Mazaris, A. D., et al. (2013). Human activities in Natura 2000 sites: A highly diversified conservation network. *Environmental Management* 51, 1025–1033.

Tsiafouli, M. A., Thébault, E., Sgardelis, S. P., et al. (2015). Intensive agriculture reduces soil biodiversity across Europe. *Global Change Biology* 21, 973–985.

Tutwiler, A., Bailey, A., Attwood, S., Remans, R., and Ramirez, M. (2017). *Agricultural biodiversity and food system sustainability*. Rome: Biodiversity International.

Uittenbroek, C. J., Janssen-Jansen, L. B., and Runhaar, H. A. C. (2013). Mainstreaming climate adaptation into urban planning: Overcoming barriers, seizing opportunities and evaluating the results in two Dutch case studies. *Regional Environmental Change* 13, 399–411.

UNFCCC (2017). Decision -/CP.23, Koronivia joint work on agriculture. Available from https://bit.ly/3GScpiG.

Van Dijk, T. C., Van Staalduinen, M. A., and Van der Sluijs, J. P. (2013). Macro-invertebrate decline in surface water polluted with imidacloprid. *PloS One* 8, e62374.

van Noordwijk, M. (Ed.). (2019). *Sustainable development through trees on farms: Agroforestry in its fifth decade*. Bogor: World Agroforestry (ICRAF).

Van Oosten, C. (2013). Forest landscape restoration: Who decides? A governance approach to forest landscape restoration. *Natureza & Conservação* 11, 119–126.

Vanhove, M. P. M., Rouchette, A., and Janssens de Bisthoven, L. (2017). Joining science and policy in capacity development for monitoring progress towards the Aichi Biodiversity Targets in the global South. *Ecological Indicators* 73, 694–697.

Velázquez Gomar, J. O. (2014). International targets and environmental policy integration: The 2010 biodiversity target and its impact on international policy and national implementation in Latin America and the Caribbean. *Global Environmental Change* 29, 202–212.

Verbruggen, P., and Havinga, T. (2017). Hybridization of food governance: An analytical framework. In *Hybridisation of food governance: Trends, types and results*. P. Verbruggen and T. Havinga (Eds.), pp. 1–27. Cheltenham: Edward Elgar.

Vijge, M. J. (2018). The (dis)empowering effects of transparency beyond information disclosure: The extractive industries transparency initiative in Myanmar. *Global Environmental Politics* 18, 13–32.

Visseren-Hamakers, I. J. (2015). Integrative environmental governance: Enhancing governance in the era of synergies. *Current Opinion in Environmental Sustainability* 14, 136–143.

Watson, J. E., Dudley, N., Segan, D. B., and Hockings, M. (2014). The performance and potential of protected areas. *Nature* 515, 67–73.

Whitehorn, P. R., Navarro, L. M., Schröter, M., et al. (2019). Mainstreaming biodiversity: A review of national strategies. *Biological Conservation* 235, 157–163.

Wilson, G. A., and Rigg, J. (2003). "Post-productivist" agricultural regimes and the South: Discordant concepts? *Progress in Human Geography* 27, 681–707.

Wunder, S., Engel, S., and Pagiola, S. (2008). Taking stock: A comparative analysis of payments for environmental services programs in developed and developing countries. *Ecological Economics* 65, 834–852. https://doi.org/10.1016/j.ecolecon.2008.03.010

Yamamuro, M., Komuro, T., Kamiya, H., et al. (2019). Neonicotinoids disrupt aquatic food webs and decrease fishery yields. *Science* 366, 620–623.

Yu, Y., Feng, K., and Hubacek, K. (2013). Tele-connecting local consumption to global land use. *Global Environmental Change* 23, 1178–1186.

Zinngrebe, Y. (2016a). Learning from local knowledge in Peru – Ideas for more effective biodiversity conservation. *Journal for Nature Conservation* 32, 10–21.

  (2016b). Conservation narratives in Peru: Envisioning biodiversity in sustainable development. *Ecology and Society* 21, 35.

  (2018). Mainstreaming across political sectors: Assessing biodiversity policy integration in Peru. *Environmental Policy and Governance* 28, 153–171.

Zinngrebe, Y., Borasino, E., Chiputwa, B., et al. (2020). Agroforestry governance for operationalising the landscape approach: Connecting conservation and farming actors. *Sustainability Science* 15, 1417–1434.

Zinngrebe, Y., Pe'er, G., Schueler, S., et al. (2017). The EU's ecological focus areas – Explaining farmers' choices in Germany. *Land-Use Policy* 65, 93–108.

14

Cities and the Transformation of Biodiversity Governance

HARRIET BULKELEY, LINJUN XIE, JUDY BUSH, KATHARINA ROCHELL,
JULIE GREENWALT, HENS RUNHAAR, ERNITA VAN WYK, CATHY OKE
AND INGRID COETZEE

14.1 Introduction

The governing of nature has been an essential part of the story of urbanization. Whether through the conversion of rivers for transportation, the creation of urban drainage systems for wastewater removal or the installation of parks for their recreational and aesthetic value (Gandy, 2004; Gleeson and Low, 2000; Rydin, 1998), nature has played a critical role in urban development. Yet, conservationist thinking, which has dominated environmental governance and policy, has tended to equate the environment as belonging to either "rural" or "wilderness" places that needed to be protected from the encroachment of (urban) society (Owens, 1992). As a result, much of the governance of biodiversity at the urban scale during the twentieth century was focused on the designation and enforcement of protected areas (Vaccaro et al., 2013). Yet such dualistic thinking has ignored the ways in which nature inhabits the city, whether intended or otherwise, from domestic gardens to public parks, urban sewers to derelict corners of the city, as well as the potential benefits that such forms of biodiversity can bring to the city.

It has only been since the late 1980s that how cities might contribute toward local, national and global sustainability has begun to be recognized. While climate change has tended to dominate this agenda, cities also have a range of different yet substantial roles in addressing the loss of nature: as habitats for biodiversity and threatened species (Aronson et al., 2014; Hall et al., 2017; Ives et al., 2016; Soanes and Lentini, 2019); as locations for people to connect with nature (Soanes and Lentini, 2019); as key jurisdictions in global and multilevel governance (Pattberg et al., 2019) and as important consumption arenas driving biodiversity loss globally (Díaz et al., 2019). Nonetheless, it was not until 2008 that the first *Global Biodiversity Summits of Local and Subnational Governments* was held in parallel to the Conference of the Parties to the Convention on Biological Diversity. These summits have since taken place biannually and are intended as a means through which to reinforce the recognition and involvement of local and subnational governments in contributing to CBD objectives and targets. While the initial version of both the *Strategic Plan for Biodiversity 2011–2020* and the *20 Aichi Biodiversity Targets* makes no direct references to cities or urban areas, a subsequent assessment of the Aichi targets and the 2030 Agenda for Sustainable Development found that Sustainable Development Goal 11 on Sustainable Cities and Communities corresponded to six (2, 4, 8, 11, 14 and 15) of the Aichi targets

(CBD, 2016). At the same time, the 2010 Decision X/22 of the Convention on Biodiversity laid out explicit terms on which the Parties to the convention were to be encouraged to recognize and facilitate the work of subnational and local authorities through the development and implementation of local biodiversity strategies and action plans (LBSAPs). Over the past decade, the urban dimension of biodiversity issues has come to be increasingly recognized.

Yet despite this, in practice biodiversity governance has yet to gain widespread traction at the local level, and local biodiversity planning has been critiqued for an overly narrow approach, the exclusion of diverse values for nature and limited effectiveness (Bomans et al., 2010; Elander et al., 2005; Evans, 2004; Wilkinson et al., 2013). In this chapter, we explore how the governance of urban nature is evolving in response to the increasing urgency of this agenda. In so doing, we follow the distinction put forward by Patterson et al. (2017) and highlighted in Chapter 1 between governance *for* transformation, where governance creates the conditions by which transformative change can emerge; governance *of* transformations, where governance is deliberately intended to advance transformative change in terms of either processes or outcomes that involve systemic or structural shifts in current socioecological orders; and transformations *in* governance, where governance regimes – their architectures, agency, power and so forth – are themselves transformed. We find that, internationally, urban biodiversity governance is being transformed both in terms of its intentions (governance *for* transformation) – moving from a concern only with reducing the threat of cities to biodiversity to also realizing their benefits (Section 14.2) – and in terms of the forms that governance is taking (transformation *in* governance) – through the growth of governance experimentation in cities and the growth in transnational governance networks (Section 14.3). These shifts are changing the *outcomes* of what biodiversity governance in the city is seeking to realize – from a focus on specific places and parts of nature to a broader engagement with multiple socio-natures and the ways in which working with nature can generate sustainability benefits for a diverse range of communities. At the same time, within urban policymaking and practice on the ground, there has yet to be a significant effort to address the ways in which cities contribute to the underlying drivers of biodiversity loss through explicitly linking their roles and responsibilities in reducing waste, combating climate change and shaping production and consumption with biodiversity agendas. We return to these points in conclusion (Section 14.4) and reflect on their implications for the ways in which cities can contribute to transformative biodiversity governance.

14.2 Transforming Biodiversity in the City: from Threat to Opportunity?

If, for the most part of modern urban development, cities were regarded as separated from nature, the global environmental challenges facing society in the twenty-first century have abruptly erased any such boundaries. As the IPBES Global Assessment makes clear, cities are a primary driver of biodiversity loss through urban expansion and pollution, as well as affecting the loss of nature globally through the consumption practices of urban residents and the global value chains of urban economies. The detailed analysis presented in the *Nature in the Urban Century* report (McDonald et al., 2018) asserts that urban growth was responsible for the loss of 190,000 km^2 of natural habitat between 1992 and 2000 and could threaten 290,000 km^2 of global natural habitat by 2030. Cities located in globally important biodiversity hotspots bear special significance in this context. Biodiversity hotspots are areas of exceptional concentrations of endemic species that are simultaneously undergoing a high rate of loss of habitat. It was estimated that in 1995, 20 percent of the world's population was living in global biodiversity hotspots, which accounts for about 12 percent of the earth's surface. Population growth in these hotspots was estimated to be 1.8 percent per annum (Cincotta et al., 2000).[1] Such impacts are not only felt in areas with particular biodiversity value: urbanization and increased impervious surfaces are also having severe impacts on urban wetlands and waterways (Booth et al., 2016). In short, even though the impact of individual cities will be highly varied, the weight of evidence suggests that urbanization processes are "catastrophic for native species, and ... a well-known threat to biodiversity worldwide" (Garrard et al., 2017: 1).

For the most part, it has been this discourse of the in-situ impacts of urbanization on biodiversity either within the city boundary or at its expanding edge that has shaped how the potential role of cities in governing biodiversity has been framed (Bulkeley et al., 2021). Over the past decade, the Convention on Biodiversity has primarily focused on the spatial planning capacities of cities as essential to managing urban encroachment on biodiversity and on the importance of protected areas for biodiversity conservation. Yet this underplays two other important ways in which cities are connected to the biodiversity challenge. First, as the IPBES Global Assessment makes clear, cities have a significant role in shaping the drivers of biodiversity loss – from climate change to consumption. Second, as urbanists have long recognized, cities are intricately connected to and dependent on nature – from water resources to urban parks (Gandy, 2002; Swyngedouw and Kaika, 2000). There is now a growing interest in the ways in which cities can benefit from both the ecosystem services that nature provides and also how urban nature and biodiversity contribute to less readily quantified values, such as heritage, well-being, stewardship and reverence, and provide an essential form of connection between nature and people in the urban milieu. Urban nature is increasingly recognized for its capacity to not only support biodiversity conservation, but also to generate additional environmental, economic and social benefits – or what are termed "nature's contribution to people" (Kabisch et al., 2016). This is reflected in a growing interest in urban nature-based solutions, an umbrella term used to encompass ecosystem-based adaptation (Geneletti and Zardo, 2016; Munang et al., 2013), green infrastructure and ecosystem services (Cohen-Shacham et al., 2016; Dorst et al., 2019; Nesshöver et al., 2017; Pauleit et al., 2017). Nature-based solutions provide a means through which cities not only have the potential to benefit directly from nature, but also contribute to addressing the global challenge of the loss of biodiversity. In the rest of this

---

[1] Urbanization does not only form a threat to nature because of the conversion of nature into built environment and because of the effects on surrounding nature areas (e.g. traffic, recreation, etc.), however. Nature *within* cities is also threatened because of competing land claims. For instance, the "compact city" paradigm and other densification strategies – aimed at preserving nature outside cities – can endanger space for nature in cities (Fischer et al., 2018).

section, we examine how cities are currently undertaking action that can contribute to three key elements of biodiversity governance – *protecting* or conserving nature, *restoring* nature and fostering the value of nature's contributions to people through *thriving* with nature. We suggest that there is significant evidence that cities can no longer be viewed simply as a *threat* to biodiversity, but are transforming their role to one of significant *opportunity*. In doing so, they are adopting new means of governing nature in the city, which in turn are leading to the transformation of biodiversity governance within and beyond its boundaries.

14.2.1 Urban Biodiversity Conservation

The main goal of conservation is to prevent further loss and degradation of natural ecosystems and resources (Young. 2000), although in practice this can include the preservation, maintenance, sustainable use and enhancement of the components of biological diversity as well as exploring how society lives in harmony with nature. Although cities have been seen to hold little conservation value, there is increasing recognition of the role that urban green spaces, waterways and wetlands play in conservation, and its wider contributions to human health and well-being (Aronson et al., 2014; Endreny et al., 2017; Parris et al., 2018). Cities also provide habitat for threatened species, and some threatened species are found exclusively in urban areas (Soanes and Lentini, 2019). Ives et al. (2016: 117) analyzed the distribution of Australia's listed threatened species and found that 30 percent are found in cities and that "Australian cities support substantially more nationally threatened animal and plant species than all other non-urban areas on a unit-area basis." Globally, while a large number of species have been disadvantaged or made locally extinct by urbanization, urban areas have also provided range expansions for other species, including fruit-eating bats (Williams et al., 2006) and nectar-feeding birds that feast on the well-watered and productive plants found in urban gardens.

It is therefore a misconception that cities cannot contribute significantly to biodiversity conservation (Soanes et al., 2019), yet even where this is recognized, there is significant debate concerning how this contribution can be realized. As the main preference is "given to conserving large, highly connected areas," "relative ambivalence [is] shown toward protecting small, isolated habitat patches" even though they are "inordinately important for biodiversity conservation" (Wintle et al., 2019: 909). Far from being delivered through systematic forms of urban (biodiversity) planning, urban reserve or park systems are often small, fragmented and disconnected, located on leftover, undevelopable land or squeezed in size due to urban development pressures and economic imperatives, but nonetheless have been shown to make important contributions to conservation (Kendal et al., 2017). While "effective conservation planning requires an understanding of species-habitat relationships" that goes beyond the simplified single species focus (Threlfall et al., 2012: 41), in urban areas the same species that may be valued as threatened species may also be labeled as pests. The grey-headed flying fox is listed as a vulnerable species in Australia, but in Melbourne, a colony of the animals was evicted from the botanic gardens for causing roosting damage to trees. Black-legged kittiwakes, a threatened gull species that nest and breed in areas along the quayside in Newcastle, UK, are blamed for mess and noise, such that while birdwatchers and wildlife enthusiasts celebrate their presence, local businesses are less enthusiastic and have used various means (spikes, nets, electric shocks) to attempt to prevent nesting. At the same time, as cities experience the impacts of climate change, what it is appropriate to conserve is also coming into question as much-loved and threatened urban species may not be able to flourish under changing conditions (Lennon, 2015; Prober et al., 2019). These complexities point to the challenges of governing biodiversity in human-dominated landscapes, suggesting that the forms of nature that are or are not valued cannot be established through scientific knowledge but, as the Introduction to this volume suggests, require the bringing together of diverse forms of knowledge often in a transdisciplinary manner.

Governing for urban conservation is therefore no straightforward matter, but rather shot through with contention over which kinds of nature should be conserved, for whom and under which conditions. The importance of fragmented urban nature and small, disconnected spaces in cities for biodiversity conservation also suggests that addressing biodiversity goals involves multiple sites and actors that are not directly engaged in the formal land-use planning or regulatory systems of local authorities. Indeed, as we discuss further below, it appears that urban conservation governance is being transformed – rather than being led by urban planning, it is now taking place through a whole host of initiatives and programs, including those conceived as nature-based solutions to diverse sustainability challenges, that are undertaken by a range of urban actors, including private and civil society organizations. In this context, rather than requiring more integrated governance, as Chapter 1 suggests, fragmented and diverse forms of governance are potentially a more viable means through which to transform the capacity of cities to address the loss of nature.

14.2.2 Urban Biodiversity Restoration

While conservation mainly focuses on preventing ongoing and future losses, restoration seeks to actively reverse such degradation (Garson, 2016). Similar to conservation activities, restoration activities differ greatly in their spatial scale and in terms of the sheer magnitude of intervention they entail (Garson, 2016). With cities' roles in biodiversity conservation being increasingly recognized, more attention is also being directed toward the restoration of urban green spaces for biodiversity habitat (Butt et al., 2018). Restoration activities have focused on habitat improvement and planting; creating artificial structures for nesting, shelter or to facilitate faunal movement and connectivity between sites; control of pest or invasive species; and community engagement and education programs, including citizen science and site or species monitoring programs (Threlfall et al., 2019). Green spaces that include understory cover and increased structural complexity of vegetation have been shown to improve biodiversity outcomes, and therefore restoration efforts that "redress the dominance of simplified and exotic vegetation ... with an increase in understorey vegetation volume and percentage of native vegetation will benefit a broad array of biodiversity" (Threlfall et al., 2017: 1874). Furthermore, with studies showing the "inordinately important" contribution of small, isolated habitat patches for biodiversity conservation, the restoration of these small patches of urban green space, wetlands and waterways should be "urgently prioritised" (Wintle et al., 2019: 909). This in turn implies a model of governance that extends beyond the capacities of local governments to include a host of actors who own and manage urban land and water systems.

Cities allow for a diversity of restored habitats that serve to improve conditions for biodiversity in public and private lands (Aronson et al., 2017). For example, practices such as the restoration of native prairie vegetation along roadsides has been shown to increase bee species richness (Hopwood, 2008). Moreover, urban green and blue spaces are being increasingly recognized for their capacity to not only support biodiversity conservation (Dunn et al., 2006; Goddard et al., 2010; Miller and Hobbs, 2002; Niemela, 1999), but also to generate additional environmental, social-cultural, and economic benefits, including managing water quality, fostering community inclusion and generating new opportunities for business (Haase et al., 2013; Kabisch et al., 2015), as well as fostering the functioning of ecosystems for climate change mitigation and adaptation (European Commission, 2015). The restoration of Merri Creek, a waterway in Melbourne, Australia, has seen the return of a range of species, including birds such as the sacred kingfisher, and pollution-sensitive insects to restored wetlands beside the creek (Bush et al., 2003; McGregor and McGregor, 2020). The restoration has provided many opportunities for community involvement in replanting, rubbish collection and so on, underpinning a remarkable community reconnection with the creek and a renewed sense of shared ownership (Bush et al., 2003). In the Netherlands, many citizen grassroots initiatives around urban nature exist, but their contribution to restoration in a classical sense (i.e. conserving rare/Red list species) is limited, not only because of their spatial scale but also because their objectives in terms of social, economic and environmental outcomes are not always aligned with such outcomes (Mattijsen et al., 2018). While restoration efforts have focused on public land, there is an increasing recognition of the potential contribution of greening the private realm. The City of Melbourne has recently joined a number of other cities, including Seattle, Helsinki and Malmo, in establishing a "Green Factor Tool" to encourage the integration of greening in new buildings and developments by private developers (Bush et al., 2021; City of Melbourne, 2017).

However, restoring urban habitat brings to the fore the potential for increasing conflicts between humans and nonhumans in these urban "shared habitats." For example, in an Australian urban creek restoration project, neighboring residents viewed the return of native birds and lizards either neutrally or favorably, but there were fears about the return of snakes, which created conflict (Maller and Farahani, 2018). In another case, we can see that while water-sensitive urban design treatments to address urban issues of flooding and stormwater management can enhance biodiversity (Parris et al., 2018), they may contain high levels of contaminants, including pesticides and heavy metals from stormwater runoff, which can potentially endanger water quality for human use (Sievers et al., 2019). At the same time, what counts as "restoration" is also continually a matter of negotiation and contestation. Global environmental change poses challenges to traditional practices of "restoring 'degraded' ecosystems to a 'natural state' of acceptable historic variability"

(Lennon, 2015) such that the end goals of restoration are far from clearly determined by science alone. Further, "novel ecosystems," which are composed of "non-historical species configurations" and dominate many urban landscapes, are rarely considered as worthy of either conservation or restoration, despite providing rich species assemblages and biodiversity habitat (Planchuelo et al., 2019).

As with the governing of urban biodiversity conservation, interventions and practices aimed at enabling the restoration of urban nature for biodiversity is fraught with conflicts, indeterminacy and the potential for exclusionary processes that revere some forms of nature at the expense of others (Tozer et al., 2020). While questions of the design and implementation of such schemes have been debated, there has been less consideration of "new principles that can help guide goal-setting for nature conservation and ecological restoration in dynamic environments" (Prober et al., 2017: 477), particularly in the face of climate change (Prober et al., 2019). Indigenous people's perspectives and knowledge, which have critical contributions for connecting past, present and future natural and cultural heritage, must be embedded in these debates for new principles as well as broader planning and implementation of conservation and restoration activities. Indigenous knowledge and perspectives are "crucial for long-term, sustained biodiversity conservation, land and water management" (Threlfall et al., 2019: 3). As has been found with conservation initiatives, recognizing the key role that restoration in cities can play toward realizing global biodiversity goals also suggests that multiple actors and modes of governing beyond traditional forms of land-use planning and regulation will need to be harnessed if its potential is to be realized. While this may take place through the development of more inclusive forms of governance, we suggest it will also involve forms of protest, contestation and conflict over whose nature should be conserved or restored.

14.2.3 Thriving with Urban Biodiversity

As the IPBES Global Assessment makes clear, in addition to seeking to conserve and restore nature, a central concern for biodiversity governance in the coming decade will be to ensure that nature's contribution to people is preserved and enhanced (Díaz et al., 2019). In short, to ensure that cities can *thrive* with nature. How, why and with what consequence it is possible to consider nature as generating a contribution to individuals and to society has been subject to intense debate, as scholars, activists and policy-makers take issue with the extent to which such contributions are framed as instrumental – a means to a human end – or as ensuing from a sense of connection, spirituality or well-being derived from knowing and being in nature (Gavin et al., 2018; Pascual et al., 2017). Attempts to identify so-called ecosystem services that contribute to societal needs and to calculate their monetary value have in particular been subject to a strong critique that doing so reduces the actual contribution that nature makes to society to a narrow range of attributes and functions that can be captured in this way (Schröter et al., 2014). Recent years have witnessed something of a move away from this position to a recognition of the multiple ways in which nature contributes toward society, as well as the continued importance of recognizing the intrinsic value of nature itself (Díaz et al., 2018).

This shift in conceiving of nature as providing singular and functional benefits for society to a position in which the multiplicity of nature's contributions is recognized can also be witnessed at the urban level. The growth and increasing prominence of the discourse of nature-based solutions, particularly in the European Union, draws explicitly on the idea that nature can contribute to addressing the challenges facing cities, for example in terms of air or water quality, while at the same time generating a wide range of benefits, such as flourishing biodiversity and enhanced well-being, that are not so readily captured in functional or economic terms. Despite the novelty of the term, it is clear that historically urban nature has played these multiple roles, offering a means through which cities could function more effectively but also creating more or less formalized spaces of connection, solidarity and spirituality for diverse communities. In Victorian Britain, for example, formalized parks were seen to provide havens from city life for reflection and recreation. In cities that experienced colonization, Indigenous communities continued to maintain and fight for rights in order to continue to access both food resources and their cultural and spiritual connections to land and water.

As cities now seek to realize diverse goals for urban sustainability, working with and for nature has come to play a vital role. In Tianjin, China, for example, the Ecological Wetland Park is a constructed, artificial wetland with an approximate size of 630,000 m^2 located in one of the largest industrial, logistics and free-trade centers of the country. Its aims are not only to enhance the environmental quality of the industrial park, but also to generate space for biodiversity, a thriving economy and enhanced social well-being. In Winnipeg, Canada, a grassroots-run neighborhood group – the Spence Neighbourhood Association – is working with Indigenous communities and local stakeholders to transform more than fifty vacant lots into edible community gardens and parks. Besides their conservation value, these urban green spaces provide important social, economic and environmental benefits. For example, in the Ogimaa Gichi Makwa Gitigaan garden, which opened in 2012, the inclusion of indigenous plants not only contributes to the conservation of local species, but also allows community members to utilize traditional knowledge while learning about horticultural practices. These examples show that cities are transforming their development approach by seeking to thrive with nature in multiple ways. Yet the multifunctionality of nature-based solutions provides both opportunities and challenges. While they are frequently asserted, the benefits, synergies and trade-offs of interventions designed to generate a contribution to society need to be better investigated (Raymond et al., 2017). Multifunctionality is also problematic in view of the organization of local governments and the private sector in specialized "silos" (Dorst et al., 2019; Kabisch et al., 2016), meaning that while in principle the idea of generating diverse contributions to society is regarded as a benefit, such interventions can lack the political champions or consistent backing required to ensure that they are taken up as part of urban development.

Understanding who benefits and how from urban nature's contribution to people is not only important from the perspective of their uptake, but also in relation to their consequences. Research has documented a persistent phenomenon of green gentrification emerging in relation to efforts to develop and enhance nature's contributions to people within cities, leading not only to forms of demographic change and displacement, but also exclusion from the very benefits that nature is supposed to generate (Anguelovski et al., 2018; Wolch et al., 2014). Such processes not only serve to reproduce and deepen urban inequalities, but also to sustain particular dominant views about which forms of nature can best contribute to society, generating elitist and often exclusionary views of what "counts" as the kinds of urban nature and biodiversity that should be conserved, restored and generated (Mattijssen et al., 2018; Tozer et al., 2020). Rather than taking for granted how nature-based solutions should intervene to contribute to biodiversity, if they are to ensure that diverse communities are to thrive with nature in the city, it is vital that the kinds of nature and biodiversity that are being generated and the auxiliary benefits they carry are subject to scrutiny by those who may need the benefits of nature most. Rather than assuming that nature and biodiversity are automatically of benefit to urban residents, it is critical that the ways in which urban nature has historically been used to repress and exclude different communities is considered in efforts to govern urban biodiversity and its wider benefits, or there is a significant risk that such interventions will contribute to, rather than transform, urban inequalities (Kuras et al., 2020). While measures to support inclusive governance, as suggested in Chapter 1, can seek to make alternative voices heard, without more fundamental changes to the structures of power within which decisions about urban futures are made, and an acknowledgment that contestation and conflict may be a necessary part of generating alternatives, inclusive governance is unlikely to be sufficient.

14.3 Transformative Urban Governance for Biodiversity?

Our analysis suggests that cities are now engaged in a vast array of efforts toward conservation, restoration and thriving with nature, both through their efforts to maintain existing forms of urban nature and through the increasing focus on nature-based solutions as interventions by which to accomplish multiple sustainability goals. Urban biodiversity governance is not confined to the actions of municipal authorities, but undertaken through a wide range of interventions. In this section, we examine how urban biodiversity governance is being transformed as a result, and with what consequences for the capacity of cities to engage in the transformative governance of biodiversity. We first examine the multiple modes of governing through which cities are mobilizing their actions on biodiversity. We then turn to examine how the urban governance of biodiversity is being transformed by the growth of transnational initiatives, generating a growing "urban biodiversity complex." We suggest that these transformations in the ways in which governing biodiversity in the city are taking place each generate new forms of transformative capacity, but that this is yet to be recognized within the global biodiversity governance landscape.

14.3.1 Transforming the Modes of Governing Biodiversity in the City

If the governance architectures envisaged by international organizations a decade ago assumed that municipal authorities might be involved in contributing to the global governance of biodiversity through the development of LBSAPs that contribute to national biodiversity strategies and action plans (NBSAPs) and global goals (Puppim de Oliveira et al., 2014), this form of vertical alignment or integration is relatively rare, with only a fraction of national plans containing urban goals, and the majority of strategies and plans developed at the urban scale operating relatively independently of national biodiversity planning (Xie and Bulkeley, 2020). In part this is due to issues of capacity and competing demands within municipalities. Planning tools and mechanisms are often limited in their coverage of, or ability to address, biodiversity. Further, making the case to invest municipal funds into natural assets is also challenging in the face of pressing city needs such as housing and poverty alleviation. Nonetheless, using the planning system to assign protected areas within and on the borders of cities has remained popular as a model to govern urban biodiversity (Vaccaro et al., 2013). However, these governance approaches have drawn criticism for their top-down character, exclusionary stipulations and the associations of this form of governance with the control of nature (Vaccaro et al. 2013). Cities located in biodiversity hotspots face different challenges, as it appears that many of them lack planning approaches that are specifically geared toward harmonizing the need to simultaneously secure globally important biodiversity and the need to accommodate growing cities (Weller et al., 2019).

As well as being shaped by the challenges of implementing biodiversity planning on the ground, the lack of alignment or integration between global, national and local policy and planning for biodiversity is a result of the increasingly complex, fragmented and multiple forms through which urban biodiversity governing takes place. Analysis of fifty-four examples of urban nature-based solutions in eighteen cities found that no fewer than twelve different modes of governing were being deployed in order to govern urban nature, ranging from those that were wholly without the involvement of municipal or other government actors, such as those undertaken by philanthropic donors or community organizations, through to those that were wholly enacted by municipalities through their capacities to finance, build and implement infrastructure projects (Bulkeley, 2019). Across these modes of governing, the forms of regulation and land-use planning associated with traditional forms of biodiversity strategies and action plans were relatively muted, in comparison to a diverse range of governing mechanisms related to incentives, persuasion, provision, enabling and so forth. This reflects a broader phenomenon now extensively documented in the literature on urban sustainability governance, which suggests that experimentation has come to be a critical means of governing the city toward sustainability (Bulkeley, 2019; Bulkeley et al., 2014; Evans et al., 2016). As Karvonen (2018: 202) explains, "experiments might not simply serve as one-off trials to provide evidence and justification for new … policies, regulations, and service provision through existing circuits of policymaking and regulation. Instead, these activities are emerging as a new mode of governance in themselves." In this view, governance by experimentation is increasingly operating alongside and indeed replacing traditional "plan-led" forms of urban governance in the face of growing fragmentation of authority and the growth of the number of actors with a stake in urban futures (Bulkeley, 2019).

Alongside the trend in the growth of biodiversity governance experimentation, analysis suggests that a specific form of intervention – nature-based solutions – is also gaining momentum (Almassy et al., 2018). The governance of nature-based solutions shows strong parallels to other forms of urban experimentation (Dorst et al., 2019), which are often characterized by participation, collaboration and learning (catalyzing local and tacit knowledge), which can contribute to inclusive, transdisciplinary and adaptive governance (Frantzeskaki, 2019; Munaretto et al., 2014; Plummer et al., 2013; Reid, 2016; Triyanti and Chu, 2018). Indeed, collaborative forms of governance dominate the design and implementation of nature-based solutions in European cities and beyond (Almassy et al., 2018; Bulkeley, 2019). While significant barriers to mainstreaming nature-based solutions remain – not least with respect to knowledge about their value and operation, the disruption they pose to existing ways of undertaking urban development, and access to finance – it is apparent that at least some forms of nature-based solutions are becoming systematically deployed. For example, as a response to changing predictions of the nature and extent of urban flooding, "sponge city" and "sustainable urban drainage" approaches are now routinely used, often creating and restoring habitat within cities and contributing to conservation goals as well as generating contributions to social well-being, health and economic development. Overall, we can suggest that the growth of urban governance experimentation is fueling the uptake of nature-based solutions, which provide forms of intervention that work across a landscape of fragmented authority and a plethora of agendas around which nature-based interventions can gather, while the multiple benefits that nature-based solutions promise serves to attract more, and more varied, actors toward governing biodiversity in the city through experimentation.

Yet despite the evident ways in which urban biodiversity governance is being transformed as a result, there is less clear evidence that urban nature-based solutions are effectively addressing issues of urban inequalities, and indeed a growing literature suggests that they could have precisely the opposite effect, casting doubt on their transformative reach. Research on the phenomenon of "green gentrification" points to the ways in which urban (re)development projects that bring nature into the city can have a significant effect on widening inequalities, displacing residents as land values and house prices rise and failing to secure access to new forms of urban nature for communities who may already suffer from multiple forms of social exclusion (Anguelovski et al., 2018; Wolch et al., 2014). For example, the now famed High-Line project in New York, while often celebrated as an economic regeneration development in the city, has also been critiqued as effectively serving the interests of business, tourists and higher income groups at the expense of the (former) residents of the neighborhood (Anguelovski et al., 2018). Equally important, efforts to bring nature into cities can serve to reproduce particular ideas about what constitutes valuable or appropriate forms of nature, failing to take account of the manifold and often contested values for nature held by diverse communities. For example, the views and values of Indigenous communities concerning the kinds of nature that should be included in urban plans are often overlooked. This suggests that the governing of urban nature can be far from transformative, serving to reproduce existing social inequalities and the systems of capitalist urban development that are in many senses responsible for driving the loss of nature globally. On the other hand, where issues of social inclusion, the multiple values of nature and justice are taken into account, there is gathering evidence that efforts at governing urban nature can be transformative. In Winnipeg, for example, an initiative has been developed to harness Indigenous knowledge to develop community gardens in vacant lots in the city to provide space for alternative nature in the city and address issues of isolation and poor mental health among these social groups. How, by and for whom urban nature is governed is therefore critical in shaping its potential to be transformative of urban inequalities. Advocating for grassroots actions, the notion of urban nature stewardship offers opportunities for scientific and policy partnerships with local communities (Connolly et al., 2014; Krasny and Tidball, 2012), highlighting the importance of the openness and inclusiveness of urban nature governance that allows the participation of different stakeholders. Yet while such approaches can be transformative for those involved, many of the issues regarding exclusion and inequality at large remain challenging to address through such interventions. This in turn suggests that alongside any efforts at more inclusive governance, there needs to be space for dissent and contestation so that the nature of ongoing inequalities and their consequences can be made visible.

14.3.2 Transnational Transformations?

In parallel to the shift from an urban planning approach to biodiversity governance at the local level toward urban experimentation and nature-based solutions, we can see that the governance of urban biodiversity is also evolving in the international arena. First, within the Convention on Biological Diversity itself there has been a renewed commitment to the importance of urban action, notably through the development of the Edinburgh Process, through which local and subnational governments have been mandated by the Secretariat of the CBD to put forward their proposals for how the post-2020 Global Biodiversity Framework should advance and support their potential contributions. To date, this constituency has focused primarily on the need to ensure that the post-2020 Framework contains an explicit mandate for local and subnational action on the goals and targets agreed internationally, to replace the previous policy architecture agreed a decade ago. Second, and often in parallel, governance arrangements and initiatives concerned with the global governance of urban development have begun to recognize the potential value of urban nature. For example, *The New Urban Agenda*,[2] adopted in 2017, refers to the value of cities and human settlements that protect ecosystems and biodiversity as well as to the importance of encouraging nature-based solutions and innovations as part of urban development processes. Cross-cutting both arenas, initiatives and arrangements that are primarily concerned with the governing of climate change have increasingly signaled the potential of urban nature-based solutions as a means through which to address climate challenges as well as the biodiversity and urban sustainability agendas, for example in the report of the *Global Commission on Adaptation* and other initiatives highlighted at the 2019 UN Climate Action Summit that took place in New York (GCA and WRI, 2019;

---

[2] https://habitat3.org/the-new-urban-agenda/.

UNCC, 2019; UNDP, 2019). Across the UN environmental governance landscape, it is evident that the potential for urban responses to play an important role in transforming biodiversity governance is increasingly recognized.

At the same time, it is critical to recognize that the global architecture for urban biodiversity governance is not confined to the workings of international conventions, but also encompasses a range of actors and networks that operate transnationally. Of these, the first to be established (in 2006) was the Cities Biodiversity Centre, part of ICLEI Local Governments for Sustainability, who were appointed in 2019 as the representative of local and subnational governments within the CBD Secretariat. Over the past two years, ICLEI's Cities Biodiversity Centre has partnered with the newly formed IUCN Urban Alliance and The Nature Conservancy to form the CitiesWithNature platform, intended to provide a focal point for urban action toward the post-2020 biodiversity agenda. To date, 174 cities from 58 countries have committed to action under the CitiesWithNature umbrella. The involvement of the IUCN and The Nature Conservancy in such initiatives is particularly significant, marking a growing interest in urban biodiversity from organizations that have traditionally been concerned with conservation and restoration in a relatively conventional sense and for whom cities have been marginal to their interests. A similar urban biodiversity initiative was launched by the Secretariat of the Ramsar Convention in 2017 – the Wetland City Accreditation scheme.[3] In Europe, a number of urban projects designed to develop and implement urban nature-based solutions are being supported under the Horizon 2020 Sustainable Cities and Communities program, with a total budget of approximately 200 million Euros. These transnational initiatives primarily seek to enhance the ways in which cities are governing biodiversity within their own territories, creating a means through which both urban biodiversity planning and the increasingly diverse forms of experimentation that cities are deploying to govern biodiversity are recognized, aggregated and shared, and learning between cities is fostered. In this way, they both benefit from the fragmentation of authority to govern urban biodiversity and, through fostering new and more varied initiatives, serve to contribute toward it.

A further, if currently embryonic, trend is the emergence of transnational initiatives that are seeking to engage cities in addressing their contribution to the underlying drivers of biodiversity loss and in so doing contributing to governance *for* transformation – primarily through taking measures either to support ecosystems on which cities depend or to improve the sustainability of production and consumption. The World Resources Institute has developed the *Cities4Forests* initiative, aimed both at improving the quality and quantity of urban forest biodiversity and enhancing the role that cities play in protecting "nearby" and "faraway" forests. One of the sixty cities who have now joined this initiative, Raleigh in North Carolina, USA, has developed a water levy to pay for partnership work with upstream landowners to protect water quality in the catchment from which it draws its own water supply. As well as taking measures to protect its surrounding forest area, Kigali, Rwanda, another member of the initiative, is partnering with the Rwandan Ministry of Environment to fulfill the aim of planting trees across 43,000 hectares of land nationwide. In

---

[3] www.ramsar.org/news/wca-applications.

October 2020, twenty-six cities came together under the *European Circular Cities Declaration*, founded by The Collaborating Centre on Sustainable Consumption and Production (CSCP) together with ICLEI, the Ellen MacArthur Foundation, Eurocities, UNEP and other partners to accelerate the transition to a circular economy at the city level in order to reduce their impact on climate change and biodiversity. The long-established *C40 Cities Climate Leadership Group* is currently promoting the use of nature-based solutions to enhance building efficiency and the adaptive capacity of cities in the face of climate change, while its Food System Networks promotes regenerative urban agriculture to decrease production emissions, close yield gaps, increase food security, support local producers, decrease food miles, mitigate urban heat island effect and reduce building energy demand (through roof and wall gardens). What is notable in these initiatives is that biodiversity is often not positioned centrally to urban actions, but rather that potentially transformative forms of governing biodiversity through urban action are emerging as a "co-benefit" of urban efforts to reconfigure their economies and address climate change. Such outcomes are therefore being generated through the fragmentation, rather than integration, of governance.

There is therefore a growing density and diversity in the multilevel governance arrangements, networks, initiatives and projects through which the urban governing of biodiversity is taking shape. Taken together, the growing *governance complex* through which the governing of urban biodiversity is taking place, as well as the diversification of modes of governing through which it is being implemented, suggest that this is an arena of biodiversity governance that has been substantially transformed over the past decade. The transformation of the architectures, arrangements, networks and substance of urban nature governing, away from a specific form of urban planning concerned primarily with nature conservation and largely isolated from wider urban sustainable development and climate change goals and toward a much more fragmented, multiple and encompassing approach, not only represents a transformation in the governance dynamics at play, but has arguably also served to shift the governing of urban biodiversity on to a more transformative footing. By bringing a whole host of new actors into the realm of urban biodiversity governance and transforming both the capacities and purpose of governing biodiversity in the city, the transformation of urban biodiversity governance is arguably paving the way for a more transformative approach to biodiversity on the ground.

14.4 Conclusions

Cities hold considerable potential for conserving and restoring biodiversity, and will be critical to ensuring that society can thrive by preserving and enhancing nature's contribution to people. As we discussed in this chapter, there is now a growing realization of the importance of urban governance for nature. Of the themes of transformative governance raised in the Introduction to this book, we find most evidence of *transformations in governance* when it comes to the role of cities in biodiversity governance. First, we have argued that biodiversity governance is being transformed within cities themselves. Rather than being confined to urban planning, we find a growth of urban experimentation as various initiatives and nature-based solutions are now being undertaken by municipal authorities and their partners, as well as a range of private and community actors, to protect, restore and thrive with nature. Second, the growing recognition of cities as key agents of change and as presenting opportunities for governing biodiversity represents a transformation in biodiversity governance internationally, which has traditionally focused on cities as a threat to biodiversity and has tended to be dominated by a focus on the nation-state. This in turn is leading to a transformation in the global architecture for biodiversity governance, such that cities are now given more prominence within the global Convention on Biological Diversity. In parallel, we witness a growth of transnational networks seeking to both advocate for cities within international fora and to foster urban responses, a phenomenon both generated by and contributing to the fragmentation of authority to govern urban nature. In short, the rise of cities on the biodiversity agenda is leading to transformations in how and by whom biodiversity is governed both within the urban arena and beyond.

However, some fundamental issues persist and form the key challenges that will need to be addressed if we are to realize a transformation in how urban biodiversity governance is pursued and to what ends – in short, if we are to generate governance *for* transformations. The first issue concerns how matters of biodiversity can become mainstream within urban development and how cities come to be positioned within biodiversity governance (and vice versa). Despite a growing recognition of its importance, biodiversity is relatively marginalized in policymaking and planning in cities. Among most of those transnational networks/ initiatives that incorporate biodiversity goals and targets, biodiversity is usually regarded as a "co-benefit" of urban efforts to reconfigure their economies and address climate change. This not only limits the attention given to biodiversity per se, but also means that the underlying drivers of biodiversity loss beyond the city limits receive limited attention – for example in terms of cities addressing the impacts of their consumption or of waste in terms of their effects on the loss of biodiversity or in terms of how they compromise the capacity of other communities to realize the benefits of nature. On the one hand, a continued emphasis on the win–win potential of initiatives for addressing biodiversity while also attending to other critical urban priorities will be necessary to maintain its position on the urban agenda, yet at the same time it will be crucial that cities come to see themselves as having a fundamental role in governing nature within and beyond their own boundaries through further embedding this issue in key policy arenas and through the actions of critical stakeholders in urban development. We suggest that it is unlikely that governance for transformative action that addresses the underlying urban drivers of biodiversity loss will be found through existing institutions, but will rather require new coalitions and partnerships that bring urban actors together with those in the business and finance sectors as well as through place-to-place partnerships. Rather than expecting this to be a fully joined-up or integrated process, as with the climate agenda, we might witness a growing fragmentation and complexity of governance in order to address the critical issue of transformative change.

Second, and related, a transformative approach to biodiversity governance would necessarily need to challenge which forms of urban nature come to count in the pursuit of urban sustainability. As nature-based solutions are gaining traction, the delicate relationship between nature and society that coexists within cities becomes particularly salient, even if such forms of "hybrid nature" are not afforded much value in terms of conservation or do not represent the restoration of previously lost ecologies. Cities are spaces for new kinds of mundane nature that bring significant worth to everyday life and also provide the space for novel ecologies that consist of what might be termed invasive or non-native species, around which forms of human and nonhuman association and community are often developed. Questioning which forms of nature are seen to belong or are to be excluded from the city, by whom and to what purpose, in turn might lead to a transformation in how urban biodiversity should be understood, conserved, restored and prioritized in order that diverse communities can thrive with nature. Such an effort will require more inclusive forms of governance, as suggested in the Introduction to this book, but it also suggests that we will need to leave space open for dissent, contestation and protest in order to realize transformative governance for biodiversity.

Last but not least, how issues of social exclusion and injustice can be addressed (rather than exacerbated) is a significant problem, but one that must be solved if biodiversity governance is to become truly transformative. While a focus on inclusive governance points to the importance of ensuring equitable processes, governance *for* transformation also requires that we focus on the outcomes that are generated through interventions for biodiversity governance and how such forms of governance either serve to reproduce or challenge existing socioeconomic and power inequalities. Given that some nature-based solutions projects risk excluding minority or Indigenous communities in the project design and implementation process, displace residents who cannot afford the resulting rising house prices and can serve urban elites at the expense of others, there is a growing concern that the governing of urban nature will entrench forms of neoliberal economic development and social exclusion. Transformative biodiversity governance will necessarily involve a fundamental reordering of structures of power and knowledge that can enable social and environmental justice to be secured and enhanced, and as such is likely to be highly contested and often contradictory and fragmented. Focusing on the underlying drivers of the loss of biodiversity and the diminished and unequal contributions that nature makes to people will, as other contributions to this volume make clear, be necessary if governance is to be transformative. This in turn suggests that it will not be sufficient for global institutions and transnational networks to promote urban action on nature, but that they will need to play a critical part in building the capacity and vision needed for cities to ensure that they take action for nature within and beyond urban boundaries that not only contributes to global biodiversity goals but also ensures social justice.

References

Almassy, D., Pinter, L., Rocha, S., et al. (2018). *Urban nature atlas: A database of nature-based solutions across 100 European cities*. NATURVATION. Available from https://bit.ly/35QI1rt.

Anguelovski, I., Connolly, J. J. T., Masip, L., and Pearsall, H. (2018). Assessing green gentrification in historically disenfranchised neighborhoods: A longitudinal and spatial analysis of Barcelona. *Urban Geography* 39, 458–491.

Aronson, M. F. J., La Sorte, F. A., Nilon, C. H., et al. (2014). A global analysis of the impacts of urbanization on bird and plant diversity reveals key anthropogenic drivers. *Proceedings of the Royal Society of London B: Biological Sciences* 281, 20133330.

Aronson, M. F. J., Lepczyk, C. A., Evans, K. L., et al. (2017). Biodiversity in the city: Key challenges for urban green space management. *Frontiers in Ecology and the Environment* 15, 189–196.

Bomans, K., Steenberghen, T., Dewaelheyns, V., Leinfelder, H., and Gulinck, H. (2010). Underrated transformations in the open space: The case of an urbanized and multifunctional area. *Landscape and Urban Planning* 94, 196–205.

Booth, D. B., Roy, A. H., Smith, B., and Capps, K. A. (2016). Global perspectives on the urban stream syndrome. *Freshwater Science* 35, 412–420.

Bulkeley, H. (2019). Taking action for urban nature: Growing effective governance solutions. NATURVATION deliverable 4. Available from https://bit.ly/3goiZ4V.

Bulkeley, H. A., Broto, V. C., and Edwards, G. A. S. (2014). *An urban politics of climate change: Experimentation and the governing of socio-technical transitions.* Routledge.

Bulkeley, H., Kok, M., and Xie, L. (2021). *Realising the urban opportunity: Cities and post-2020 biodiversity governance.* PBL Briefing. Available from https://bit.ly/3rpBQCR.

Bush, J., Ashley, G., Foster, B., and Hall, G. (2021). Integrating green infrastructure into urban planning: Developing Melbourne's Green Factor Tool. *Urban Planning* 6, 20–31.

Bush, J., Miles, B., and Bainbridge, B. (2003). Merri Creek: Managing an urban waterway for people and nature. *Ecological Management and Restoration* 4, 170–179.

Butt, N., Shanahan, D. F., Shumway, N., et al. (2018). Opportunities for biodiversity conservation as cities adapt to climate change. *Geo: Geography and Environment* 5, e00052.

CBD (2016). Technical Note on Biodiversity and the 2030 Agenda for Sustainable Development. CBD/FAO/WBG/UNEP/UNDP. Available from https://bit.ly/3Gu8HLf.

Cincotta, R., Wisnewski, J., and Engelman, R. (2000). Human population in the biodiversity hotspots. *Nature* 404, 990–992.

City of Melbourne (2017). *Green our city strategic action plan 2017–2021: Vertical and rooftop greening in Melbourne.* Melbourne: City of Melbourne. Available from https://bit.ly/332LIJG.

Cohen-Shacham, E., Walters, G., Janzen, C., and Maginnis, S. (2016). *Nature-based solutions to address societal challenges.* Gland: International Union for Conservation of Nature.

Connolly, J. J. T., Svendsen, E. S., Fisher, D. R., and Campbell, L. K. (2014). Networked governance and the management of ecosystem services: The case of urban environmental stewardship in New York City. *Ecosystem Services* 10, 187–194.

Díaz, S., Pascual, U., Stenseke, M., et al. (2018). Assessing nature's contributions to people. *Science* 359, 270–272.

Díaz, S., Settele, J., Brondízio, E. S., et al. (2019). *Summary for policymakers of the global assessment report on biodiversity and ecosystem services of the Intergovernmental Science-Policy Platform on Biodiversity and Ecosystem Services.* Bonn: IPBES Secretariat.

Dorst, H., Raven R., van der Jagt, A., and Runhaar, H. (2019). Urban greening through nature-based solutions – Key characteristics of an emerging concept. *Sustainable Cities and Society* 49, 101620.

Dunn, R. R., Gavin, M. C., Sanchez, M. C., and Solomon, J. N. (2006). The pigeon paradox: Dependence of global conservation on urban nature. *Conservation Biology* 20, 1814–1816.

Elander, I., Alm, E. L., Malbert, B., and Sandström, U. G. (2005). Biodiversity in urban governance and planning: Examples from Swedish cities. *Planning Theory & Practice* 6, 283–301.

Endreny, T., Santagata, R., Perna, A., et al. (2017). Implementing and managing urban forests: A much needed conservation strategy to increase ecosystem services and urban wellbeing. *Ecological Modelling* 360, 328–335.

European Commission. (2015). *Towards an EU research and innovation policy agenda for nature-based solutions and re-naturing cities.* Final Report of the Horizon 2020 expert group on "nature-based solutions and re-naturing cities." Brussels: European Commission.

Evans, J. (2004). What is local about local environmental governance? Observations from the local biodiversity action planning process. *Area* 36, 270–279.

Evans, J., Karvonen, A., and Raven, R. (2016). *The experimental city.* London; New York: Routledge.

Fischer, L. K., Honold, J., Cvejić, R., et al. (2018). Beyond green: Broad support for biodiversity in multicultural European cities. *Global Environmental Change* 49, 35–45.

Frantzeskaki, N. (2019). Seven lessons for planning nature-based solutions in cities. *Environmental Science and Policy* 93, 101–111.

Gandy, M. (2002). *Concrete and clay: Reworking nature in New York City*. Cambridge, MA: MIT Press.

(2004). Rethinking urban metabolism: Water, space and the modern city. *City* 8, 363–379.

Garrard, G. E., Williams, N. S. G., Mata, L., Thomas, J., and Bekessy, S. A. (2017). Biodiversity sensitive urban design. *Conservation Letters* 11, 1–10.

Garson, J. (2016). Ecological restoration and biodiversity conservation. In *The Routledge handbook of philosophy of biodiversity*. J. Garson, A. Plutynski and S. Sarkar (Eds.), pp. 326–337. New York: Routledge.

Gavin, M., McCarter, J., Berkes, F., et al. (2018). Effective biodiversity conservation requires dynamic, pluralistic, partnership-based approaches. *Sustainability* 10, 1846.

Geneletti, D., and Zardo, L. (2016). Ecosystem-based adaptation in cities: An analysis of European urban climate adaptation plans. *Land Use Policy* 50, 38–47.

Gleeson, B., and Low, N. (2000). Cities as consumers of the world's environment. In *Consuming cities: The urban environment in the global economy after the Rio declaration*. N. Low, B. Gleeson, I. Elander and R. Lidskog (Eds.), pp. 1–29. London: Routledge.

Global Commission on Adaptation (GCA) and World Resources Institute (WRI). (2019). *Adapt now: A global call for leadership on climate resilience*. Available from https://bit.ly/3HvITQ9.

Goddard, M. A., Dougill, A. J., and Benton, T. G. (2010). Scaling up from gardens: Biodiversity conservation in urban environments. *Trends in Ecology & Evolution* 25, 90–98.

Haase, D., Kabisch, N., and Haase, A. (2013). Endless urban growth? On the mismatch of population, household and urban land area growth and its effects on the urban debate. *PLoS ONE* 8, e66531.

Hall, D. M., Camilo, G. R., Tonietto, R. K., et al. (2017). The city as a refuge for insect pollinators. *Conservation Biology* 31, 24–29.

Hopwood, J. L. (2008). The contribution of roadside grassland restorations to native bee conservation. *Biological Conservation* 141, 2632–2640.

Ives, C. D., Lentini, P. E., Threlfall, C. G., et al. (2016). Cities are hotspots for threatened species. *Global Ecology and Biogeography* 25, 117–126.

Kabisch, N., Frantzeskaki, N., Pauleit, S., et al. (2016). Nature-based solutions to climate change mitigation and adaptation in urban areas: Perspectives on indicators, knowledge gaps, barriers, and opportunities for action. *Ecology and Society* 21, 39.

Kabisch, N., Qureshi, S., and Haase, D. (2015). Human environment interactions in urban green spaces—A systematic review of contemporary issues and prospects for future research. *Environmental Impact Assessment Review* 50, 25–34.

Karvonen, A. (2018). The city of permanent experiments? In *Innovating climate governance: Moving beyond experiments*. B. Turnheim, P. Kivimaa and F. Burkhout (Eds.), pp. 201–215. Cambridge: Cambridge University Press.

Kendal, D., Zeeman, B., Ikin, K., et al. (2017). The importance of small urban reserves for plant conservation. *Biological Conservation* 213, 146–153.

Krasny, M. E., and Tidball, K. G. (2012). Civic ecology: A pathway for Earth stewardship in cities. *Frontiers in Ecology and the Environment* 10, 267–273.

Kuras, E. R., Warren, P. S., Zinda, J. A., et al. (2020). Urban socioeconomic inequality and biodiversity often converge, but not always: A global meta-analysis. *Landscape and Urban Planning* 198, 103799.

Lennon, M. (2015). Nature conservation in the Anthropocene: Preservation, restoration and the challenge of novel ecosystems. *Planning Theory and Practice* 16, 285–290.

Maller, C., and Farahani, L. M. (2018). *Snakes in the city: Understanding residents' responses to greening interventions for biodiversity*. Adelaide: State of Australian Cities Conference.

Mattijssen, T., Buijs, A., Elands, B., and Arts, B. (2018). The "green" and "self" in green self-governance – A study of 264 green space initiatives by citizens. *Journal of Environmental Policy and Planning* 20, 96–113.

McDonald, R. I., Colbert, M. L., Hamann, M., et al. (2018). *Nature in the urban century: A global assessment of where and how to conserve nature for biodiversity and human wellbeing.* The Nature Conservancy. Available from https://apo.org.au/node/204131.

McGregor, B. A., and McGregor, A. M. (2020). Communities caring for land and nature in Victoria. *Journal of Outdoor and Environmental Education* 23, 153–171. doi:10.1007/s42322-020-00052-9

Miller, J. R., and Hobbs, R. J. (2002). Conservation where people live and work. *Conservation Biology* 16, 330–337.

Munang, R., Thiaw, I., Alverson, K., et al. (2013). Climate change and ecosystem-based adaptation: A new pragmatic approach to buffering climate change impacts. *Current Opinion in Environmental Sustainability* 5, 67–71.

Munaretto, S., Siciliano, G., and Turvani, M. E. (2014). Integrating adaptive governance and participatory multicriteria methods: A framework for climate adaptation governance. *Ecology and Society* 19, 74. http://dx.doi.org/10.5751/ES-06381-190274

Nesshöver, C., Assmuth, T., Irvine, K. N., et al. (2017). The science, policy and practice of nature-based solutions: An interdisciplinary perspective. *Science of the Total Environment* 579, 1215–1227.

Niemela, J. (1999). Ecology and urban planning. *Biodiversity & Conservation* 8, 119–131.

Owens, S. E. (1992). Energy, environmental sustainability and land use planning. In *Sustainable development and urban form.* M. Breheny (Ed.), pp. 79–105. London: Pion.

Parris, K. M., Amati, M., Bekessy, S. A., et al. (2018). The seven lamps of planning for biodiversity in the city. *Cities* 83, 44–53.

Pascual, U., Balvanera, P., Díaz, S., Pataki, G., and O'Farrell, P. (2017). Valuing nature's contribution to people. *Current Opinion in Environmental Sustainability* 26–27, 7–16.

Pattberg, P., Widerberg, O., and Kok, M. T. J. (2019). Towards a global biodiversity action agenda. *Global Policy* 10, 385–390. doi:10.1111/1758-5899.12669

Patterson, J., Schulz, K., Vervoort, J., et al. (2017). Exploring the governance and politics of transformations towards sustainability. *Environmental Innovation and Societal Transitions* 24, 1–16. https://doi.org/10.1016/j.eist.2016.09.001

Pauleit, S., Zölch, T., Hansen, R., Randrup, T. B., and Konijnendijk van den Bosch, C. (2017). Nature-based solutions and climate change – Four shades of green. In *Nature-based solutions to climate change adaptation in urban areas. Theory and practice of urban sustainability transitions.* N. Kabisch, H. Korn, J. Stadler and A. Bonn (Eds.), pp. 29–49. Cham: Springer.

Planchuelo, G., von Der Lippe, M., and Kowarik, I. (2019). Untangling the role of urban ecosystems as habitats for endangered plant species. *Landscape and Urban Planning* 189, 320–334.

Plummer, R., Armitage, D. R., and de Loë, R. C. (2013). Adaptive comanagement and its relationship to environmental governance. *Ecology and Society* 18, 21. http://dx.doi.org/10.5751/ES-05383-180121.

Prober, S. M., Doerr, V. A. J., Broadhurst, L. M., Williams, K. J., and Dickson, F. (2019). Shifting the conservation paradigm: A synthesis of options for renovating nature under climate change. *Ecological Monographs* 89, e01333. https://doi.org/10.1002/ecm.1333.

Prober, S. M., Williams, K. J., Broadhurst, L. M., and Doerr, V. A. J. (2017). Nature conservation and ecological restoration in a changing climate: What are we aiming for? *Rangeland Journal* 39, 477–486.

Puppim de Oliveira, J. A., Doll, C. N. H., Moreno-Peñaranda, R., and Balaban, O. (2014). Urban biodiversity and climate change. In *Global environmental change.* B. Freedman (Ed.), pp. 461–468. Dordrecht: Springer Netherlands.

Raymond, C. M., Frantzeskaki, N., Kabisch, N., et al. (2017). A framework for assessing and implementing the co-benefits of nature-based solutions in urban areas. *Environmental Science & Policy* 77, 15–24.

Reid, H. (2016). Ecosystem- and community-based adaptation: Learning from community-based natural resource management. *Climate and Development* 8, 4–9.

Rydin, Y. (1998). *Urban and environmental planning in the UK. Planning, environment, cities.* London: Palgrave.

Schröter, M., van der Zanden, E. H., van Oudenhoven, A. P. E., et al. (2014). Ecosystem services as a contested concept: A synthesis of critique and counter-arguments. *Conservation Letters* 7, 514–523.

Sievers, M., Hale, R., Swearer, S. E., and Parris, K. M. (2019). Frog occupancy of polluted wetlands in urban landscapes. *Conservation Biology* 33, 389–402.

Soanes, K., and Lentini, P. E. (2019). When cities are the last chance for saving species. *Frontiers in Ecology and the Environment* 17, 225–231.

Soanes, K., Sievers, M., Chee, Y. E., et al. (2019). Correcting common misconceptions to inspire conservation action in urban environments. *Conservation Biology* 33, 300–306.

Swyngedouw, E., and Kaika, M. (2000). The environment of the city...or the urbanisation of nature. In *A companion to the city*. G. Bridge and S. Watson (Eds.), pp. 96–107. Oxford: Blackwell.

Threlfall, C. G., Law, B., and Banks, P. B. (2012). Sensitivity of insectivorous bats to urbanization: Implications for suburban conservation planning. *Biological Conservation* 146, 41–52.

Threlfall, C. G., Mata, L., Mackie, J. A., et al. (2017). Increasing biodiversity in urban green spaces through simple vegetation interventions. *Journal of Applied Ecology* 54, 1874–1883.

Threlfall, C. G., Soanes, K., Ramalho, C. E., et al. (2019). *Conservation of urban biodiversity: A national summary of local actions*. Report prepared by the Clean Air and Urban Landscapes Hub. Melbourne: Clean Air and Urban Landscapes Hub.

Tozer, L., Hörschelmann, K., Anguelovski, I., Bulkeley, H., and Lazova, Y. (2020). Whose city? Whose nature? Towards inclusive nature-based solution governance. *Cities* 107, 102892.

Triyanti, A., and Chu, E. (2018). A survey of governance approaches to ecosystem-based disaster risk reduction: Current gaps and future directions. *International Journal of Disaster Risk Reduction* 32, 11–21.

UNCC. (2019). *Climate action and support trends*. Bonn: United Nations Climate Change. Available from https://bit.ly/3umMe0l.

UNDP. (2019). *The heat is on – Taking stock of global climate ambition*. NDC Global Outlook Report 2019. Available from https://bit.ly/3orA1Ul.

Vaccaro, I., Beltran, O., and Paquet, P. A. (2013). Political ecology and conservation policies: Some theoretical genealogies. *Journal of Political Ecology* 20, 255–272.

Weller, R., Drozdz, Z., and Kjaersgaard, S. P. (2019). Hotspot cities: Identifying peri-urban conflict zones. *Journal of Landscape Architecture* 14, 8–19.

Williams, N. S. G., McDonnell, M. J., Phelan, G. K., Keim, L. D., and Van Der Ree, R. (2006). Range expansion due to urbanization: Increased food resources attract grey-headed flying-foxes (Pteropus poliocephalus) to Melbourne. *Austral Ecology* 31, 190–198.

Wilkinson, C., Sendstad, M., Parnell, S., and Schewenius, M. (2013). Urban governance of biodiversity and ecosystem services. In *Urbanization, biodiversity and ecosystem services: Challenges and opportunities*. T. Elmqvist, M. Fragkias, J. Goodness, et al. (Eds.), pp. 539–587. Dordrecht: Springer.

Wintle, B. A., Kujala, H., Whitehead, A., et al. (2019). Global synthesis of conservation studies reveals the importance of small habitat patches for biodiversity. *Proceedings of the National Academy of Sciences of the United States of America* 116, 909–914.

Wolch, J., Byrneb, J., and Newell, J. P. (2014). Urban green space, public health, and environmental justice: The challenge of making cities "just green enough." *Landscape and Urban Planning* 125, 234–244.

Xie, L., and Bulkeley, H. (2020). Nature-based solutions for urban biodiversity governance. *Environmental Science & Policy* 110, 77–87.

Young, T. P. (2000). Restoration ecology and conservation biology. *Biological Conservation* 92, 73–83.

15
Transformative Governance for Ocean Biodiversity

BOLANLE ERINOSHO, HASHALI HAMUKUAYA, CLAIRE LAJAUNIE, ALANA MALINDE S. N. LANCASTER, MITCHELL LENNAN, PIERRE MAZZEGA, ELISA MORGERA AND BERNADETTE SNOW

15.1 Introduction

The ocean's enormity and depth are illustrated by the limited ability of humankind to comprehend it. The current science and policy seascape remains largely fragmented, and as a result the integrity of marine life and the well-being of those (human and nonhuman) dependent on a healthy ocean is being negatively impacted. Fragmented governance is an indirect driver of ocean biodiversity loss due to its inability to provide synergistic solutions to address simultaneously multiple direct drivers for such loss (overfishing, land-based and marine pollution, and climate change). This governance problem is well known (Kelly et al., 2019; Watson-Wright and Valdés, 2018), and to some extent it is being addressed in ongoing international negotiations on an international instrument on marine biodiversity of areas beyond national jurisdiction (A/RES/72/249, 2017).

This chapter will shed new light on these well-known problems by applying the lens of "transformative governance," understood as "formal and informal (public and private) rules, rule-making systems and actor-networks at all levels of human society (from local to global) that enable transformative change ... towards biodiversity conservation and sustainable development more broadly," with a view to "respond[ing] to, manag[ing], and trigger[ing] regime shifts in coupled socio-ecological systems at multiple scales" (Visseren-Hamakers et al., 2021: 21; see also Chaffin et al., 2016 and Chapter 1 of this volume). We share the editors' views that there is a need to shift away "from the technocratic and regulatory fix of environmental problems to more fundamental and transformative changes in social-political processes and economic relations" (Otsuki (2015: 1; see also Chapter 1 of this volume). This can also help us to better understand how ocean biodiversity can contribute to "other environmental and social justice issues"[1] that are interwoven with the ocean in less visible ways than terrestrial biodiversity, such as poverty (Singh et al., 2018) and resource-grabbing (Virdin et al., 2021).[2]

All the authors are part of the One Ocean Hub, a collaborative research for sustainable development project funded by UK Research and Innovation (UKRI) through the Global Challenges Research Fund (GCRF) (Grant Ref: NE/S008950/1). GCRF is a key component in delivering the UK AID strategy and puts UK-led research at the heart of efforts to tackle the United Nations Sustainable Development Goals. In addition, Mr. Hamukuaya was financially supported by the National Research Foundation (NRF) toward this research: Opinions expressed and conclusions arrived at are those of the author and are not necessarily to be attributed to the NRF.

[1] Chapter 1 in this volume.
[2] The term "ocean-grabbing" is increasingly utilized to refer to a situation "[w]here the benefits from use of finite ocean space and resources characterized as public goods are captured by a few, while traditional ocean users (who are often politically

In particular, the chapter will illustrate the broad recognition of the vital need for *integrative* and *inclusive* governance of ocean biodiversity, to ensure that solutions also have sustainable impacts at other scales and in other sectors, and to empower those whose interests are currently not being met and represent transformative sustainability values.[3] The complementary roles of adaptive governance (enabling learning, experimentation, reflexivity, monitoring and feedback) and anticipatory (precautionary) governance will also be touched upon. The latter has been extensively debated in international legal scholarship (Guston, 2014; Birnie et al., 2009), so we will reflect on how the former can contribute to the latter. Fundamentally, however, the chapter will focus on the role of *transdisciplinary* governance (the recognition of different knowledge systems and the inclusion of underrepresented types of knowledge) in supporting integration, inclusion and learning in ocean affairs for transformative change.

Accordingly, this chapter will first engage in a brief analysis of the major underlying causes of marine biodiversity loss, by drawing on global synthesis reports. Second, considering the extensive literature assessing existing regulatory mechanisms and their effects on the status and uses of marine biodiversity, this chapter proposes to focus specifically on the lessons learned for transformative ocean governance in the context of area-based management and spatial planning from the international to the local level. Finally, an alternative governance approach will be proposed as a possible way forward, building on the factual and legal interdependencies between human rights and marine biodiversity. The chapter will suggest taking a broader approach to fair and equitable benefit-sharing to shift toward transformative governance for the ocean at different scales.

15.2 Marine Biodiversity Loss: Causes and Consequences

The ocean is an integrated physical and biological system that provides a multitude of planetary services. These include the provision of half of the oxygen we breathe, absorption of 26 percent of anthropogenic CO_2 emissions from the atmosphere, and rich and diverse life (UNGA, 2016: A/70/112). The full extent of the ocean's biodiversity is not fully known or understood, but there is sufficient knowledge indicating that marine life is declining dramatically, albeit not yet irreversibly (Serrao-Neumann et al., 2016). Additionally, we have limited understanding of the intrinsic, as well as the social and cultural, values of marine biodiversity, and its multiple contributions to human identity and well-being (IPCC, 2019).

The causes of marine biodiversity loss are numerous, pervasive and interconnected. Globally, the major direct drivers include overexploitation, climate change and pollution. The increasing number of zoonotic pathogens associated with biodiversity loss is also affecting marine life, as well as humans (Morand and Lajaunie, 2017). Examples include outbreaks of influenza in seabird populations, and distemper morbillivirus in seal colonies

---

marginalized) lose access to resources and a just operating space within the ocean economy. For example, loss of access for small-scale fisheries, which are by far the ocean's largest employers, has threatened human rights and exacerbated inequity" (Virdin et al., 2021).

[3] Chapter 1 in this volume.

(Bogomolni et al., 2008; Morand and Lajaunie, 2017; Waltzek et al., 2012). This led to calls for a more comprehensive global approach in 2020 as the COVID-19 pandemic raged (Corlett, 2020; Ostfeld, 2009), and serves as a reminder of the links between human well-being and healthy, resilient ecosystems. The following subsections will explore threats to marine biodiversity on the basis of seminal global scientific assessments (UNGA, 2016: A/70/112; FAO, 2020; IBPES, 2019; IPCC, 2019).

15.2.1 Exploitation of Living and Nonliving Marine Resources

The exploitation of marine resources has brought about the largest relative impact on biodiversity since 1970 (IPBES, 2019). Illustrative examples may be drawn from fisheries and aquaculture, as well as the projected impacts of commercial mining activities in the deep seabed, all of which can contribute to habitat and biodiversity loss in the ocean.

Fishing has had the most impact on marine biodiversity in the past fifty years, including impacts across scales on target and nontarget species, habitats and ecosystems (IPBES, 2019). Combined with the effects of climate change, fishing is expected to remain a leading driver in worsening the state of marine biodiversity (IPBES, 2019). Funded by harmful government subsidies, commercial fishing fleets have expanded geographically and into deeper waters that were previously not financially viable to exploit (IPBES, 2019; Sumaila et al., 2019), directly contributing to a global decline in fish stocks (FAO, 2020). Fishing above sustainable levels causes negative impacts on marine biodiversity and reduces fish productivity and ecosystem functioning (FAO, 2020). Bycatch caused by nonselective fishing methods impacts marine biodiversity, and some fishing gear, such as bottom trawls and pelagic drift nets, also cause damage to habitats and biodiversity. The United Nations has recognized that the threat of illegal, unreported and unregulated (IUU) fishing goes beyond the depletion of fish populations, and there is a close nexus between the illegal activities in fisheries and transnational organized criminal activity, known as fisheries crime (A/63/111, 2008).[4] Fisheries crime threatens fish stocks and undermines the international goal to conserve and use the ocean for sustainable development (A/RES/70/1, 2015; A/RES/60/31/2006). Finally, the impacts of fisheries crime are being exacerbated by climate change (Cheung, 2016; IPBES, 2019; NIC, 2016).

Aquaculture, whether it is coastal farming or offshore aquaculture (Holmer, 2010), has been promoted as a means to address both overfishing and food security, but may have a negative impact on the environment and biodiversity, mainly arising from excess feed, pesticides and medicines leaching into the marine environment (Tovar et al., 2000). Aquaculture may affect ecosystems and biodiversity with the loss of critical habitats like mangrove or wetlands, with consequences for coastal protection (Páez-Osuna, 2001), or the alteration of hydrologic regimes by the use of structures such as fish cages (Eng et al., 1989).

---

[4] There is no universally accepted definition of fisheries crime, and different organizations describe this concept differently. The United Nations Office on Drugs and Crime (UNODC), for example, describes fisheries crime as "[a]n ill-defined legal concept referring to a range of illegal activities in the fisheries sector. These activities – frequently transnational and organised in nature – include illegal fishing, document fraud, trafficking, and money laundering. Criminal activities in the fisheries sector are often regarded as synonymous with illegal fishing, which many States do not view or prosecute as criminal offences, but rather as a fisheries management concern." Refer to the UNODC *Fisheries Crime*, at https://bit.ly/3GYAGUv.

The intensification of aquaculture has a dramatic effect on seabed fauna and their abundance (Diana, 2009; Tsutsumi et al., 1991). In turn, coastal pollution (agriculture, hydrocarbon, heavy metals) and marine pollution affect the success of aquaculture (Eng et al., 1989).

15.2.2 Pollution

Pollution is the direct or indirect introduction by humans of substances that result or are likely to result in deleterious effects to the environment (UNCLOS, Art. 1(4)). Marine and coastal areas are highly vulnerable to pollution from activities on land or at sea, which have a direct impact on marine biodiversity. Land-based pollution comes in many forms, including nutrient run-off (untreated sewage), agricultural and industry run-off such as pesticides, heavy metals or oils entering river systems and then the open ocean (UNEP/EA.4/Res.11, 2019). Marine pollution can come from a variety of activities at sea, including plastics from discarded fishing gear, dumping from vessels and underwater noise (UNEP/EA.3/L.19, 2018).[5] Marine environmental pollution has gathered international attention, as captured in Sustainable Development Goal (SDG) 14.1: "By 2025, prevent and significantly reduce marine pollution of all kinds, in particular from land-based activities, including marine debris and nutrient pollution."

Plastic pollution is pervasive in the marine environment, and the widespread impacts of macro- and microplastics on marine biodiversity at all levels are sobering. Addressing plastic pollution presents a complex governance challenge and is subject to intensified international attention. For example, the UN has highlighted the pervasive nature of plastic pollution, highlighting that between 4.8–12.7 million tons of plastic enters the ocean annually (UNEP/EA.3/L.19, 2017). The vast majority of this (~80 percent) is from land-based sources,[6] while the rest comes from maritime activities, including fishing (Isensee and Valdes, 2015), which requires stronger monitoring and control by states to prevent plastic entering ocean systems (Haward, 2018) and potentially new measures at the international level (Borrelle, et al., 2017).

Deep-seabed mining for minerals and rare-earth metals at a commercial scale occurs in areas within national jurisdiction and may soon be a reality in the Area (which is the seabed beyond the jurisdiction of any state; one of the two areas outside national jurisdiction, together with the high seas) (Casson et al., 2020).[7] Noise and light pollution, as well as sediment plumes, may have a harmful effect on marine species, while the mining itself may permanently destroy deep-sea habitats and may impact communities relying on fish stocks, with potential human rights implications (Miller et al., 2018). Deep-sea sediments act as long-term stores of atmospheric carbon, meaning mining activities may pose an additional climate risk by releasing carbon through sediment disturbance (Sala et al., 2021).[8] Climate

---

[5] See also, for example, https://bit.ly/3tSSBYU.

[6] Isensee and Valdes (2015) estimated that around 4.8–12.7 million tonnes of plastic is dumped in the ocean from land-based sources.

[7] Article 1(1)(1) of UNCLOS defines the "Area" to be "the seabed and ocean floor and subsoil thereof, beyond the limits of national jurisdiction." Within Namibia's jurisdiction, commercial seabed mining activities for diamonds occur and may soon expand to mining the seabed for phosphate. (Casson et al., 2020).

[8] Seabed disturbance can remineralize carbon stored in the seabed into CO_2 which can be subsequently dissolved into the ocean or released into the atmosphere; the following study suggests protecting the carbon-rich seabed as a nature-based solution to climate change (Sala et al., 2021).

change is also predicted to alter deep-ocean environments and to be exacerbated by other deep-sea extractive activities such as oil and gas extraction and bottom fishing (Levin et al., 2020).

15.2.3 Climate Change

There is scientific consensus that human-induced climate change is altering the physical and chemical makeup of the ocean (Stocker et al., 2013). The main impacts of climate change on the ocean are warming (IPCC, 2019), acidification and deoxygenation, which simultaneously occur due to increasing carbon dioxide (CO_2) and other greenhouse gas emissions (Beaugrand et al., 2015; Molinos et al., 2016). These changes are expected to persist throughout this century, as levels of CO_2 increase to those unseen in human times (Gattuso et al., 2015). Transformative governance has thus been recommended by the Intergovernmental Panel on Climate Change to address and adapt to these issues (IPCC, 2018).

The consequences of climate change on marine biodiversity include species extinction, local changes in species richness, proliferation of invasive species, ecosystem collapse, and disruption of ecosystem functioning and services (Beaugrand et al., 2015; Cheung et al., 2009; FAO, 2018; IPCC, 2019; Molinos et al., 2016). In addition, climate change is projected to decrease net ocean primary production and fish biomass (IPBES, 2019). Changes in the distribution of fish populations from historical locations can affect livelihoods, income and food security (IPCC, 2019), and increase conflicts between fishers, communities, authorities and states, highlighting a need for *adaptive* governance in the conservation and management of marine species (Spijkers et al., 2019; SROCC, 2019).

Roughly half of the CO_2 emitted by anthropogenic activities between 1800 and 1994 is stored in the deep ocean as organic matter from absorption by planktonic organisms (Sabine et al., 2004). Since 1980, this uptake has been between 20 percent and 30 percent of total anthropogenic CO_2 emissions, causing an increase in ocean acidification (IPCC, 2019). Acidification of the ocean decreases its ability to uptake and store carbon (IPBES, 2019), and leads to habitat destruction, with coral reef ecosystems particularly under threat (IPCC, 2019), alteration of marine food webs (Feely et al., 2004; Kleypas et al., 1999) and sensory perception changes in marine species (Dixson et al., 2010; Munday et al., 2009; 2010).

As a result of both climate change and pollution, ocean deoxygenation has become a pervasive yet overlooked issue. Deoxygenation is caused by the warming of ocean waters, from agricultural run-off into rivers and from the atmosphere from the burning of fossil fuels (Laffoley and Baxter, 2019). This causes species loss, resulting in changes in ecosystem structure and function (Laffoley and Baxter, 2019). There has been a marked loss in ocean oxygen levels from the surface to 1000 m depth since 1970, leading to the prevalence of oxygen minimum zones, which are uninhabitable for many marine species (IPCC, 2019).

15.2.4 Lessons Learned

While our global understanding of the multiple threats to marine biodiversity is growing, ocean science is "still weak in most countries" due to limited holistic approaches for understanding cumulative impacts of various threats, and lack of capacity to conduct science (A/71/733, 2017). Low- and middle-income countries face the greatest challenges in this regard: to prevent and mitigate negative development impacts connected to the ocean, participate in traditional and emerging ocean activities (Blasiak, 2018), and predict and harness the socioeconomic benefits of ocean conservation (Blasiak, 2018). As a result, scientific understanding of the effectiveness of conservation and management responses is poor, meaning it is more difficult to predict the productivity limits and recovery time of marine ecosystems in these countries. Meanwhile, the negative social, economic and cultural impacts of degraded mangroves and corals on local communities are increasingly noted (CBD, Decision XII/23, 2014), as are the negative impacts of declining fisheries on the human rights to food and culture (A/67/268, 2012). The urgency of advancing ocean science, in and to the benefit of all countries, is expected to take centerstage globally, with the UN declaring 2021–2030 as the Decade of Ocean Science for Sustainable Development (UNESCO, 2020).

This situation is compounded by limited efforts to bridge different knowledge systems (notably Indigenous and local knowledge), which contributes to marginalizing these knowledge holders from relevant decision-making, even if these groups are disproportionally affected by the negative consequences. Furthermore, limited understanding of the benefits that derive from a healthy ocean for society and the economy fuels a "disconnect" between some communities and the ocean (Jamieson et al., 2021). In effect, only recently have global scientific reviews highlighted the multiple dependencies of people's right to health on the marine environment (WHO/CBD, 2015; A/HRC/34/49, 2017; A/75/161, 2020).

From a transdisciplinary governance perspective, all the facts observed and anticipated scenarios in the global reports analyzed above are not equally known, and even less equally predictable. For instance, if the recent rate of fishing capture is maintained, the collapse of some fisheries is almost certain, while others, especially close to the shores of the more important fishing nations, have already collapsed, leading these states to travel greater distances, thereby replicating the process elsewhere. It is also projected that the warming and acidification of the ocean will exacerbate this. In contrast, the severity and the intensity of the impacts that will result from deep-sea mining is very difficult to evaluate, as are as the effects of all the occurring changes that are cascading through unpredictable interactions. Here, the limited predictability of changes in the state of the ocean and marine resources is not a matter of observation, monitoring techniques or models (Mazzega, 2018). Rather, unpredictability is intrinsic to the complex dynamics of the ocean system, emphasizing the need for ocean governance to be anticipatory and adaptive.[9]

Furthermore, while the main trends summarized above represent scientific consensus, these global syntheses of current knowledge are based on a small fraction of the volume of

---

[9] Chapter 1 in this volume.

articles annually published on these themes.[10] The limitation of these systematic reviews is of particular concern because the impacts of human activities and environmental changes on biodiversity are for the vast majority manifesting at relatively local scales, in specific ecosystems or biomes. They require careful observations and analysis in context (Allan et al., 2013).

15.3 An Assessment of Existing Mechanisms for Ocean Governance

The international legal framework for the ocean is considered "critical" to make progress in all target areas of SDG 14: "life below water" (A/71/733, 2015). The international framework, though, is notoriously so complex and fragmented (sectorally and geographically) that it presents colossal challenges to effective, let alone transformative, ocean governance. To an extent, fragmentation is the result of historical processes of international lawmaking. The earliest marine treaties focused on clarifying the rights and obligations of states over portions of the ocean,[11] establishing safeguards,[12] regulating discharge of wastes and pollution from shipping,[13] and managing fishing resources. The next wave of treaties prioritized specific objectives, including the protection of (marine) species.[14] However, the narrow scope and diverse approaches encapsulated within these instruments often failed to consider the impacts on ecosystems in a holistic and integrated manner (Kimball, 2001; Mossop, 2007). As these treaties resulted in a patchwork approach to marine management, early attempts at integrated ocean governance began with the negotiations of the 1982 United Nations Convention of the Law of the Sea (UNCLOS).[15]

UNCLOS, commonly referred to as the "constitution of the oceans," firmly embodies elements of customary international law, as well as several innovative features for a more comprehensive approach to the regulation of ocean activities, including on the basis of a general obligation to protect and preserve the marine environment. UNCLOS, however, heavily relies on other international instruments and mechanisms, thereby confirming the continued relevance of sectoral and regional governance approaches.

For instance, the UN Fish Stocks Agreement (UNFSA) implements UNCLOS Articles 63–68, and 116–120 on straddling and highly migratory fish, and sets out obligations to ensure sustainable fishing activities and mitigate the impacts of fishing on the marine environment and biodiversity, applying the precautionary principle when scientific information is inadequate or absent (Art. 6). UNFSA, in turn, is significantly underpinned by regional, collaborative approaches (Arts. 9 and 15). Arguably, therefore, UNFSA both

---

[10] This situation should be compared with the synthesis of knowledge on the climate (see Minx et al., 2017).
[11] For example, the Byzantine *Lex Rhodia,* the *Rolls of Oléron* and the *Laws of Wisby.*
[12] For example, the General Treaty for the Cessation of Plunder and Piracy by Land and Sea, Dated February 5, 1820 and the 1914 International Convention for the Safety of Life at Sea.
[13] International Convention for the Prevention of Pollution from Ships 1973/38; Convention on the Prevention of Marine pollution by Dumping of Wastes and other Matter 1972; 1996 Protocol (London Protocol).
[14] For example, the International Convention for the Regulation of Whaling (ICRW), Washington DC, December 2, 1946, in force November 10, 1948; 161 UNTS 17, 338 UNTS 336; Convention on the Conservation of Migratory Species of Wild Animals (CMS), Bonn, June 23, 1979, in force November 1, 1983, 19 ILM (1980) 15; Convention on International Trade in Endangered Species of Wild Fauna and Flora (CITES), Washington DC, March 3, 1973, in force July 1, 1975, 993 UNTS 243.
[15] United Nations Convention on the Law of The Sea (UNCLOS), Montego Bay, December 10, 1982, in force November 16, 1994, 21 ILM 1261.

Table 15.1 *Main biodiversity-related changes*

Biodiversity Related Changes	Direct Drivers* CC F E			Local	Regional	Global	International instruments					Organisations	
							UNCLOS	FSA	CBD	CMS	CITES	ISA	IMO
Species Extinction							x	x	x	x	x	?	
Decrease Net Primary Production								?	x				
Decrease Fish Biomass								x	x				
Perturbation of Ecosystem Functioning								x	x		x		
Protection of Marine Habitats							x	x	x	x	x	x	x
Perturbation of Life Cycles								?	x	x			
Proliferation Invasive Species							x		x	x	?		x
Community Recomposition								x	x				
Loss of Species Richness								x	x	?			

*CC: climate change ; F: Fisheries ; E: exploitation of non-living resources

Direct drivers (climate change: CC; fisheries: F; exploitation of nonliving resources: E), spatial scales (local, regional, global), concerned conventions and organizations analyzed in the chapter. An x indicates that the authors understand the conventions concerned, or the instruments deployed by the organizations have sought to address these changes and drivers. A question mark indicates the conventions or their decisions may be applicable to these changes and drivers, but need further study. The table is meant as a basis for discussion with other legal and nonlegal experts, as the understanding of governance landscape may be subject to differing interpretations.

requires, and sets the conditions for, an integrative, anticipatory and inclusive approach at the regional level, which, with the correct synergies, may be scaled up to the global level. Examples of such approaches will be discussed in Section 15.2.3.

While UNCLOS reflects to some extent the evolution of natural sciences and ecosystem management by referring to the interrelatedness of the problems of ocean spaces and the need to consider them as a whole, a parallel legal development under international environmental law has also contributed to a more integrative and inclusive approach to ocean governance. This is the case of the Convention on Biological Diversity (CBD)[16] and its objectives of conservation, sustainable use, and fair and equitable benefit-sharing (Morgera and Razzaque, 2017). Over the years, the CBD has provided integrative tools to complement earlier biodiversity-related treaties, including the Convention on International Trade in Endangered Species of Wild Fauna and Flora (CITES) and the Convention on the Conservation of Migratory Species of Wild Animals (CMS) (UNEP-WCMC, 2012), and contributed to addressing the nexus between the ocean, climate change and biodiversity (Morgera, 2011; Diz, 2017). It has also addressed an increasing number of new and emerging human activities that pose challenges to biodiversity conservation and sustainable use, such as renewables development, which can increase demands for ocean space (UNCTAD/DITC/TED/2014/5). In doing so, the CBD has also addressed the specific concerns of Indigenous peoples and local communities (IPLC), and highlighted the importance of their knowledge (Morgera, 2020), thereby contributing to defining inclusive and transdisciplinary ocean governance.

These developments have occurred under the CBD ecosystem approach (CBD Decisions V/6, 2000; VII/11, 2004), which aims at integrating the management of land, water and living resources, and balancing the three objectives of the Convention, as well as integrating different legal and management strategies, depending on local, national, regional or global conditions (CBD Decision V/6, 2000, Annex, para. 5), through adaptive management and precaution (thereby contributing to adaptive and anticipatory governance) (Morgera, 2011). The ecosystem approach also aims to integrate modern science and Indigenous and local knowledge (CBD Decision V/6, 2000, Principle 11), as well as equity concerns, recognizing that human beings and their cultural diversity are an integral component of many ecosystems (CBD Decision V/6, 2000, para. 2). Under this umbrella, one of the key obligations under the CBD is to establish a system of protected areas (CBD, Art 8[a]). This was complemented with a target of a 10 percent increase in marine protected areas (MPA) coverage by 2020 among the Aichi Biodiversity Targets[17] by implementing effective and equitable protection of marine and coastal areas, particularly those important for biodiversity and ecosystem services (Aichi target 11).[18] Scientific guidance for the development of representative MPA networks had been previously adopted by CBD Parties in 2008 (CBD Decision X/2, 2010, target 11),[19] and "ecologically or biologically significant marine areas"

---

[16] Convention on Biological Diversity (CBD) 1992, 1760 UNTS 79 (CBD), Art 1.   [17] See www.cbd.int/sp/targets/.
[18] It is estimated that there are 15,292 MPAs covering 6.4 percent of the global ocean area or 14.4 percent of coastal and marine areas under national jurisdiction, as of July 2017; see www.unep-wcmc.org/; See also SDG 14.2 Update source: https://mpatlas.org/.
[19] The criteria for describing "ecologically or biologically significant marine areas in need of protection and guidance" for designing representative networks of MPA required sites to reflect at least one of the listed criteria of uniqueness or rarity;

(EBSAs) have been described by states as meeting the scientific requirements to benefit from enhanced conservation and management measures, protected status and impact assessments.[20] That said, commentators (Diz et al., 2018) have underscored that while progress has been made toward the 10 percent target in quantitative terms, the qualitative elements of the MPA target (effectively and equitably managed, ecologically representative and well connected systems), which would contribute to inclusive and integrative governance, have received far less attention (Rees et al., 2018).

Also linked to the ecosystem approach, the guidance elaborated under the CBD in relation to marine spatial planning places a focus on the need to identify stakeholder roles and interests, promoting a deeper understanding of their dependence on ecosystem services, enhancing collaboration across different cultures, and demonstrating fairness, transparency and inclusiveness, including by employing a long-term historical perspective on how current conditions and issues evolved in a given area (CBD Decision XIII/9, 2016). This approach can address one of the main sources of opposition to the creation of MPAs: rather than pitting conservation against fisheries as competing interests, it could support the co-development of MPAs as integral components of ecosystem-based fisheries management (Rees et al., 2020). This approach can also support the fair and equitable sharing of benefits arising from the establishment of MPA networks (discussed in Section 15.3.1) with ecosystem stewards and traditional knowledge holders, thereby contributing to integrative, inclusive and transdisciplinary governance (Ntona and Morgera, 2018).

15.3.1 A Common but Differentiated Strategy: The Use of Area-Based Management Tools in Achieving Integrative Governance of the Ocean

UNCLOS,[21] as well as treaties aimed at improving safety at sea,[22] support area-based management tools (ABMTs)[23] such as MPAs (Baxter et al., 2016; De Santo, 2018; Warner, 2019), and previous experiences led by regional organizations serve to illuminate key opportunities and challenges (De Santo, 2018). ABMTs have in effect been promoted from early on in the regional context, most notably through the Regional Seas Programme, which was birthed from early attempts by UNEP to catalyze a more specialized and integrated methodology at the regional level (Akiwumi and Melvasalo, 1998).[24] Described as one of UNEP's most significant achievements in the past thirty-five years,[25] the concept's linchpin is to engage neighboring countries in comprehensive and specific

---

special importance for life history stages of species; importance for threatened, endangered or declining species and/or habitats; vulnerability, fragility, sensitivity or slow recovery; biological productivity; biological diversity and naturalness.

[20] Areas described as EBSA range from relatively small sites to very extensive oceanographic features representative of a full range of ecosystem habitats, biotic diversity and ecological processes.

[21] E.g. Articles 61(2), (3) and (4).

[22] Such as those under the International Maritime Organization (IMO) that give rise to special areas and particularly sensitive sea areas.

[23] Area-Based Management Tools (ABMTs) could be defined as "regulations of human activity in a specified area to achieve conservation or sustainable resource management objectives." Examples include marine protected areas, ridge to reef, marine spatial planning, areas of particular environmental interest, pollution control zones or fisheries closure (https://bit.ly/33DJlgJ).

[24] UNEP, *Regional Seas Programme* (online) at https://bit.ly/3IyiiCg; refer also to the Strategic Action Plan document available at https://bit.ly/3GW5EN2.

[25] https://sustainabledevelopment.un.org/partnership/?p=7399.

actions for the sustainable management and use of the marine and coastal environment (A/ 9625, 1974). An additional advantage of the framework is the opportunity that the Regional Seas Programme provides stakeholders to share experiences and support more integrative ocean governance. For instance, relevant states participating in the regional seas Abidjan Convention in West Africa have cooperated with the Benguela Current Commission[26] for the management of the Benguela Large Marine Ecosystem (Cochrane et al., 2009), and the OSPAR Convention for the Protection of the Marine Environment of the North-East Atlantic[27] provides almost complete coverage of the Eastern Atlantic.[28] This has led to exchanging knowledge and capacity, as well as ensuring coherent implementation of the ecosystem approach, beyond the scope of the respective conventions. That said, there is widespread understanding that UNCLOS provides limited guidance on MPA networks, and progress has been too limited in areas beyond national jurisdiction. For these reasons, ABMTs are currently being addressed in international negotiations on a new international instrument on marine biodiversity of areas beyond national jurisdiction (De Santo, 2018).

Regional fisheries management organizations (RFMOs) have also established ABMTs. The advantage of RFMOs is that they can adopt targeted management measures that are adapted to the political and ecological characteristics of a given region. The key difference with regional seas organizations is that RFMOs can adopt measures that are binding on their member states. Many RFMOs now include an ecosystem and precautionary approach to fisheries.

While such provisions do not confer upon RFMOs the mandate to regulate activities other than fisheries, they generally allow them to conduct cumulative impact assessments to evaluate the aggregate effects of human activities on the ecosystems in their regulatory area.

(Diz and Ntona, 2018: 19)

Nevertheless, RFMOs are still not cooperating with other organizations to the extent necessary to ensure cross-sectoral cooperation for MPAs, other area-based management and risk assessments "in adopting integrated and coherent conservation and management measures within ecologically meaningful boundaries (or ecosystem-based units/ functional units)" (Diz and Ntona, 2018: 19; Kenny et al., 2018). Thus, their sector-focused approach to management still poses an obstacle to the integrated management of fisheries (Leroy and Morin, 2018; Pentz et al., 2018).

For that reason, synergies between the Regional Seas Programme and RFMOs have been pursued. One approach has been to focus on large marine ecosystems (LMEs),[29] wide areas of ocean space along the planet's continental margins, spanning 200,000 km² or more. LMEs are another type of ABMT that include both ocean space and connected coastal land areas, such as river basins and estuaries (Sherman and Alexander, 1986), to maintain and restore ecosystem functions. As discussed in Section 15.3.2, the establishment of the Benguela Current Commission between Angola, Namibia and South Africa, as the three

---

[26] https://bit.ly/3fSgF61.
[27] Convention for the protection of the marine environment of the North-East Atlantic, Paris, September 22, 1992, in force March 25, 1998, 2354 UNTS 67. www.ospar.org/convention/text.
[28] See https://bit.ly/3AtW8hq.   [29] See www.lmehub.net.

states that border the LME, is an example of transformative ocean governance. The connection between the Regional Seas Programme, RFMOs and LMEs is being deepened by the Sustainable Ocean Initiative, led by the CBD (CBD, 2016).

Against this background, a case study will serve to illustrate progress and continued challenges in creating MPAs as a leading ABMT methodology that is integral to marine spatial planning for balancing ocean uses to support sustainable development and enhance ocean governance. (Finke et al., 2020a; Kirkman et al., 2019). The next subsection will thus identify lessons learned in ensuring integrative and inclusive ocean governance, understood as inclusivity of diverse representative species and biodiversity hotspots, as well as of varied human dependences on marine ecosystems through stakeholder engagement, securing of resource rights, and the recognition of Indigenous and local knowledge systems that can contribute to biodiversity conservation goals (MacKinnon et al., 2015).

15.3.2 Experiments in Integrated and Inclusive Approaches: The Benguela Current Commission and South Africa's MSP Process

The Benguela Current Commission is a notable example of integrating and upscaling efforts between the Regional Seas Programme, RFMO and a large marine ecosystem (CBD, 2016).[30] The establishment of the Commission resulted from the cooperation over two decades in ocean governance between Angola, Namibia and South Africa toward a multi-sectoral ocean governance approach.[31] Cooperation culminated in several international instruments, including the 1999 Strategic Action Programme for the Ecosystem, which was given effect through a voluntary 2007 Interim Agreement on the Establishment of the Benguela Current Commission.[32] This was to ensure effective longstanding transboundary cooperation and the sustainable management and protection of the LME (O'Toole and Shannon, 2003). In 2013, the Interim Agreement was replaced by the Benguela Current Convention (BCC), cementing the legal status of the Benguela Current Commission.[33]

Several remarkable features of the BCC make it a good basis for more inclusive and integrative ocean governance. First, the BCC addresses the complex legacy of fragmented governance left by colonial and political histories (Cochrane et al., 2009), including Angola's independence and forty years of debilitating war (Cochrane et al., 2009), Namibia's independence from South Africa,[34] and the end of apartheid in South Africa (Finke et al., 2020a), with the social impacts spilling over into the establishment and effectiveness of South Africa's MPA system (Sowman and Sunde, 2018).

---

[30] See https://bit.ly/33J4FRT.
[31] Two noteworthy regional cooperative initiatives were the Benguela-Environment-Fisheries Interaction & Training (BENEFIT) Programme and the BCLME Programme. The BENEFIT Programme goal was to increase the science capability required for the optimal and sustainable utilization of marine living resources of the BCLME. The BCLME Programme's goal was "to sustain the ecological integrity of the BCLME through integrated transboundary ecosystem management." For more information refer to O'Toole, and Shannon (2003).
[32] See https://bit.ly/3IyuJOw.
[33] Adopted March 18, 2013; in force December 10, 2015. Available at https://bit.ly/3GXwUL3.
[34] In regard to the complex legacy between South Africa and Namibia, which was formally known as South West Africa, for more detail refer to Devine (1986); Security Council Resolution 276 (1970); and *Advisory Opinion on Legal Consequences for States of the Continued Presence of South Africa in Namibia/ South West Africa*, ICJ Rep. 16, 1970.

Secondly, the Commission links the Benguela Current Large Marine Ecosystem with the neighboring Agulhas and Somali LMEs, which is vital, as these boundaries are highly dynamic and the neighboring warmer waters directly influence the Benguela ecosystem and its living marine resources (Heileman and O'Toole, 2001).

Thirdly, the arrangement reinforces the framework under the Abidjan Regional Seas Convention, as well as relevant regional fisheries arrangements.[35] Finally, there is an established linkage between the Benguela Current Commission and the Orange-Senqu Commission that comprises the four riparian states[36] fed by the largest river discharging into the Benguela LME (Finke et al., 2020b). This in turn allows a link between ocean management and a wetland of international importance under the Ramsar Convention.[37]

The BCC allows its members to manage transboundary resources holistically while balancing different ocean users' needs with conservation imperatives. Its objective is "to promote a coordinated regional approach to the long-term conservation, protection, rehabilitation, enhancement and sustainable use of the [LME], to provide economic, environmental and social benefits" (BCC, Art. 2). According to the BCC, member states must be guided by principles on sustainable use and management, precautionary and prevention (BCC, Art. 4; Vrancken, 2011), thereby providing the legal basis for integrative and anticipatory governance.

Member states and the Commission are guided by a five-year Strategic Action Programme (Hamukuaya et al., 2016), which addresses the following eight themes: living marine resources; nonliving marine resources; productivity and environmental variability; pollution; ecosystem health and biodiversity; human dimensions; enhance the economic development potential; and governance (Hamukuaya, 2020). The Strategic Action Programme is based on a transboundary diagnostic analysis, consisting of a scientific and technical assessment to identify important transboundary issues related to the marine environment and their impacts on the environment and socioeconomy of the region (Hamukuaya et al., 2016). Both instruments are reviewed and updated every five years.[38] The Commission included marine spatial planning into its 2015–2019 Strategic Action Programme (Finke et al., 2020a) to support a variety of ecosystems and sectors, make contributions to the existing economies of member states and tackle increasing demands on the region's marine space (Finke et al., 2020b). This is in line with the progress already made under the Benguela Ecologically or Biologically Significant Areas Project (Kirkman et al., 2019), the Second National Biodiversity Strategy and Action Plan (to implement the CBD) of Namibia[39] and Angola,[40] and the three countries' commitment to implementing an ecosystem approach to fisheries (Kirkman et al., 2016).

Through the Benguela Current Commission, a regional working group for MSP was established to foster cooperation between different stakeholders (Finke et al., 2020a), including government officials, technical experts and representatives of civil society, supporting the implementation of MSP within the three states and enabling information

---

[35] See FAO, Regional Fisheries Bodies Map Viewer: www.fao.org/figis/geoserver/factsheets/rfbs.html.
[36] The whole of Lesotho and parts of Botswana, Namibia and South Africa.   [37] See https://bit.ly/33DPtFL.
[38] The Benguela Current Commission has undertaken to update the Strategic Action Programme document as the current one "expired" in 2019.
[39] See http://extwprlegs1.fao.org/docs/pdf/nam169118.pdf.   [40] See www.cbd.int/doc/world/ao/ao-nbsap-v2-en.pdf.

exchange, mutual learning and capacity-building in the form of expertise (Finke et al., 2020a). These are not limited to the region. The regional working group has engaged with the European Commission, the Baltic Marine Environment Protection Commission and the Baltic Sea Spatial Planning Organization (Finke et al., 2020a). A valuable output from the regional working group is enabling a uniform approach to MSP in the region (Finke et al., 2020a). For the successful implementation of MSP within the region, however, extensive data is required on the state of the marine area, the impact of human activities and the effect of external pressures such as climate change.

To date, the Benguela Current Commission has undertaken projects to inform the regional MSP process, such as the spatial biodiversity assessment of marine and coastal biodiversity in the ecosystem, focusing on the ecosystem threat status, ecosystem protection levels and priority areas for protection (Holness et al., 2012). In addition, through the Marine Spatial Management and Governance Project (MARISMA), member states have been supported in describing the region's EBSAs, in line with the CBD, as part of MSP.

The main challenge facing the Benguela Current Commission, in addition to lack of long-term funding for the MSP process, is how to engage with stakeholders across different sectors as part of its efforts to strategically organize the use of the marine space, to avoid conflicts and limit threats while ensuring the long-term sustainable development of the blue economy in the region.[41] The challenge facing the Commission is, therefore, encompassing inclusive, transdisciplinary and adaptive governance.

Regarding national efforts, there are currently no MPAs legislated in Angola.[42] In Namibia, the Namibian Islands are currently the sole MPA, but will be one of seven marine areas that have been described as an EBSA under the CBD (Finke et al., 2020b). South Africa has legislated forty-four MPAs in line with Operation Phakisa MPA Network.[43] Of the three states, only South Africa promulgated legislation specifically on marine spatial planning (Marine Spatial Planning Act of 2018). Nevertheless, Namibia and Angola have established similar institutional structures to South Africa, enabling different government agencies to work together to implement MSP through the National Working Groups by using experts of the MARISMA project (Finke et al., 2020b). The three states are thus developing plans sequentially to focus on one marine area at a time to integrate learning from one planning process into the next (Finke et al., 2020b).

In South Africa, researchers and government partners have identified Algoa Bay in the Eastern Cape as a case-study area for developing the first marine spatial plan, with a view to using lessons learned for the development of marine area plans as set out in the Marine

---

[41] For example, the successful implementation of MSP in South Africa hinges upon elaborating marine spatial plans within the framework of South Africa's MPAs, based on increasing representation of marine habitats, benchmarking and precaution. Sowman and Sunde (2018), however, underscored that a failure to address social impacts under Operation Phakisa, including historical injustices experienced by communities in the establishment of MPAs, has led to growing discontent among coastal fishing communities. The *Gongqose and Others v Minister of Agriculture, Forestry and Others, Gongqose and S* (1340/16, 287/17) [2018] ZASCA 87 is an example of South African case law where these conflicts were present.

[42] Even though Angola has no MPAs at present, the government has recognized the potential of the blue economy and expanded the mandate of the Ministry of Fisheries. It launched a marine spatial plan to address conflicting uses of marine resources and is planning to set up the first MPA contiguous with Angola's largest national park. These plans are coupled with the doubling of terrestrial protected areas, which are impacted by illegal occupation of the vulnerable Quiçama coastline as a consequence of the Angolan war, but also after the peace in 2002.

[43] The Network is a unique initiative, developed in a unique context, with participation from seventeen ministries as part of the Operation Phakisa Oceans Economy Lab.

Spatial Planning Act (Dorrington et al., 2018). Algoa Bay has been extensively researched and is home to government-funded research platforms, therefore providing a substantial body of data, allowing an understanding and management of the complexity of legal and socioeconomic requirements, on one hand, and environmental (physical, chemical and biological) considerations, on the other (Dorrington et al., 2018). The development of the Algoa Bay marine spatial plan is following the Intergovernmental Oceanographic Commission of UNESCO (IOC-UNESCO) ten-step approach, underpinned by the CBD ecosystem approach principles, which include recognition of Indigenous knowledge systems (CBD Decision V/6, 2000, Principle 11). This case study can, therefore, become an entry point for recognizing human rights as part of the governance of the ocean and its resources, integrating different systems of knowledge. In addition, the case study is viewed through a systems approach lens and the development of system dynamic tools/models that provide opportunities for scenario-planning and determining possible inter-sectorial impacts and environmental impacts (Lombard et al., 2019). Algoa Bay, therefore, entails a research-stakeholder-led "enabling approach" to developing capacities for the "governance of transformations" (i.e. governance to actively trigger and steer a transformation process).[44] It aims to bring together natural science findings and methods across fisheries, marine ecology and oceanography, with social sciences, law and art to support transdisciplinary, integrative, adaptive and inclusive ocean governance. Algoa Bay provides an example that could be scaled up not only to the national but also the regional level, including with a view to supporting the Benguela Current Commission and the Western Indian Ocean in constructively engaging with stakeholders over trade-offs, by expanding their current integrative and anticipatory governance approaches to include inclusive, adaptive and transdisciplinary approaches. Lessons learned are providing guidance for the development of the Western Indian Marine Spatial Planning Strategy (Lombard et al., 2021). This is for marine planning at a regional scale, rather than at local levels, which is considered key for the development of a sustainable blue economy (Friess and Grémaud-Colombier, 2021).

15.3.3 Ways Forward

Among the possible ways forward for transformative ocean governance in all its dimensions at different scales, this section will investigate the potential of the interdependence between human rights and marine biodiversity to address indirect drivers of biodiversity loss, including power dynamics.

From an international law perspective, even if the CBD and its guidelines do not use explicit human rights language, they have made significant conceptual and normative contributions to the relationship with human rights, specifically with regard to Indigenous peoples' rights to natural resources (Morgera, 2018a). As a result, the CBD and its instruments have been increasingly relied upon by international human rights bodies (A/HRC/37/59, 2018). This recognition has implications both for national-level action, as well as for international cooperation, at the global and regional levels (A/HRC/34/49, 2017, paras. 36–

---

[44] See Chapter 1 in this volume.

48), and can have a bearing on the inclusiveness and integration of ocean governance. Notably, human rights can help address, from a legal perspective, the "politics of transformative change,"[45] preventing a shifting of the burden of response onto the vulnerable; paying attention to social differentiation, through the lens of nondiscrimination; and addressing issues of power and legitimacy. In other words, human rights can serve to address questions of justice[46] in ocean governance. The integration of international human rights law into the interpretation and application of the law of the sea, however, is not very advanced (Barnes, 2018).

One way in which human rights considerations can be put into practice in the context of ocean governance, with a view to making it more integrated and inclusive, is reliance on the international legal concept of fair and equitable benefit-sharing, which is already included in the law of the sea and international human rights law, and has been elaborated upon under the CBD (Morgera, 2018b). As will be argued below, fair and equitable benefit-sharing can support transformative governance in terms of framing and agenda-setting, leadership, financial investment, capacity for learning and increasing institutionalization.[47]

Fair and equitable benefit-sharing norms in the law of the sea are conceived narrowly in relation to deep-seabed mining and marine scientific cooperation (UNCLOS, Arts. 82(1) and (4), 242–244 and Part XI; Noyes, 2011; Salpin, 2013), and they are currently being developed with regard to bioprospecting in areas beyond national jurisdiction as part of the negotiations of a new legally binding instrument on marine biodiversity of these areas (Morgera, 2018–19). Benefit-sharing has, however, become a broader obligation in international biodiversity law (Morgera, 2016) arising from the conservation and sustainable use of natural resources (both within and outside national jurisdiction, beyond access to genetic resources) to address equity and sustainability issues as part of the ecosystem approach (*Contra* Baslar, 1998).[48] Along parallel lines, under international human rights law, benefit-sharing has been identified as a safeguard to protect the human rights of Indigenous peoples (A/HRC/27/59, 2018, Principle 15; Morgera, 2019), small-scale fishing communities (A/RES/73/165, 2019; Morgera and Nakamura, forthcoming) and rural women (CEDAW/C/GC/34, 2016), including in connection with their effective participation in the creation and management of protected areas. In addition, benefit-sharing is part and parcel of the human right to science (the right of everyone to benefit from scientific advancements), which reveals the human rights dimensions of interstate obligations related to scientific cooperation, capacity-building and technology transfer (International Covenant on Economic, Social and Cultural Rights, Art. 15(3); Morgera, 2015).

That said, benefit-sharing implementation is often dominated by a transactional logic to obtain a "green light" for conservation or development projects, rather than redress power asymmetries that threaten biodiversity conservation and sustainable use (Martin et al., 2014). A different interpretation, however, emerges from CBD guidance that is more aligned with human rights standards. This interpretation focuses on the active participation of beneficiaries in the identification of benefits, which relies on an iterative, concerted and

---

[45] Chapter 1 in this volume.   [46] Chapter 8 in this volume.   [47] Chapter 1 in this volume, referring to Chaffin et al., 2016.
[48] Who instead suggested that common heritage as such should be applied to other natural resources of different international legal status as a functional rather than territorial concept.

good-faith dialogue to develop a common understanding as part of mutual learning and an adaptive approach. Based on a combined reading of interpretative materials, "sharing" principally conveys the idea of agency, as opposed to the passive enjoyment of benefits (Mancisidor, 2015), and therefore a shift away from unidirectional (likely, top-down) or one-off flows of benefits. In addition, benefit-sharing usually relies on a menu of benefits, the nature of which can be economic and noneconomic. This arguably allows taking into account, through the concerted, dialogic process of sharing, the beneficiaries' needs, values and priorities through a contextual selection of the combination of benefits that may best serve to lay the foundation for partnership (Morgera, 2016). The expressions "fair and equitable," which is generally left to subsequent negotiations, can be interpreted to express the rationale of balancing competing rights and interests (Burke, 2014), with a view to integrating both procedural and substantive dimensions of justice (Kläger, 2011) into a relationship regulated by international law that is characterized by power imbalances (Kläger, 2011).

Applied at the multilateral level, this interpretation of benefit-sharing can support the voice of developing countries in co-identifying the benefits and needs for transformative ocean governance through the integrated implementation of capacity-building, technology transfer, scientific cooperation and information-sharing obligations (Morgera, 2016). In particular, this can be applied to the creation and management of MPA networks, with a focus on equity and power imbalances in ocean science production and area-based management and impacts at local levels. It could also support the co-development of MPAs as integral components of ecosystem-based fisheries management based on better understanding of the dependence on ecosystem services for different actors and sectors. As the Post-2020 Global Biodiversity Framework indicates, this would be aligned with the broader goal of valuing and maintaining nature's contributions to people through conservation and sustainable use "for the benefit of all" and would take into account the importance of spatial approaches to this end:

The number of people who can benefit from nature's contributions to people depends not only on nature's ability to provide the benefit, but also on societies' ability to manage their distribution, fairly and equitably, within and between generations.   *(CBD/SBSTTA/24/3.Add.2, 2021, para. 36)*

This approach is aligned with the innovative theory of change in the Global Biodiversity Framework, which emphasizes "a whole-of-government and society approach" for transformative change and the role of a rights-based approach and cross-scale partnerships for ensuring that "biodiversity is used sustainably in order to meet people's needs," notably gender equality, youth inclusion, and the full and effective participation of Indigenous peoples and local communities in the implementation of this framework (CBD/POST2020/PREP/2/1, 2020).

This co-identification and delivery of benefits can be supported by a process of institutionalization:[49] multilateral facilitative and brokering arrangements can serve to operationalize relevant duties of cooperation with a view to ensuring equitable distribution

---

[49] Chapter 1 in this volume.

across different regions, monitoring of effectiveness, and learning from experience. The need for such an approach has already been demonstrated in other international processes, such as the International Seabed Authority (ISA) and the International Maritime Organization (IMO) (Morgera and Ntona, 2018). In addition, benefit-sharing is a key element to recognizing Indigenous peoples and local communities for their global contributions to the conservation and sustainable use of biodiversity, and to respectfully integrate their knowledge systems[50] in relation to MPA creation and management at different levels. This could allow for the co-identification of benefits and needs for transformative ocean governance beyond the current state-centric model, with a view to enhancing both transdisciplinary and inclusive ocean governance.

The key elements of a benefit-sharing inspired multilateral approach to transformative ocean governance would then be the following:

- Joined-up thinking on the implementation of various international obligations on scientific cooperation and information-sharing, financial and technological solidarity, capacity-building and their human rights dimensions (integrative and transdisciplinary governance);
- Dialogue to enhance collaboration across sectors, among duty-bearers and among human rights-holders, to contribute to the achievement of international biodiversity, ocean, climate change and human rights objectives (integrative governance);
- Deliberation and mutual learning with a view to setting priorities to the benefit of the most vulnerable (inclusive governance);
- The provision of international institutional support for facilitating and brokering scientific cooperation opportunities;
  o Co-identifying information-sharing, technology transfer and regulatory and institutional capacity-building needs and available assistance; and
  o Building, and assessing the effects of partnerships, including public–private partnerships (adaptive governance);
- Multistakeholder identification and assessment of obstacles, co-development of proposals for enhancement, joint monitoring and reflection on lessons learned on emerging transformative approaches (inclusive and adaptive governance); and
- Transparency about, and assessment of, the distribution of benefits across regions, as well as good practices and lessons learned at the local, national and regional levels, with a view to ensuring fairness and equity in benefit-sharing (arising from the dialogue and incrementally shaping funding and governance across scales – adaptive governance).[51]

15.4 Conclusions

These elements could be applied in the context of area-based management and spatial approaches under the ongoing negotiations of an international instrument on marine biodiversity of areas beyond national jurisdiction (Morgera, 2022), and under the

---

[50] This is inspired and adapted from Morgera et al. (2020).   [51] This is inspired and adapted from Morgera et al. (2020).

Sustainable Ocean Initiative. This chapter focuses on the latter, as an already institutionalized opportunity for transformative governance. The Initiative has become a regular process to facilitate the exchange of experiences, to identify options and opportunities to enhance cross-sectoral collaboration toward internationally agreed goals and to discuss the need for specific tools, guidelines or other initiatives to strengthen collaboration among not only regional seas conventions and RFMOs, but also sectoral international organizations like the Food and Agriculture Organization of the United Nations, the IMO and the ISA (Diz and Ntona, 2018). The Initiative could take the approach outlined above to understand the reasons why "many protected areas are not effectively or equitably managed," as well as "the importance of focusing on biodiversity outcomes rather than spatial area" included within MPAs, and the "provision of ecosystem services and to maintain integrity of planetary ecological processes" (CBD/SBSTTA.24/3/Add.2, 2021, paras. 54–56). Equally, the Initiative could provide a forum to reflect on equity issues across scales in interregional scientific cooperation, notably in relation to carrying out fisheries assessments in data-poor environments (Kenny et al., 2018), implementation of the precautionary approach to fisheries (UNFSA, Art. 6 and Annex II; A/Conf.210/2016/5, 2016, para. 36), habitat protection in the context of conflicts of use (i.e. fishing or fishing survey activities vs seismic activities) (NAFO, 2016), and the effects of climate change and ocean acidification on marine ecosystems (A/RES/72/73, 2018, para. 196). Furthermore, scientific and participatory methodologies for assessing coastal communities' and coastal and marine ecosystems' vulnerabilities to climate change and ocean acidification are a crucial area of scientific cooperation and capacity-building to identify adaptation measures in most vulnerable regions (Cochrane et al., 2017).

A reflection has already been started on the role of the Regional Seas Programme for contributing to the Post-2020 Biodiversity Framework (CBD/SBSTTA/24/INF/24, 2021). Based on the key challenge facing the Benguela Current Commission and the findings from the Algoa Bay case study in South Africa, the SOI could share learning across scales on integrating social and natural sciences insights, as well as different knowledge systems. This could support regional seas organizations to engage in complex stakeholder engagements and deliberations on trade-offs in a constructive manner, to maximize the potential for transformation, by expanding their current integrative and anticipatory governance approaches to inclusive, adaptive and transdisciplinary approaches. The Initiative could also provide a forum to engage with the increasing concentration of businesses in the blue economy and explore how to build fair partnerships with the private sector in the context of MPA networks at different scales (Virdin, 2021). These efforts could contribute to strengthening the adaptive and transdisciplinary governance dimensions of efforts on EBSAs and ABMTs across scales, contributing to implementing CBD obligations to monitor biodiversity components that require urgent conservation measures and those that offer the best potential for sustainable use through international technical and scientific cooperation on conservation and the sustainable use of biodiversity (CBD, Arts. 7 and 17–18). It could also support CBD Parties in providing the evidence base to identify processes with (likely) significant adverse impacts on biodiversity conservation and sustainable use (CBD, Art 7 (c)), as well as to assess and minimize adverse impacts (CBD, Art. 14), while building capacity by sharing cross-regional learning on transboundary MSP approaches (CBD, Art. 12; CBD/EBSA/EM/2017/1/INF/1, 2017).

At the national level, this rights-based interpretation of benefit-sharing could be explored as part of marine spatial planning processes. It could support bottom-up forms of deliberations (Cotula and Webster, 2020), characterized by the agency of beneficiaries, the respect of human rights, and mutual understanding of different benefits and priorities in MPA creation and other area-based management tools, as well as in the sustainable use of marine resources and the advancement of ocean science. Such dialogues could be informed by interdisciplinary and transdisciplinary research (Morgera et al., 2021) to assist different actors in the respectful and constructive engagement with beneficiaries' choice and capabilities, knowledge systems, and different worldviews of nature and development, and an understanding of different benefits and risks across scales (Ntona and Morgera, 2018). The partnership that is being built among researchers from different disciplines, different sectors of government and different knowledge holders could also contribute to the contextual application of the precautionary principle and new technologies (anticipatory governance), through learning, experimentation and reflexivity (adaptive governance). Research is equally needed to document good practices in integrating the evidence base across marine sciences and social sciences through inclusive approaches, with a view to understanding barriers and opportunities to scaling up to the national, regional and international levels.

References

Akiwumi, P., and Melvasalo, T. (1998). UNEP's regional seas programme: Approach, experience and future plans. *Marine Policy* 22, 229–234.

Allan, E., Weisser, W. W., Fischer, M., et al. (2013). A comparison of the strength of biodiversity effects across multiple functions. *Oecologica* 173, 223–237.

Barnes, R. A. (2018). Environmental rights in marine spaces. In *Environmental rights in Europe and beyond*. S. Bogojević and R. G. Rayfuse (Eds.), pp. 49–84. London: Hart.

Baslar, K. (1998). The concept of the common heritage of mankind in international law. Leiden: Martinus Nijhoff.

Baxter, J. M., Laffoley, D., and Simard, F. (2016). *Marine protected areas and climate change*. Gland: IUCN. Available from https://bit.ly/362bezX.

Beaugrand, G., Edwards, M., Raybaud, V., et al. (2015). Future vulnerability of marine biodiversity compared with contemporary and past changes. *Nature Climate Change* 5, 695–701.

Birnie, P., Boyle, A. E., and Redgwell, C. (2009). *International law and the environment*, 3rd ed. Oxford: Oxford University Press.

Blasiak, R., Jouffray, J.-B., Wabnitz, C. C., et al. (2018). Corporate control and global governance of marine genetic resources. *Science Advances* 4, eaar5237.

Bogomolni, A. L., Gast, R. J., Ellis, J. C., et al. (2008). Victims or vectors: A survey of marine vertebrate zoonoses from coastal waters of the Northwest Atlantic. *Diseases of Aquatic Organisms* 81, 13–38.

Borrelle, S. B., Rochman, C. M., Liboiron, M., et al. (2017). Opinion: Why we need an international agreement on marine plastic pollution. *Proceedings of the National Academy of Sciences USA* 114, 9994–9997.

Burke, C. (2014). *An equitable framework for humanitarian intervention*. Portland, OR: Hart.

Casson, L., Alexander, J., Miller, K., et al. (2020). Deep trouble: The murky would of the deep sea mining industry 2020. Greenpeace International. Available from https://bit.ly/3rONA0N.

CBD (2016). Outcome of the Sustainable Ocean Initiative Global Dialogue with Regional Seas Organizations and Regional Fisheries Bodies on Accelerating Progress towards the Aichi Biodiversity Targets, held in Seoul from 26 to 28 September 2016 ("Seoul Outcome"). Available from https://oceanconference.un.org/commitments/?id=14827.

Chaffin, B. C., Garmestani, A. S., Gunderson, L. H., et al. (2016). Transformative environmental governance. *Annual Review of Environment and Resources* 41, 399–423.

Cheung, W. W. L. (2016). Climate change effects on illegal, unreported and unregulated fishing. Available from https://bit.ly/3FXkC3P.

Cheung, W. W. L., Lam, W. Y. V., Sarmiento, J. L., et al. (2009). Projecting global marine biodiversity impacts under climate change scenarios. *Fish and Fisheries* 10, 235–251.

Cochrane, K. L., Augustyn, C. J., Fairweather, T., et al. (2009). Benguela Current large marine ecosystem – Governance and management for an ecosystem approach to fisheries in the region. *Coastal Management* 37, 235–254.

Cochrane, K. L., Rakotondrazafy, H., Aswani, S., et al. (2017). Report of the GLORIA Workshop, Antananarivo, Madagascar, June 14–16, 2016. Available from https://bit.ly/3H2DakA.

Corlett, R. T. (2020). Impacts of the coronavirus pandemic on biodiversity conservation. *Biological Conservation* 246, 108571.

Cotula, L., and Webster, E. (2020). COVID-19 and the sites of rights resilience. Strathclyde Law School blogpost. Available from https://bit.ly/3r0qu8e.

De Santo, E. M. (2018). Implementation challenges of area-based management tools (AMBTs) for biodiversity beyond national jurisdiction (BBNJ). *Marine Policy* 97, 34–43.

Devine, D. J. (1986). Some thoughts on the interim preservation of the Namibian fishing heritage. *Verfassung und Recht in Übersee* 19, 379–381.

Diana, J. S. (2009). Aquaculture production and biodiversity conservation. *BioScience* 59, 27–38.

Dixson, D. I., Munday, P. L., and Jones, G. P. (2010). Ocean acidification disrupts the innate ability of fish to detect predatory olfactory cues. *Ecology Letters* 13, 68–75.

Diz, D. (2017). Marine biodiversity: Unravelling the intricacies of global frameworks and applicable concepts. In *Encyclopaedia of environmental law: Biodiversity and nature protection law*. E. Morgera and J. Razzaque (Eds.), pp. 123–144. Northampton, MA: Edward Elgar.

Diz, D., Johnson, D., and Riddel, M. (2018). Mainstreaming marine biodiversity into the SDGs: The role of other effective area-based conservation measures (SDG 14.5). *Marine Policy* 93, 251–261.

Diz, D., and Ntona, M. (2018). Background report for the Second Meeting of Sustainable Ocean Initiative Global Dialogue with Regional Seas Organizations and Regional Fisheries Bodies on Accelerating Progress towards the Aichi Biodiversity Targets and Sustainable Development Goals. On file with authors.

Dorrington, R. A., Lombard, A. T., Bornman, T. G., et al. (2018). Working together for our oceans: A marine spatial plan for Algoa Bay, South Africa. *South African Journal of Science* 114, 1–6.

Eng, C. T., Paw, J. N., and Guarin, F. Y. (1989). The environmental impact of aquaculture and the effects of pollution on coastal aquaculture development in Southeast Asia. *Marine Pollution Bulletin* 20, 335–343.

FAO (2018). *Impacts of climate change of fisheries and aquaculture: Synthesis of current knowledge, adaptation and mitigation options. FAO Fisheries and Aquaculture Technical Paper 627*. Rome: FAO.

(2020). *State of world fisheries and aquaculture 2020. Sustainability in action*. Rome: FAO.

Feely, R. A., Sabine, C. L., Lee, K., et al. (2004). Impact of anthropogenic CO_2 on the $CaCO_3$ system in the oceans. *Science* 305, 362–366.

Finke, G., Gee, K., Gxaba, T., et al. (2020a). Marine spatial planning in the Benguela Current large marine ecosystem. *Environmental Development* 36, 100569.

Finke, G., Gee, K., Kreiner, A., Amunyela, M., and Braby, R. (2020b). Namibia's way to marine spatial planning – Using existing practices or instigating its own approach? *Marine Policy* 121, 104107.

Friess, B., and Grémaud-Colombier, M. (2021). Policy outlook: Recent evolutions of maritime spatial planning in the European Union. *Marine Policy* 132, 103428. https://doi.org/10.1016/j.marpol.2019.01.017

Gattuso, J. P., Magnan, A., Billé, R., et al. (2015). Contrasting futures for ocean and society from different anthropogenic CO_2 emissions scenarios. *Science* 349, aac4722.

Guston, D. H. (2014). Understanding "anticipatory governance." *Social Studies of Science* 44, 218–242.

Hamukuaya, H. (2020). Benguela Current Convention supports ecosystem assessment and management practice. *Environmental Development* 36, 100574.

Hamukuaya, H., Attwood, C., and Willemse, N. (2016). Transition to ecosystem-based governance of the Benguela Current Large Marine Ecosystem. *Environmental Development* 17, 310–321.

Haward, M. (2018). Plastic pollution of the world's seas and oceans as a contemporary challenge in ocean governance. *Nature Communications* 9, 667.

Heileman, S., and O'Toole, M. J. (2001). Benguela Current LME. In *LMEs and regional seas*. AIS. Available from www.ais.unwater.org/ais/aiscm/getprojectdoc.php?docid=3920.

Holmer, M. (2010). Environmental issues of fish farming in offshore waters: Perspectives, concerns and research needs. *Aquaculture Environment Interactions* 1, 57–70.

Holness, S., Wolf, T., Lombard, M., et al. (2012). Spatial biodiversity assessment and spatial management, including marine protected areas: Report on progress. Available from https://bit.ly/3IFCYIo.

IPBES (2019). *Summary for policymakers of the global assessment report on biodiversity and ecosystem services of the Intergovernmental Science-Policy Platform on Biodiversity and Ecosystem Services*. Bonn: IPBES Secretariat.

IPCC (2018). *Global warming of 1.5 C – Summary for policymakers*. Geneva: World Meteorological Organization.

(2019). *Special report on the ocean and cryosphere in a changing climate, summary for policy makers*. In press. Available from https://bit.ly/35YMIj2.

Isensee, K., and Valdes, L. (2015). *GSDR 2015 brief: Marine litter: Microplastics*. IOC/UNESCO. Available online at https://bit.ly/32DGt2U.

Jamieson, A. J., Singleman, G., Linley, T. D., and Casey, S. (2021). Fear and loathing of the deep ocean: Why don't people care about the deep sea? *ICES Journal of Marine Science* 78, 797–809.

Kelly, C., Ellis, G., and Flannery, W. (2019). Unravelling persistent problems to transformative marine governance. *Frontiers in Marine Science* 6, 213.

Kenny, A. J., Campbell, N., Koen-Alonso, M., et al. (2018). Delivering sustainable fisheries through adoption of a risk-based framework as part of an ecosystem approach to fisheries management. *Marine Policy* 93, 232–240.

Kimball, L. A. (2001). *International ocean governance: Using international law and organizations to manage marine resources sustainably*. Gland: IUCN.

Kirkman, S. P., Blamey, L., Lamont, T., et al. (2016). Spatial characteristics of the Benguela ecosystem for ecosystem-based management. *African Journal of Marine Science* 38, 7–22.

Kirkman, S. P., Holness, S., Harris, L. R., et al. (2019). Using systematic conservation planning to support marine spatial planning and achieve marine protection targets in the transboundary Benguela ecosystem. *Ocean & Coastal Management* 168, 117–129.

Kläger, R. (2011). *Fair and equitable treatment' in international investment law*. Cambridge: Cambridge University Press.

Kleypas, J. A., Buddemeier, R. W., Archer, D., et al. (1999). Geochemical consequences of increased atmospheric carbon dioxide on coral reefs. *Science* 284, 118–120.

Laffoley, D., and Baxter, J. M. (Eds.). (2019). Oxygen deoxygenation: Everyone's problem: Causes, impacts, consequences and solutions. Gland: IUCN. https://doi.org/10.2305/IUCN.CH.2019.13.en

Leroy, A., and Morin, M. (2018). Innovation in the decision-making process of the RFMOs. *Marine Policy* 97, 156–162.

Levin, L. A., Wei, C.-L., Dunn, D. C., et al. (2020). Climate change considerations are fundamental to management of deep-sea resource extraction. *Global Change Biology* 26, 4664–4678.

Lombard, A. T., Ban, N. C., Smith, J. L., et al. (2019). Practical approaches and advances in spatial tools to achieved multi-objective marine spatial planning. *Frontiers in Marine Science* 6. https://doi.org/10.3389/fmars.2019.00166

Lombard, A. T., Clifford-Holmes, J. K., Snow, B., et al. (2021). *A regional marine spatial planning strategy for the Western Indian Ocean.* Nairobi Convention, 2021. Available from www.nairobiconvention.org/clearinghouse/taxonomy/term/6445.

MacKinnon, C. J., Lemieux, K., Beazley, S., et al. (2015). Canada and Aichi biodiversity target 11: Understanding "other effective area-based conservation measures" in the context of the broader target. *Biodiversity and Conservation* 24, 3559–3581.

Mancisidor, M. (2015). Is there such a thing as a human right to science in international law? *European Society of International Law* 4, 1–6.

Martin, A., Akol, A., and Phillips, J. (2014). Just conservation? On the fairness of sharing benefits. In The justices and injustices of ecosystem services. T. Sikor (Ed.), pp. 69–89. London: Routledge.

Mazzega, P. (2018). On the ethics of biodiversity models, forecasts and scenarios. *Asian Bioethics Review* 10, 295–312.

Miller, K. A., Thompson, K. F., Johnston, P., and Santillo, D. (2018). An overview of seabed mining including the current state of development, environmental impacts, and knowledge gaps. *Frontiers in Marine Science* 4, 418.

Minx, J. C., Callaghan, M., Lamb, W. F., Garard, J., and Edenhofer, O. (2017). Learning about climate change solutions in the IPCC and beyond. *Environmental Science and Policy* 77, 252–259.

Molinos, J. G., Halpern, B. S., Schoeman, D. S., et al. (2016). Climate velocity and the future global redistribution of marine biodiversity. *Nature Climate Change* 6, 83–88.

Morand, S., and Lajaunie, C. (2017). Loss of biological diversity and emergence of infectious diseases. In *Biodiversity and health*, 1st ed. S. Morand and C. Lajaunie (Eds.), pp. 29–47. Amsterdam: Elsevier.

Morgera, E. (2011). Far away, so close: A legal analysis of the increasing interactions between the Convention on Biological Diversity and climate change law. *Climate Law* 2, 85–115.

(2015). Fair and equitable benefit-sharing at the crossroads of the human right to science and international biodiversity law. *Laws* 4, 803–831.

(2016). The need for an international legal concept of fair and equitable benefit-sharing. *European Journal of International Law* 27, 353–383.

(2018a). Dawn of a new day? The evolving relationship between the Convention on Biological Diversity and international human rights law. *Wake Forest Law Review* 54, 691–712.

(2018b). Fair and equitable benefit-sharing. In *Encyclopedia of environmental law: Principles of environmental law*. E. Orlando and L. Krämer (Eds.), pp. 323–337. Cheltenham: Edward Elgar.

(2018–19). Fair and equitable benefit-sharing in a new international instrument on marine biodiversity: A principled approach towards partnership building? *Maritime Safety and Security Law Journal* 5, 48–77.

(2019). Under the radar: Fair and equitable benefit-sharing and the human rights of Indigenous peoples and local communities connected to natural resources. *International Journal of Human Rights* 23, 1098–1139.

(2020). Biodiversity as a human right and its implications for the EU's external action. Report to the European Parliament. Available from https://bit.ly/3H3VutP.

(2022). The relevance of the human right to science for the conservation and sustainable use of marine biodiversity of areas beyond national jurisdiction: A new legally binding instrument to support co-production of ocean knowledge across scales. In *International law and marine areas beyond national jurisdiction*. V. De Lucia, L. Nguyen and A. G. Oude Elferink (Eds.), pp. 242–274. Leiden: Brill.

Morgera, E., and Nakamura, J. (Forthcoming). Shedding a light on the human rights of small-scale fisherfolk: complementarities and contrasts between the UN Declaration on Peasants' Rights and the Small-Scale Fisheries Guidelines. In *Commentary on the Declaration on the Rights of Peasants*. M. Bruonori et al. (Eds.).

Morgera, E., and Ntona, M. (2018). Linking small-scale fisheries to international obligations on marine technology transfer. *Marine Policy* 93, 214–222.

Morgera, E., Parks, L., and Schroeder, M. (2021). Methodological challenges of transnational environmental law. In *Research handbook on transnational environmental law*. V. Heyvaert and L. A. Duvic-Paoli (Eds.), pp. 48–65. Cheltenham: Edward Elgar.

Morgera, E., and Razzaque, J. (Eds.). (2017). *Biodiversity and nature protection law.* Elgar Encyclopedia of Environmental Law volume III. Cheltenham: Edward Elgar Publishing.

Morgera, E., Switzer, S., and Geelhoed, M. (2020). Study for the European Commission on 'Possible ways to address digital sequence information – legal and policy aspects'. Available from https://bit.ly/3Jb0tJQ.

Mossop, J. (2007). Protecting marine biodiversity on the continental shelf beyond 200 nautical miles. *Ocean Development & International Law* 38, 283–304.

Munday, P. L., Dixson, D. L., Donelson, J. M., et al. (2009). Ocean acidification impairs olfactory discrimination and homing ability of a marine fish. *Proceedings of the National Academy of Sciences of the United States of America* 106, 1846–1852.

Munday, P. L., Dixson, D. L., McCormick, M. I., et al. (2010). Replenishment of fish populations is threatened by ocean acidification. *Proceedings of the National Academy of Sciences of the United States of America* 107, 12930–12934.

NAFO (2016). Report of the NAFO Joint Fisheries Commission–Scientific Council Working Group on Ecosystem Approach Framework to Fisheries Management (WG-EAFFM). FC-SC Doc. 16/03. Available from https://bit.ly/3AABM6c.

NIC (2016). Global implications of illegal, unreported and unregulated (IUU) fishing. Available from https://fas.org/irp/nic/fishing.pdf.

Noyes, J. E. (2011). The common heritage of mankind: Past, present and future. *Denver Journal of International Law & Policy* 40, 447–471.

Ntona, M., and Morgera, E. (2018). Connecting SDG 14 with the other sustainable development goals through marine spatial planning. *Marine Policy* 93, 214–222.

Ostfeld, R. S. (2009). Biodiversity loss and the rise of zoonotic pathogens. *Clinical Microbiology and Infection* 15, 40–43.

O'Toole, M., and Shannon, V. (2003). Sustainability of the Benguela: Ex Africa semper aliquid novi. In Large marine ecosystems of the world: Trends in exploitation, protection and research. G. Hemper and K. Sherman (Eds.), pp. 227–253. Amsterdam: Elsevier.

Otsuki, K. (2015). *Transformative sustainable development: Participation, reflection and change.* New York: Routledge.

Páez-Osuna, F. (2001). The environmental impact of shrimp aquaculture: A global perspective. *Environmental Pollution* 112, 229–231.

Pentz, B., Klenk, N., Ogle, S., and Fisher, J. A. D. (2018). Can regional fisheries management organizations (RFMOs) manage resources effectively during climate change? *Marine Policy* 92, 13–20.

Rees, S., Foster, N. L., Langmead, O., Pittman, S., and Johnson, D. (2018). Defining the qualitative elements of Aichi Biodiversity Target 11 with regard to the marine and coastal environment in order to strengthen global efforts for marine biodiversity conservation outlined in the United Nations Sustainable Development Goal 14. *Marine Policy* 93, 241–250.

Rees, S., Sheehan, E. V., Stewart, B. D., et.al. (2020). Emerging themes to support ambitious UK marine biodiversity conservation. *Marine Policy* 117, 1–10.

Sabine, C. L., Feely, R. A., Gruber, N., et al. (2004). The ocean sink for anthropogenic CO_2. *Science* 305, 367–371.

Sala, E., Mayorga, J., and Bradley, D. et al. (2021). Protecting the global ocean for biodiversity, food and climate. *Nature* 592, 397–402.

Salpin, C. (2013). The law of the sea: A before and an after Nagoya? In *The 2010 Nagoya Protocol on Access and Benefit-Sharing in perspective: Implications for international law and national implementation.* E. Morgera, M. Buck and E. Tsioumani (Eds.), pp. 149–183. Leiden: Brill/Martinus Nijhoff.

Serrao-Neumann, S., Davidson, J. L., and Baldwin, C. L. (2016). Marine governance to avoid tipping points: Can we adapt the adaptability envelope? *Marine Policy* 65, 56–67.

Sherman, K., and Alexander, L. M. (Eds.). (1986). *Variability and management of large marine ecosystems. AAAS Selected Symposium 99.* Boulder, CO: Westview Press.

Singh, G. G., Cisneros-Montemayor, A. M., Swartz, W., et al. (2018). A rapid assessment of co-benefits and trade-offs among sustainable development goals. *Marine Policy* 93, 223–231.

Sowman, M., and Sunde, J. (2018). Social impacts of marine protected areas in South Africa on coastal fishing communities. *Ocean & Coastal Management* 157, 168–179.

Spijkers, J., Singh, G., Blasiak, R., et al. (2019). Global patterns of fisheries conflict: Forty years of data. *Global Environmental Change* 57, 101921.

Stocker, T. F., Qin, D., Plattner, G.-K., et al. (Eds.). 2013. *AR5 Climate Change 2013: The physical science basis. Contribution of Working Group I to the Fifth Assessment Report of the Intergovernmental Panel on Climate Change.* Cambridge:Cambridge University Press.

Sumaila, U. R., Ebrahim, N., Schuhbauer, A., et al. (2019). Updated estimates and analysis of global fisheries subsidies. *Marine Policy* 109, 103695.

Tovar, A., Moreno, C., Mánuel-Vez, M. P., and García-Vargas, M. (2000). Environmental impacts of intensive aquaculture in marine waters. *Water Research* 34, 334–342.

Tsutsumi, H., Kikuchi, T., Tanaka, M., et al. (1991). Benthic faunal succession in a cove organically polluted by fish farming. *Marine Pollution Bulletin* 23, 233–238.

UNEP-WCMC (2012). *Promoting synergies within the cluster of biodiversity-related multilateral environmental agreements.* Cambridge: UNEP-WCMC. Available from https://bit.ly/3rSpNx5.

UNESCO (2020). United Nations decade of ocean science for sustainable development, implementation plan, Version 2.0. Available from www.oceandecade.org/decade-publications/.

United Nations General Assembly (UNGA). (2016). *First global integrated marine assessment.* Available from www.un.org/Depts/los/global_reporting/WOA_RegProcess.htm.

Virdin, J., Vegh, R., Jouffray, J.-B., et al. (2021). The Ocean 100: Transnational corporations in the ocean economy. *Scientific Advances* 7, eabc8041.

Visseren-Hamakers, I. J., Razzaque, J., McElwee, P., et al.(2021). Transformative governance of biodiversity: Insights for sustainable development. *Current Opinion in Environmental Sustainability* 53, 20–28.

Vrancken, H. G. (2011). *South Africa and the law of the sea.* Leiden: Brill.

Waltzek, T. B., Cortés-Hinojosa, G., Wellehan Jr., J. F. X., and Gray, G, C. (2012). Marine mammal zoonoses: A review of disease manifestations. *Zoonoses and Public Health* 59, 521–535.

Warner, R. (2019). Area-based management tools developing regulatory frameworks for areas beyond national jurisdiction. *Asia-Pacific Journal of Ocean Law and Policy* 4, 142–157.

Watson-Wright, W., and Valdés, J. L. (2018). Fragmented governance of our one global ocean. In *The future of ocean governance and capacity development – Essays in honor of Elisabeth Mann Borgese (1918–2002).* D. Werle, P. Boudreau, M. R. Brookes, et al. (Eds.), pp. 16–22. Leiden: Brill Nijhoff.

World Health Organization and Secretariat of the Convention on Biological Diversity (WHO/CBD). (2015). State of knowledge review on biodiversity and health, connecting global priorities: Biodiversity human health. Available from www.cbd.int/health/doc/Summary-SOK-Final.pdf.

Part V
Strategic Reflections

16

Enabling Transformative Biodiversity Governance in the Post-2020 Era

MARCEL T. J. KOK, ELSA TSIOUMANI, CEBUAN BLISS, MARCO
IMMOVILLI, HANS KEUNE, ELISA MORGERA, SIMON R. RÜEGG, ANDREA
SCHAPPER, MARJANNEKE J. VIJGE, YVES ZINNGREBE AND INGRID J.
VISSEREN-HAMAKERS

16.1 Introduction

While there are increasing calls for transformative change and transformative governance, what this means in the context of addressing biodiversity loss remains debated. The *aim* of this edited volume *Transforming Biodiversity Governance* is to open up this debate and identify ways forward in the context of the implementation of the Post-2020 Global Biodiversity Framework (GBF) of the Convention on Biological Diversity (CBD). To become transformative, biodiversity governance needs to be transformed: yet how and by whom? These questions are urgent, given the fact that around one million species are threatened with extinction (Díaz et al., 2019), despite over half a century of global efforts to avoid this tragedy. By bringing together insights from previous chapters, we here reflect on these questions.

The *research questions* that guided this book were:

a) What are the lessons learned from existing attempts to address the underlying causes of biodiversity loss?
b) What are the lessons learned from different approaches to, and instruments for, transformative governance as operationalized below?

We turn to question a) in Section 16.2, where we provide specific reflections on the theoretical and conceptual insights from the chapters in the book. In Section 16.3 we address opportunities and challenges for transformative biodiversity governance in the context of the Post-2020 GBF and its further implementation. We end with a final section with concluding remarks (Section 16.4).

16.2 Theoretical and Conceptual Insights

In this section, we summarize some of the insights from the various chapters regarding the operationalization of the main concepts of the book.

16.2.1 Our Starting Point

In Chapter 1, we defined *transformative governance* as the formal and informal (public and private) rules, rule-making systems and actor-networks at all levels of human society (from local to global) that enable transformative change, in our case, toward biodiversity conservation and sustainable development more broadly. We argued that governance becomes transformative if it:

a) Focuses on addressing underlying causes (indirect drivers) of sustainability issues;
b) Implements the five governance approaches below in conjunction; and
c) Operationalizes these approaches in the following specific manners:
   1. *Integrative*, operationalized in ways that ensure solutions also have sustainable impacts at other scales and locations, on other issues, and in other sectors;
   2. *Inclusive*, in order to empower and emancipate those whose interests are currently not being met and who represent values that constitute transformative change toward sustainability;
   3. *Adaptive*, since transformative change and governance, and our understanding of them, are moving targets, so governance needs to enable learning, experimentation, reflexivity, monitoring and feedback;
   4. *Transdisciplinary*, in ways that recognize different knowledge systems, and support the inclusion of sustainable and equitable values by focusing on types of knowledge that are currently underrepresented; and
   5. *Anticipatory*; utilizing the precautionary principle when governing in the present for uncertain future developments, and especially the development or use of new technologies (Visseren-Hamakers et al., 2021; Chapter 1).

16.2.2 Revisiting the Concept of Transformative Change

In Chapter 1 we used the following definition of *transformative change*, which was inspired by the Intergovernmental Science-Policy Platform on Biodiversity and Ecosystem Services (IPBES) definition (Díaz et al., 2019): "transformative change [is] a fundamental, society-wide reorganization across technological, economic and social factors and structures, including paradigms, goals and values." In comparison to IPBES, this definition emphasizes the need for society-wide, structural change (instead of systemic change through specific transitions, as elaborated below). It includes both the indirect drivers of biodiversity loss[1] and the values underlying these indirect drivers. Building on insights from the various chapters, we here further refine this conceptualization of transformative change to represent change of *the underlying causes* of biodiversity loss, which includes both the indirect drivers *and* the paradigms, goals and values underlying societies that determine the behavior of individuals and society at large.

Highlighting the inclusion of changing paradigms, goals and values is pivotal for transformative biodiversity governance. How and to what extent can changes in paradigms, goals and values be governed? To date, the literature on (governing) transformative change, transformations or transitions (see Chapters 1 and 4) has paid relatively little attention to this

---

[1] According to the IPBES GA, indirect drivers can be demographic (e.g. human population dynamics), sociocultural (e.g. consumption patterns), economic (e.g. production and trade) or technological, or can relate to institutions, governance, conflicts and epidemics (Díaz et al., 2019; Chapter 1).

question. In particular, the transition literature zooms in on transitions in specific regimes, for example the transitions on food, energy, animal-free innovation and mobility. While this focus makes analyses and governance more tangible and relevant for practitioners working in a specific regime, it diverts attention from more generic societal structures, including paradigms, goals and values. In this sense, biodiversity governance needs to be transformed in order to include explicit attention to all underlying causes, including those generic for societies at large. As sustainability issues, such as climate change and environmental justice, share many of the same underlying causes, this shift in attention implies the need to take a broader perspective beyond traditional conservation and mainstreaming policies.

Based on these insights, we propose further specifying the concept of transformative change by combining the concepts of transformations and transitions as follows. Transformations refer to changing the generic societal underlying causes, including institutions, governance structures, developments, power relationships, paradigms, goals and values (e.g. globalization, the paradigm of economic growth, values on the relationships between humans and nonhumans). Transitions focus on regime-specific underlying causes (e.g. the discourse of having to feed almost 10 billion people in 2050, thereby arguing for the intensification and expansion of agricultural production). Specific transitions are thus embedded in, and an integral part of, more generic, society-wide transformations. Together, transformations and transitions represent transformative change. Combining insights from the transformations and transitions literatures in such a manner, transformative biodiversity governance focuses both on the generic and regime-specific underlying causes of sustainability problems. This means governance mixes need to include instruments designed to realize transformative change both within specific regimes and in society more broadly (see Chapter 4 for more details).

16.2.3 Deepening Our Understanding of Transformative Biodiversity Governance

Based on the contributions of the book, we have also further nuanced the definition of transformative governance (see above for the definition as introduced in Chapter 1), especially by approaching the concept of sustainable development more broadly. Several chapters extend the idea of what transformative change for biodiversity entails, including a focus on just transitions, animal rights, rights of nature and human rights (see Chapters 8, 9 and 15), and Chapter 4 argues that transformative biodiversity governance is about prioritizing biodiversity concerns (instead of compromise or optimization approaches). Based on these insights, this book suggests that transformative biodiversity governance means prioritizing ecological, justice and equity concerns over economic ones, with a view to enabling *ecocentric, compassionate and just sustainable development.*

This notion of prioritizing biodiversity concerns in biodiversity governance seems obvious, but in practice it is not. Most biodiversity governance initiatives over the past decades have been based on deliberative, compromise approaches, in which biodiversity represents one of many interests, or optimization approaches that apply economic logic to decide whether addressing biodiversity loss "is worth it" and mostly use market-based solutions (see Chapter 4 for an introduction to the four problem conceptions). In this sense, biodiversity governance needs to be transformed to actually prioritize biodiversity concerns. This does not mean deliberative or market-based solutions are obsolete, but they need to be applied in a manner prioritizing biodiversity concerns. As Chapter 4 highlights, governance mixes need to change over time as transformative governance is evolving, with the role of market-based instruments shrinking as the underlying causes are increasingly addressed. Deliberative approaches remain needed throughout the transformation in order for stakeholders to reflect on whether transformative governance is still on track.

16.2.4 Plurality versus Priority and Inclusiveness versus Emancipation

Authors have different views on the best ways forward to conserve and sustainably and equitably use biodiversity, and this book includes these different perspectives. Some highlight the need for plurality in biodiversity governance (Chapters 2 and 6), and argue that transformative biodiversity governance means embracing a plurality of values, including intrinsic, instrumental and relational values, as well as a plurality of worldviews and epistemologies. These different values not only represent different ways of looking at human–nature relationships but also entail different views on what the problem of biodiversity loss is and the most appropriate and effective solutions to that problem. They can in some ways be mapped onto the three main objectives of the CBD (intrinsic – conservation; instrumental – sustainable use; and relational – equitable sharing of the benefits), although the different values would also interpret the other aims differently (e.g. those holding intrinsic values would perhaps have more ambitious definitions of what sustainable use would entail, or would actually be against certain forms of sustainable use, such as trophy hunting).

Interestingly, those proposing pluralism are often not complete in the values they describe as relevant for biodiversity governance, often omitting animal rights and post-humanist values (as highlighted in Chapter 9). Moreover, the question is whether those promoting pluralism are actually making the case for including all different views. Do they really aim to defend the right of those actors responsible for large-scale habitat destruction to participate in biodiversity governance? It seems that instead they are promoting the emancipation of relational values and the rights of Indigenous people and local communities (IPLC). This is a legitimate position, but using the concept of pluralism for this purpose blurs the discussion.

Others promote the problem conception of prioritization, as opposed to compromise or optimization conceptions (Chapter 4), and actually see the call for pluralism as a suboptimal solution, representing a compromise problem conception. While recognizing and deliberating values is vital, actors have to be clear on what the problem is they are prioritizing, whether it be emancipation of certain groups of humans or nonhumans, promoting economic development, or conserving (certain types of) biodiversity. Therefore, a crucial part of transformative biodiversity governance is to explicitly discuss the values and problem conceptions of different actors – not with the aim to find compromise, but to achieve clarity on different priorities.

Also, the call for plurality is sometimes used as a call for including actors holding different values, and sometimes for including different types of knowledge, with the former relating to the concept of inclusive governance and the latter to the concept of transdisciplinary governance. These two calls are obviously related, since knowledge is value-laden, and the call for inclusiveness entails including different knowledge-holders, such as IPLC. So both calls aim to emancipate IPLC and recognize their values and knowledge systems. This is an important societal goal, but a different priority from addressing biodiversity loss per se (although they are related since IPLC play an important role in conserving and sustainably using biodiversity). Also, these calls have different implications for biodiversity governance, since basing biodiversity governance on integrated bodies of knowledge is different to facilitating the participation of different types of stakeholders. This difference is not always clear in calls for plurality.

Moreover, value plurality is different to diversity in problem conceptions. Values inform and underlie problem conceptions. So it is possible for coalitions striving for the same priority (e.g. addressing biodiversity loss or promoting IPLC rights) to include actors representing different values. However, in practice these differentiations between different values and between different priorities is not made explicit. In transformative biodiversity governance, actors should discuss these differentiations to see whether they really represent one, or several, perhaps overlapping discourse coalitions, and deliberating values should precede discussing priorities.

This touches upon the definition of inclusive governance as part of the operationalization of transformative governance in Chapter 1. We stated there that it should be operationalized in ways that empower those whose interests are currently not being met and represent values embodying transformative change for sustainability. This means a strategic approach toward participatory processes: so not including all stakeholders for the purpose of compromise, but designing the participatory process in such a manner that it emancipates those who prioritize transformative sustainability. Obviously, all stakeholders should be heard to design a legitimate process, but this does not mean a process of compromise. The ambition for prioritizing transformative sustainability, or ecocentric, compassionate and just sustainable development, should be leading for the design of the participatory process. We need inclusive governance that contributes to changing power dynamics from the domination of unsustainable politics and practices to sustainable ones. These insights are in line with more critical perspectives that incorporate politics, power and equity issues in the debates on transformations.

So, while this book set out to include analytical, normative and critical approaches to study transformations (Burch et al., 2019), we have come to the conclusion that any such analysis is, in essence, normative, since analyses that do not incorporate issues of power and justice could be seen to implicitly accept current power relationships. Transformative change and governance – or lack thereof – and their analysis are, therefore, in essence political and normative.

16.2.5 Emerging Values for the Governance of Transformative Change

Rights of nature, animal rights, Buen Vivir, degrowth and convivial conservation are some of the alternative approaches that this book has covered. Despite comprising different normative visions (for a comparison see Escobar, 2015), they commonly share criticisms of the current neoliberal socioeconomic system, capitalism and/or focus on instrumental values of nature as the underlying causes of ecological crises (Acosta, 2013; Büscher and Fletcher, 2020; Escobar, 2015; Gudynas, 2019). These approaches often advocate replacing the dominant paradigm of economic growth and capital accumulation and suggest broader cultural, political and social transformations of institutions and practices (Büscher and Fletcher, 2020; Demaria et al., 2013; Escobar, 2015). The adoption of rights of nature, animal rights and the rising popularity of the "Buen Vivir" notion, for instance, can precipitate new forms of transformative biodiversity governance in which the modern human–nature dichotomy and anthropocentrism are no longer the dominant ontological assumptions, human, nature and animal well-being are not subordinate to economic reasoning, and the relationships between humans and nonhumans are redefined. Moreover, there is growing interest in understanding how alternatives can unfold over time and space to enact transformative change, away from neoliberal logic and practices, and breaking current lock-ins (Schmid, 2019). For instance, Feola et al. (2021) illustrate how creating space for a postcapitalist alternative requires "unmaking" current capitalist structures that are at the root of the current ecological crisis.

The recognition of alternative values, beliefs, worldviews and approaches can serve to form new conceptualizations of transformative change toward multispecies justice (Celermajer et al., 2021) – or actually represent such transformative change. These new approaches can be understood as a reconfiguration of justice, recognizing rights of ecosystems holistically, including nature, animals and human beings, and calling for the establishment of alternative structures, institutions and processes. Multispecies justice requires rethinking liberalism as the dominant political ideology, rethinking the social contract tradition, and rethinking democracy and representation (Kopnina et al., 2021). Reconfigurations of justice also include paying increased attention to intergenerational justice concerns and the rights of future generations (Hiskes, 2009; Shue, 2014). Considering that these (justice) alternatives are grounded in (and inspired by) a dense network of social mobilizations, civil society, activists and new forms of transnational actor constellations, discussing them sheds light on the importance of bottom-up processes, since these processes enable transformations (in contrast to specific transitions) of the society-wide underlying causes of our current unsustainability. Particularly, it allows for stronger consideration of local knowledge and experiences for transformation, in parallel with traditional top-down conservation practices. Realizing the right to participate in transformation processes for these representatives of transformative values will, therefore, be a core concern in the transformation toward ecocentric, compassionate and just sustainable development.

16.2.6 Can Transformative Change be Governed?

Yes it can, to a certain extent. "Coalitions of the willing," including governmental, civil society and market actors sharing the same priorities, can together develop governance mixes focused on accelerating transformative change, addressing the main (generic and transition-specific) underlying causes of sustainability issues through a process of transformative governance, including the five governance approaches introduced above and in Chapter 1. Over time, the governance mixes will need to be adjusted to reflect what the change process requires during its evolution. Various competing coalitions representing different priorities will emerge and the process will inevitably be complex. Nevertheless, governance can progressively become transformative, since governing transformative change becomes easier as the underlying causes are increasingly addressed.

16.3 Challenges and Opportunities for Transformative Governance through the Post-2020 GBF

As noted in Chapter 1, the focus of biodiversity policy has broadened over time from conservation to mainstreaming. Now the call for transformative change and addressing indirect drivers adds a new dimension to biodiversity governance. Based on insights from the chapters in this book and a review of emerging literature on the Post-2020 GBF, we here examine challenges and opportunities for the GBF and its further implementation to contribute to transformative change for biodiversity. We address how the transformative character of the GBF and its further implementation can be harnessed using the overall conceptualization of transformative governance applied in this book. We especially look at the governance mechanisms that the GBF puts forward in more or less explicit terms: the whole-of-government approach, the whole-of-society approach, ensuring just transitions, implementation support mechanisms, and the responsibility and transparency mechanism. First, we address the question of how to understand the role of the GBF in achieving transformative change.

16.3.1 What Makes the Post-2020 GBF Transformative?

In various negotiation drafts of the GBF, its stated ambition has been that the framework should be transformative. This, first and foremost, requires the framework to focus on addressing the underlying causes of biodiversity loss in an equitable manner and be part of the broader sustainability agenda of the Sustainable Development Goals (SDGs). The GBF is built around a theory of change that recognizes that urgent policy action globally, regionally and nationally is required to transform economic, social and financial models. It assumes that whole-of-government and whole-of-society approaches are necessary. The framework's theory of change *"assumes* that transformative actions are taken to (a) put in place tools and solutions for implementation and mainstreaming, (b) reduce the threats to biodiversity and (c) ensure that biodiversity is used sustainably in order to meet people's needs and that these actions are supported by enabling conditions, and adequate means of implementation, including financial resources, capacity and technology." It also *"assumes* that progress is monitored in a transparent and accountable manner with adequate stocktaking exercises to ensure that, by 2030, the world is on a path to reach the 2050 Vision for biodiversity" (CBD, 2021a: paras. 5–7, italics added).

While the GBF has the ambition of galvanizing urgent and transformative action, it provides little detail on how to achieve this, beyond setting ambitious goals that form the core of the GBF (Díaz et al., 2020). The first draft of the GBF addresses indirect drivers, such as "reduce negative impacts" from businesses and "full sustainability for extraction and production practices," as well as "harmful subsidies." However, it does this without giving guidance on how to identify what type of action (and by whom) is needed to successfully implement it. The GBF also contains provisions for implementation support mechanisms, enabling conditions (including finance), the responsibility and transparency mechanism, and a mechanism for outreach, awareness and uptake (CBD, 2021a). The GBF is meant to be a voluntary international governance mechanism to achieve transformative change for biodiversity. To realize its goals and targets it will depend on, among others, the mainstreaming, capacity building and resource mobilization strategies that the CBD is developing in support of the GBF, and consequently the domestic implementation of whole-of-government (see King, 2020 and Yang et al., 2019) and whole-of-society approaches for transformative biodiversity governance (Pattberg et al., 2019).

Different views exist about the role the Post-2020 Framework can play in achieving transformative change. As also noted by Bulkeley et al. (2020), transformative change in the GBF is mostly defined in terms of its outcomes, and not how goals and targets will be achieved. Some refer to the GBF as the blueprint or roadmap for global biodiversity governance (Phang et al., 2020), while others suggest that the GBF provides a set of shared principles that can act as a guiding "compass," establishing a common direction of travel (Birdlife International, 2019; Bulkeley et al., 2020; 2021a; Franks, 2020; Grumbine and Xu, 2021).

Yet, based on literature on the governance of transformations (Burch et al., 2019; Masarella et al., 2021; Patterson et al., 2017, Visseren et al., 2021; see also Chapters 1, 3 and 4), we suggest that both governance *for* transformative change – the vision and conditions that enable others to take action on this agenda – and transformation *in* governance arrangements is needed if we are to realize these outcomes. Setting ambitious goals is not sufficient. If biodiversity governance seeks to galvanize transformative change, it must embrace transformation in its working arrangements, mechanisms and institutions (Bulkeley et al., 2020; 2021a). Grumbine and Xu (2021: 638) highlight that "fulfilling the goals of the CBD will not occur without strategic learning about societal change, explicit incorporation of climate concerns into conservation; future-forward reframing of what protected means, mainstreaming environmental values into multiple rules and regulations, and finding the money to pay for it all." So the GBF needs to include not only ambition for the "what" but also the "how" – in other words, not only ambition for transformative change, but also transformative governance.

16.3.2 The Whole-of-Government Approach

The "whole-of-government approach" advocated by the GBF points at the dimensions of *integrative* and *adaptive* governance. Biodiversity governance needs active support from a range of other policy domains to address the indirect and direct drivers of biodiversity loss. *Integrative governance* can become transformative if "solutions also have sustainable impacts at other scales and locations, on other issues and in other sectors" (Chapter 1). This requires that policy domains such as trade and finance, climate change, agriculture and development take into account biodiversity in implementing sustainable transformation pathways, and that dependencies, risks and benefits of nature in these policy domains are recognized and prioritized. Integration of biodiversity concerns in other policy areas in turn will have implications for biodiversity policy, forcing it to go beyond its traditional conservation approach to deal with competing priorities and a plurality of values (Fougeres et al., 2020; Pascual et al., 2021). Next to questions of such horizontal policy coherence, the analysis of the role of cities (Chapter 14) shows the need to include multilevel governance approaches, among others, in the sense that cities, as subnational actors, increasingly play an independent role in transforming biodiversity governance beyond implementing national policies.

As analyses in this book have also shown, transformative change necessitates new ways of doing things, including creating *new spaces for transformative action* and new institutions. Hence *adaptive governance* is required to "enable learning, experimentation, reflexivity, monitoring and feedback" (Chapter 1) in developing and engaging with transformation pathways to achieve transformative change. Contestation and politics is inevitable when creating such pathways. It also requires breaking down business-as-usual approaches, as we need to consider the structures and conditions causing biodiversity loss. It is thus urgent to open up a space for alternatives and consider possible pathways and futures that are currently neglected or marginalized in sustainability debates because they are considered "unfeasible" or not "cost-efficient" (Beck and Oomen, 2021). Considering this "space for alternatives" helps to address underlying causes of sustainability issues and hence explore "radical" alternatives that literally go to the root causes of current societal problems (Meadows, 1999) and imagine desirable futures (Stålhammar, 2021) that, for example, could inform the mainstreaming discussions from a transformative change perspective. Chapter 13 identifies an inclusive vision linking biodiversity to national development, social capital for integrative governance among governmental and private actors, as well as adaptive learning as key elements for governance realizing transformative change.

The mainstreaming agenda is a long-standing discussion in the CBD focusing on integration and policy coherence (EMG, 2021). It is clear that integrative approaches need to overcome multiple challenges. These range from legal challenges related to the fragmentation of international law and the different mandates and memberships of various bodies and conventions, to epistemological challenges linked to the sheer difficulty of biodiversity as a subject matter and the lack of, and differences in, understanding between different scientific and policy communities. This last point also shows the importance of transdisciplinary governance to "recognize different knowledge systems, and support the inclusion of ... types of knowledge that are currently underrepresented" (see Chapter 1).

Various chapters (e.g. Chapters 5 and 9) have argued for the need to broaden the regime complex for biodiversity through a One Health approach and to strengthen links between the CBD, World Health Organization (WHO), Food and Agriculture Organization (FAO) and the World Organisation for Animal Health (OIE). This resonates with the experience of the global community regarding the COVID-19 pandemic. In the early days, One Health mainly focused on nature-related health risks, taking potential nature-related health benefits far less into account and, similar to biodiversity governance, not addressing indirect drivers and structural change. Later, One Health was given a broader perspective, including nature-related health benefits (e.g. WHO and CBD, 2015) and incorporating a more systemic approach with *structural One Health* (Wallace et al., 2015). Gradually, over time, more and more professional communities were convinced of the importance of a One Health approach, but before COVID-19 it was far from *mainstream*. The pandemic strongly highlighted the interlinkages between biodiversity, wildlife and human health. It has underscored the urgency of tackling the root causes of biodiversity loss and promoting fair and equitable policies when tackling global challenges, with a focus on the vulnerable and the disenfranchised, who often happen to be biodiversity stewards. This led to broader support for the One Health approach, beyond the One Health expert communities, yet still with a diversity of conceptualizations and practical strategies (Chapter 5).

Another important issue for consideration in a "whole-of-government" approach is the link between biodiversity and oceans. Chapters 10 and 15 on bioprospecting and ocean governance elaborate rights-based proposals to link the biodiversity and oceans regimes by focusing on issues of access and benefit-sharing. As negotiations on oceans continue under the UN General Assembly, the equity question concerns how to secure benefits from global common resources for all, not only for politically, financially or technologically strong actors. Chapter 10 proposes a way to deal with bioprospecting to serve public interests. Chapter 15 proposes an alternative governance approach for oceans, building on the interdependencies between human rights and marine biodiversity, and a broader approach to fair and equitable benefit-sharing to support transformative governance for oceans at various scales. Enhancing the interdependency between human rights and marine biodiversity is suggested to address the indirect drivers of biodiversity loss, including power dynamics. These are also examples of how changing the essence of resource use (moving from market logic to public logic) is essential to enable such synergies between ocean governance and biodiversity governance, an example of how addressing the indirect drivers can support conservation and equitable use.

The fundamental question is how to make the "whole-of-government agenda" transformative. This attention to other sectors does not imply lowering the ambitions for conservation. Ecosystem- and species-focused conservation remains vital in the Post-2020 era. The whole-of-government approach should be seen as an additional priority, not a replacement of conservation, as discussed elsewhere in this book (e.g. Chapter 11). Possible starting points range from intensified cooperation between scientific bodies to recognize the multiple values of nature in various policy domains (e.g. the IPBES and IPCC workshop report on biodiversity and climate change [Pörtner et al., 2021]) to promoting high-level recognition of biodiversity's contribution to all SDGs (Erdelen, 2020), the further development of rights-based approaches to sustainable use and benefit-sharing, taking into account the biodiversity footprint of consumption and production (Chapters 8,

12, 14 and 15), as well as governmental initiatives to promote biodiversity considerations and biodiversity safeguards within relevant sectors (Chapter 13). It finally requires seriously considering biodiversity as a political priority that needs to be dealt with in coherence with climate change, and understood as a socioeconomic development issue that requires us to reshape our economic system (Dasgupta, 2021; Otero et al., 2020; World Bank, 2021; see also Chapter 4).

16.3.3 The Whole-of-Society Approach

Next to the whole-of-government approach, the GBF advocates a stronger engagement with actor groups beyond the state through a "whole-of-society" approach. This includes civil society, cities and subnational governments, IPLC, business, finance and youth. The GBF argues that "all relevant stake- and rightsholders need to be involved in realizing its objectives" (CBD, 2021a: para. 2).

Transformative biodiversity governance must be inclusive, strategic and purposeful, with an aim of focusing on actors that want to influence the indirect drivers of biodiversity loss. In Chapter 1, the dimension of inclusive governance suggests focusing on "empower[ing] and emancipat[ing] those whose interests are currently not being met and who represent values that constitute transformative change toward sustainability." Through the UN major-group system, the CBD has a long tradition of involving various stakeholders in its formal processes, specifically promoting the participation of IPLC. Beyond that, processes are in place to strengthen the position of, for example, cities and subnational governments (Edinburgh process) and businesses that want to contribute to nature-positive strategies. However, more imaginative inclusion processes are needed: the recent shift in international policy domains like climate change (the UNFCCC), oceans and the SDGs encouraging a stronger role for nonstate actors marks a shift in international environmental governance that goes beyond traditional representation of major groups as in the CBD processes (Pattberg et al., 2019).

This development toward stronger involvement of nonstate and subnational actors is not uncontested and has at least two dimensions (see also Chapter 3). It requires working with nonstate actors with the power and ability to induce ownership and leadership to work for biodiversity (Bull and Brownlie, 2017; Bull et al., 2020; Smith et al., 2019), as well as addressing vested interests that may resist transformative change. Such vested interests may include sectors that are based on the (often unsustainable) use of natural resources, including biodiversity. Examples of the latter are provided in Chapters 10 and 13 in industry responses to the evolving regulation of marine bioprospecting in polar regions and biodiversity policy integration in agricultural landscapes. These businesses are seldomly engaged in biodiversity governance and may use domestic implementation of international agreements to create room to maneuver. Political will is needed to address regulatory and implementation gaps in current legislation, power asymmetries and trade-offs between different policy objectives.

As illustrated by Chapter 14, cities provide a case in point of how the involvement of nonstate and subnational actors provides opportunities for the Post-2020 GBF (see also Bulkeley et al., 2021b; Xie and Bulkeley, 2020). Urban biodiversity governance is recently being transformed both in terms of its focus – moving from only a concern with reducing the threat of cities to biodiversity to also realizing their benefits – and in terms of the forms that governance is taking – through governance experimentation in cities and the growth in transnational governance networks. The growing recognition of cities as key agents of change and as presenting both opportunities and challenges for governing biodiversity is also relevant for business, finance and other nonstate actors (Meijer et al., 2021; Smith at al., 2019; van Oorschot et al., 2020).

Some of the challenges that the urban agenda illustrates include the need to go beyond biodiversity and nature-based solutions as win–win solutions to also addresses the underlying causes of biodiversity loss beyond city boundaries, among others through unsustainable production and consumption (i.e. the biodiversity footprint). It also requires answers to the questions of how to address injustice, and the risk that governing urban nature will entrench forms of neoliberal economic development and social exclusion. The way inclusiveness is taking shape continuously needs to be examined. And, lastly, urban governance needs to respond to such challenges through new institutional mechanisms, since existing institutions will most likely not be able to do so. This in turn shows the limitations of integrative forms of governance, as it suggests that it will not be sufficient for global institutions and transnational networks to promote urban action on nature; they will need to play a critical part in building the capacity and vision needed for cities to ensure they take action for nature within and beyond urban boundaries, not only contributing to global biodiversity goals but also ensuring social justice. That is why our operationalization of transformative governance (see above) highlights that integrative and inclusive governance need to be implemented in conjunction.

The "whole-of-society" approach can contribute to a transformative GBF, if fully embedded throughout its theory of change. This implies that nonstate actors will be included in the goals and targets to address indirect drivers, and that the mainstreaming and capacity-building strategy is extended to nonstate actors and social movements to empower them in enhancing nature-inclusive transitions and broader society-wide transformative change, and include them in the responsibility and transparency mechanism for the GBF.

16.3.4 Ensuring Just Transitions

Another element of transformative biodiversity governance, and especially pertinent for the further implementation of the Post-2020 GBF, is the issue of justice and equity (see Chapters 4, 7, 8, 9, 10, 11, 12, 14 and 15). This relates to the *inclusiveness* dimension of transformative governance (Chapter 1). The depth, scale and urgency of transformative change require heightened attention to both existing injustices and the advancement of multiple dimensions of justice, including procedural justice, recognition and distributive justice. Various chapters suggest proposals for combining conservation and justice objectives.

A strong access and benefit-sharing (ABS) regime, for example, can support conservation while promoting equity and justice considerations. This represents an example of synergies between justice and equity concerns one the one hand and conservation concerns on the other. Technological developments such as bioinformatics and synthetic biology, addressed in the CBD negotiations under the umbrella term "digital sequence information" (DSI), can both enable or disable this trend. Through these developments, harvesting for bioprospecting may be less necessary, since the information derived from genetic resources can be publicly available in biobanks long-term. On the other hand, unless such public access to data is accompanied by strong provisions to ensure fair and equitable benefit-sharing, including capacity building to analyze it, these technological developments risk reinforcing global asymmetries in bio-based research and development (Chapter 7). Thus, broad ABS rules, in addition to a radical restructuring of the intellectual property rights system, are needed to move toward transformative biodiversity governance that is inclusive and emancipatory (see also Chapters 10 and 15).

The efforts in the GBF to expand protected areas and other effective conservation measures also opens up questions of justice; namely, its redistributive effects and issues of procedural justice and recognition in decision-making (Chapters 8, 11 and 12). Although the redistributive effects of protected area expansion are often understood in human terms (for an example, see Schleicher et al., 2019), an ecological justice perspective – which extends compassion and rights to the entire living community – draws attention to the ways in which protected area expansion redistributes the Earth's resources between humans and nonhumans (Bhola et al., 2020; Fougeres et al., 2020; Kopnina et al., 2018). A perspective on justice that encompasses both human and nonhuman concerns could highlight possible areas of convergence between ecocentric conservation and social justice activists. Chapter 11 specifically addressed new approaches for protected and conserved areas to ensure that positive biodiversity outcomes are accompanied by equitable outcomes for IPLC. Here, especially, inclusive and transdisciplinary governance becomes relevant for recognizing different knowledge systems, and supporting the inclusion of multiple values by focusing on types of knowledge that are currently underrepresented in conservation. In this regard, the broadening and pluralizing of ways of understanding nature (Chapter 2) is fundamental for creating a space that focuses on the inclusion of currently underrecognized knowledge systems. Conservation, therefore, should recognize and enforce the rights of IPLC (Armitage et al., 2020), animal rights and rights of nature as part of a vision of ecocentric, compassionate and just sustainable development.

It is also important to take distributional justice into account, to address consumption and production in developed and newly industrialized countries, which have the largest impact on global biodiversity loss. This requires the GBF to take a differentiated approach regarding responsibilities in addressing the loss of biodiversity. Developed and newly industrialized countries and relevant nonstate and subnational actors such as business and cities need to address the underlying drivers of biodiversity loss linked to unsustainable production, consumption and global trade, which negatively impacts biodiversity in low-income countries. This latter point is also stressed in Chapter 12 on convivial conservation and structural transformation (see also Buscher and Fletcher, 2020). This chapter argues that fundamental changes in consumption patterns, global trade and the world economy cannot be achieved through mainstream institutional and societal structures. Instead, transformative governance will need to take a "whole Earth" approach and address the indirect drivers of biodiversity loss, including land use, economic development and economic growth. The chapter proposes "Biodiversity Impact Chains" (BIC) as a potential political methodology and a transformative governance mechanism. The basic idea behind BICs is to better understand and politicize the relationships among different actors and the impacts that their livelihoods and consumption choices have on biodiversity elsewhere. BICs challenge many of the embedded assumptions in biodiversity policy by refocusing attention on those with the largest footprints.

Underlining the need to strengthen equity in biodiversity governance, various chapters (5, 9, 10, 15; see also Bernstein et al., 2021) argue for upholding a rights-based approach in the GBF to promote embedding justice and equity concerns in its enabling conditions, equitable access to finance and intergenerational equity. Moreover, rights-based approaches are critical for groups such as IPLC who, despite being at the forefront of biodiversity conservation and sustainable use, are often left behind due to power asymmetries. These "traditional" rights-based approaches can be complemented by more novel approaches to biodiversity governance, including rights of nature and animal rights. Chapters 2 and 9, for instance, argue that integrating animal rights and rights of nature approaches is necessary to fully enable ecocentric approaches in biodiversity governance, and that such an integrated approach should be included in the (implementation of the) GBF to enable transformative change. In particular, extending the agenda with animal rights perspectives would be a novel step from a biodiversity governance perspective that would enable compassionate sustainable development.

Chapter 8 summarizes how principles of justice and equity could be interpreted and upheld in efforts to pursue transformative biodiversity governance. The chapter suggests the following policy options: further development of international norms of justice and equity in global sustainability governance and across all three objectives of the CBD; better compliance with or fulfillment of existing norms; and stronger integration of justice concerns and procedural rights in biodiversity policy-making, implementation and review at all levels of governance. This can also build on SDG implementation that includes goals on equity. Alongside more conventional measures to alleviate the impacts of conservation initiatives on marginalized groups (including social impact assessment and financial transfers), just transformation is likely to require strengthening broad-based social safety nets, international recognition of Indigenous and community conserved areas (ICCAs) and other measures to remedy unjust asymmetries of power in political systems (e.g. land reform and recognition of indigenous rights).

16.3.5 Implementation Support Mechanisms: Mainstreaming and Finance

The national implementation challenge has long been recognized in the CBD (see Chapter 3). In addition to the Post-2020 GBF itself, the CBD is developing support mechanisms for domestic implementation, including a resource mobilization strategy, a strategic framework for capacity building, a mainstreaming strategy, a gender plan of action and a communication strategy. Building also on the analysis regarding whole-of-government, whole-of-society and just transitions in this section so far, we here address the issues of mainstreaming and resource mobilization in the context of national implementation.

Mainstreaming of biodiversity, as a form of integrative governance, is one of the main strategies of the CBD, as exemplified by the new long-term strategy for mainstreaming that is being developed as a complement to the Post-2020 GBF (CBD, 2021b). The recognition of the need for a whole-of-society approach, as discussed above, has major implications for mainstreaming strategies and will need to include nonstate actors (Milner-Gulland et al., 2021). Chapter 13 on agriculture provides an example of how a transformative change lens is relevant for mainstreaming biodiversity in other policy domains. The chapter finds that biodiversity policies are predominantly "add-on" and agricultural policies so far neither directly address biodiversity-threatening agricultural practices nor specifically support more "nature-inclusive" agriculture. Thus, existing knowledge on biodiversity-sound agriculture is not reflected in dominant agricultural policies and practices. The authors argue that political will can target the following leverage points to transform existing governance structures for agriculture: a) working toward a clear vision for sustainable agriculture (Wanger et al., 2020); b) building social capital; c) integrating private sector initiatives and d) better integrating knowledge and learning in policy development and implementation. The Post-2020 GBF should focus on the transformation of agricultural governance systems by concretely addressing key leverage points and providing specific guidance for Parties to address country-specific drivers and potential for sustainable innovation and change through biodiversity policy integration in the agricultural sector. Since the agricultural sector especially touches upon many different sustainability issues, including climate change, water use, animal welfare, pollution and biodiversity, such mainstreaming of biodiversity should be seen as part of a broader agenda for ecocentric, compassionate and just sustainable development.

A resource mobilization strategy is under development (CBD, 2021c) to realize the financial resources required, as put forward in the specific targets on finance in the GBF. Chapters 6 and 8 address these issues. Chapter 6 critically examines the transformative potential of biodiversity finance. This addresses part of the challenge put forward by IPBES to reform the current economic and financial system. The chapter argues that biodiversity finance has not yet challenged the foundations of the capitalist system that has often been argued to undergird many of the known drivers of biodiversity loss, because it reproduces the existing (skewed) power relations that this system builds on. According to the authors it seems implausible that, on their own, innovative financial instruments can bring about the fundamental transformation that is advocated in this book, although they can contribute to catalyzing it. Financial instruments represent the market-based instruments that, as Chapter 4 argues, will have an increasingly smaller role as the sustainability transformation progresses. In this respect, they are rather transitory facilitators of the transformative changes required for effective biodiversity conservation and, therefore, a component of a broader system of transformative governance. With respect to resource mobilization, Chapter 8 argues that this requires credible, time-bound, multilateral, national and nonstate commitments to scale-up resource mobilization to support biodiversity policy in developing countries – including meaningful progress on the multilateral benefit-sharing mechanisms in the context of the ABS framework.

16.3.6 Responsibility and Transparency

The GBF states that "its successful implementation requires responsibility and transparency, which will be supported by effective mechanisms for planning, monitoring, reporting and review" (CBD, 2021a: paras. 18–20). A responsibility and transparency mechanism is key to ensuring that countries and society can change course when ambition and implementation gaps become evident. This relates to the idea of adaptive governance: "transformative change and governance, and our understanding of them, are moving targets, so governance needs to enable learning, experimentation, reflexivity, monitoring and feedback" (Chapter 1). Such a mechanism is largely missing within the CBD and countries have to date been unwilling to implement a legally binding compliance mechanism, and for the GBF now seem to opt for a nonpunitive, voluntary system for accountability (Chapter 3).

Accountability could be strengthened through more transparency in reporting on the progress of Parties and nonstate actors, especially on addressing indirect drivers (Milner-Gulland et al., 2021). Furthermore, meaningful ways of monitoring and evaluating equity in conservation, sustainable use and benefit-sharing need to be put in place. A potential pledge and review mechanism shows promise and could be accepted by Parties, if accompanied by a robust resource mobilization mechanism (Chapter 8). In addition, peer-review mechanisms could be strengthened to facilitate learning. The main challenge here is to make the contribution and commitments of countries to the Post-2020 targets more directly visible and attributable, and if needed to step up ambitions and actions.

16.4 Concluding Remarks

The GBF deliberations have the ambition to develop a transformative framework for a new phase in biodiversity governance, and shape the agenda for new and more effective biodiversity policies across governments and society at large for the coming decade. Over half a century of conservation efforts around the world have failed to bend the curve for biodiversity – in fact the downward curve has steepened despite our efforts. We need to essentially transform the ways in which we govern biodiversity – tweaking the system will not be enough. Transformative biodiversity governance means prioritizing ecological, justice and equity concerns through addressing the root causes of biodiversity loss.

This book has developed ideas to make biodiversity governance transformative. One of the main aspects of our operationalization of transformative governance is the implementation of five governance approaches: integrative, inclusive, adaptive, transdisciplinary and anticipatory, operationalized in a specific manner, in conjunction and focused on the underlying causes of biodiversity loss and unsustainability. The individual approaches themselves cannot become transformative without being implemented with the other dimensions in mind. In this sense, these approaches serve as a heuristic and guidance to further develop and implement transformative governance. The GBF and its implementation, therefore, must be continuously evaluated and adapted as a system of approaches that only together can become transformative.

In order to do so, the global community can apply the guidance on transformative governance as suggested in this volume in the further development and implementation of the GBF, and the SDGs more broadly. The GBF should, therefore, not only be transformative but also be governed transformatively, and should:

- Prioritize halting biodiversity loss through actions across all levels of governance, around the world, in all sectors and on all issues, including biodiversity impacts elsewhere (integrative governance);
- Strategically design the participatory processes in order to empower and emancipate those whose interests are currently not being met and who represent values that constitute transformative change toward sustainability (local stewards of biodiversity, rights of nature, animal rights);
- Regularly evaluate whether implementation is still transformative by addressing indirect drivers and prioritizing ecocentric, compassionate and just sustainable development (adaptive governance);
- Ensure all knowledge systems are respected and all necessary types of knowledge are being used and facilitated (transdisciplinary governance);
- Apply the precautionary principle, not only in relation to new technological developments, but also more broadly in policy (anticipatory governance).

Parties to the CBD, and all other actors and stakeholders, can continuously reflect on the extent to which the process and governance mixes are truly transformative. Through this process of prioritization and learning, global biodiversity governance in the Post-2020 era can become increasingly transformative in order to achieve the goal of halting biodiversity loss and restoring nature. If the global community truly wants to transform our societies and achieve the Sustainable Development Goals by 2030, and the goals and targets of the GBF, we urgently need to change our priorities toward ecocentric, compassionate and just sustainable development – and the ways in which we govern the transformation toward those priorities.

References

Acosta, A. (2013). *El Buen Vivir: Sumak Kawsay, una oportunidad para imaginar otros mundos*. Barcelona: Icaria.

Armitage, D. Mbatha, P., Muhl, E.-K., Rice, W., and Sowman, M. (2020). Governance principles for community-centered conservation in the post-2020 global biodiversity framework. *Conservation Science and Practice* 2, e160. https://doi.org/10.1111/csp2.160

Beck, S., and Oomen, J. (2021). Imagining the corridor of climate mitigation – What is at stake in IPCC's politics of anticipation? *Environmental Science & Policy* 123, 169–178.

Bernstein, J., Heinz, V., Schouwink, R., et al. (2021). *Strengthening equity in the post-2020 Global Biodiversity Framework*. London: IIED.

Bhola, N., Klimmek, H., Kingston, N., et al. (2020). Perspectives on area-based conservation and what it means for the post-2020 biodiversity policy agenda. *Conservation Biology* 35,168–178. https://doi.org/10.1111/cobi.13509

Birdlife International. (2019). Achieving transformative change for the post-2020 global biodiversity framework. Available from https://bit.ly/3B6fUQI.

Bulkeley, H., Kok, M., and van Dijk, J. (2020). *Harnessing the potential of the Post-2020 Global Biodiversity Framework. Report prepared by an Eklipse Expert Working Group*. Wallingford: UK Centre for Ecology & Hydrology.

Bulkeley, H., Kok, M., and van Dijk, J. (2021a). Embedding transformative change in global biodiversity governance, Policy Brief, Post2020 Biodiversity Framework-EU Support project. Available from https://bit.ly/3GzauyQ.

Bulkeley, H., Kok, M., and Xie, L. (2021b). *Realising the urban opportunity: Cities and Post-2020 biodiversity governance*. The Hague: PBL Netherlands Environmental Assessment Agency.

Bull, J. W., and Brownlie, S. (2017). The transition from no net loss to a net gain of biodiversity is far from trivial. *Oryx* 51, 53–59. https://doi.org/10.1017/S0030605315000861

Bull, J. W., Milner-Gulland, E. J., Addison, P. F. E., et al. (2020). Net positive outcomes for nature. *Nature Ecology and Evolution* 4, 4–7.

Burch, S., Gupta, A., Inoue, C. Y., et al. (2019). New directions in earth system governance research. *Earth System Governance* 1, 100006.

Büscher, B., and Fletcher, R. (2020). *The conservation revolution: Radical ideas for saving nature beyond the Anthropocene*. London; New York: Verso Trade.

CBD (2010). *The Strategic Plan for Biodiversity 2011-2010 and the Aichi Biodiversity Targets*. UNEP/CBD/COP/DEC/X/2. Montreal: Convention on Biological Diversity.

(2021a). First draft of the post-2020 Global Biodiversity Framework, CBD/WG2020/3/3, Montreal. Available from https://bit.ly/3r4XV9J.

(2021b). Mainstreaming of biodiversity within and across sectors and other strategic actions to enhance implementation, CBD/SBI/3/13, Montreal. Available from https://bit.ly/35n1Nuh.

(2021c). Resource mobilization. CBD/SBI/3/5, Montreal. Available from https://bit.ly/34epODb.

Celermajer, D., Schlosberg, D., Rickards, L., et al. (2021). Multispecies justice: Theories, challenges, and a research agenda for environmental politics. *Environmental Politics* 30, 119–140.

Dasgupta, P. (2021). *The economics of biodiversity: The Dasgupta review*. London: HM Treasury.

Demaria, F., Schneider, F., Sekulova, F., and Martinez-Alier, J. (2013). What is degrowth? From an activist slogan to a social movement. *Environmental Values* 22, 191–215.

Díaz, S., Settele, J., Brondízio, E. S., et al. (2019). Summary for policymakers of the global assessment report on biodiversity and ecosystem services of the Intergovernmental Science-Policy Platform on Biodiversity and Ecosystem Services. Intergovernmental Science-Policy Platform on Biodiversity and Ecosystem Services. Available from https://bit.ly/3onTu88.

Díaz, S., Zafra-Calvo, N., Purvis, A., et al. (2020). Set ambitious goals for biodiversity and sustainability. *Science* 370, 411–413. DOI: 10.1126/science.abe1530.

EMG (2021). Supporting the global biodiversity agenda. Available from https://bit.ly/3IHgW87.

Erdelen, W. R. (2020). Shaping the fate of life on Earth: The Post-2020 Global Biodiversity Framework. *Global Policy* 11, 347–359. https://doi.org/10.1111/1758-5899.12773

Escobar, A. (2015). Degrowth, postdevelopment, and transitions: A preliminary conversation. *Sustainability Science* 10, 451–462.

Feola, G., Koretskaya, O., and Moore, D. (2021). (Un) making in sustainability transformation beyond capitalism. *Global Environmental Change* 69, 102290.

Fougeres, D., Andrade, A., Jones, M., and McElwee, P. D. (2020). *Transformative conservation in social-ecological systems. Discussion paper for the 2021 World Conservation Congress*. Commission on Ecosystem Management, International Union for the Conservation of Nature. Available from https://bit.ly/3Hb3DfW.

Franks, P. (2020). Truly transformative change is key to combating the biodiversity crisis. Available from https://bit.ly/3AJ9lDl.

Grumbine, R. E., and Xu, J. (2021). Five steps to inject transformative change into the Post-2020 Global Biodiversity Framework. *BioScience* 71, 637–646.

Gudynas, E. (2019). Value, growth, development: South American lessons for a new ecopolitics. *Capitalism Nature Socialism* 30, 234–243.

Hiskes, R. (2009). *The human right to a green future: environmental rights and intergenerational justice*. Cambridge: Cambridge University Press.

King, N. (2020). *Key African priorities for a post-2020 global biodiversity framework*. South African Institute of International Affairs. Available from https://bit.ly/3LbYlDe.

Kopnina, H., Spannring, R., Hawke, S., et al. (2021). Ecodemocracy in practice: Exploration of debates on limits and possibilities of addressing environmental challenges within democratic systems. *Visions for Sustainability* 15, 9–23.

Kopnina, H., Washington, H., Gray, J., and Taylor, B. (2018). The "future of conservation" debate: Defending ecocentrism and the Nature Needs Half movement. *Biological Conservation* 217, 140–148.

Masarella, K., Nygren, A., Fletcher, R., et al. (2021). Transformation beyond conservation: How critical social science can contribute to a radical new agenda in biodiversity conservation. *Current Opinion in Environmental Sustainability* 49, 79–87.

Meadows, D. H. (1999). *Leverage points: Places to intervene in a system*. Hartland Four Corners, VT: Sustainability Institute.

Meijer, J., van Oosten, C., Subramanian, S. M., Yiu, E., and Kok, M. T. J. (2021). *Seizing the landscape opportunity to catalyse transformative biodiversity governance. A contribution to the CBD post-2020 Global Biodiversity Framework*. The Hague: PBL Netherlands Environmental Assessment Agency.

Milner-Gulland, E. J., Addison, P., Arlidge, W. N. S., et al. (2021). Four steps for the Earth: mainstreaming the post-2020 Global Biodiversity Framework. *ONEEAR* 4, 75–87.

Otero, I., Farrell, K. N., Pueyo, S., et al. (2020). Biodiversity policy beyond economic growth. *Conservation Letters* 13, e12713.

Pascual, U., Adams, W. M., Díaz, S., et al. (2021). Biodiversity and the challenge of pluralism. *Nature Sustainability* 4, 567–572.

Pattberg, P., Widerberg, O., and Kok, M. T. (2019). Towards a global biodiversity action agenda. *Global Policy* 10, 385–390.

Patterson, J., Schulz, K., Vervoort, J., et al. (2017). Exploring the governance and politics of transformations towards sustainability. *Environmental Innovation and Societal Transitions* 24, 1–16.

Phang, S. C., Failler, P., and Bridgewater, P. (2020). Addressing the implementation challenge of the global biodiversity framework. *Biodiversity and Conservation* 29, 3061–3066. https://doi.org/10.1007/s10531-020-02009-2.

Pörtner, H. O., Scholes, R. J., Agard, J., et al. (2021). IPBES-IPCC co-sponsored workshop report on biodiversity and climate change; IPBES and IPCC. DOI: 10.5281/zenodo.4782538.

Schleicher, J., Zaehringer, J. G., Fastré, C., et al. (2019). Protecting half of the planet could directly affect over one billion people. *Nature Sustainability* 2, 1094–1096.

Schmid, B. (2019). Degrowth and postcapitalism: Transformative geographies beyond accumulation and growth. *Geography Compass* 13, e12470.

Shue, H. (2014). *Climate justice: Vulnerability and protection*. Oxford: Oxford University Press.

Smith, T. et al. (2019). Biodiversity means business: Reframing global biodiversity goals for the private sector. *Conservation Letters* 13, e12690. https://doi.org/10.1111/conl.12690

Stålhammar, S. (2021). Assessing people's values of nature: Where is the link to sustainability transformations? *Frontiers in Ecology and Evolution* 9, 145. https://doi.org/10.3389/fevo.2021.624084

Van Oorschot, M. M. P., Kok, M. T. J., and Van Tulder, R. (2020). *Business for biodiversity. Mobilising business towards net positive impact.* The Hague: PBL Netherlands Environmental Assessment Agency.

Visseren-Hamakers, I. J., Razzaque, J., McElwee, P., et al. (2021). Transformative governance of biodiversity: Insights for sustainable development. *Current Opinion in Environmental Sustainability* 53, 20–28. https://doi.org/10.1016/j.cosust.2021.06.002.

Wallace, R. G., Bergmann, L., Kock, R., et al. (2015). The dawn of structural One Health: A new science tracking disease emergence along circuits of capital. *Social Science and Medicine* 129, 68–77.

Wanger, T. C., DeClerck, F., Garibaldi, L. A., et al. (2020). Integrating agroecological production in a robust post-2020 Global Biodiversity Framework. Correspondence. *Nature Ecology & Evolution* 4, 1150–1152. www.nature.com/articles/s41559-020-1262-y

WHO and CBD. (2015). Connecting global priorities: Biodiversity and human health. Available from https://bit.ly/3goq71b.

World Bank. (2021). *The economic case for nature: A global Earth-economy model to assess development policy pathways*. Washington, DC: The World Bank.

Xie, L., and Bulkeley, H. (2020). Nature-based solutions for urban biodiversity governance. *Environmental Science & Policy* 110, 77–87. https://doi.org/10.1016/j.envsci.2020.04.002

Yang, R., Peng, Q., Cao, Y., et al. (2019). Transformative changes and paths toward biodiversity conservation in China. *Biodiversity Science* 27, 1032–1040. DOI: 10.17520/biods.2019217

Index

Aarhus Convention (United Nations Economic Commission for Europe [UNECE] Convention on Access to Information, Public Participation in Decision-Making and Access to Justice in Environmental Matters), 161, 172
access and benefit-sharing (ABS) – equitable, 146, 350, 353
access without benefit-sharing (ABS), 147
accountability, 52, 57, 58–59, 120, 172, 278, 356
agricultural landscapes, 17, 33, 264–265, 269, 272, 274, 278–283, 351
agricultural policy, 269, 275, 277, 278, 282–283
agriculture, 266–268
agrihealth, 97
Aichi Biodiversity Targets, 229
Aichi Targets (ATs), **4**, 17, 33, 44, 138, 161, 169, 221–222, 223, 225–226, 228, 229–230, 233, 264, 293, 321
   cost of, 166
   equitable management component of, 227–228
   missed, 53
   overview of, **4**
   transformative governance and, 230
amplification effect, 96
animal rights, 16, 76, 82, 165, 182–183, 184–185, 186, 191, 343, 344, 345, 346, 353, 357
   academic debate about, 180–181
   rights of nature and, 179, 192–194
   rights-based approaches to, 354
animal welfare, 8, 165, 179, 182, 183–186, 190–195, 264, 282, 355
   animal rights versus conservation and, 183
Anthropocene, 30, 96
anthropocentric, 16, 30, 34, 79, 104, **131**, 179, 184–185, 186, 192, 195
anthropocentrism, 123, 125, 181, 346
architecture, 43, 128, 204, 274, 294, 301, 304, 305, 306, 307
ArcticZymes / Biotec Pharmacon, 212, **213**
area-based management, 18, 314, 322, 323, 329, 330, 332
Atrato River, 35
Āyurveda, 98

benefit-sharing, 18, 138, 147–149, 151, 156, 172, 202, 204, 314, 321, 328–330, 332, 356
Benguela Current Commission, 324–327
Benguela Current Convention (BCC), 324
biocultural diversity, 14, 26, 31–34, 37
biodiversity, 26–28, 122, 137, 157–158, 255–260, 270, 273–274, 277–278, 294–296
   financeable, 121–126
   governing in the city, 301–304
   institutional capacity and, 276–277
   urban, 299–301
biodiversity and health, 93, 97–98
biodiversity conservation, 10, 13, 26, 29, 37, 56, 75–76, 79, 82, 115, 117–118, 123, 138, 161, 201, 223, 224, 234, 235, 236, 244–245, 247, 250–251, 259, 268, 270, 274–275, 278, 280, 296–297, 313, 321, 324, 328, 331, 342, 354, 356
   agricultural landscapes and, 17
   Aichi Targets (ATs) and, 161, 222, 264
   application of synthetic biology and, 143
   biocultural diversity and, 33
   biological diversity and, 27
   BPI and, 270
   business case for, 120
   CBD objectives of, 47
   Earth jurisprudence for, 187
   economic importance of, 118, 229
   financing instrument for, 124, 128
   funding gap and, 116, 118, 121
   incentives for, 204
   influence of cities on, 51
   management plans for, 274
   objectives of, 44–47, 224
   prioritizing, 77, 79, 83
   promotion of, 265
   public finance for, 119
   reason for failing objectives of, 164
   resources for, 15
   threat to, 272
   transformative change and, 13, 69, 75, 78, 85, 221
   transformative governance agenda, 230

biodiversity finance, 15, 124, 126–130, 167, 168, 170, 355
  assessment of, 120–121
  developments in, 116
  diversity of instruments and, 116–121
  literature about, 167
  sources and needs, **119**
  transformative change and, 121
  transformative governance and, 115–121
  unlocking, 118–120
biodiversity govern, **159**
biodiversity governance, 3–5, 9, **14**, 14, 18, 25, 26, 28, 29, 31, 34, 36–37, 43, 44, 47, 50, 51, 55–56, 60, 76–77, 84, 93–94, **108**, 108, 115, 120, 123, 130, 137–142, 150, 151, 155–159, 164, 165, 166, 169, 170, 179–180, 184, 188, 194–195, 200, 274, 304, 341–357
  external actors and, 164
  justice and equity in, 157, 159–163
  transformative, 13, 15, 18, 26, 35, 51, 85, 122, 171–173, 343–344
  urban, 18
biodiversity health, 27, 94–97
biodiversity impact, 17, 140, 244–246, 253–259, 267, 281, 357
biodiversity intactness, 27
biodiversity offsets, **119**, 122, 124, 129
Biodiversity Policy Integration (BPI), **14**, 17, 264–265, 269, 270–278, 282
  framework for, 269–273
biodiversity-loss-as-material-risk, 130
bioinformatics, 15, 137–138, 146, 353
biological diversity, 3, 27, 31–32, 44–47, 139, 141, 151, 296
biopiracy, 202, 203, 210, 212, 214
bioprospecting, 16, 200–202, 211, 215, 328, 350, 351, 353
  ABS and, 204, 214
  corporate strategies and, 210
  governing, 206–210
  in the polar regions, 205–210
  Norway and, 208–210
  Novozymes', 211, 214
bioprospectors, 208, 211, 214, 215
biotechnology, 137, 143–144, 145, 201, 205, 207, 208, 215
black-boxes, 37
blended finance, 127, 130
buen vivir, 34, 35, 123, 186, 345, 346

capitalism, 9, 247, 249, 250, 346
carbon dioxide removal (CDR), 15, 137, 139
Cities4Forest initiative, 305
CitiesWithNature, 305
Clearing House Mechanism, 207, 215
climate change, 6–7, 17, 28, 50, 55, 58, 82, 94, 137, 139–140, 155, 247, 266, 280, 293, 294, 295, 297, 298, 299, 304, 306, 315, 317, 321, 330, 331, 343, 350, 351
  deep ocean environments and, 317
collaboration, 26, 50, 70, 83, 98, 102, 107, 109, 123, 143, 185, 191, 195, 203, 208, 211, 212, 214, 274, 303, 322, 330, 331
Communities of Practice, 109
complex adaptive systems approach, 103
complexity, 15, 36, 53, 83, 104, 109, 126, 148, 151, 155, 224, 229, 237, 307, 327
  structural societal drivers, 297
  system/systemic, 68
compliance, 46, 53–55, 58, 149, 172, 203, 210, 211, 232, 234, 273, 354, 356
  lack of, 150
conservation, 44, **45**, 47, 49, 51, 52, 191–192, 233–237, 274–276
conservation medicine, 97
conservation strategies, 28, 37
conserved areas, 17, 168, 221, 222–225, 226, 230, 233–236, 237, 353
  designation of, 228–229
Convention on Biological Diversity (CBD), 3, 44–47, 52–53, 59–60, 137, 156, 167, 221, 264, 354
Convention on International Trade in Endangered Species of Wild Fauna and Flora (CITES), 49
Convention on the Conservation of Migratory Species of Wild Animals (CMS), 3
Convention on Wetlands, the Convention on Migratory Species (CMS), 49
convivial conservation, 17, 35, 244–246, 249–253, 255, 257, 258, 345, 354
  BICs as a tool in, 259–260
  transformative governance of, 255
COVID-19, 4–5, 15, 93–94, 102, 103, 104, 106–108, 115, 191, 200, 315, 350
cultural landscapes, 29, 33–34

decision-making, 164–165
decision-making (biodiversity) – justice and equity in, 163–165
determinants of health, 97
digital sequence information (DSI), 15, 137, 146–149, 204, 206, 207, 209, 211, 215, 353
dilution effect, 96, 103

early warning and prevention, 107
Earth jurisprudence, 16, 186–190, 192
ecocentrism, 35
EcoHealth, 97, 109
ecologically or biologically significant marine areas (EBSAs), 322, 326
ecosystem services, 6, 14, 26, 29–31, 33, 37, 94, 99, 122, 125, 126, 170, 188, 190, 222, 331
Edinburgh Process, 51, 304, 351
emerging infectious diseases, 93, 97, 104, 107
enabling environment, 106, 109

Index

environmental justice, 4, 156, 158, 162, 163, 251, 308, 343
environmental policy-making, 36
Environmental Reserve Quota, 123
enzymes (marine), 204, 205–206, 209–210, 212–214
epistemic pact, 35
equity, 79, *See also* justice
Escazú Agreement, 161
ethical and sociocultural valuations, 31
experimentation, 12, 18, 59, 70, 224, 237, 294, 302–303, 304, 305, 307, 314, 332, 342, 349, 352, 356

finance, 355–356
finance (biodiversity)
  allocation, 167–168
  effort-sharing, 167
financeable objects, 122–123, 131
financeable' biodiversity, 121–126
five freedoms, 183–186
Free, Prior and Informed Consent (FPIC), 165
funding gap, 116, 118–120, 121

gene drives, 15, 137, 138, 142–147, 150
geoengineering, 137–138, 150–151
  climate-related, 15, 138, 139–142
Global Biodiversity Framework (GBF), 347–348, 355
global health, 97
Global Health Ethics, 103
governance, 7–12, 50–52, 56–57, 68–69, 79–86, 105, 106, 168–173, 228–230, 249–253, 345–351
  transformations and transitions and, 67–68
  transformative, 70
  transitional, 69–70
  transnational, 72
governance *for* transformation, 305, 308
green bonds, 117, 122, 126
Green Factor Tool, 298
green gentrification, 300
green prescription, 94

hard law, 53
health, 99–102, 103–104, 105
  human, 102, 104–105
  integrative concepts of, 97
Health in All Policies, 97
health pluralism, 97
Health Promotion, 94, 97
  biodiversity conservation, 56
human-natural systems, 37

illegal, unreported and unregulated fishing (IUU), 315
implementation, 53–55, 57–59, 172
Indigenous and Community Conserved Areas (ICCAs), 169
Indigenous peoples and local communities (IPLC), 150, 162, 329–330
  justice and equity, 34, 156, 276, 321

innovative financial instruments, 15, 130, 355
Integrated Conservation and Development Projects (ICDPs), 159, 171
integrative governance, 15, 15, 16, 50, 51, 55–56, 72, 79, 85, 97, 107, 116, 129, 130, 131, 179, 195, 223, 236, 280–281, 283, 322–324, 330, 348–349, 355, 357
interdisciplinary, 102, 107, 146, 191
intergenerational justice, 158, 183, 346
Intergovernmental Science-Policy Platform on Biodiversity and Ecosystem Services (IPBES), 6, 31, 55, 93, 139, 190, 227, 244, 342
international institutions, 43, 48, 146
international law, 53, 141, 319, 327, 329, 349
interspecies justice, 180, 193, 195

just sustainabilities, 158
just transition, 163, 343, 347, 352–355
justice
  corrective or remedial, 158
  distributive, 16, 158, 164, 167, 173, 352
  environmental, 163, 251, 308, 343
  procedural, 16, 156, 158, 163, 164, 165–166, 168, 352, 353
  recognition, 16, 164, 352
justice and equity, 16, 18, 79, 86, 94, 103, 157, 161–163
  theories of, 157–158
  transformative biodiversity governance, 155–157

land use change, 266, 281
large marine ecosystem (LMEs), 323
Local Biodiversity Strategies and Action Plans (LBSAPs)., 294

mainstreaming, 4, 16, 47, 49, 50, 51, 52–53, 54, 59, 98, 120, 179–180, 303, 355–356
  justice and equity in, 170–171
marine biobanks, 202, 209
marine biodiversity, 18, 147, 314, 323, 330, 350
  causes of, 314
  impact of fishing on, 315–316
  international instruments on, 313
  loss of, 314–315
  threats to, 318–319
marine genetic resources, 200–201, 205, 207, 211, 215
marine protected areas (MPAs), 117, 223–224, 226, 321
marine resources, 315–316
Marine Spatial Management and Governance Project (MARISMA), 326
marine spatial planning, 322, 324, 325, 326–327, 332
market-based instruments, 116, 120, 344, 356
markets, 13, 50, 51, 84, 116, 117, 188, 248, 252, 258, 266, 274, 278, 281
medicines, 6
Mother Nature, 36

multi-functionality, 278, 300
multi-level governance, 59, **61**, 349
multinational corporations, 201, 214, 215
multi-perspective scenario exercises, 37
multispecies justice, 158, 195, 346

nature, 4, 6
　definition of, 25, 26, 28, 30, 31, 34, 36, 94, 99
　definitions of, 14
Nature Futures Framework, 36
nature-related health risks and benefits, 109
nature's contribution to people, 295, 299, 300, 306
nature-as-natural-capital, 116, 118, 122, 126
nature-based solutions, 295, 297, 300–301, 303, 352
neoliberal/neoliberalism, 17, 50, 56, 57, 72
networking, 51
non-anthropocentric, 192
novel ecosystems', 299
Novozymes, 210–212

objectification, 122
ocean governance, 18, 314, 319–322, 323, 327, 329–330, 350
　need for, 318
　transformative, 323–324, 327–328
One Health, 15, 93–102, 105, 108–110, 191–192, 350
　challenges in, 102–104
　COVID-19 lessons and, 106–108
　methodological gaps in, 106
　understanding concept of, 94–97
openness, 107, 109, 304
other effective area-based conservation measures (OECMs), 33, 168, 221
outcomes, 233

Pachamama, 34, 35, 186
pandemic, 4, 5, 15, 93–94, 102, 103, 104, 107, 108, 155, 200, 245, 259, 350, *See also* COVID-19
patent applications, 148, 201, 203, 205, 206, 211, 214
Payments for Ecosystem Services (PES), 117
planetary health, 97, 98
Planetary Health ethics, 103
pluralism, 14, 25, 26, 37, 103, 115, 125, 344–345
pluralistic approaches, 14, 25, 37
polar regions, 200–202, 205–206, 207, 208, 351
policy coherence, 47, 50, 275–276, 349
policy learning, 282
policy mix, 128–129
pollution, 316–317
polycentric governance, 223
practical implementation, 109
preparedness, 107, 109
primary health care, 94, 101
procedural justice, 168, 173
protected areas, 224, 230
　Aichi Target 11, 169, 226, 236

qualitative descriptions, 36
quantitative descriptions, 36

REDD+, 122, 128, 132, 228–229, 230–231, 236
reductionist approaches, 14, 37
Regime Complex, 14, 43, 47–50, 55–56, 349
regional fisheries management organizations (RFMOs), 323
resilience, 70, 104
　de-growth, 85, 106
　ecosystem services, 229
resources, budget, 166–167, 277
restoration, 235, 297–299, 301, 305, 308
rewilding, 33, 183
rights of nature, 14, 16, 26, 34–35, 37, 76, 78, 179–180, 183, 186–191, 192, 194, 343, 345–346, 353, 354, 357
risk management, 126, 145
rules-in-use, 128, 129, 131

salutogenesis, 99, 101, 104
scenarios, 7, 14, 26, 36, 70, 139, 280
scientific analysis, **108**
silo/silos, 3, 55, 107, 109, 300
singularization, 122, 124
social exclusion, 303, 308
societal deliberation, **108**
soft law, 44, 47, 54, 57
soft skills, 109
solar radiation modification (SRM), 15, 139
sponge city, 303
Structural One Health, 99, 100, 107, 109, 350
structural societal drivers, 15
Sustainable Development Goals (SDGs), 4, **5**, 56, 74, 77–78, 102, 172, 249, 265, 313, 319, 347, 357
Sustainable Ocean Initiative (SOI), 324, 331
sustainable urban drainage, 303
sustainable use, 3, 30, 33, 45, 47, 51–52, 56, 60, 74, 116, 137, 138, 144, 169, 172, 231, 296, 325, 328–330, 331–332, 344, 350, 356
synthetic biology, 15, 137–139, 142–147, 150–151, 353

traditional knowledge systems of medicine, 98
transdisciplinary, 10, 15, 18, 59, 69, 72, 79, 85, 102, 107, 109, 115, 121, 122–123, 125, 126–127, 353, 357
transformative biodiversity governance, 94
transformative change, 3, 4, 13, 14–15, 16, 18, 36, 43, 52, 56, 60, 70, 72, 75, 77–78, 84–86, 99, 103, 104, 105, 109, 115–116, 120, 138, 144, 155, 157, 162, 163, 169, 172–173, 187, 195, 201, 221, 222, 237, 244–246, 253, 259, 265, 294, 307, 313–314, 328, 329, 341, 342–343, 346–347, 356
　biodiversity finance and, 121
　biodiversity loss and, 6–9
　governance and, 7–12, 68–69, 84–85, 106
　integration and reflection, 67–68

need for, 156
  prioritization and, 67–69
transformative governance, 9–12, 84, 149–151
  justice and equity in, 161–163
  role of science in, 82–85
transnational governance, 18, 50, 72, 294, 352
transnational networks, 307, 308, 352
transparency, 14, 37, 52, 57, 58, 59, 167, 207, 322, 330, 347, 348, 352, 356
truth-claims, 37

UN Declaration on the Rights of Indigenous Peoples (UNDRIP), 161, 193
UN Fish Stocks Agreement (UNFSA), 319
UN Framework Convention on Climate Change - equity in, 142, 156, 228, 265
United Nations Convention of the Law of the Sea (UNCLOS), 319
urban biodiversity governance, 294, 301, 303, 305, 306, 307, 352
urban nature, 18, 99, 294, 295, 297, 298–299, 300–301, 302–304, 307–308, 352
urbanisation, 295

valuation, 31, 57, 94, 123, 125, 126, 130, **131**, 190, 249
value pluralism, 14, 26, 115, 125
value plurality, 345
values
  instrumental, 14, 34, 36, 122, 346
  intrinsic, 33, 125, 344
  relational, 26, 33, 344

whole-of-government approach, 348–351
whole-of-society approach, 120–121, **131**, 347–348, 351–352, 355
wilderness, 14, 26, 28–30, 32, 33, 37, 248, 293
wilderness protection, 33, 249
World Health Organization (WHO), 93, 147, 191, 350
World Organisation for Animal Health, 98, 185, **193**, 350
worldviews, 25, 31, 34–35, 37, 98, 103, 155, 158, 162, 332, 344, 346

zoonoses, 200